AF356542

Wide Bandgap Based Devices: Design, Fabrication and Applications, Volume II

Wide Bandgap Based Devices: Design, Fabrication and Applications, Volume II

Editor

Giovanni Verzellesi

MDPI • Basel • Beijing • Wuhan • Barcelona • Belgrade • Manchester • Tokyo • Cluj • Tianjin

Editor
Giovanni Verzellesi
Department of Sciences and
Methods for Engineering
University of Modena and
Reggio Emilia
Reggio Emilia
Italy

Editorial Office
MDPI
St. Alban-Anlage 66
4052 Basel, Switzerland

This is a reprint of articles from the Special Issue published online in the open access journal *Micromachines* (ISSN 2072-666X) (available at: www.mdpi.com/journal/micromachines/special_issues/Wide_Bandgap_Devices_Volume_II).

For citation purposes, cite each article independently as indicated on the article page online and as indicated below:

LastName, A.A.; LastName, B.B.; LastName, C.C. Article Title. *Journal Name* **Year**, *Volume Number*, Page Range.

ISBN 978-3-0365-3994-2 (Hbk)
ISBN 978-3-0365-3993-5 (PDF)

© 2022 by the authors. Articles in this book are Open Access and distributed under the Creative Commons Attribution (CC BY) license, which allows users to download, copy and build upon published articles, as long as the author and publisher are properly credited, which ensures maximum dissemination and a wider impact of our publications.

The book as a whole is distributed by MDPI under the terms and conditions of the Creative Commons license CC BY-NC-ND.

Contents

About the Editor

Giovanni Verzellesi

Giovanni Verzellesi received the "Laurea" degree in Electrical Engineering from the University of Bologna, Bologna, Italy, in 1989, and the Ph.D. degree also in Electrical Engineering from the University of Padova, Padova, Italy, in 1994. In 1993-1994, he was a visiting graduate student with the University of California, Santa Barbara (CA, USA). From 1994 to 1999, he was with the University of Trento, Italy, as an Assistant Professor of Electronics. Since 1999, he has been with the University of Modena and Reggio Emilia, Italy, where he became Associate Professor in 2000 and Professor in 2006.

His research activity has been concerned with the modeling, simulation, and characterization of semiconductor devices and sensors and has covered, over the years, the following topics: impact-ionization effects in Si bipolar transistors, Si- and SiC-based radiation detectors, GaN LEDs, and GaAs and GaN based field-effect transistors. His current research interests are mainly in field of GaN field-effect transistors for RF and power switching applications and GaN LEDs.

He has coauthored more than 100 papers in international journals and more than 130 papers in proceedings of international conferences.

micromachines

MDPI

Editorial

Editorial for the Special Issue on Wide Bandgap Based Devices: Design, Fabrication and Applications, Volume II

Giovanni Verzellesi 

Department of Sciences and Methods for Engineering (DISMI), University of Modena and Reggio Emilia, 42122 Reggio Emilia, Italy; giovanni.verzellesi@unimore.it

Citation: Verzellesi, G. Editorial for the Special Issue on Wide Bandgap Based Devices: Design, Fabrication and Applications, Volume II. *Micromachines* **2022**, *13*, 403. https://doi.org/10.3390/mi13030403

Received: 27 February 2022
Accepted: 27 February 2022
Published: 1 March 2022

Publisher's Note: MDPI stays neutral with regard to jurisdictional claims in published maps and institutional affiliations.

Copyright: © 2022 by the author. Licensee MDPI, Basel, Switzerland. This article is an open access article distributed under the terms and conditions of the Creative Commons Attribution (CC BY) license (https://creativecommons.org/licenses/by/4.0/).

Wide bandgap (WBG) semiconductors are becoming a key enabling technology for several strategic fields of human activities. SiC- and GaN-based transistors are finding their way to market and are expected to become the technology of choice for high-power-density RF amplifiers and high-efficiency power converters, the latter being indispensable elements for the electrification of transports and of energetic systems in industry and buildings. GaN LEDs are the light source technology dominating all segments of the illumination market today. III-nitrides are being evaluated as materials for sensor and transducers. WBG semiconducting oxides are emerging as new materials having potentially superior properties for different applications, such as power conversion, displays and illumination.

Research is still required for all of the above technologies at different levels, from materials to devices and from circuits to systems, so the success of this Special Issue is not surprising. There are 23 papers published, including 20 articles and 3 review papers providing contributions within the full spectrum of the WBG semiconductor applications delineated above. Not surprisingly, one-third of the papers [1–7] focuses on GaN device technologies, which are important for next-generation high-efficiency power converters and their impellent contribution to the decarbonization of human activities. In addition, three papers [8–10] address GaN and SiC circuital applications in power conditioning systems. One paper deals with GaN LEDs [11], whereas three contributions [12–14] are concerned with III-nitride-based devices for sensing applications. Two papers [15,16] cover advanced processing techniques. The remaining papers [17–23] explore the properties and growth techniques of emerging WBG materials.

Regarding GaN device technologies, Jorudas et al. [1] presents results from buffer-free, AlGaN/GaN Schottky barrier diodes and HEMTs on SiC substrates, showing uncompromised performance for high-frequency and high-power applications. Kim et al. [2] propose a normally off, p-GaN/AlGaN/GaN HFET, allowing for unidirectional operation by means of a p-GaN drain electrode shorted to the ohmic drain electrode and avoiding the need for a separate reverse blocking device. Huang et al. [3] characterize normally off, p-GaN gate, AlGaN/GaN HEMTs they fabricated on a low-resistivity SiC substrate, guaranteeing efficient heat removal in high-power performance at a price that is lower than high-resistivity SiC. Alim et al. [4] report on a systematic study based on the measurements, an equivalent-circuit model and sensitivity analysis of the temperature-dependent DC and microwave characteristics of 0.15-μm ultra-short gate-length AlGaN/GaN HEMTs over a wide temperature range from $-40\ ^\circ$C to 150 $^\circ$C. A novel method to achieve AlGaN/GaN MIS-HEMTs in a Si-CMOS platform along with a process for repairing interface defects by a supercritical NH_3 fluid treatment are reported by Liu et al. [5]. Zagni et al. [6] investigated the compensation ratio between the densities of donors and acceptors introduced by carbon doping in the buffer of GaN power HEMTs, assumed to correctly simulate breakdown voltage and current collapse effects. A comprehensive review of the current status of GaN-on-Si transistor technologies is provided by Hsu et al. [7], along with recent different substrate structures, including silicon-on-insulator, engineered substrates and the 3D hetero-integration of GaN and CMOS technologies.

A power conditioning system is designed and built using SiC MOSFETs as switching devices by Ma et al. in [8], which, by leveraging the excellent thermal and voltage capability of SiC MOSFETs, is suitable for grid-level energy storage systems based on vanadium redox flow batteries. A digitally controlled photovoltaic emulator based on an advanced GaN power converter is developed by Ma et al. in [10], whereas in [9], the driving requirements of SiC MOSFETs and GaN HEMTs are illustrated, and the driving circuits designed for WBG switching devices are surveyed.

In [11], Kim et al. demonstrate that InGaN/GaN MQW LEDs on Si substrates with an AlN buffer layer grown with NH_3 interruption show improved crystal quality and enhanced optical output compared to LEDs with conventional AlN buffer. On the sensing application side, AlN is exploited by Chiu et al. [12] to fabricate piezoelectric micromachined ultrasonic transducers that are used to build a high-accuracy time-of-flight ranging system. Nguyen et al. [13] investigate the sensing characteristics of NO_2 gas sensors based on Pd-AlGaN/GaN HEMTs at high temperatures, while Thalhammer et al. [14] describe a novel class of X-ray sensors based on AlGaN/GaN HEMTs offering superior sensitivity and the opportunity for dose reduction in medical applications.

On the advanced processing technique side, laser micromachining on the frontside of SiC and sapphire wafers and the conditions by which the degradation of the performance of GaN HEMT electronics on the backside can be avoided are investigated by Indrišiūnas et al. in [15]. A novel dual laser beam asynchronous dicing method is proposed by Zhang et al. in [16] to improve the cutting quality of SiC wafers.

Regarding the properties and growth of emerging WBG materials, a methodology to synthesize gallium nitride nanoparticles by combining crystal growth with thermal vacuum evaporation is proposed by Fathy et al. in [22]. AlN is explored as an ultra WBG material in three papers: annealing Ni/AlN/SiC Schottky barrier diodes in an atmosphere of nitrogen and oxygen is shown to lead to a significant improvement in the electrical properties of the structures by Kim et al. in [19]; the effect of high-temperature nitridation and a buffer layer on semi-polar AlN films grown on sapphire by hydride vapor phase epitaxy is studied by Zhang et al. [21]; and the thermal annealing of AlN films with different polarities and its impact on crystal quality are studied by in Yue et al. in [23]. The effect of the annealing temperature on the microstructure and performance of sol-gel-prepared NiO films for electrochromic applications is analyzed by Shi et al. in [17]. Solution-processed In_2O_3 thin films and TFTs are fabricated, and the factors affecting the stability of these devices are investigated by Yao et al. in [18]. The electronic structure and the optical properties of Sr-doped β-Ga_2O_3 are studied by Kean Ping et al. [20] using DFT first-principles calculations.

I would like to take this opportunity to thank all the authors for submitting their manuscripts to this Special Issue and all the reviewers for their time and their fundamental help in improving the quality of the accepted papers.

Conflicts of Interest: The author declares no conflict of interest.

References

1. Jorudas, J.; Šimukovič, A.; Dub, M.; Sakowicz, M.; Prystawko, P.; Indrišiūnas, S.; Kovalevskij, V.; Rumyantsev, S.; Knap, W.; Kašalynas, I. AlGaN/GaN on SiC Devices without a GaN Buffer Layer: Electrical and Noise Characteristics. *Micromachines* **2020**, *11*, 1131. [CrossRef] [PubMed]
2. Kim, T.; Jang, W.; Yim, J.; Cha, H. Unidirectional Operation of p-GaN Gate AlGaN/GaN Heterojunction FET Using Rectifying Drain Electrode. *Micromachines* **2021**, *12*, 291. [CrossRef]
3. Huang, Y.; Chiu, H.; Kao, H.; Wang, H.; Liu, C.; Huang, C.; Chen, S. High Thermal Dissipation of Normally off p-GaN Gate AlGaN/GaN HEMTs on 6-Inch N-Doped Low-Resistivity SiC Substrate. *Micromachines* **2021**, *12*, 509. [CrossRef] [PubMed]
4. Alim, M.; Gaquiere, C.; Crupi, G. An Experimental and Systematic Insight into the Temperature Sensitivity for a 0.15-μm Gate-Length HEMT Based on the GaN Technology. *Micromachines* **2021**, *12*, 549. [CrossRef]
5. Liu, M.; Yang, Y.; Chang, C.; Li, L.; Jin, Y. Fabrication of All-GaN Integrated MIS-HEMTs with High Threshold Voltage Stability Using Supercritical Technology. *Micromachines* **2021**, *12*, 572. [CrossRef]
6. Zagni, N.; Chini, A.; Puglisi, F.; Pavan, P.; Verzellesi, G. On the Modeling of the Donor/Acceptor Compensation Ratio in Carbon-Doped GaN to Univocally Reproduce Breakdown Voltage and Current Collapse in Lateral GaN Power HEMTs. *Micromachines* **2021**, *12*, 709. [CrossRef]

7. Hsu, L.; Lai, Y.; Tu, P.; Langpoklakpam, C.; Chang, Y.; Huang, Y.; Lee, W.; Tzou, A.; Cheng, Y.; Lin, C.; et al. Development of GaN HEMTs Fabricated on Silicon, Silicon-on-Insulator, and Engineered Substrates and the Heterogeneous Integration. *Micromachines* **2021**, *12*, 1159. [CrossRef]

8. Ma, C.; Tian, Y. Design and Implementation of a SiC-Based VRFB Power Conditioning System. *Micromachines* **2020**, *11*, 1099. [CrossRef]

9. Ma, C.; Gu, Z. Review on Driving Circuits for Wide-Bandgap Semiconductor Switching Devices for Mid- to High-Power Applications. *Micromachines* **2021**, *12*, 65. [CrossRef]

10. Ma, C.; Tsai, Z.; Ku, H.; Hsieh, C. Design and Implementation of a Flexible Photovoltaic Emulator Using a GaN-Based Synchronous Buck Converter. *Micromachines* **2021**, *12*, 1587. [CrossRef]

11. Kim, S.; Oh, S.; Lee, K.; Kim, S.; Kim, K. Improved Performance of GaN-Based Light-Emitting Diodes Grown on Si (111) Substrates with NH_3 Growth Interruption. *Micromachines* **2021**, *12*, 399. [CrossRef]

12. Chiu, Y.; Wang, C.; Gong, D.; Li, N.; Ma, S.; Jin, Y. A Novel Ultrasonic TOF Ranging System Using AlN Based PMUTs. *Micromachines* **2021**, *12*, 284. [CrossRef] [PubMed]

13. Nguyen, V.; Kim, K.; Kim, H. Performance Optimization of Nitrogen Dioxide Gas Sensor Based on Pd-AlGaN/GaN HEMTs by Gate Bias Modulation. *Micromachines* **2021**, *12*, 400. [CrossRef] [PubMed]

14. Thalhammer, S.; Hörner, A.; Küß, M.; Eberle, S.; Pantle, F.; Wixforth, A.; Nagel, W. GaN Heterostructures as Innovative X-ray Imaging Sensors-Change of Paradigm. *Micromachines* **2022**, *13*, 147. [CrossRef] [PubMed]

15. Indrišiūnas, S.; Svirplys, E.; Jorudas, J.; Kašalynas, I. Laser Processing of Transparent Wafers with a AlGaN/GaN Heterostructures and High-Electron Mobility Devices on a Backside. *Micromachines* **2021**, *12*, 407. [CrossRef]

16. Zhang, Z.; Wen, Z.; Shi, H.; Song, Q.; Xu, Z.; Li, M.; Hou, Y.; Zhang, Z. Dual Laser Beam Asynchronous Dicing of 4H-SiC Wafer. *Micromachines* **2021**, *12*, 1331. [CrossRef]

17. Shi, M.; Qiu, T.; Tang, B.; Zhang, G.; Yao, R.; Xu, W.; Chen, J.; Fu, X.; Ning, H.; Peng, J. Temperature-Controlled Crystal Size of Wide Band Gap Nickel Oxide and Its Application in Electrochromism. *Micromachines* **2021**, *12*, 80. [CrossRef]

18. Yao, R.; Fu, X.; Li, W.; Zhou, S.; Ning, H.; Tang, B.; Wei, J.; Cao, X.; Xu, W.; Peng, J. Bias Stress Stability of Solution-Processed Nano Indium Oxide Thin Film Transistor. *Micromachines* **2021**, *12*, 111. [CrossRef]

19. Kim, D.; Schweitz, M.; Koo, S. Effect of Gas Annealing on the Electrical Properties of Ni/AlN/SiC. *Micromachines* **2021**, *12*, 283. [CrossRef]

20. Kean Ping, L.; Mohamed, M.; Kumar Mondal, A.; Mohamad Taib, M.; Samat, M.; Berhanuddin, D.; Menon, P.; Bahru, R. First-Principles Studies for Electronic Structure and Optical Properties of Strontium Doped β-Ga_2O_3. *Micromachines* **2021**, *12*, 348. [CrossRef]

21. Zhang, Q.; Li, X.; Zhao, J.; Sun, Z.; Lu, Y.; Liu, T.; Zhang, J. Effect of High-Temperature Nitridation and Buffer Layer on Semi-Polar (10–13) AlN Grown on Sapphire by HVPE. *Micromachines* **2021**, *12*, 1153. [CrossRef] [PubMed]

22. Fathy, M.; Gad, S.; Anis, B.; Kashyout, A. Crystal Growth of Cubic and Hexagonal GaN Bulk Alloys and Their Thermal-Vacuum-Evaporated Nano-Thin Films. *Micromachines* **2021**, *12*, 1240. [CrossRef] [PubMed]

23. Yue, Y.; Sun, M.; Chen, J.; Yan, X.; He, Z.; Zhang, J.; Sun, W. Improvement of Crystal Quality of AlN Films with Different Polarities by Annealing at High Temperature. *Micromachines* **2022**, *13*, 129. [CrossRef] [PubMed]

 *micromachines*

Article

AlGaN/GaN on SiC Devices without a GaN Buffer Layer: Electrical and Noise Characteristics

Justinas Jorudas [1,*], Artūr Šimukovič [1], Maksym Dub [2,3], Maciej Sakowicz [2,3], Paweł Prystawko [2], Simonas Indrišiūnas [1], Vitalij Kovalevskij [1], Sergey Rumyantsev [2,3], Wojciech Knap [2,3] and Irmantas Kašalynas [1,*]

[1] Center for Physical Sciences and Technology (FTMC), Saulėtekio 3, 10257 Vilnius, Lithuania; arturas.simukovic@ftmc.lt (A.Š.); simonas.indrisiunas@ftmc.lt (S.I.); vitalij@ftmc.lt (V.K.)

[2] Institute of High Pressure Physics PAS, ul. Sokołowska 29/37, 01-142 Warsaw, Poland; mdub@unipress.waw.pl (M.D.); sakowicz400@gmail.com (M.S.); pprysta@unipress.waw.pl (P.P.); roumis4@gmail.com (S.R.); knap.wojciech@gmail.com (W.K.)

[3] CENTERA Laboratories, Institute of High Pressure Physics PAS, ul. Sokołowska 29/37, 01-142 Warsaw, Poland

[*] Correspondence: justinas.jorudas@ftmc.lt (J.J.); irmantas.kasalynas@ftmc.lt (I.K.); Tel.: +370-5-231-2418 (I.K.)

Received: 29 November 2020; Accepted: 16 December 2020; Published: 20 December 2020

Abstract: We report on the high-voltage, noise, and radio frequency (RF) performances of aluminium gallium nitride/gallium nitride (AlGaN/GaN) on silicon carbide (SiC) devices without any GaN buffer. Such a GaN–SiC hybrid material was developed in order to improve thermal management and to reduce trapping effects. Fabricated Schottky barrier diodes (SBDs) demonstrated an ideality factor n at approximately 1.7 and breakdown voltages (fields) up to 780 V (approximately 0.8 MV/cm). Hall measurements revealed a thermally stable electron density at $N_{2DEG} = 1 \times 10^{13}$ cm^{-2} of two-dimensional electron gas in the range of 77–300 K, with mobilities $\mu = 1.7 \times 10^3$ cm^2/V·s and $\mu = 1.0 \times 10^4$ cm^2/V·s at 300 K and 77 K, respectively. The maximum drain current and the transconductance were demonstrated to be as high as 0.5 A/mm and 150 mS/mm, respectively, for the transistors with gate length $L_G = 5$ μm. Low-frequency noise measurements demonstrated an effective trap density below 10^{19} cm^{-3} eV^{-1}. RF analysis revealed f_T and f_{max} values up to 1.3 GHz and 6.7 GHz, respectively, demonstrating figures of merit $f_T \times L_G$ up to 6.7 GHz × μm. These data further confirm the high potential of a GaN–SiC hybrid material for the development of thin high electron mobility transistors (HEMTs) and SBDs with improved thermal stability for high-frequency and high-power applications.

Keywords: AlGaN/GaN; SiC; high electron mobility transistor; Schottky barrier diode; breakdown field; noise; charge traps; radio frequency

1. Introduction

Aluminium gallium nitride/gallium nitride (AlGaN/GaN) high electron mobility transistors (HEMTs) are widely used in high-power and high-frequency applications due to their superior characteristics based on the unique physical properties of III-nitride materials. The AlGaN/GaN heterostructures can be grown on sapphire, silicon, silicon carbide, and native GaN substrates [1–7]. While sapphire and silicon substrates are the most cost-effective, the best characteristics are achieved on transistors fabricated on silicon carbide (SiC) and GaN substrates. Considerable improvements in electrical performance including the low-frequency noise were demonstrated on the AlGaN/GaN/sapphire platform [8,9]. The advantage of the SiC over GaN substrates is higher SiC thermal conductivity and therefore potentially better thermal management of the transistors fabricated using AlGaN/GaN/SiC structures. The common approach to compensate for lattice mismatch and to reduce the dislocation density in these structures is to grow the aluminium gallium nitride (AlN) nucleation layer (NL) with reduced crystalline quality followed by a several-micrometres-thick GaN buffer doped with deep acceptors such as Fe or C which compensate for residual doping of an *n*-type GaN [10–12]. However, this approach deteriorates the overall thermal resistance of the structure and diminishes the advantage of a SiC substrate operating as a heatsink [13,14]. Also, the acceptor-type impurities in a thick GaN buffer introduce deep charge trapping centres, resulting in the increase of low-frequency noise, and facilitate the current collapse effects in HEMTs [9,15,16].

A new heteroepitaxy approach based on thin GaN–AlN–SiC heterostructures without a GaN buffer has been developed recently [17,18]. Although this approach has already been demonstrated to be promising, there are only a few studies on realistic devices such as transistors [17,19]. Thin GaN–AlN–SiC structures should provide better thermal management of the devices and could potentially reduce short channel effects. One expects also that this technology will reduce the effects of traps from a GaN:C buffer. However, the GaN:C buffer helps in reducing the number of threading dislocations. Therefore, GaN–AlN–SiC structures with the thin buffer may exhibit a higher concentration of threading dislocations, which may deteriorate the characteristics of devices. Indeed, it is well known that the dislocations may act as traps increasing low-frequency noise and current collapse effects and/or lowering maximum voltage breakdown of the devices.

In this work, the GaN–AlN–SiC hybrid material was used to develop thin Schottky barrier diodes (SBDs) and thin HEMTs (T-HEMTs) to study realistic devices under high DC voltages and in radio frequency (RF) regimes. We show that all the devices fabricated on this material have good thermal stability and demonstrate good DC as well as radio frequency (RF) characteristics. By systematic low-frequency noise measurements, we estimated the trap density, showing that avoiding a GaN:C buffer in the GaN–AlN-SiC material does not lead to an increase in active (dislocation related) trap density. We also show that deep trap-related current collapse phenomena are avoided and that all the fabricated devices demonstrate good DC, high voltage, as well as radio frequency (RF) characteristics. This way, we confirm the high potential of a GaN–SiC hybrid material in the development of improved thermal stability HEMTs and SBDs for high-frequency and high-power applications.

2. Materials and Methods (Experimental Details)

The heterostructures with the sequence of layers shown in Figure 1a were obtained commercially from the "SweGaN" company. They were grown on a 4″ diameter, 500-μm-thick semi-insulating SiC substrate. The layers consisted of a 2.4-nm GaN cap, a 20.5-nm $Al_{0.25}Ga_{0.75}N$ barrier, and a 255-nm GaN channel grown directly on a 62-nm high-quality AlN NL on SiC. The sheet resistance (R_{Sh}) of the as-grown T-HEMT structure determined from contactless eddy current measurements was $380 \pm 10\ \Omega/\square$. The band diagram and electron distribution were calculated by a 1D Poisson simulator using the nominal thickness of all layers [20,21]. The results are shown in Figure 1b. The density of the two dimensional electron gas (2DEG) was calculated by integrating an electron distribution in the quantum well. Its value was found to be about $1 \times 10^{13}\ cm^{-2}$.

The devices were fabricated using standard ultraviolet (UV) photolithography [8,22]. Mesas of 140 nm depth were formed by inductively coupled plasma reactive ion etching (ICP-RIE) (Oxford Instruments, Bristol, UK) using Cl plasma and chemical treatment in tetramethylammonium hydroxide (TMAH) solution (Microchemicals, Ulm, Germany). For ohmic contacts, Ti/Al/Ni/Au metal stacks of thicknesses 30/90/20/150 nm were deposited and annealed in nitrogen ambient for 30 s at 850 °C. The resistance (R_c), and the specific resistivity (ρ_c) of ohmic contacts were determined by transmission line method (TLM), demonstrating average values of about $1\ \Omega \times mm$ and $2 \times 10^{-5}\ \Omega \times cm^2$, respectively. Schottky contacts were formed from Ni/Au (25/150 nm).

The Schottky diodes (Figure 2) and HEMTs of two different designs (see Figures 3 and 4) were fabricated. Circular SBDs were used by depositing an inner Schottky contact with radius $r = 40\ \mu m$ and an outer ohmic contact of a variable radius in such a way that the distance between the electric contact,

L, ranged from 5 µm to 40 µm (see Figure 2). The designs of the Schottky diodes and transistors shown in Figures 2 and 4, respectively, do not require mesa isolation. For testing at RF, the transistor design shown in Figure 3 was used (RF T-HEMT).

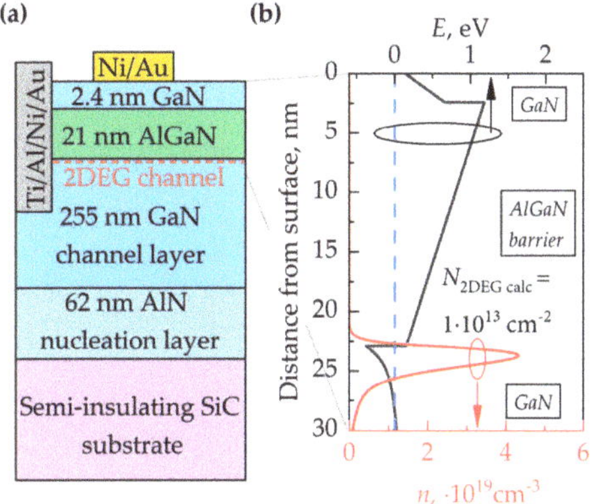

Figure 1. (**a**) Schematic of the thin high electron mobility transistor (T-HEMT) structure cross section with ohmic and Schottky contacts and (**b**) the calculated band diagram and electron density distribution in the upper layers of the heterostructure.

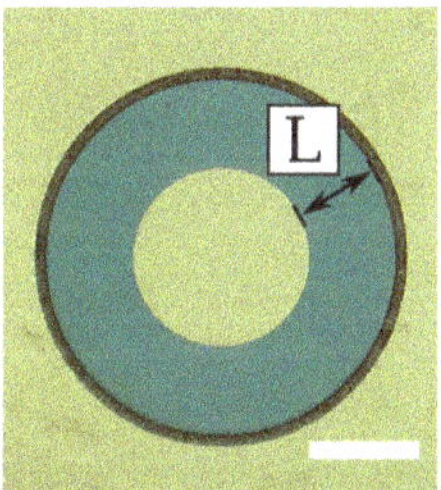

Figure 2. Microscope image of the fabricated Schottky barrier diode (SBD): L is the separation between the Ohmic and Schottky contacts. The scale bar is 50 µm.

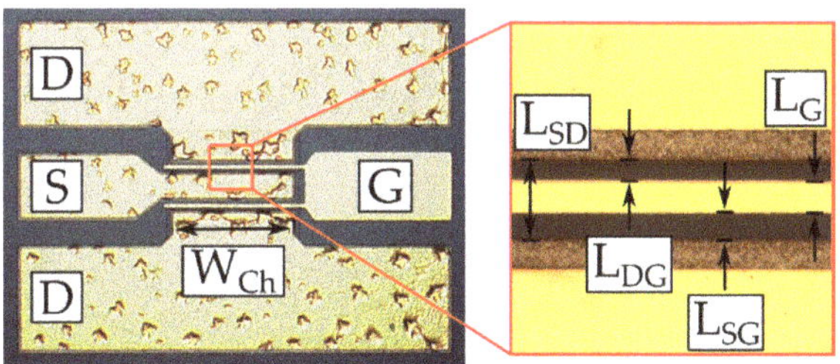

Figure 3. Microscope image of the radio frequency (RF) T-HEMT (left hand side) and details of the design parameters (right hand side) illustrating the Gate (G), Source (S), and Drain (D) electrodes in a 150-µm pitch implementation.

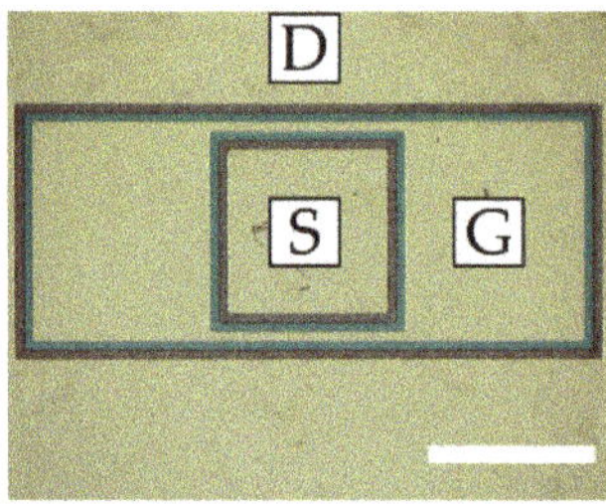

Figure 4. Microscope image of the fabricated DC T-HEMT: the scale bar is 100 μm.

These RF T-HEMTs consisted of 200 μm (RF T-HEMT-1) or 300 μm (RF T-HEMT-2)-wide two-finger transistors, each of drain-source distance L_{SD} = 14 μm, gate length L_G = 5 μm, and gate-source distance L_{SG} = 5 μm. For comparison, the T-HEMTs with rectangular-type electrodes (see Figure 4), labelled here as DC T-HEMT, were also investigated (see also reference [9]). Similar to RF T-HEMTs, all DC T-HEMTs had the same gate length and gate-source distance of 5 μm, but the channel width was of 200 μm and the drain-source distances were 17.5 μm, 15 μm, and 12.5 μm for three sample transistors labelled DC T-HEMT-1, DC T-HEMT-2, and DC T-HEMT-3, respectively.

All transistors were measured on the wafer in DC and RF regimes by using Süss Microtech probe station PM8 (SUSS MicroTec SE, Garching, Germany). For the RF measurements, the G-S-G (ground–signal–ground) 150-μm pitch high frequency probes, Agilent E8364B PNA Network Analyzer (Agilent, Santa Clara, CA, USA), and E5270B Precision IV Analyzer with IC-CAP software were used (Keysight Technologies, Santa Rosa, CA, USA). The two-step open-short de-embedding method was implemented, and small signal S-parameters were obtained. The unity current gain cut-off frequency (f_T) and the unity maximum unilateral power gain frequency (f_{max}) were found from de-embedded S-parameter frequency characteristics.

The SBDs were investigated using EPS150 probe station (Cascade Microtech, Beaverton, OR, USA), high voltage source-meter Keithley 2410 (Tektronix, Beaverton, OR, USA), and impedance analyser Agilent 4294A (Agilent, Santa Clara, CA, USA).

The low-frequency noise in transistors was measured in the linear regime with the source grounded. The voltage fluctuations from the drain load resistor, R_L, were amplified by a low-noise amplifier and analysed using "PHOTON" spectrum analyser (Bruel & Kjaer, Nærum, Denmark). The spectral noise density of drain current fluctuations was calculated in the usual way with $S_I = S_V((R_L + R_{DS})/R_L R_{DS})^2$, where S_V is the drain voltage fluctuations and R_{DS} is the total drain to source resistance.

3. Experimental Results and Discussions

The 2DEG density (N_{2DEG}), mobility (μ_{2DEG}), and sheet resistance (R_{Sh}) were determined in the Hall experiments using Van der Pauw (VdP) geometry. The results are summarized in Table 1. Good agreement between the calculated carrier density, an integral of electron distribution in the quantum well (see Figure 1b), measured sheet resistance using contactless eddy current method, and the results of the Hall experiment were found within a deviation interval of 7%.

Table 1. Parameters of 2DEG in T-HEMT heterostructures at 300 K and 77 K.

Parameter	Hall Measurements		Simulation	Eddy Current Measurements
	300 K	77 K	300 K	300 K
N_{2DEG}, $\times 10^{13}$ cm^{-2}	1.00	0.96	1.0	-
μ_{2DEG}, cm^2/V·s	1.7×10^3	1.0×10^4	-	-
R_{Sh}, Ω/□	375	64	-	380 ± 10

These values are typical for the state-of-the-art AlGaN/GaN heterostructures [23–27]. Therefore, we can conclude that elimination of the buffer layer did not worsen the parameters of the 2DEG.

3.1. Performance of SBDs

Typical capacitance–voltage (*C-V*) characteristics of SBD measured at frequencies 100 kHz and 1 MHz are shown in Figure 5a. One can see that frequency dispersion is negligible, indicating that deep levels do not affect the *C-V* characteristics. The pinch-off voltage (V_{po}) needed to fully deplete a 2DEG channel was found to be about −3.1 V. The density of 2DEG under Schottky contact was calculated using the integral capacitance technique [28]:

$$N_{G-2DEG} = \frac{1}{eA} \int_{V_{po}}^{0} C_P(V)dV,$$ (1)

where *e* is the elementary charge, *A* is the area of Schottky contact, and $C_P(V)$ is the capacitance. The carrier density *N* dependence on the distance from the surface *W* was found from *C-V* data using the following formulas [28]:

$$W = \frac{\varepsilon\varepsilon_0 A}{C_P},$$ (2)

$$N = \frac{C_P{}^3}{e\varepsilon\varepsilon A^2}\left(\frac{dC_P(V)}{dV}\right)^{-1},$$ (3)

where $\varepsilon = 8.9$ is the relative permittivity of GaN and ε_0 is the vacuum permittivity. The obtained *N* dependence on the parameter *W* is shown in Figure 5b. The density of 2DEG was found to be N_{G-2DEG} = 0.69 × 10^{13} cm^{-2} at 300 K. This density is smaller than that found from the Hall measurements due to depletion by the Schottky barrier built-in voltage [29,30].

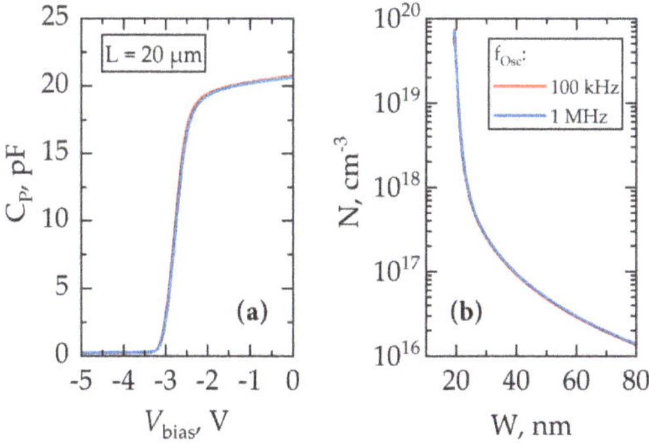

Figure 5. (**a**) Capacitance-voltage (*C-V*) characteristics of SBD with *L* = 20 μm at modulation frequencies of 100 kHz (red line) and 1 MHz (blue line), and (**b**) carrier distribution *N(W)* calculated from *C-V* data using Equations (2) and (3).

Figure 6 shows examples of the forward and reverse current–voltage characteristics of SBDs. The forward current–voltage characteristics demonstrated an ideality factor of $n \cong 1.7$. The barrier height found based on the thermionic emission (TE) model was $\varphi = 0.75$ eV. These values are typical for Ni/AlGaN Schottky barriers [31]. Under reverse bias, leakage currents were saturated at approximately −5 V and remained constant until the breakdown (see Figure 6b). Moreover, SBDs demonstrated a sufficiently high j_{ON}/j_{OFF} ratio; for example, for SBD with *L* = 40 μm, the highest achieved value was found to be more than three orders of magnitude, $j_{ON}/j_{OFF} \geq 3200$, taking into account also the reverse-current densities prior to a breakdown which occurred at a voltage of −780 V. Furthermore, a 2.5 times improvement in the maximum current density was obtained in comparison with previously reported SBDs fabricated on standard AlGaN/GaN HEMT structures with a thick GaN:C buffer [8]. Note the dependence of forward current on the distance between ohmic and Schottky contacts indicating good performance of the fabricated ohmic contacts with negligible losses.

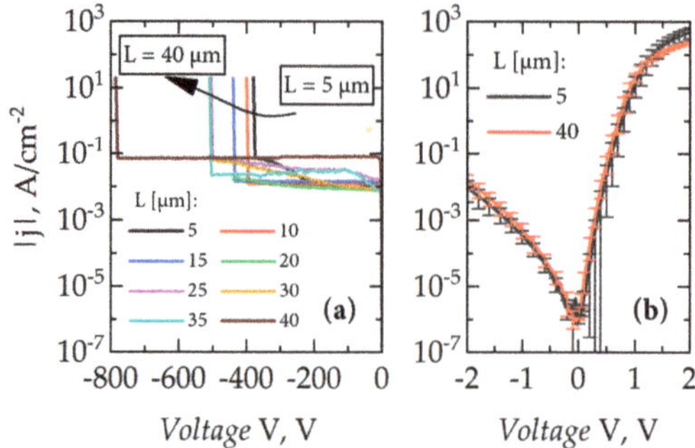

Figure 6. (**a**) Reverse current-voltage characteristics of SBDs and (**b**) current-voltage characteristics of SBDs with L = 5 and 40 µm at low voltages.

GaN–AlN–SiC buffer-free structures with a thin AlN layer may potentially exhibit a higher concentration of the threading dislocations, which may deteriorate the breakdown characteristics. On the other hand, as discussed in References [17,18], high-quality AlN NL in a T-HEMT structure can serve as a back barrier which enhances the critical breakdown field. Figure 7 shows the breakdown voltage and critical electric field dependences on the distance between ohmic and Schottky contacts.

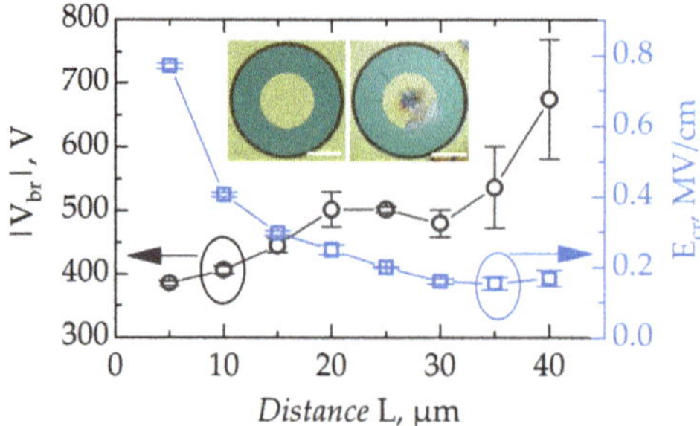

Figure 7. Breakdown voltage and critical electric field dependences on the distance between ohmic and Schottky contacts: *error bars in the critical field data are depicted by the size of the symbols. Inset:* images of L = 40 µm SBD before and after breakdown (scale bar is 50 µm).

As seen in Figure 7, the breakdown voltage depends on the distance, L, between contacts and ranges from 800 V to 400 V for L = 40 µm and L = 5 µm, respectively. The average breakdown field for L = 5 µm devices was found to be 0.8 MV/cm. It is worth noting that the maximum critical field asymptotically decreased down to 0.2 MV/cm with distance increasing from 5 µm to 30 µm and was independent of the distance for larger L values. The inset in Figure 7 shows the optical microscope images of a Schottky diode before and after breakdown. One can see that the inner contact is mostly damaged. Lateral breakdown occurs close to the inner Schottky contact, where the electric field has its maximum. A similar reverse breakdown field dependence on the distance between two ohmic contacts fabricated on the T-HEMT with locally removed 2DEG was reported previously in Reference [18]. There, the critical breakdown field values reached 2 MV/cm for a short distance of L = 5 µm between two isolated devices. In our work, realistic devices—SBDs—were investigated in the reverse bias regime, demonstrating similar behaviour for the breakdown field with maximum values close to 0.8 MV/cm for the short distance (5 μm) between Schottky and ohmic contacts. Therefore, we conclude that the actual breakdown field is higher than 0.8 MV/cm and the absence of the thick GaN buffer does not deteriorate the breakdown characteristics by much.

3.2. Performance of T-HEMTs

Typical DC characteristics of representative T-HEMT are shown in Figure 8. As seen in Figure 8a, RF T-HEMT demonstrated drain current saturation at the level of 266 mA/mm under DC biases of $V_D = 10$ V and $V_G = +1$ V. This translates into an input power value of 2.6 W/mm for T-HEMT with a channel width of 0.4 mm. The drain current in the saturation region fell by 1–2% only. This indicates the advantages of efficient heat removal from the 2DEG channel in AlGaN/GaN with AlN NL that exploits the absence of the buffer layer and high thermal conductivity of the SiC substrate.

The transfer and transconductance (g_m) characteristics at $V_D = 5$ V for various T-HEMTs are shown in Figure 8b,c. The impact of mesa on the device performance can be identified from the transfer characteristics. Indeed, the circular DC T-HEMT devices demonstrated up to two orders of magnitude larger leakage currents in comparison to those measured for RF T-HEMTs. Both the maximum drain current and the transconductance values were found to be higher for the DC T-HEMTs demonstrating values up to 507 mA/mm and 154 mS/mm, respectively. Meanwhile, RF T-HEMTs demonstrated only 266 mA/mm and 77 mS/mm. The pinch-off region is observed beyond a gate bias of −3 V, which is in good agreement with V_{po} obtained from *C-V* measurements.

One of the most effective ways to evaluate the quality of the material and the deep level traps is the low-frequency noise measurements. It is well known that low-frequency noise may differ significantly for the devices with almost identical DC characteristics. Elevated noise level is an indication of lower quality of the material, higher concentration of the deep level traps, lower reliability, and reduced lifetime of the devices. In the majority of cases, the low-frequency noise in field effect transistors complies with the McWhorter model [32,33]. In accordance with the model, the $1/f$ low-frequency noise is a result of tunnelling of the carriers to the layers adjacent to the channel. The model allows for estimation of the effective trap density responsible for noise, which is a good figure of merit for the noise level and overall quality of the material.

The spectra of the drain current fluctuations had the form of $1/f^{\gamma}$ noise with exponent $\gamma = 0.9$–1.1. The dependences of the noise S_I/I^2 on the gate voltage swing (V_G-V_T) at $f = 10$ Hz for three representative devices are shown in Figure 9a (here, V_T is the threshold voltage determined from the transfer current voltage characteristics in the linear regime). As seen, noise depends on the gate voltage as $(V_G$-$V_T)^2$ or steeper. It is known that, in many cases, this dependence at high gate voltages may become flat, indicating a contribution of the contact noise. It is seen from Figure 9a that this is not the case for the studied devices and that contacts do not contribute to noise significantly. The effective trap density N_T in the McWhorter model can be estimated from gate voltage noise as follows [9]:

$$S_{V_G} = \frac{S_I/I^2}{\left(g_m/I\right)^2} \tag{4}$$

$$S_{V_G} = \frac{kTN_T e^2}{\gamma f W_{Ch} L_G C^2}, \tag{5}$$

where k is the Boltzmann constant, T is the temperature, W_{Ch} and L_G is the channel area, C is the gate capacitance per unit area, and γ is the attenuation coefficient of the electron wave function under the barrier, taken to be 10^8 cm^{-1}.

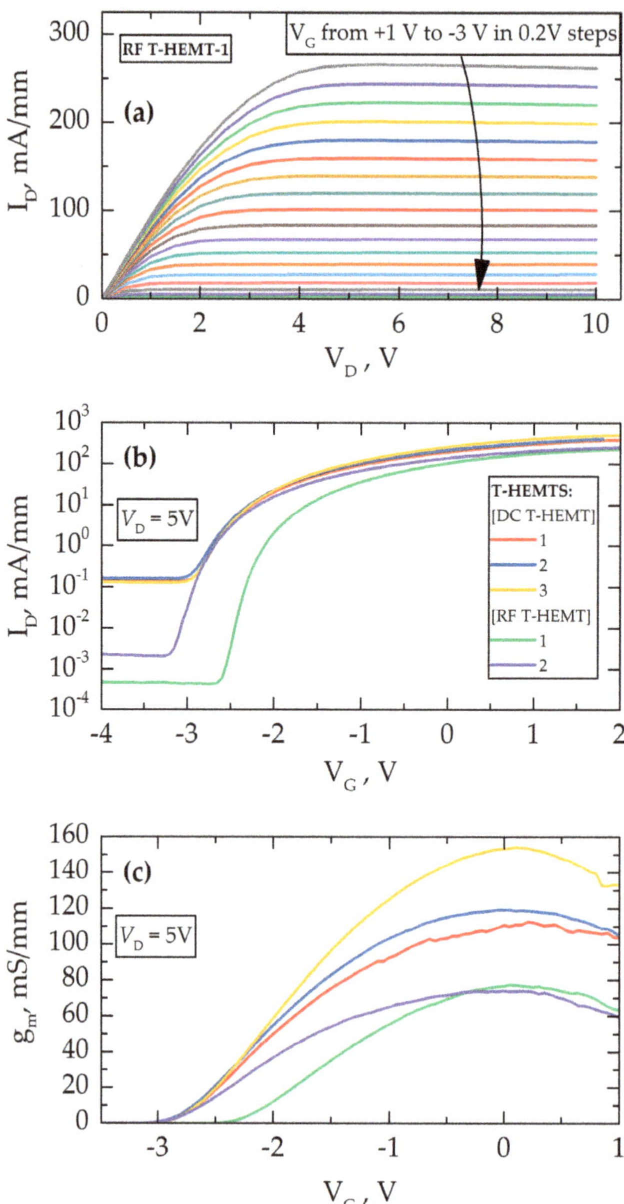

Figure 8. DC characteristics of T-HEMTs under study: (**a**) DC output characteristics of 0.4 mm wide RF T-HEMT-1 and comparisons of transfer (**b**) and transconductance (**c**) characteristics of the RF T-HEMTs and DC T-HEMTs with various values of the channel widths W_{Ch}. The gate length for all devices is 5 μm.

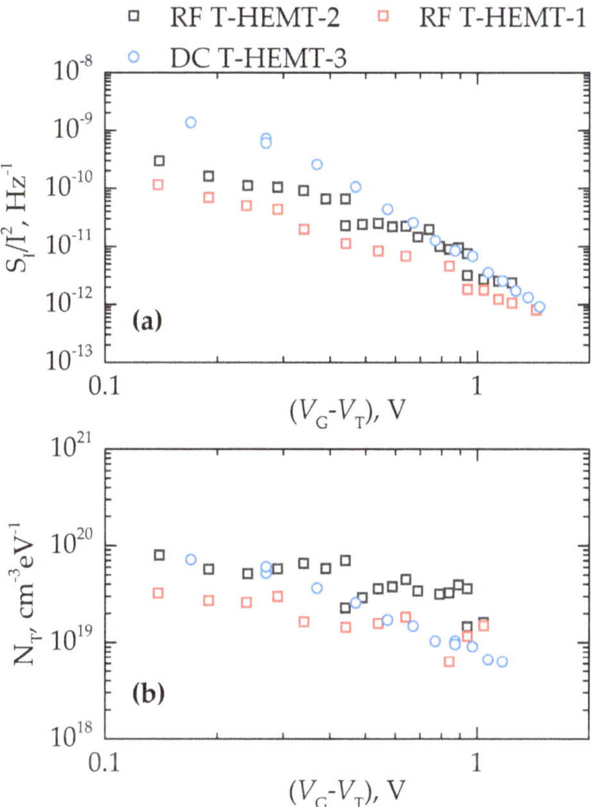

Figure 9. (**a**) Drain current noise S_I/I^2 at frequency f = 10 Hz for T-HEMTs of different channel widths ranging from 0.2 mm to 0.6 mm and (**b**) the effective trap density N_T as a function of the gate voltage swing (V_G-V_T) for the same transistors.

According to the McWhorter model, input gate voltage noise does not depend on access resistance and carrier concentration in the channel [9]. The dependence of the effective trap density on the gate voltage in Figure 9b can be attributed to the dependence of the trap density on energy. The number of traps in this T-HEMT structure was found to be in the range 10^{19}–10^{20} cm^{-3} eV^{-1}. Some of the devices demonstrated $N_T < 10^{19}$ cm^{-3} eV^{-1}. These values are of the same order or even smaller than those reported earlier for AlGaN/GaN HEMTs with a thick buffer layer [9]. Therefore, we conclude that studied T-HEMTs are characterized by the same quality as or even better quality than regular devices with thick buffers.

The unity current gain cut-off frequency (f_T) and the unity maximum unilateral power gain frequency (f_{max}) were found at various voltages down to the threshold voltage. The results are shown in Figure 10. The RF T-HEMTs with a 0.4-mm channel width demonstrated the highest operational frequencies, with values reaching f_T = 1.33 GHz at V_{GS} = 0 V with V_D = 5 V and f_{max} = 6.7 GHz at the bias of V_G = −0.8 V and V_D = 7 V. These results revealed a figure of merit (FOM) factor $f_T \times L_G$ up to 6.7 GHz $\times$ μm, which is comparable with the best value of 9.2 GHz $\times$ μm reported for the T-HEMTs in Reference [19]. The performance of RF T-HEMTs can be further improved in our processing via optimization of ohmic contact/access resistance and the reduction of channel length L_{SD} in tandem with gate length L_G [34,35]. Note that there is up to 3 times difference between the FOM factor of T-HEMTs and that of standard HEMTs, which requires more detailed investigations in the future [36].

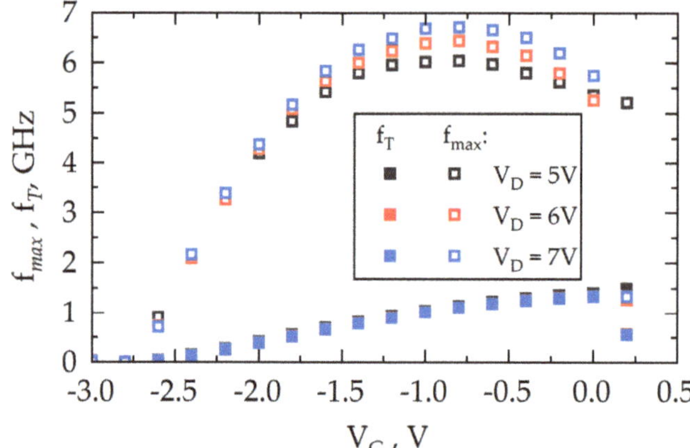

Figure 10. Frequencies f_T and f_{max} at different biasing conditions extracted from S-parameters measurements of RF T-HEMT-1 with W_{Ch} = 0.4 mm and L_G = 5 μm.

4. Conclusions

AlGaN/GaN SBDs and HEMTs without GaN buffer layers have been fabricated on SiC substrates. 2DEG densities of 1×10^{13} cm^2 with mobility of 1.7×10^3 cm^2/V·s and 1.0×10^4 cm^2/V·s at 300 K and 77 K, respectively, were found from the Hall measurements. The unterminated and unpassivated SBDs fabricated on these heterostructures exhibited high breakdown voltages up to −780 V, with the critical breakdown field reaching 0.8 MV/cm. Transistors on these heterostructures, so-called T-HEMTs, demonstrated maximum current density and transconductance values up to 0.5 A/mm and 150 mS/mm, respectively, with a negligible reduction in the drain current. This indicates improved thermal management due to a heterostructure design on the SiC substrate without a GaN buffer layer. By systematic low-frequency noise measurements, we estimated the effective trap density, which in T-HEMT structures was below the level of 10^{19} cm^{-3} eV^{-1}. This value is similar to or even smaller than previously reported trap densities in heterostructures with thick GaN:C buffers. This means that avoiding a GaN:C buffer in GaN–AlN-SiC material does not lead to an increase in active (dislocation-related) trap density. The unity current gain cut-off and unity maximum unilateral power gain were measured to be 1.3 GHz and 6.7 GHz, respectively. Using this data, the figure of merit $f_T \times L_G$ is estimated at 6.7 GHz × μm. Therefore, we conclude that a buffer-free design did not compromise the quality of the structures or the performance of the devices. Our results confirm the potential of a GaN–SiC hybrid material for the development of HEMTs and SBDs for high-frequency and high-power applications with improved thermal stability.

Author Contributions: I.K., J.J., and S.R. conceived the idea. J.J., P.P., S.I., and V.K. characterized the material and prepared the samples. J.J., A.Š., and I.K. conducted the electrical characterization. J.J., M.D., and S.R. performed the noise characterization. M.S., S.R., W.K., and I.K. acquired funding. All authors contributed to the discussion, data analysis, and manuscript preparation. All authors have read and agreed to the published version of the manuscript.

Funding: The work was supported by the Research Council of Lithuania (Lietuvos mokslo taryba) under the "TERAGANWIRE" project (grant No. S-LL-19-1), by the National Science Centre of Poland (grant no. 2017/27/L/ST7/03283), and by the "International Research Agendas" program of the Foundation for Polish Science co-financed by the European Union under the European Regional Development Fund (Nos. MAB/2018/9).

Conflicts of Interest: The authors declare no conflict of interest.

References

1. Koksaldi, O.S.; Haller, J.; Li, H.; Romanczyk, B.; Guidry, M.; Wienecke, S.; Keller, S.; Mishra, U.K. N-Polar GaN HEMTs Exhibiting Record Breakdown Voltage Over 2000 V and Low Dynamic On-Resistance. *IEEE Electron. Device Lett.* **2018**, *39*, 1014–1017. [CrossRef]
2. Choi, U.; Kim, H.-S.; Lee, K.; Jung, D.; Kwak, T.; Jang, T.; Nam, Y.; So, B.; Kang, M.-J.; Seo, K.-S.; et al. Direct Current and Radio Frequency Characterizations of AlGaN/AlN/GaN/AlN Double-Heterostructure High-Electron Mobility Transistor (DH-HEMT) on Sapphire. *Phys. Status Solidi* **2020**, *217*, 1900695. [CrossRef]
3. Chandrasekar, H.; Uren, M.J.; Eblabla, A.; Hirshy, H.; Casbon, M.A.; Tasker, P.J.; Elgaid, K.; Kuball, M. Buffer-Induced Current Collapse in GaN HEMTs on Highly Resistive Si Substrates. *IEEE Electron. Device Lett.* **2018**, *39*, 1556–1559. [CrossRef]
4. Romanczyk, B.; Mishra, U.K.; Zheng, X.; Guidry, M.; Li, H.; Hatui, N.; Wurm, C.; Krishna, A.; Ahmadi, E.; Keller, S. W-Band Power Performance of SiN-Passivated N-Polar GaN Deep Recess HEMTs. *IEEE Electron. Device Lett.* **2020**, *41*, 349–352. [CrossRef]
5. SaifAddin, B.K.; Almogbel, A.S.; Zollner, C.J.; Wu, F.; Bonef, B.; Iza, M.; Nakamura, S.; DenBaars, S.P.; Speck, J.S. AlGaN Deep-Ultraviolet Light-Emitting Diodes Grown on SiC Substrates. *ACS Photonics* **2020**, *7*, 554–561. [CrossRef]
6. Alshahed, M.; Heuken, L.; Alomari, M.; Cora, I.; Toth, L.; Pecz, B.; Wachter, C.; Bergunde, T.; Burghartz, J.N. Low-Dispersion, High-Voltage, Low-Leakage GaN HEMTs on Native GaN Substrates. *IEEE Trans. Electron. Devices* **2018**, *65*, 2939–2947. [CrossRef]
7. Pashnev, D.; Kaplas, T.; Korotyeyev, V.; Janonis, V.; Urbanowicz, A.; Jorudas, J.; Kašalynas, I. Terahertz time-domain spectroscopy of two-dimensional plasmons in AlGaN/GaN heterostructures. *Appl. Phys. Lett.* **2020**, *117*, 051105. [CrossRef]
8. Jakštas, V.; Jorudas, J.; Janonis, V.; Minkevičius, L.; Kašalynas, I.; Prystawko, P.; Leszczynski, M. Development of AlGaN/GaN/SiC high-electron-mobility transistors for THz detection. *Lith. J. Phys.* **2018**, *58*, 135–140. [CrossRef]
9. Sai, P.; Jorudas, J.; Dub, M.; Sakowicz, M.; Jakštas, V.; But, D.B.; Prystawko, P.; Cywinski, G.; Kašalynas, I.; Knap, W.; et al. Low frequency noise and trap density in GaN/AlGaN field effect transistors. *Appl. Phys. Lett.* **2019**, *115*, 183501. [CrossRef]
10. Chang, S.-J.; Bhuiyan, M.A.; Won, C.-H.; Lee, J.-H.; Jung, H.W.; Shin, M.J.; Do, J.-W.; Cho, K.J.; Lee, J.-H.; Ma, T.P.; et al. Investigation of GaN channel thickness on the channel mobility in AlGaN/GaN HEMTs grown on sapphire substrate. In Proceedings of the 2017 IEEE International Symposium on Radio-Frequency Integration Technology (RFIT), Seoul, Korea, 30 August–1 September 2017; pp. 87–89.
11. Selvaraj, S.L.; Suzue, T.; Egawa, T. Breakdown Enhancement of AlGaN/GaN HEMTs on 4-in Silicon by Improving the GaN Quality on Thick Buffer Layers. *IEEE Electron. Device Lett.* **2009**, *30*, 587–589. [CrossRef]
12. Poblenz, C.; Waltereit, P.; Rajan, S.; Heikman, S.; Mishra, U.K.; Speck, J.S. Effect of carbon doping on buffer leakage in AlGaN/GaN high electron mobility transistors. *J. Vac. Sci. Technol. B Microelectron. Nanom. Struct.* **2004**, *22*, 1145. [CrossRef]
13. Manoi, A.; Pomeroy, J.W.; Killat, N.; Kuball, M. Benchmarking of Thermal Boundary Resistance in AlGaN/GaN HEMTs on SiC Substrates: Implications of the Nucleation Layer Microstructure. *IEEE Electron. Device Lett.* **2010**, *31*, 1395–1397. [CrossRef]
14. Cho, J.; Bozorg-Grayeli, E.; Altman, D.H.; Asheghi, M.; Goodson, K.E. Low Thermal Resistances at GaN–SiC Interfaces for HEMT Technology. *IEEE Electron. Device Lett.* **2012**, *33*, 378–380. [CrossRef]
15. Uren, M.J.; Moreke, J.; Kuball, M. Buffer Design to Minimize Current Collapse in GaN/AlGaN HFETs. *IEEE Trans. Electron. Devices* **2012**, *59*, 3327–3333. [CrossRef]
16. Fang, Z.-Q.; Claflin, B.; Look, D.C.; Green, D.S.; Vetury, R. Deep traps in AlGaN/GaN heterostructures studied by deep level transient spectroscopy: Effect of carbon concentration in GaN buffer layers. *J. Appl. Phys.* **2010**, *108*, 063706. [CrossRef]
17. Chen, J.-T.; Bergsten, J.; Lu, J.; Janzén, E.; Thorsell, M.; Hultman, L.; Rorsman, N.; Kordina, O. A GaN–SiC hybrid material for high-frequency and power electronics. *Appl. Phys. Lett.* **2018**, *113*, 041605. [CrossRef]

18. Lu, J.; Chen, J.-T.; Dahlqvist, M.; Kabouche, R.; Medjdoub, F.; Rosen, J.; Kordina, O.; Hultman, L. Transmorphic epitaxial growth of AlN nucleation layers on SiC substrates for high-breakdown thin GaN transistors. *Appl. Phys. Lett.* **2019**, *115*, 221601. [CrossRef]

19. Chen, D.-Y.; Malmros, A.; Thorsell, M.; Hjelmgren, H.; Kordina, O.; Chen, J.-T.; Rorsman, N. Microwave Performance of 'Buffer-Free' GaN-on-SiC High Electron Mobility Transistors. *IEEE Electron. Device Lett.* **2020**, *41*, 828–831. [CrossRef]

20. Tan, I.; Snider, G.L.; Chang, L.D.; Hu, E.L. A self-consistent solution of Schrödinger—Poisson equations using a nonuniform mesh. *J. Appl. Phys.* **1990**, *68*, 4071–4076. [CrossRef]

21. Snider, G.L.; Tan, I.-H.; Hu, E.L. Electron states in mesa-etched one-dimensional quantum well wires. *J. Appl. Phys.* **1990**, *68*, 2849–2853. [CrossRef]

22. Kruszewski, P.; Prystawko, P.; Kasalynas, I.; Nowakowska-Siwinska, A.; Krysko, M.; Plesiewicz, J.; Smalc-Koziorowska, J.; Dwilinski, R.; Zajac, M.; Kucharski, R.; et al. AlGaN/GaN HEMT structures on ammono bulk GaN substrate. *Semicond. Sci. Technol.* **2014**, *29*, 75004. [CrossRef]

23. Narang, K.; Bag, R.K.; Singh, V.K.; Pandey, A.; Saini, S.K.; Khan, R.; Arora, A.; Padmavati, M.V.G.; Tyagi, R.; Singh, R. Improvement in surface morphology and 2DEG properties of AlGaN/GaN HEMT. *J. Alloys Compd.* **2020**, *815*, 152283. [CrossRef]

24. Yang, C.; Luo, X.; Sun, T.; Zhang, A.; Ouyang, D.; Deng, S.; Wei, J.; Zhang, B. High Breakdown Voltage and Low Dynamic ON-Resistance AlGaN/GaN HEMT with Fluorine Ion Implantation in SiNx Passivation Layer. *Nanoscale Res. Lett.* **2019**, *14*, 191. [CrossRef] [PubMed]

25. Nifa, I.; Leroux, C.; Torres, A.; Charles, M.; Reimbold, G.; Ghibaudo, G.; Bano, E. Characterization and modeling of 2DEG mobility in AlGaN/AlN/GaN MIS-HEMT. *Microelectron. Eng.* **2019**, *215*, 110976. [CrossRef]

26. Geng, K.; Chen, D.; Zhou, Q.; Wang, H. AlGaN/GaN MIS-HEMT with PECVD SiNx, SiON, SiO2 as Gate Dielectric and Passivation Layer. *Electronics* **2018**, *7*, 416. [CrossRef]

27. Lutsenko, E.V.; Rzheutski, M.V.; Vainilovich, A.G.; Svitsiankou, I.E.; Tarasuk, N.P.; Yablonskii, G.P.; Alyamani, A.; Petrov, S.I.; Mamaev, V.V.; Alexeev, A.N. Investigation of Photoluminescence, Stimulated Emission, Photoreflectance, and 2DEG Properties of Double Heterojunction AlGaN/GaN/AlGaN HEMT Heterostructures Grown by Ammonia MBE. *Phys. Status Solidi* **2018**, *215*, 1700602. [CrossRef]

28. Schroder, D.K. Carrier and Doping Density. In *Semiconductor Material and Device Characterization*; John Wiley & Sons, Inc.: Hoboken, NJ, USA, 2005; pp. 61–125.

29. Pashnev, D.; Korotyeyev, V.V.; Jorudas, J.; Kaplas, T.; Janonis, V.; Urbanowicz, A.; Kašalynas, I. Experimental evidence of temperature dependent effective mass in AlGaN/GaN heterostructures observed via THz spectroscopy of 2D plasmons. *Appl. Phys. Lett.* **2020**, *117*, 162101. [CrossRef]

30. Hoon Shin, J.; Je Jo, Y.; Kim, K.-C.; Jang, T.; Sang Kim, K. Gate metal induced reduction of surface donor states of AlGaN/GaN heterostructure on Si-substrate investigated by electroreflectance spectroscopy. *Appl. Phys. Lett.* **2012**, *100*, 111908. [CrossRef]

31. Kruszewski, P.; Grabowski, M.; Prystawko, P.; Nowakowska-Siwinska, A.; Sarzynski, M.; Leszczynski, M. Properties of AlGaN/GaN Ni/Au-Schottky diodes on 2°-off silicon carbide substrates. *Phys. Status Solidi* **2017**, *214*, 1600376. [CrossRef]

32. Handler, P.; Farnsworth, H.E.; Kleiner, W.H.; Law, J.T.; Garrett, C.G.B.; Autler, H.; McWhorter, A.L. Electrical Properties of a Clean Germanium Surface. In *Semiconductor Surface Physics*; Kingston, R.H., Ed.; University of Pennsylvania Press: Philadelphia, PA, USA, 1957; pp. 23–52.

33. Christensson, S.; Lundström, I.; Svensson, C. Low frequency noise in MOS transistors—I Theory. *Solid State Electron.* **1968**, *11*, 797–812. [CrossRef]

34. Chavarkar, P.; Mishra, U.K. High Electron Mobility Transistors. In *RF and Microwave Semiconductor Device Handbook*; CRC Press: Boca Raton, FL, USA, 2017; pp. 8.1–8.32.

35. Schwierz, F.; Liou, J.J. RF transistors: Recent developments and roadmap toward terahertz applications. *Solid State Electron.* **2007**, *51*, 1079–1091. [CrossRef]

36. Shinohara, K.; Regan, D.C.; Tang, Y.; Corrion, A.L.; Brown, D.F.; Wong, J.C.; Robinson, J.F.; Fung, H.H.; Schmitz, A.; Oh, T.C.; et al. Scaling of GaN HEMTs and Schottky Diodes for Submillimeter-Wave MMIC Applications. *IEEE Trans. Electron. Devices* **2013**, *60*, 2982–2996. [CrossRef]

Publisher's Note: MDPI stays neutral with regard to jurisdictional claims in published maps and institutional affiliations.

 © 2020 by the authors. Licensee MDPI, Basel, Switzerland. This article is an open access article distributed under the terms and conditions of the Creative Commons Attribution (CC BY) license (http://creativecommons.org/licenses/by/4.0/).

Article

Unidirectional Operation of p-GaN Gate AlGaN/GaN Heterojunction FET Using Rectifying Drain Electrode

Tae-Hyeon Kim, Won-Ho Jang, Jun-Hyeok Yim and Ho-Young Cha *

School of Electrical and Electronic Engineering, Hongik University, 94 Wausan-ro, Mapo-gu, Seoul 04066, Korea; jxajxa@g.hongik.ac.kr (T.-H.K.); jwh8904@mail.hongik.ac.kr (W.-H.J.); jhgjhg4@g.hongik.ac.kr (J.-H.Y.)
* Correspondence: hcha@hongik.ac.kr; Tel.: +82-2-320-3062

Abstract: In this study, we proposed a rectifying drain electrode that was embedded in a p-GaN gate AlGaN/GaN heterojunction field-effect transistor to achieve the unidirectional switching characteristics, without the need for a separate reverse blocking device or an additional process step. The rectifying drain electrode was implemented while using an embedded p-GaN gating electrode that was placed in front of the ohmic drain electrode. The embedded p-GaN gating electrode and the ohmic drain electrode are electrically shorted to each other. The concept was validated by technology computer aided design (TCAD) simulation along with an equivalent circuit, and the proposed device was demonstrated experimentally. The fabricated device exhibited the unidirectional characteristics successfully, with a threshold voltage of ~2 V, a maximum current density of ~100 mA/mm, and a forward drain turn-on voltage of ~2 V.

Keywords: AlGaN/GaN heterojunction; p-GaN gate; unidirectional operation; rectifying electrode

Citation: Kim, T.-H.; Jang, W.-H.; Yim, J.-H.; Cha, H.-Y. Unidirectional Operation of p-GaN Gate AlGaN/GaN Heterojunction FET Using Rectifying Drain Electrode. *Micromachines* **2021**, *12*, 291. https://doi.org/10.3390/mi12030291

Academic Editor: Giovanni Verzellesi

Received: 23 February 2021
Accepted: 8 March 2021
Published: 10 March 2021

Publisher's Note: MDPI stays neutral with regard to jurisdictional claims in published maps and institutional affiliations.

Copyright: © 2021 by the authors. Licensee MDPI, Basel, Switzerland. This article is an open access article distributed under the terms and conditions of the Creative Commons Attribution (CC BY) license (https://creativecommons.org/licenses/by/4.0/).

1. Introduction

AlGaN/GaN heterojunction field-effect transistors (HFETs) have been extensively studied for high-efficiency power switching and high-frequency applications owing to their properties, such as wide energy bandgap, high critical electric field, and two-dimensional electron gas (2DEG) channels with high electron mobility and electron density [1–7]. While the power switching devices must be operated in a normally-off mode, conventional AlGaN/GaN HFETs exhibit normally-on characteristics. A widely adopted device structure for the normally-off mode is a p-GaN gate AlGaN/GaN HFET, where the gate region has a p-GaN layer to deplete the area underneath the AlGaN/GaN channel [4,8–13]. Such device types have been successfully commercialized and they are currently used in various power modules for different electronic devices, such as fast chargers, switching mode power supplies, and lighting drivers. Some applications of switching devices are to prevent reverse conduction in order to protect the circuit, so-called unidirectional switching characteristics. A reverse blocking device or circuit must be added to the switching device to achieve unidirectional characteristics, which enlarges the chip size and increases the manufacturing cost. Some studies have reported the unidirectional operation of GaN devices without adding extra components [14–18]. In this study, we proposed a unidirectional switching device that is based on a normally-off p-GaN gate AlGaN/GaN HFET in which a drain electrode consisted of a rectifying gating electrode and an ohmic electrode. The proposed device requires no separate blocking device or additional manufacturing costs.

2. Device Structure and TCAD Simulation

2.1. Simulation Details

The epitaxial structure used for device simulation consists of a 70 nm p-GaN layer with a p-type doping concentration of 3×10^{17} cm^{-3}, a 15 nm unintentionally-doped Al$_{0.2}$Ga$_{0.8}$N barrier layer with an n-type doping concentration of 1×10^{16} cm^{-3}, a 35 nm unintentionally-doped GaN channel layer with an n-type doping concentration of 1×10^{16} cm^{-3}, and a

1.95 μm $Al_{0.05}Ga_{0.95}N$ buffer layer. Figure 1a,b demonstrate the cross-sectional schematics of a conventional p-GaN gate AlGaN/GaN HFET and a proposed unidirectional device, respectively, with a gate length of 2 μm for both of the structures. The length of the p-GaN drain region was 1 μm for the unidirectional device, which was separated from the drain electrode by 0.5 μm.

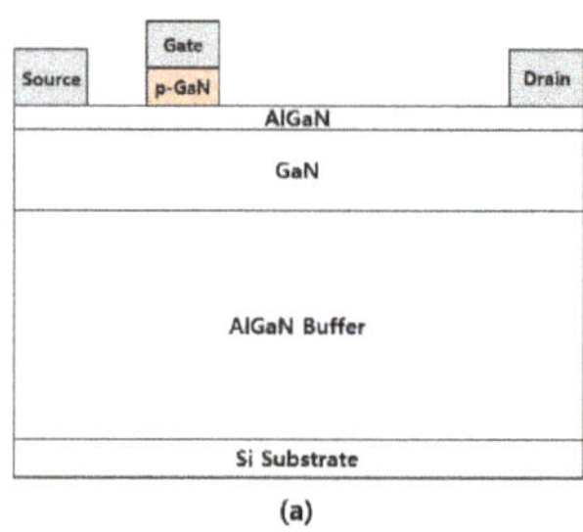

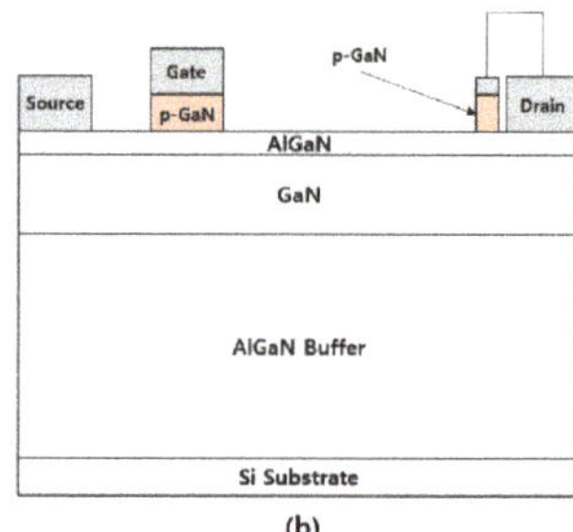

Figure 1. Cross-sectional schematics of (**a**) p-GaN gate AlGaN/GaN heterojunction field-effect transistor (HFET) and (**b**) unidirectional HFET.

The simulations were carried out using SILVACO ATLAS (Silvaco, Silicon Valley, CA, USA). Figure 2 shows the models used in the simulation code, which was adopted from an example file provided by SILVACO (ganfetex07.in). A detailed explanation of the simulation models can be found in ref [19], which includes a polarization model, a temperature dependent low field mobility model, a nitride specific high field dependent mobility model, a lattice heating model, and a trap model.

===

```
models consrh auger fermi print temp=300
mobility GaNsat.n
models lat.temp ni.fermi
mobility albrct.n bn.albrct=3e-05 an.albrct=3e-05
mobility region=5 albrct.p bp.albrct=1e04 ap.albrct=1e04
model ten.piezo psp.scale=0.67 piezo.scale=0.67 calc.strain
model region=5 pch.ins

trap region=2 donor e.level=3.2 density=1.27e18 sign=1e-15 sigp=1e-15 degen=2
trap region=2 acceptor e.level=0.36 density=7e17 sign=1e-15 sigp=1e-15 degen=4

thermcontact num=1 name=substrate alpha=2500
```

===

Figure 2. Physical models and parameters used in simulation code.

2.2. Simulation Result and Disscussion

Figure 3 compares the simulation results of the forward and reverse characteristics for two different structures. The conventional device exhibited a typical normally-off operation with reverse conduction characteristics, whereas the proposed structure exhibited the same normally-off operation with reverse blocking characteristics. The threshold voltage was 1.8 V for both devices, which was determined by the p-GaN gate electrode. A positive shift in the forward drain turn-on characteristics was observed for the proposed unidirectional device, which is the same as the gate threshold voltage of the device. The positive shift and reverse blocking characteristics can be explained while using the equivalent circuit that is shown in Figure 4. The p-GaN gate can be represented by a gate electrode of the HFET in conjunction with a PN heterojunction diode. When the p-GaN gate voltage exceeds the threshold voltage (1.8 V), the 2DEG channel is formed between the AlGaN barrier layer and GaN channel layer, creating a conduction path between the source and drain. As the p-GaN gate voltage becomes higher than the forward turn-on voltage of the p-GaN/AlGaN/GaN heterojunction diode, the current flows from the p-GaN gate to the source. On the drain side, the p-GaN region acts as a "gate" electrode, which is electrically shorted to the ohmic electrode. Therefore, the current can flow from the ohmic drain electrode to the source electrode by creating the 2DEG channel under the p-GaN region, as the drain voltage becomes higher than the gate threshold voltage (1.8 V). That is, no current flows when the drain voltage is lower than the gate threshold voltage, which is why the device has forward drain turn-on characteristics that are similar to the gate threshold characteristics. As the drain voltage becomes higher than the forward turn-on voltage of the p-GaN/AlGaN/GaN heterojunction diode, the current can flow from both the p-GaN drain and ohmic drain regions to the source electrode. In the reverse region, when the drain voltage is negative, the p-GaN drain region is reverse-biased and it further depletes the channel, blocking the current flow from the drain. Therefore, the device exhibits reverse blocking characteristics. The electron concentration distributions under forward and reverse modes are compared in Figure 5a,b, respectively. The electron channel exists under the p-GaN drain region in the forward mode that is shown in Figure 5a, where both gate and drain voltages were +5 V. On the other hand, the channel under the p-GaN drain region was depleted in the reverse mode that is shown in Figure 5b where the gate and drain voltages were +5 V and −5 V, respectively.

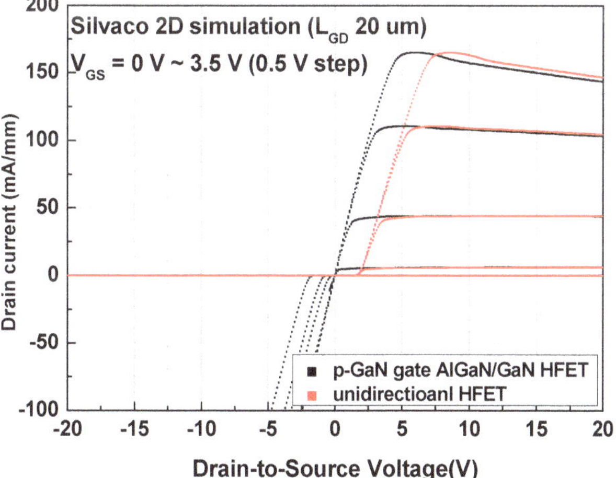

Figure 3. Forward and reverse characteristics of p-GaN gate AlGaN/GaN HFET (black lines) and unidirectional HFET (red lines).

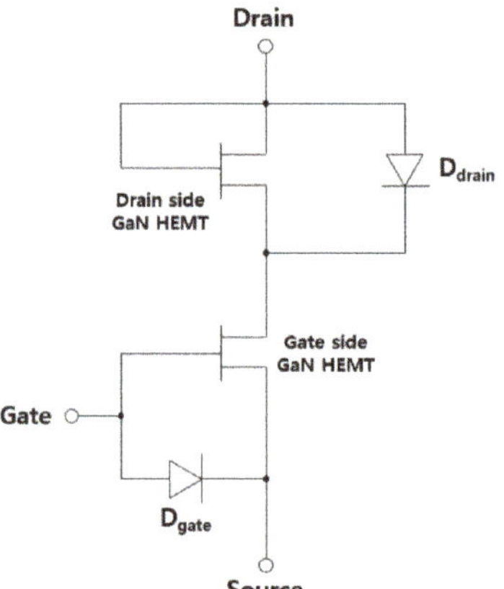

Figure 4. Equivalent circuit of unidirectional HFET.

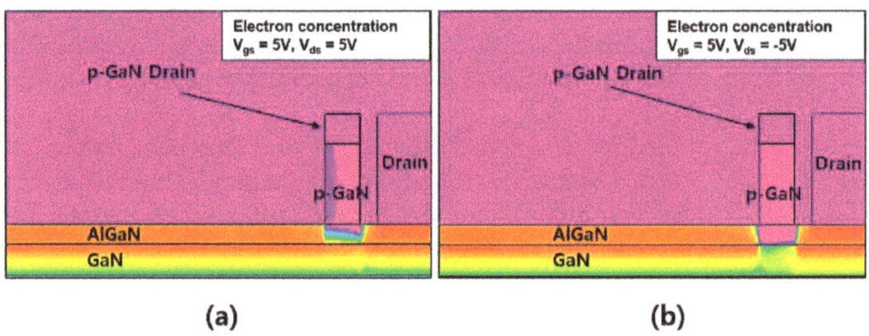

(a)   (b)

Figure 5. Electron concentration under the p-GaN drain region at (**a**) V_{gs} = 5 V and V_{ds} = 5 V and (**b**) V_{gs} = 5 V and V_{ds} = −5 V. Two electrodes (p-GaN drain and drain electrodes) are shorted electrically to each other in the simulation.

3. Fabrication

3.1. Device Structure and Fabrication

Two device structures were fabricated to validate the proposed concept, as follows. Figure 6a,b shows the cross-sectional schematics of the conventional p-GaN gate AlGaN/GaN HFET and the unidirectional HFET, respectively. The epitaxial structure consisted of a 70 nm p-GaN layer, a 15 nm $Al_{0.2}Ga_{0.8}N$ barrier layer, a 320 nm GaN layer, and a 3.6 μm buffer layer grown on a Si (111) substrate. After solvent and acid cleaning of the surface, the p-GaN layer was etched while using a two-step etching process, during which the gate and p-GaN drain regions were covered by photoresist. First, the p-GaN layer was partially etched by a low-damage plasma etching process using Cl_2/BCl_3-based inductively coupled plasma reactive ion etching (ICP-RIE) with an etch depth target of 45 nm. A source RF power of 250 W, a bias RF power of 5 W, a gas flow rate of Cl_2/BCl_3 = 18/2 sccm, and a chamber pressure of 5 mTorr were used, which resulted in an etch rate of ~1 Å/s. Subsequently, the remaining p-GaN layer was etched by a selective etching process using $Cl_2/N_2/O_2$-based ICP-RIE to minimize the plasma-induced damage on the surface. A source RF power of 2000 W, a bias RF power of 25 W, a gas flow rate of $Cl_2/N_2/O_2$ = 40/10/2 sccm, and a chamber pressure of 20 mTorr were used with a chuck temperature of 60 °C [20]. The selectivity between p-GaN and AlGaN was approximately 50:1 with a p-GaN etch rate of 3.6 Å/s. After the p-GaN layer was completely removed, the oxidized AlGaN surface was treated for 30 s using a buffered oxide etchant (30:1). Subsequently, damage recovery annealing was performed at 500 °C for 5 min. in an N_2 ambient. The ohmic contact region was etched down to the GaN channel layer while using the low-damage BCl_3/Cl_2-based ICP-RIE with an etch depth of 15 nm, after which an additional photolithography process defined the ohmic metallization area with an overhang structure. The overhang region was extended to the p-GaN drain region for the unidirectional device, as shown in Figure 6b. A Ti/Al/TiN (=30/100/20 nm) metal stack was used for the Au-free ohmic contact, which was annealed at 550 °C for 1 min. in N_2 ambient. The transfer contact resistance was 0.56 $\Omega\cdot$mm. MESA isolation was then carried out using the BCl_3/Cl_2-based RIE with an etch depth of 450 nm. A forward power of 100 W, a gas flow rate of $Cl_2/BCl_3 = 18/6$ sccm, and a chamber pressure of 75 mTorr were used for the RIE process. Subsequently, a 170-nm TiN film was sputtered for the gate and pad electrode regions. The surface was passivated with a 180-nm SiNx film using ICP chemical vapor deposition (ICP-CVD). A RF power of 200 W, a gas flow rate of $SiH_4(5\%)/N_2/NH_3 = 25/400/12$ sccm, and a chamber pressure of 2000 mTorr were used with a chuck temperature of 350 °C. Finally, SF_6-based ICP-RIE was used to open the probe contact region. Notably, the unidirectional device does not require an additional process step. The source-to-drain distance, p-GaN length for the gate region, and gate-to-drain distance were 3, 4, and 6 µm, respectively, where the gate metal length was 2 µm, and the ohmic overhang extension was 1 µm. The length of the p-GaN drain region was 2 µm in the unidirectional device.

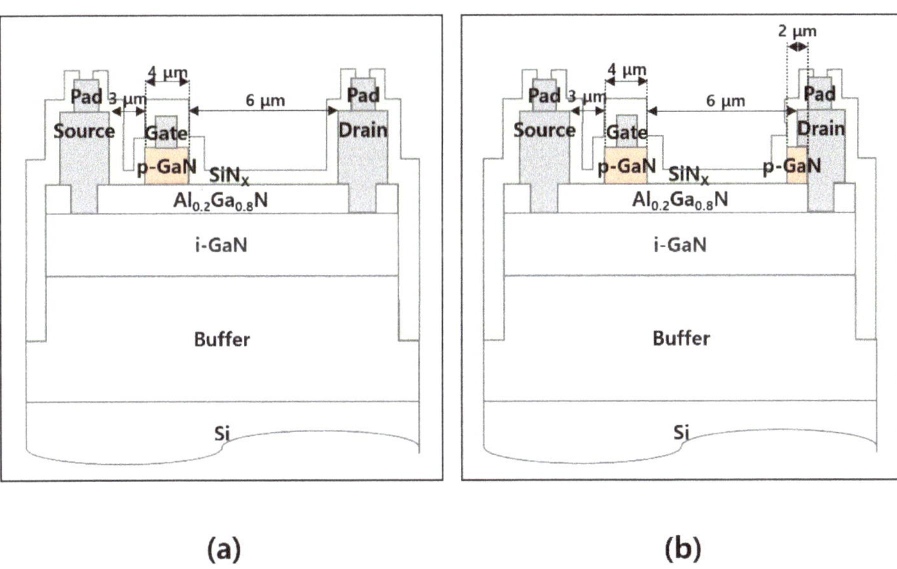

(a)   **(b)**

Figure 6. Cross-sectional schematics of (**a**) fabricated p-GaN gate AlGaN/GaN HFET and (**b**) unidirectional HFET.

3.2. Device Characteristics

Figure 7 shows the transfer current–voltage characteristics of the fabricated p-GaN gate AlGaN/GaN HFET (black lines) and unidirectional device (red lines) that were measured at a drain voltage of 10 V. No significant difference was observed between the two devices, in which the gate threshold voltage was ~2 V at 1 mA/mm.

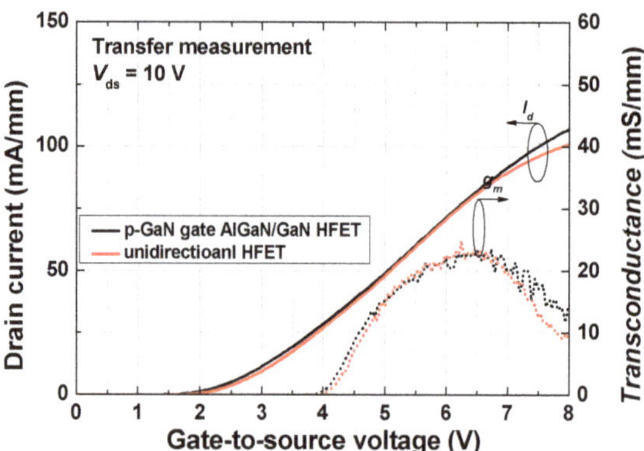

Figure 7. Transfer current–voltage characteristics of fabricated p-GaN gate AlGaN/GaN HFET (black line) and unidirectional HFET (red line).

Figure 8 shows the forward and reverse output current–voltage characteristics. The p-GaN gate AlGaN/GaN HFET (black lines) exhibited bidirectional characteristics, whereas the proposed device exhibited unidirectional characteristics (red lines). The forward drain turn-on voltage for the unidirectional device was ~2 V, which is the same as the gate threshold voltage, as discussed previously. A potential drawback of the proposed device is the forward drain turn-on characteristic. However, the overall device unit would have similar forward turn-on characteristics when an additional reverse blocking device is added to achieve the unidirectional characteristics. It is suggested that the p-GaN drain region be etched partially and/or a different metal contact be used for the p-GaN drain region in order to reduce the forward drain turn-on voltage.

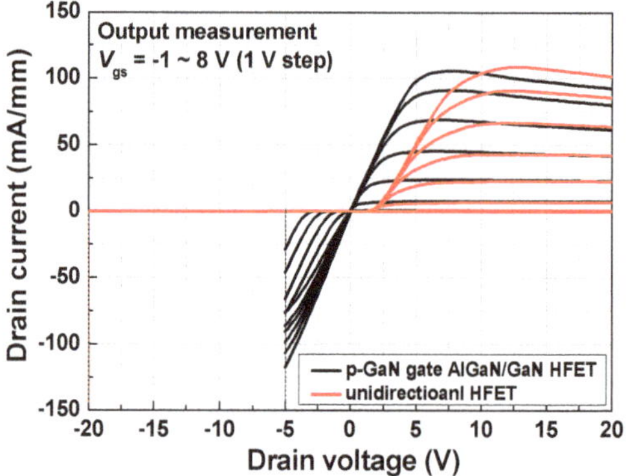

Figure 8. Output current–voltage characteristics of fabricated p-GaN gate AlGaN/GaN HFET (black line) and unidirectional HFET (red line).

4. Conclusions

A unidirectional p-GaN/AlGaN/GaN HFET was proposed to implement a normally-off, unidirectional operation, which was validated by both simulation and device demonstration. A p-GaN drain electrode was embedded in front of the ohmic drain electrode, in which they were electrically shorted to each other. The p-GaN drain region acted as a gate in the forward mode and as a reverse-biased rectifier in the reverse mode, which resulted in reverse blocking characteristics. The proposed device would be a cost-effective solution for achieving unidirectional operation, because it requires no separate reverse blocking device or an additional process step. The fabricated device exhibited a threshold voltage of

~2 V, a maximum current density of ~100 mA/mm, and a drain forward turn-on voltage of ~2 V. It is suggested that the drain turn-on voltage in the forward operation mode can be further reduced by the process engineering for the p-GaN drain contact.

Author Contributions: Simulation, device fabrication, data curation, visualization, writing-original draft preparation, T.-H.K.; device fabrication, data curation, W.-H.J. and J.-H.Y.; resources supervision, project administration, conceptualization, writing-review and editing, funding acquisition, H.-Y.C.; All authors have read and agreed to the published version of the manuscript.

Funding: This work was supported by the Basic Science Research Programs (2015R1A6A1A03031833 and 2019R1A2C1008894) and the Korea Institute for Advancement of Technology (KIAT) grant funded by the Korea Government (MOTIE) (P07820002101, The Competency Development Program for Industry Specialist).

Conflicts of Interest: The authors declare no conflict of interest.

References

1. Mishra, U.; Parikh, P.; Wu, Y.-F. AlGaN/GaN HEMTs-an overview of device operation and applications. *Proc. IEEE* **2002**, *90*, 1022–1031. [CrossRef]
2. Higashiwaki, M.; Mimura, T.; Matsui, T. AlGaN/GaN Heterostructure Field-Effect Transistors on 4H-SiC Substrates with Current-Gain Cutoff Frequency of 190 GHz. *Appl. Phys. Express* **2008**, *1*, 021103. [CrossRef]
3. Lee, M.-S.; Kim, D.; Eom, S.; Cha, H.-Y.; Seo, K.-S. A Compact 30-W AlGaN/GaN HEMTs on Silicon Substrate with Output Power Density of 8.1 W/mm at 8 GHz. *IEEE Electron. Device Lett.* **2014**, *35*, 995–997. [CrossRef]
4. Jones, E.A.; Wang, F.; Ozpineci, B. Application-based review of GaN HFETs. In Proceedings of the 2014 IEEE Workshop on Wide Bandgap Power Devices and Applications, Knoxville, TN, USA, 13–15 October 2014; pp. 24–29. [CrossRef]
5. Zeng, F.; An, J.X.; Zhou, G.; Li, W.; Wang, H.; Duan, T.; Jiang, L.; Yu, H. A Comprehensive Review of Recent Progress on GaN High Electron Mobility Transistors: Devices, Fabrication and Reliability. *Electronics* **2018**, *7*, 377. [CrossRef]
6. Ambacher, O.; Foutz, B.; Smart, J.; Shealy, J.R.; Weimann, N.G.; Chu, K.; Murphy, M.; Sierakowski, A.J.; Schaff, W.J.; Eastman, L.F.; et al. Two dimensional electron gases induced by spontaneous and piezoelectric polarization in undoped and doped AlGaN/GaN heterostructures. *J. Appl. Phys.* **2000**, *87*, 334–344. [CrossRef]
7. Ambacher, O.; Smart, J.; Shealy, J.R.; Weimann, N.G.; Chu, K.; Murphy, M.J.; Schaff, W.J.; Eastman, L.F.; Dimitrov, R.; Wittmer, L.L.; et al. Two-dimensional electron gases induced by spontaneous and piezoelectric polarization charges in N- and Ga-face AlGaN/GaN heterostructures. *J. Appl. Phys.* **1999**, *85*, 3222–3233. [CrossRef]
8. Efthymiou, L.; Longobardi, G.; Camuso, G.; Chien, T.; Chen, M.; Udrea, F. On the physical operation and optimization of the p-GaN gate in normally-off GaN HEMT devices. *Appl. Phys. Lett.* **2017**, *110*, 123502. [CrossRef]
9. Greco, G.; Iucolano, F.; Roccaforte, F. Review of technology for normally-off HEMTs with p-GaN gate. *Mater. Sci. Semicond. Process.* **2018**, *78*, 96–106. [CrossRef]
10. Meneghini, M.; Hilt, O.; Wuerfl, J.; Meneghesso, G. Technology and Reliability of Normally-Off GaN HEMTs with p-Type Gate. *Energies* **2017**, *10*, 153. [CrossRef]
11. Tapajna, M.; Hilt, O.; Bahat-Treidel, E.; Wurfl, J.; Kuzmik, J. Gate Reliability Investigation in Normally-Off p-Type-GaN Cap/AlGaN/GaN HEMTs Under Forward Bias Stress. *IEEE Electron. Device Lett.* **2016**, *37*, 385–388. [CrossRef]
12. Xu, N.; Hao, R.; Chen, F.; Zhang, X.; Zhang, H.; Zhang, P.; Ding, X.; Song, L.; Yu, G.; Cheng, K.; et al. Gate leakage mechanisms in normally off p-GaN/AlGaN/GaN high electron mobility transistors. *Appl. Phys. Lett.* **2018**, *113*, 152104. [CrossRef]
13. Hilt, O.; Knauer, A.; Brunner, F.; Bahat-Treidel, E.; Würfl, J. Normally Off AlGaN/GaN HFET with p-Type Ga Gate and AlGaN Buffer. In Proceedings of the 22nd International Symposium on Power Semi-conductor Devices & IC's (ISPSD), Hiroshima, Japan, 6–10 June 2010; pp. 347–350.
14. Bahat-Treidel, E.; Lossy, R.; Wurfl, J.; Trankle, G. AlGaN/GaN HEMT With Integrated Recessed Schottky-Drain Protection Diode. *IEEE Electron. Device Lett.* **2009**, *30*, 901–903. [CrossRef]
15. Zhou, C.; Chen, W.; Piner, E.L.; Chen, K.J. Schottky-Ohmic Drain AlGaN/GaN Normally Off HEMT With Reverse Drain Blocking Capability. *IEEE Electron. Device Lett.* **2010**, *31*, 668–670. [CrossRef]
16. Lee, J.-G.; Han, S.-W.; Park, B.-R.; Cha, H.-Y. Unidirectional AlGaN/GaN-on-Si HFETs with reverse blocking drain. *Appl. Phys. Express* **2013**, *7*, 014101. [CrossRef]
17. Zhao, S.-L.; Mi, M.-H.; Hou, B.; Luo, J.; Wang, Y.; Dai, Y.; Zhang, J.-C.; Ma, X.-H.; Hao, Y. Mechanism of improving forward and reverse blocking voltages in AlGaN/GaN HEMTs by using Schottky drain. *Chin. Phys. B* **2014**, *23*, 107303. [CrossRef]
18. Wang, H.; Mao, W.; Zhao, S.; Du, M.; Zhang, Y.; Zheng, X.; Wang, C.; Zhang, C.; Zhang, J.; Hao, Y. 1.3 kV Reverse-Blocking AlGaN/GaN MISHEMT With Ultralow Turn-On Voltage 0.25 V. *IEEE J. Electron. Devices Soc.* **2021**, *9*, 125–129. [CrossRef]
19. Silvaco Webpage. Available online: https://silvaco.co.kr/tech_lib_TCAD/simulationstandard/2012/jan_feb_mar/a2/a2.html (accessed on 10 March 2021).
20. Jang, W.-H.; Seo, K.-S.; Cha, H.-Y. P-GaN Gated AlGaN/GaN E-mode HFET Fabricated with Selective GaN Etching Process. *J. Semicond. Technol. Sci.* **2020**, *20*, 485–490. [CrossRef]

 *micromachines*

Article

High Thermal Dissipation of Normally off p-GaN Gate AlGaN/GaN HEMTs on 6-Inch N-Doped Low-Resistivity SiC Substrate

Yu-Chun Huang, Hsien-Chin Chiu *, Hsuan-Ling Kao, Hsiang-Chun Wang, Chia-Hao Liu, Chong-Rong Huang and Si-Wen Chen

Department of Electronics Engineering, Chang Gung University, Taoyuan 33324, Taiwan; james19961202@gmail.com (Y.-C.H.); snoopy@mail.cgu.edu.tw (H.-L.K.); smallflgt@hotmail.com (H.-C.W.); r3287133@gmail.com (C.-H.L.); gain525252@gmail.com (C.-R.H.); swchen@mail.cgu.edu.tw (S.-W.C.)
* Correspondence: hcchiu@mail.cgu.edu.tw

Abstract: Efficient heat removal through the substrate is required in high-power operation of Al-GaN/GaN high-electron-mobility transistors (HEMTs). Thus, a SiC substrate was used due to its popularity. This article reports the electrical characteristics of normally off p-GaN gate AlGaN/GaN high-electron-mobility transistors (HEMTs) on a low-resistivity SiC substrate compared with the traditional Si substrate. The p-GaN HEMTs on the SiC substrate possess several advantages, including electrical characteristics and good qualities of epitaxial crystals, especially on temperature performance. Additionally, the price of the low-resistivity SiC substrate is three times lower than the ordinary SiC substrate.

Keywords: p-GaN gate HEMT; normally off; low-resistance SiC substrate; temperature

Citation: Huang, Y.-C.; Chiu, H.-C.; Kao, H.-L.; Wang, H.-C.; Liu, C.-H.; Huang, C.-R.; Chen, S.-W. High Thermal Dissipation of Normally off p-GaN Gate AlGaN/GaN HEMTs on 6-Inch N-Doped Low-Resistivity SiC Substrate. *Micromachines* **2021**, *12*, 509. https://doi.org/10.3390/mi12050509

Academic Editor: Giovanni Verzellesi

Received: 31 March 2021
Accepted: 26 April 2021
Published: 1 May 2021

Publisher's Note: MDPI stays neutral with regard to jurisdictional claims in published maps and institutional affiliations.

Copyright: © 2021 by the authors. Licensee MDPI, Basel, Switzerland. This article is an open access article distributed under the terms and conditions of the Creative Commons Attribution (CC BY) license (https://creativecommons.org/licenses/by/4.0/).

1. Introduction

Gallium nitride high-electron-mobility transistors (HEMTs) have attracted increasing attention in the field of high-frequency and high-power device applications due to their high breakdown field, high mobility, and good thermal properties. However, they are naturally normally on devices. For high-power applications, off devices are normally desirable for system reliability [1]. Thus, several research works have been proposed to realize the normally off operation characteristics of AlGaN/GaN HEMTs such as fluorine base plasma treatment [2,3], an ultrathin barrier (UTB) [4], and gate-recessed structures [5,6].

Recently, GaN HEMTs with a p-GaN gate stack (p-GaN gate HEMTs) have been suggested as one of the candidates, in which a p-GaN layer on top of the AlGaN barrier depletes the 2D electron gas carriers in the channel [7–9]. The normally off p-GaN gate AlGaN/GaN high-electron-mobility transistor (HEMT) on a SiC substrate is expected to be a good choice for high-power switching components due to its high thermal conductivity, low resistivity, and high-voltage capability. Another advantage of using a SiC substrate is its lower lattice mismatch of ~3% for GaN (that of Si is ~17%). Owing to the high material properties of gallian nitride and the SiC substrate, these devices are expected to operate in high-temperature environments [10].

Here, we analyzed the DC, breakdown, pulsed, and thermal measurement performances of AlGaN/GaN HEMTs with a p-GaN gate between low-resistivity SiC and ordinary Si substrates. Finally, the heat removal through the SiC substrate had the most outstanding performance.

2. Experimental Procedures

In this work, an epitaxy wafer was grown by metal organic chemical vapor deposition on 6-in n-doped low-resistivity SiC substrates, as shown in Figure 1a. A 650-nm-thick undoped GaN channel layer was grown on top of a 3.8-µm-thick undoped GaN buffer layer. A 17.5-nm-thick undoped barrier (1.5-nm-AlN/15-nm-AlGaN/1-nm-AlN) layer was sandwiched between the GaN channel layer and a 75-nm p-type GaN cap layer. The fabrication process started with device isolation by an Ar implantation. The 5-µm-long p-GaN gate island was removed by $Cl_2/BCl_3/SF_6$ dry etching and the etching depth was stopped by the 1 nm AlN etching stop layer. After dry etching, the source and drain ohmic contacts were prepared using the electron beam evaporation of a multilayered Ti/Al/Ni/Au (25 nm/120 nm/25 nm/150 nm) sequence, patterned by a lift-off process, and annealed by a 30-s rapid thermal annealing (RTA) at 875 °C in ambient N_2. Finally, a Ti/Au (25/150 nm) gate metal stack was deposited, and 100 nm of SiN was passivated by plasma-enhanced chemical vapor deposition (PECVD). The descriptions above were made by our own laboratory research process.

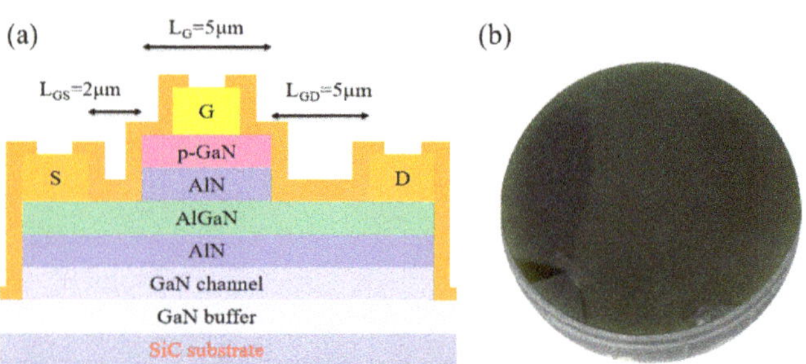

Figure 1. (**a**) Structure of p-GaN gate HEMT on low-resistance substrate. (**b**) Outward appearance of low-resistance SiC wafer.

3. Results and Discussion

Figure 2a,b show the log-scale transfer (I_{DS}-V_{GS}) and output (I_{DS}-V_{DS}) characteristics of GaN on a low-resistivity SiC substrate HEMT (LRSiC-HEMT) and a Si substrate HEMT (Si-HEMT). As shown in Figure 2a, the off-state currents for the LRSiC-HEMT and Si-HEMT are 1.37×10^{-5} and 5.2×10^{-5} mA/mm at $V_{GS} = 0$ V, the I_{on}/I_{off} ratios are 1.5×10^8 and 1.85×10^6, and they deliver the normally off operation with a positive V_{TH} of 3.2 V and 1.8 V defined at $I_D = 1$ mA/mm, respectively. In Figure 2b, the corresponding maximum drain current density (I_{Dmax}) values are 131 mA/mm and 108 mV/mm at a gate-to-source voltage (V_{GS}) = 8 V and a drain-to-source voltage (V_{DS}) = 10 V. The I_{Dmax} value of the LRSiC-HEMT was 28% higher than that of the Si-HEMT. Additionally, the LRSiC-HEMT also exhibits a lower on resistance (R_{ON}) of 16 Ω·mm.

The breakdown voltage, BV, of the devices is determined by the drain leakage current reaching 1 mA/mm. As shown in Figure 3, the off-state breakdown voltages and vertical breakdown voltages of LRSiC-HEMT and Si-HEMT are 325 V, 310 V, 413 V, and 319 V, respectively. Vertical breakdown voltage measurements were performed on both wafers by the grounded substrate, and the ohmic contact pattern was swept from 0 V up to the breakdown voltage; the size of the ohmic pattern is about 100×100 µm. Although the epitaxy technology using the low-resistivity SiC substrate was not as stable, the on device performance and the substrate's breakdown voltages were still better than the traditional Si substrate.

Having confirmed the importance of the surface temperature distribution measurements, the temperature–time curve is shown in Figure 4 by using an infrared (IR) thermographic system with micro-Raman spectroscopy. On account of having much higher thermal conductivity, SiC could achieve an outstanding performance on heat dissipation. For the set of device measurements, V_{DS} was held constant at 10 V, while I_{DS} was kept at 100 mA/mm for 60 s and then waited for the device cool down for about 50 s. As expected, the center of the gate area reached the highest temperature. As shown in Figure 4, following the time, the temperature gradient appears to increase. The temperature of LRSiC-HEMT was heated up to 34.4 °C and cooled down to 30 °C rapidly. On the contrary, Si-HEMT accumulated much more heat to reach 37.5 °C and removed it slowly.

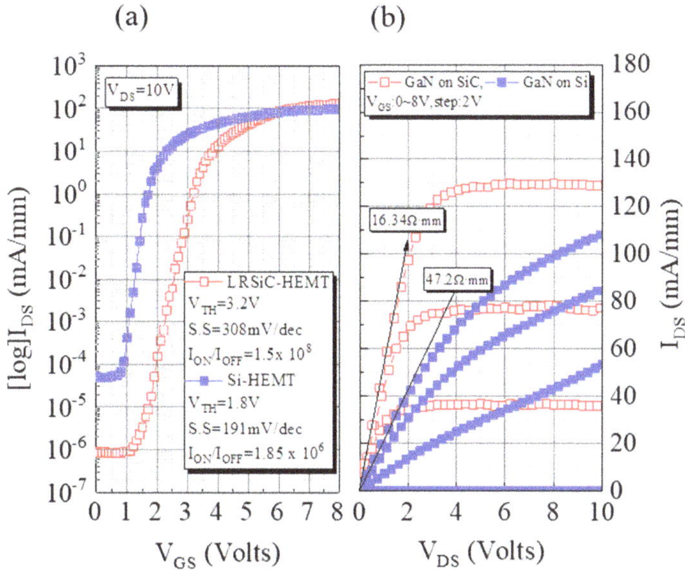

Figure 2. I–V characteristics of LRSiC-HEMT and Si-HEMT with $L_{GS}/L_G/L_{GD}/W_G = 2/5/5/100$ µm. (**a**) Transfer I_{DS}-V_{GS} characteristic. (**b**) Output I_{DS}-V_{DS} characteristic.

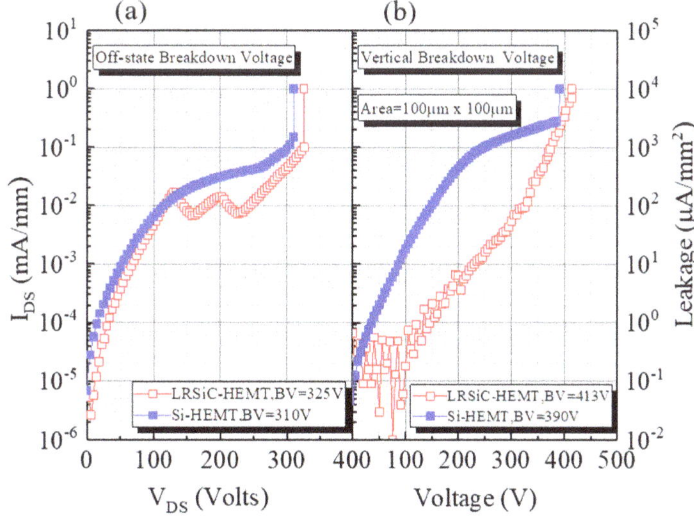

Figure 3. (**a**) Off-state breakdown voltage and (**b**) vertical breakdown voltage measurement.

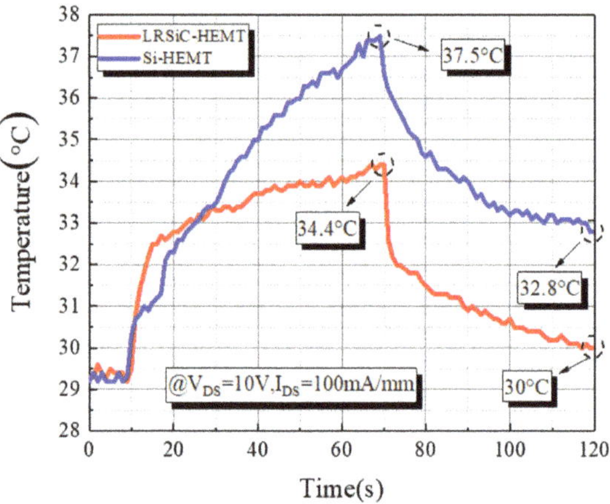

Figure 4. Temperature following the time, with the device operating for 60 s and cooling down for 50 s.

To investigate the thermal stability of LRSiC-HEMT, transfer characteristics were measured from room temperature (25 °C) to 175 °C with a 50 °C step (Figure 5a). The device shows a good thermal stability with the V_{TH} shifting less than 0.4 V up to 175 °C (Figure 5b). R_{ON} increases by about 3.3 times, due to stronger phonon scattering at higher temperature [11].

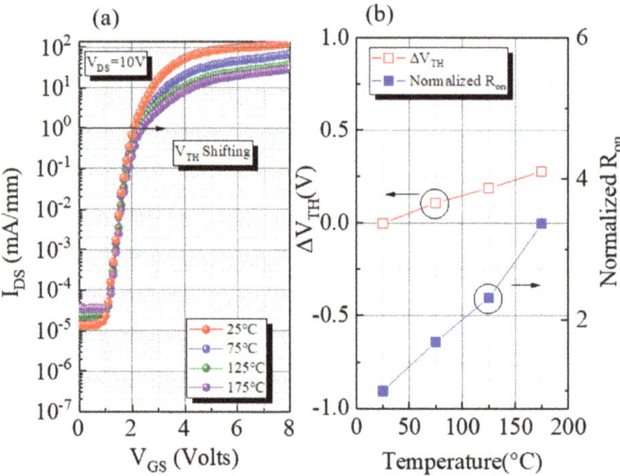

Figure 5. (a) Transfer characteristics of LRSiC-HEMT from 25 to 175 °C; (b) T-dependence of V_{TH} and R_{ON}.

To analyze the trapping/detrapping effect, the pulsed I–V characteristic and the dynamic R_{on} ratio of LRSiC-HEMT were measured using a pulse width of 2 μs and a period of 200 μs. The carriers will be trapped in the buffer layer or near the surface which is in the AlGaN layer or passivation interface during the pulse measurement. Therefore, it will reduce the carrier density and make the resistance higher [12]. In this work, the V_{DSQ} for LRSiC-HEMT is swept from 0 to 80 V with a step of 20 V in Figure 6. Clearly, the current collapse and the dynamic R_{ON} ratio of Si-HEMT are both worse than LRSiC-HEMT. Additionally, being under relatively high stress led to an I–V slope decrease because of the defect trap density in the buffer layer and surface. In Figure 7, the dynamic R_{ON} ratios are 6.8 times and 2.5 times at V_{DSQ} = 80 V.

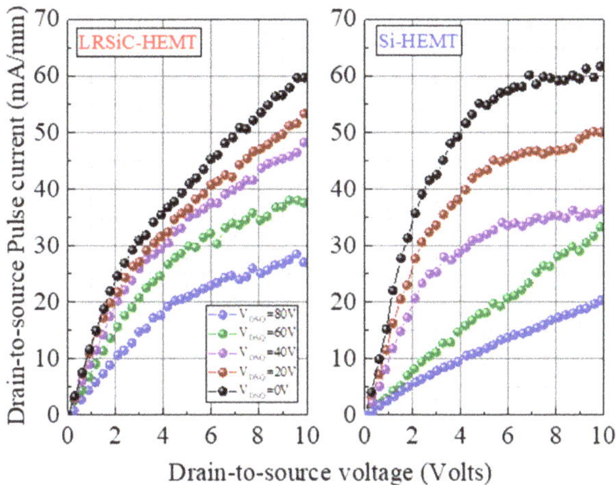

Figure 6. Pulsed I_{DS}–V_{DS} characteristics from quiescent gate bias (V_{GSQ}) point of 0 V with 2 µs pulse width and 200 µs pulse period. The quiescent drain bias (V_{DSQ}) was then swept from 0 to 80 V (in 20-V increments).

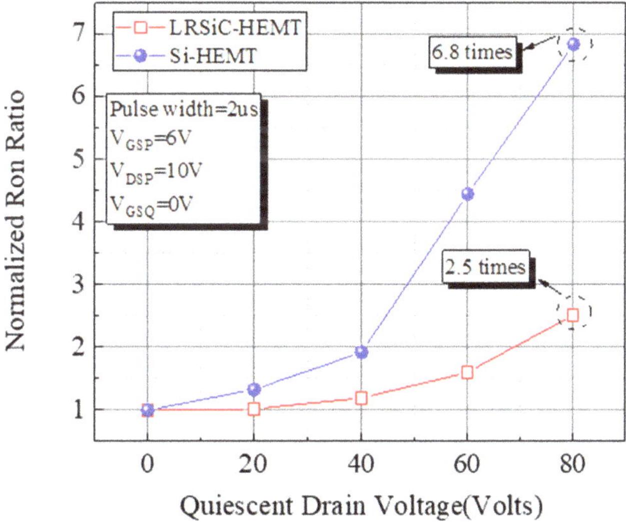

Figure 7. Dynamic R_{on} ratio of LRSiC-HEMT and Si-HEMT.

4. Conclusions

In this work, normally off AlGaN/GaN high-electron-mobility transistors (HEMTs) with a p-GaN gate on a low-resistivity SiC substrate (LRSiC-HEMT) were developed. Comparing to Si-HEMT, LRSiC-HEMT obtained many advantages, such as a higher output current, higher off-state and vertical breakdown voltages, and a lower dynamic Ron ratio, especially in thermal performance. In addition, the price of the low-resistivity SiC substrate is much lower than the ordinary SiC substrate, which is shown in Table 1. Therefore, it holds promise to be an excellent choice to solve the heat problem and cost consideration for power devices.

Table 1. Reference price and resistivity of low-resistivity SiC substrate and high-resistivity SiC substrate.

	Reference Price	Resistivity ($\Omega\cdot$cm)
LRSiC (6 inch)	USD1000	0.015~0.025
HRSiC (6 inch)	USD3000	>1E5

Author Contributions: Data curation: Y.-C.H., C.-H.L.; formal analysis: H.-C.C.; funding acquisition: H.-L.K.; investigation: C.-R.H.; methodology: H.-C.W.; supervision: H.-C.C. and S.-W.C. All authors have read and agreed to the published version of the manuscript.

Funding: This research was funded by the Ministry of Science and Technology (MOST), Taiwan, R.O.C., grant number MOST 109-2218-E-182-001.

Conflicts of Interest: The authors declare no conflict of interest.

References

1. Khan, M.A.; Chen, Q.; Sun, C.J.; Yang, J.W.; Blasingame, M.; Shur, M.S.; Park, H. Enhancement and depletion mode GaN/AlGaN heterostructure field effect transistors. *Appl. Phys. Lett.* **1996**, *68*, 514–516. [CrossRef]
2. Chen, W.; Wong, K.-Y.; Chen, K.J. Monolithic integration of lateral field-effect rectifier with normally-off HEMT for GaN-onSi switch-mode power supply converters. In Proceedings of the IEEE IEDM, San Francisco, CA, USA, 15–17 December 2008; pp. 1–4.
3. Cai, Y.; Zhou, Y.; Chen, K.J.; Lau, K.M. Highperformance enhancement-mode AlGaN/GaN HEMTs using fluoridebased plasma treatment. *IEEE Electron. Device Lett.* **2005**, *26*, 435–437. [CrossRef]
4. Ohmaki, Y.; Tanimoto, M.; Akamatsu, S.; Mukai, T. Enhancementmode AlGaN/AlN/GaN high electron mobility transistor with low ON-state resistance and high breakdown voltage. *Jpn. J. Appl. Phys.* **2006**, *45*, 42–45. [CrossRef]
5. Oka, T.; Nozawa, T. AlGaN/GaN recessed MIS-gate HFET with high-threshold-voltage normally-off operation for power electronics applications. *IEEE Electron. Device Lett.* **2008**, *29*, 668–670. [CrossRef]
6. Kim, K.-W.; Jung, S.-D.; Kim, D.-S.; Kang, H.-S.; Im, K.-S.; Oh, J.-J.; Ha, J.-B.; Shin, J.-K.; Lee, J.-H. Effects of TMAH treatment on device performance of normally off Al2O3/GaN MOSFET. *IEEE Electron. Device Lett.* **2011**, *32*, 1376–1378. [CrossRef]
7. Uemoto, Y.; Hikita, M.; Ueno, H.; Matsuo, H.; Ishida, H.; Yanagihara, M.; Ueda, T.; Tanaka, T.; Ueda, D. Gate injection transistor (GIT) normally-off AlGaN/GaN power transistor using conductivity modulation. *IEEE Trans. Electron. Devices* **2007**, *54*, 3393–3399. [CrossRef]
8. Meneghini, M.; de Santi, C.; Ueda, T.; Tanaka, T.; Ueda, D.; Zanoni, E.; Meneghesso, G. Time- and field-dependent trapping in GaN-based enhancement-mode transistors with p-gate. *IEEE Electron. Device Lett.* **2012**, *33*, 375–377. [CrossRef]
9. Hilt, O.; Knauer, A.; Brunner, F.; Bahat-Treidel, E.; Wurfl, J. Normallyoff AlGaN/GaN HFET with p-type GaN gate and AlGaN buffer. In Proceedings of the ISPSD, Hiroshima, Japan, 6–10 June 2010; pp. 347–350.
10. Fleury, C. High temperature performances of normally-off p-GaN gate AlGaN/GaN HEMTs on SiC and Si substrates for power applications. *Microelectron. Reliab.* **2015**, *55*, 1687–1691. [CrossRef]
11. Wang, C. E-Mode p-n Junction/AlGaN/GaN (PNJ) HEMTs. *IEEE Electron. Device Lett.* **2020**, *41*, 545–548. [CrossRef]
12. Hwang, I.; Kim, J.; Chong, S.; Choi, H.S.; Hwang, S.K.; Oh, J.; Shin, J.K.; Chung, U.I. Impact of channel hot electrons on current collapse in AlGaN/GaN HEMTs. *IEEE Electron. Device Lett.* **2013**, *34*, 1494–1496. [CrossRef]

 *micromachines*

Article

An Experimental and Systematic Insight into the Temperature Sensitivity for a 0.15-µm Gate-Length HEMT Based on the GaN Technology

Mohammad Abdul Alim [1], Christophe Gaquiere [2] and Giovanni Crupi [3,*]

[1] Department of Electrical and Electronic Engineering, University of Chittagong, Chittagong 4331, Bangladesh; mohammadabdulalim@cu.ac.bd

[2] Institute of Electronic, Microelectronic and Nanotechnology (IEMN), The University of Lille, F-59000 Lille, France; christophe.gaquiere@iemn.univ-lille1.fr

[3] Department of Biomedical and Dental Sciences and Morphofunctional Imaging, University of Messina, 98125 Messina, Italy

* Correspondence: crupig@unime.it

Abstract: Presently, growing attention is being given to the analysis of the impact of the ambient temperature on the GaN HEMT performance. The present article is aimed at investigating both DC and microwave characteristics of a GaN-based HEMT versus the ambient temperature using measured data, an equivalent-circuit model, and a sensitivity-based analysis. The tested device is a 0.15-µm ultra-short gate-length AlGaN/GaN HEMT with a gate width of 200 µm. The interdigitated layout of this device is based on four fingers, each with a length of 50 µm. The scattering parameters are measured from 45 MHz to 50 GHz with the ambient temperature varied from $-40\,°C$ to $150\,°C$. A systematic study of the temperature-dependent performance is carried out by means of a sensitivity-based analysis. The achieved findings show that by the heating the transistor, the DC and microwave performance are degraded, due to the degradation in the electron transport properties.

Keywords: gallium nitride (GaN); high electron-mobility transistor (HEMT); equivalent-circuit modeling; microwave frequency; scattering-parameter measurements; temperature

Citation: Alim, M.A.; Gaquiere, C.; Crupi, G. An Experimental and Systematic Insight into the Temperature Sensitivity for a 0.15-µm Gate-Length HEMT Based on the GaN Technology. *Micromachines* **2021**, *12*, 549. https://doi.org/10.3390/mi12050549

Academic Editor: Giovanni Verzellesi

Received: 26 April 2021
Accepted: 11 May 2021
Published: 12 May 2021

Publisher's Note: MDPI stays neutral with regard to jurisdictional claims in published maps and institutional affiliations.

Copyright: © 2021 by the authors. Licensee MDPI, Basel, Switzerland. This article is an open access article distributed under the terms and conditions of the Creative Commons Attribution (CC BY) license (https://creativecommons.org/licenses/by/4.0/).

1. Introduction

As well-known, high electron-mobility transistors (HEMTs) based on the aluminum gallium nitride/gallium nitride (AlGaN/GaN) heterojunction are outstanding candidates for high-frequency, high-power, and high-temperature applications, owing to the unique physical properties of the GaN material. Throughout the years, many studies have been dedicated to the investigation of how the temperature impacts the performance of GaN-based HEMT devices. To this end, both electro-thermal simulations [1–6] and measurement-based analysis [7–26] have been developed. Although the electro-thermal device simulation is undoubtedly a very powerful and costless tool to deeply understand the underlying physics behind the operation of the transistor in order to improve the device fabrication, the measurement-based investigation is a step of crucial importance for achieving a reliable validation of a transistor technology prior to its use in real applications. Typically, measurements are coupled with the extraction of a small-signal equivalent-circuit model, which can be used as cornerstone for building both large-signal [27–29] and noise [30–32] transistor models that are essential for a successful design of microwave high-power [33–36] and low-noise amplifiers [36–38]. Compared to the effective modeling approach based on using artificial neural networks (ANNs) [39,40], the equivalent-circuit model allows a physically meaningful description [41–43], thereby enabling development of a sensitivity-based investigation.

To gain a comprehensive insight, the present article focuses on the impact of the ambient temperature (T_a) on the behavior of an on-wafer GaN HEMT using DC and microwave measurements coupled with a small-signal equivalent-circuit model and a sensitivity-based analysis. The device under test (DUT) is an ultra-short gate-length HEMT based on an AlGaN/GaN heterojunction grown on a silicon carbide (SiC) substrate. The DUT has a gate length of 0.15 µm and a gate width of 200 µm. The interdigitated layout consists of four fingers, each being 50-µm long. The DC characteristics and the scattering parameters from 45 MHz to 50 GHz are measured at nine different ambient temperature conditions by both cooling and heating the device, spanning the $-40\,°C$ to $150\,°C$ temperature range. The measured data are used for equivalent-circuit extraction and sensitivity-based analysis, enabling one to assess the impact of the variation in the ambient temperature on the transistor performance. Basically, the main goal of this work is to extend the results of a previous article focused on the same DUT [15] by developing a sensitivity-based analysis, thus enabling a quantitative and systematic investigation of the effects of changes in the ambient temperature on the DC and microwave characteristics. Nevertheless, it should be pointed out that the obtained results are not of general validity, as they may strongly depend on the selected device and operating bias condition.

The paper is structured with the following sections. Section 2 describes the DUT and the experimental characterization, Section 3 reports and discusses the achieved findings, and Section 4 presents the conclusions.

2. Device under Test and Experimental Details

The metal organic chemical vapor deposition (MOCVD) technique is used to grow the $Al_{0.253}Ga_{0.747}N/GaN$ heterostructure on a 400-µm-thick SiC substrate. The schematic cross-sectional view and the photograph of the tested GaN HEMT are illustrated in Figure 1. The epitaxial layer structure of the device is made up of a 25-nm-thick undoped (UD) AlGaN barrier and a 1.5-µm-thick UD GaN buffer layer. A 300-nm-thick graded AlN relaxation layer was grown between the GaN buffer and the SiC substrate. The device was capped with a 5-nm-tick n+-GaN layer. The evaporation process was employed to create the source and drain ohmic contacts (Ti/Al/Ni/Au with thicknesses of 12/200/40/100 nm, respectively) and followed by 30 s of thermal annealing at $900\,°C$. The Schottky mushroom-shaped gate was formed through Pt/Ti/Pt/Au evaporation and the subsequent lift-off process. Finally, a Si_3N_4 layer with a thickness of 240 nm was deposited to passivate the device. The gate length of the tested GaN device is 0.15 µm. The interdigitated architecture of the device is based on the parallel connection of four 50-µm long fingers, resulting in a total gate width of 200 µm. The source-to-gate distance (L_{SG}) and the gate-to-drain distance (L_{GD}) are 1 µm and 2.85 µm, respectively. The DUT was fabricated at the University of Lille, France.

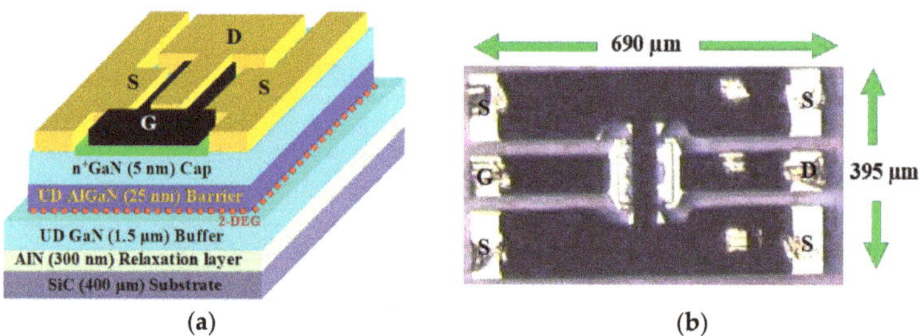

Figure 1. (**a**) Schematic drawing of the epitaxial structure and (**b**) photograph of the tested 0.15 µm $\times\,(4 \times 50)$ µm GaN HEMT.

The microwave experiments consist of DC and S-parameters measured from 45 MHz to 50 GHz at nine different ambient temperatures: $-40\,°C$, $-25\,°C$, $0\,°C$, $25\,°C$, $50\,°C$, $75\,°C$, $100\,°C$, $125\,°C$, and $150\,°C$. The analysis is performed using the DC characteristics and the S-parameters at a bias point in the saturation region: $V_{ds} = 15$ and $V_{gs} = -5$ V. The device parameters were measured with a thermal probe station connected to an HP8510C vector network analyzer (VNA) and with the aid of commercially available software to guarantee that the data are free of human error. The DC and frequency-dependent measurements were performed at each temperature after the sample reached uniform steady-state temperature. Figure 2 shows the measurement process, model extraction, and sensitivity-based analysis.

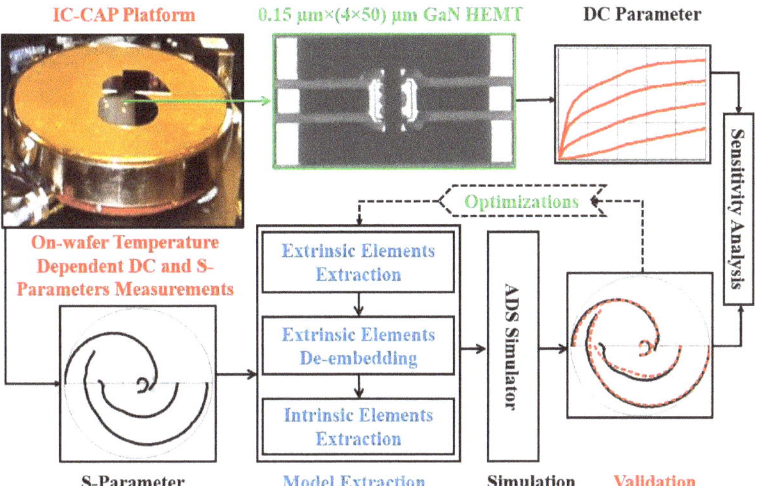

Figure 2. The flow diagram of the measurement process, model extraction, and sensitivity-based analysis for the tested on-wafer GaN HEMT.

3. Experimental Results and Systematic Analysis

The systematic sensitivity-based analysis at the selected bias voltages is accomplished using the dimensionless relative sensitivity of each parameter (*RSP*) with respect to T_a, which is calculated by normalizing the relative change in P to the relative change in T_a:

$$RSP = \frac{\Delta P}{P_0}\frac{T_{a0}}{\Delta T_a} = \frac{(P - P_0)}{P_0}\frac{T_{a0}}{(T_a - T_{a0})} \tag{1}$$

where P_0 is the value of the selected parameter P at the reference temperature (T_{a0}) of 25 °C.

The remainder of this section is divided into two subsections: the first part is focused on the impact of the ambient temperature on the DC characteristics, whereas the second part is dedicated to the effects of the variations in the ambient temperature on the microwave performance.

3.1. Sensitivity-Based Analysis of DC Characteristics

The DC output characteristics for the tested GaN HEMT at $V_{gs} = -4$ V and -5 V under different temperature conditions are illustrated in Figure 3. As can be clearly observed, I_{ds} is considerably reduced with increasing temperature. This might be attributed to the degradation in the carrier transport properties as a consequence of the enhancement of the phonon-scattering processes at higher temperatures. Analogously, the reduction in I_{ds} at higher temperatures can be observed by plotting the DC transcharacteristics of the studied device at $V_{ds} = 15$ V (see Figure 4). Similar fashion of degradation can be seen in the transconductance by plotting the g_m-V_{gs} curves at $V_{ds} = 15$ V (see Figure 5a). As a matter of fact, by heating the device, the transconductance is significantly reduced. However, it should be underlined that a higher temperature leads to a wider and flatter curve of g_m versus V_{gs}, thus implying a better linearity. Over the years, many studies have been devoted at improving the flatness of g_m versus V_{gs}, in order to yield to an improved transistor linearity and then to a more linear power amplifier [44,45]. For the sake of completeness, the behavior of g_m is plotted also as a function of I_{ds} (see Figure 5b). At the selected bias point: $V_{ds} = 15$ V and $V_{gs} = -5$ V, both I_{ds} and g_m are significantly degraded when the temperature is raised, as illustrated in Figure 6a. The interesting feature found in the $g_m - V_{gs}$ curves of Figure 5a is that, by heating the device, the peak value of g_m is not only greatly reduced but also shifted toward less negative values of V_{gs}. As shown in Figure 6b, the value of V_{gs} at which the peak in g_m occurs (V_{gm}) is increased from -5.2 V at $-40\,°$C to -4.8 V at $150\,°$C. It is worth noting that also the threshold voltage (V_{th}) shifts toward less negative values at higher T_a. As illustrated in Figure 6b, V_{th} is increased from -6.24 V at $-40\,°$C to -5.64 V at $150\,°$C.

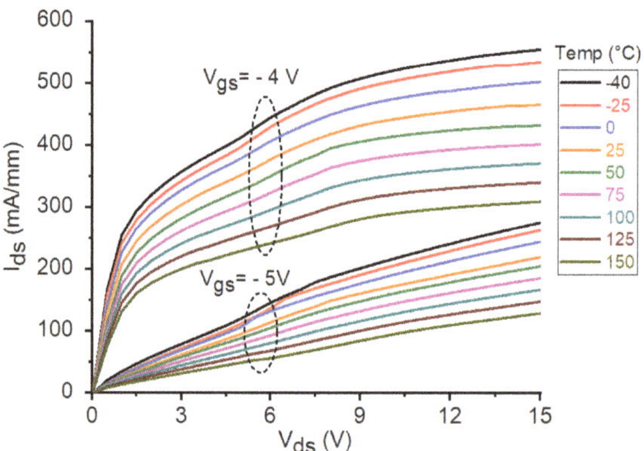

Figure 3. DC output characteristics of the studied GaN HEMT at $V_{gs} = -4$ V and -5 V under different temperature conditions.

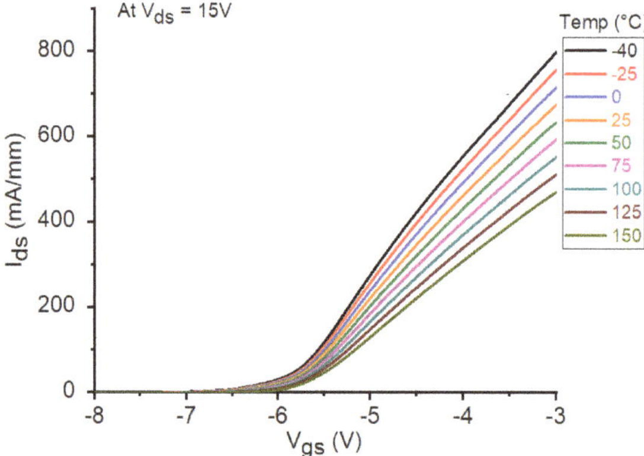

Figure 4. DC transcharacteristics of the studied GaN HEMT at $V_{ds} = 15$ V under different temperature conditions.

Using Equation (1), the relative sensitivities of I_{ds}, g_m, V_{gm}, and V_{th} with respect to T_a are calculated and reported in Figure 7. As can be observed, RSI_{ds}, RSg_m, RSV_{th}, and RSV_{gm} are negative for the studied device, as a consequence of the fact that an increase in T_a leads to a reduction in the values of I_{ds}, g_m, V_{th}, and V_{gm}.

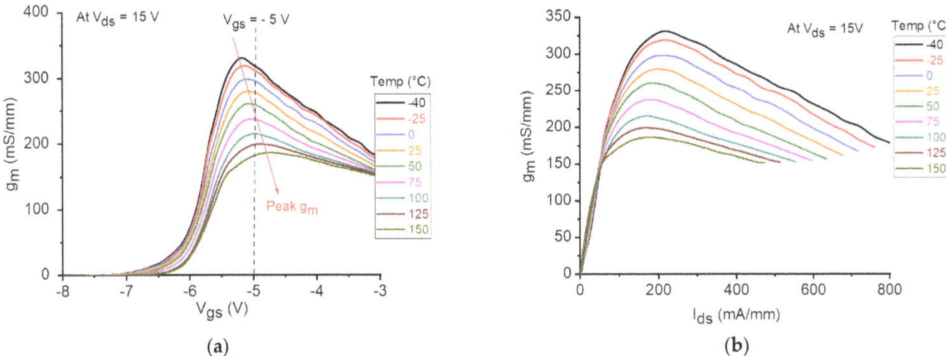

Figure 5. DC transconductance of the studied GaN HEMT at $V_{ds} = 15$ V under different temperature conditions. The data are reported as a function of (**a**) V_{gs} and (**b**) I_{ds}.

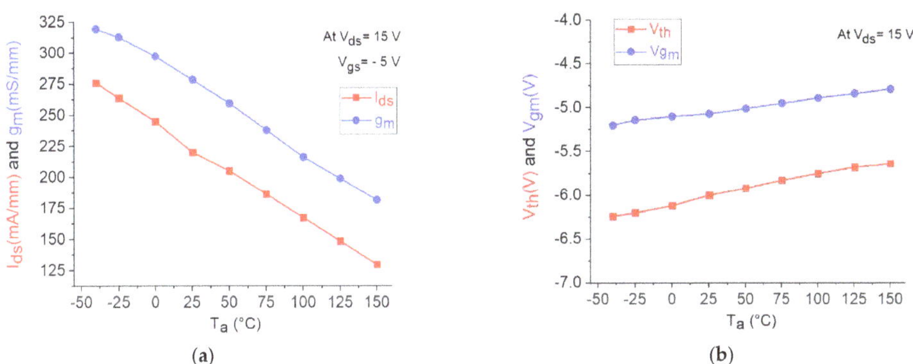

Figure 6. Temperature dependence of the DC parameters of the studied GaN HEMT: (**a**) I_{ds} and g_m at $V_{ds} = 15$ V and $V_{gs} = -5$ V; (**b**) V_{th} and V_{gm} (i.e., V_{gs} at which g_m shows its peak value) at $V_{ds} = 15$ V.

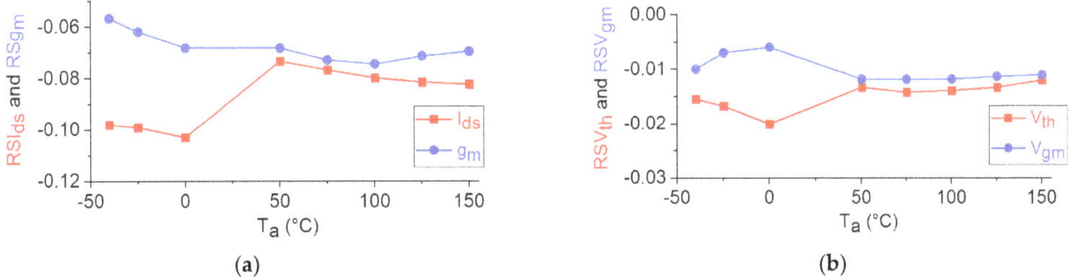

Figure 7. Temperature dependence of the relative sensitivities of the DC parameters of the studied GaN HEMT: (**a**) RSI_{ds} and RSg_m at $V_{ds} = 15$ V and $V_{gs} = -5$ V; (**b**) RSV_{th} and RSV_{gm} at $V_{ds} = 15$ V.

3.2. Sensitivity-Based Analysis of Small-Signal Parameters and RF Figures of Merit

The equivalent-circuit model in Figure 8 was used to model the measured S-parameters of the studied device. The equivalent-circuit parameters (ECPs) were extracted as described in [15], using the well-known "cold" pinch-off approach that has been widely and successfully applied to the GaN technology over the years [46–50]. The effect of T_a on the measured S-parameters at the selected bias point is shown in Figure 9. It should be highlighted that as the carrier transport properties deteriorate with increasing T_a, the low-frequency magnitude of S_{21} is reduced. This is in line with the degradation of the DC g_m at higher T_a (see Figure 5). As can be observed, the tested device is affected by the kink effect in S_{22}. As well-known, the GaN HEMT technology is prone to be affected by this phenomenon, owing to the relatively high transconductance [51–54]. In accordance with this, the observed kink effect in S_{22} is more pronounced at lower T_a, due to the higher g_m. The DC parameters, ECPs, intrinsic input and feedback time constants (i.e., $\tau_{gs} = R_{gs}C_{gs}$ and $\tau_{gd} = R_{gd}C_{gd}$), the unity current gain cut-off frequency (f_t), and the maximum frequency of oscillation (f_{max}) are reported at 25 °C in Table 1. The three intrinsic time constants (τ_m, τ_{gs}, and τ_{gd}), which emerge from the inertia of the intrinsic transistor in reacting to rapid signal changes, are meant to represent the intrinsic non-quasi-static (NQS) effects, which play a more significant role at higher frequencies. The values of f_t and f_{max} are obtained from the frequency-dependent behavior of the measured short-circuit current gain (h_{21}) and maximum stable/available gain (MSG/MAG), respectively (see Figure 10).

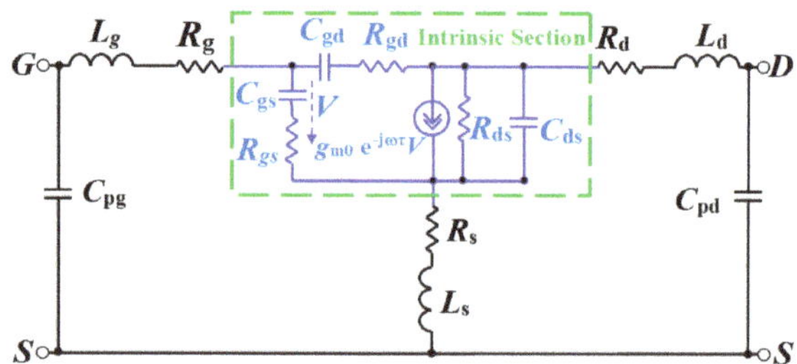

Figure 8. The equivalent-circuit model for the GaN HEMT under investigation.

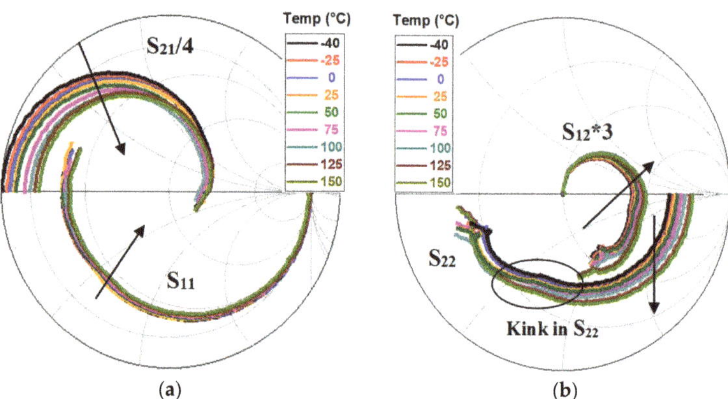

(a)　　　　　(b)

Figure 9. Measured S-parameters of the studied GaN HEMT at V_{ds} = 15 V and V_{gs} = −5 V under different temperature conditions: (**a**) S_{11}, S_{21}, (**b**) S_{12}, and S_{22}.

Similarly, to what was done for the DC parameters, the relative sensitivities of the other parameters in Table 1 are calculated using equation 1 and then shown in Figure 11. Because of their low dependence on the temperature, the relative sensitivities of the extrinsic capacitances and inductances are almost nil, as depicted in Figure 11a,b. It can be observed in Figure 11c–e that the relative sensitivities of the extrinsic and intrinsic resistances are positive, reflecting the fact that the resistive contributions increase at higher temperatures. Figure 11f illustrates that unlike the resistances, the transconductance has a negative relative sensitivity, as this parameter is degraded when heating the device.

Table 1. Analyzed parameters for the studied GaN HEMTs at 25 °C for the bias condition of $V_{ds} = 15$ V and $V_{gs} = -5$ V.

Parameters	Value	Parameters	Value
I_{ds} (mA)	44	C_{gs} (fF)	154
g_m (mS)	54	C_{gd} (fF)	28
V_{th} (V)	−6	C_{ds} (fF)	121
V_{gm} (V)	−5.1	R_{gs} (Ω)	1.7
C_{pg} (fF)	50	R_{gd} (Ω)	11.3
C_{pd} (fF)	86	R_{ds} (Ω)	225
L_g (pH)	141	g_{mo} (mS)	63
L_s (pH)	1.8	τ_m (ps)	3.0
L_d (pH)	63	τ_{gs} (ps)	1.6
R_g (Ω)	2.7	τ_{gd} (ps)	2.0
R_s (Ω)	3.0	f_t (GHz)	47.3
R_d (Ω)	5.8	f_{max} (GHz)	97

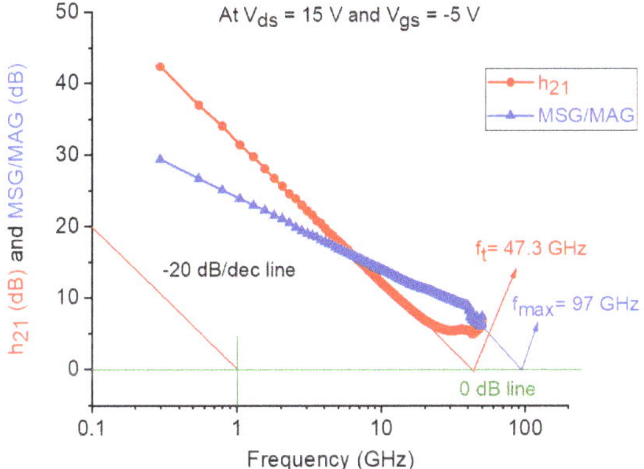

Figure 10. Behavior of the magnitude of the h_{21} and MAG/MSG versus the frequency of the studied GaN HEMT at $V_{ds} = 15$ V, $V_{gs} = -5$ V, and $T_a = 25$ °C.

As illustrated in Figure 11d, the relative sensitivity of C_{gs} is negative, while the relative sensitivities of C_{gd} and C_{ds} are positive. Figure 11g shows that the relative sensitivities of the intrinsic time constants are positive, indicating that they increase when the temperature is raised. This finding implies that the NQS effects occur at lower frequencies when the device is heated. As can be observed in Figure 11f, the relative sensitivities of f_t and f_{max} are negative, implying lower operating frequencies at higher temperatures. Figure 11h shows that the relative sensitivities of the magnitude of S_{21} and h_{21} at 45 MHz are negative, in line with the reduction of the transconductance at higher temperatures, while the stability factor (K) shows a positive temperature sensitivity as illustrated at 1 GHz.

For the tested device, a good agreement between measured and simulated S-parameters was achieved. As an example, Figure 12 depicts the comparison between measurements and S-parameter simulations at two different T_a for the tested GaN HEMT at the selected bias condition. The simulations are obtained using the equivalent-circuit model depicted in Figure 8 by means of the commercial microwave simulation software advanced design system (ADS). The small-signal ECPs extracted for different T_a from the measured S-parameters are used as inputs to the schematic.

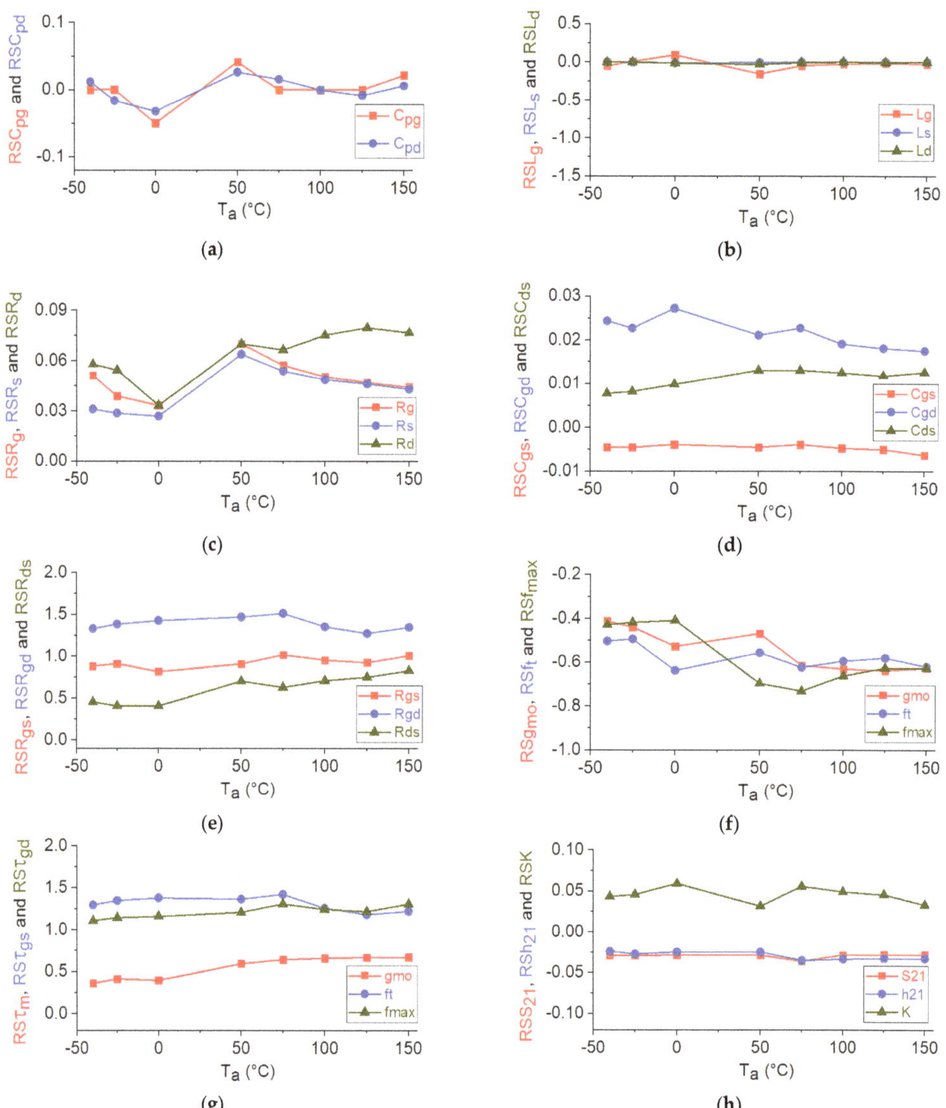

Figure 11. The relative sensitivities of the analyzed parameters versus ambient temperature for the studied GaN HEMT: (**a**) RSC_{pg} and RSC_{pd}; (**b**) RSL_g, RSL_s, and RSL_d; (**c**) RSR_g, RSR_s, and RSR_d; (**d**) RSC_{gs}, RSC_{gd}, and RSC_{ds}; (**e**) RSR_{gs}, RSR_{gd}, and RSR_{ds}; (**f**) RSg_{mo}, RSf_t, and RSf_{max}; (**g**) $RS\tau_m$, $RS\tau gs$, and $RS\tau gd$; (**h**) RSS_{21} and RSh_{21} at 45 MHz and RSK at 1 GHz The illustrated bias points for intrinsic parameters and RF figures of merits is: $V_{ds} = 15$ V and $V_{gs} = -5$ V.

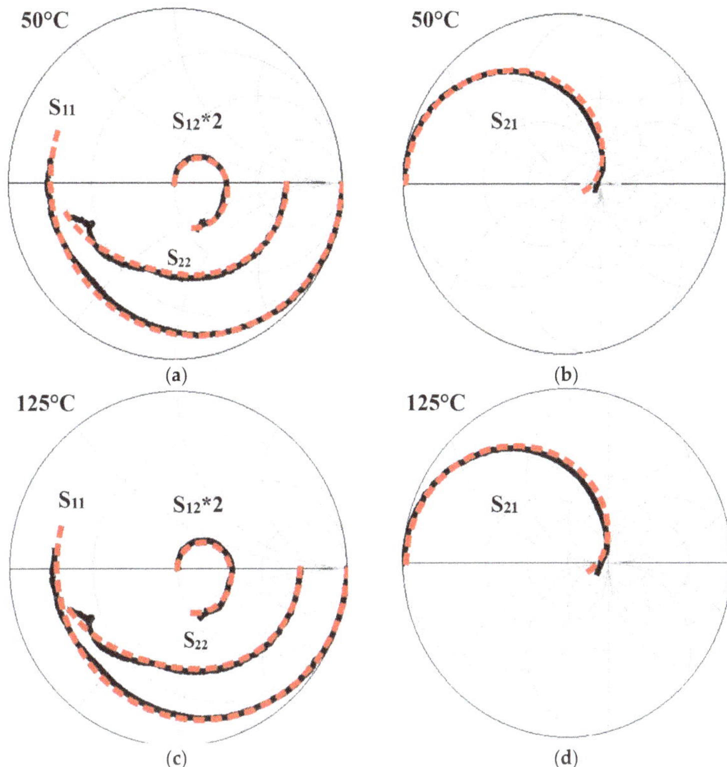

Figure 12. Measured (solid black lines) and simulated (dashed red lines) S-parameters from 45 MHz to 50 GHz for the studied GaN HEMT at 50 °C (**a**,**b**) and 125 °C (**c**,**d**). The illustrated bias point is: $V_{ds} = 15$ V and $V_{gs} = -5$ V.

4. Conclusions

We have reported an experimental investigation on the impact of the ambient temperature on the DC and microwave performance of a transistor based on an ultra-short 0.15-μm GaN HEMT technology. Measurements have been coupled with an equivalent-circuit model and a sensitivity-based study to assess the thermal effects on device performance over the wide temperature range going from -40 °C to 150 °C. The relative sensitivity was used as the evaluation indicator for this study because it enables investigation of the effects of the ambient temperature on the device performance in a quantitative, systematic, and simple way. The measurement-based findings show that both DC and microwave performance of the studied device are remarkably degraded with increasing temperature.

Author Contributions: Conceptualization, M.A.A. and G.C.; methodology, M.A.A. and G.C.; validation, M.A.A.; investigation, M.A.A.; writing—original draft preparation, M.A.A. and G.C.; writing—review and editing, C.G. and G.C.; supervision, C.G. and G.C. All authors have read and agreed to the published version of the manuscript.

Funding: This research received no external funding.

Institutional Review Board Statement: Not applicable.

Informed Consent Statement: Not applicable.

Data Availability Statement: The data presented in this study are available on request from the authors.

Conflicts of Interest: The authors declare no conflict of interest. The funders had no role in the design of the study; in the collection, analyses, or interpretation of data; in the writing of the manuscript, or in the decision to publish the results.

References

1. Heller, E.R.; Crespo, A. Electro-thermal modeling of multifinger AlGaN/GaN HEMT device operation including thermal substrate effects. *Microelectron. Reliab.* **2008**, *48*, 45–50. [CrossRef]
2. Bertoluzza, F.; Delmonte, N.; Menozzi, R. Three-dimensional finite-element thermal simulation of GaN-based HEMTs. *Microelectron. Reliab.* **2009**, *49*, 468–473. [CrossRef]
3. Zhang, Y.; Sun, M.; Liu, Z.; Piedra, D.; Lee, H.-S.; Gao, F.; Fujishima, T.; Palacios, T. Electrothermal Simulation and Thermal Performance Study of GaN Vertical and Lateral Power Transistors. *IEEE Trans. Electron Devices* **2013**, *60*, 2224–2230. [CrossRef]
4. Chvála, A.; Marek, J.; Príbytný, P.; Šatka, A.; Donoval, M.; Donoval, D. Effective 3-D device electrothermal simulation analysis of influence of metallization geometry on multifinger power HEMTs properties. *IEEE Trans. Electron Device* **2017**, *64*, 333–336. [CrossRef]
5. Gupta, M.P.; Vallabhaneni, A.K.; Kumar, S. Self-consistent electrothermal modeling of passive and microchannel cooling in AlGaN/GaN HEMTs. *IEEE Trans. Compon. Packag. Manuf. Technol.* **2017**, *7*, 1305–1312. [CrossRef]
6. Hao, Q.; Zhao, H.; Xiao, Y.; Kronenfeld, M.B. Electrothermal studies of GaN-based high electron mobility transistors with improved thermal designs. *Microelectron. Reliab.* **2018**, *116*, 496–506. [CrossRef]
7. Akita, M.; Kishimoto, S.; Mizutani, T. High-frequency measurements of AlGaN/GaN HEMTs at high temperatures. *IEEE Electron. Device Lett.* **2001**, *22*, 376–377. [CrossRef]
8. Nuttinck, S.; Gebara, E.; Laskar, J.; Harris, H.M. Study of self-heating effects, temperature-dependent modeling, and pulsed load-pull measurements on GaN HEMTs. *IEEE Trans. Microw. Theory Tech.* **2001**, *49*, 2413–2420. [CrossRef]
9. Lee, J.W.; Webb, K.J. A temperature-dependent nonlinear analytic model for AlGaN-GaN HEMTs on SiC. *IEEE Trans. Microw. Theory Tech.* **2004**, *52*, 2–9. [CrossRef]
10. Thorsell, M.; Andersson, K.; Fagerlind, M.; Südow, M.; Nilsson, P.-A.; Rorsman, N. Thermal study of the high-frequency noise in GaN HEMTs. *IEEE Trans. Microw. Theory Tech.* **2009**, *57*, 19–26. [CrossRef]
11. Darwish, A.M.; Huebschman, B.D.; Viveiros, E.; Hung, H.A. Dependence of GaN HEMT millimeter-wave performance on temperature. *IEEE Trans. Microw. Theory Tech.* **2009**, *57*, 3205–3211. [CrossRef]
12. Darwish, A.M.; Ibrahim, A.A.; Hung, H.A. Temperature Dependence of GaN HEMT Small Signal Parameters. *Int. J. Microw. Sci. Technol.* **2011**, *2011*, 945189. [CrossRef]
13. Marinković, Z.; Crupi, G.; Caddemi, A.; Avolio, G.; Raffo, A.; Marković, V.; Vannini, G.; Schreurs, D.M.M.-P. Neural approach for temperature-dependent modeling of GaN HEMTs. *Int. J. Numer. Model. Electron. Netw. Devices Fields* **2014**, *28*, 359–370. [CrossRef]
14. Crupi, G.; Raffo, A.; Avolio, G.; Schreurs, D.M.M.-P.; Vannini, G.; Caddemi, A. Temperature influence on GaN HEMT equivalent circuit. *IEEE Microw. Wirel. Comp. Lett.* **2016**, *26*, 813–815. [CrossRef]
15. Alim, M.A.; Rezazadeh, A.A.; Gaquiere, C. Temperature effect on DC and equivalent circuit parameters of 0.15-μm gate length GaN/SiC HEMT for microwave applications. *IEEE Trans. Microw. Theory Tech.* **2016**, *64*, 3483–3491. [CrossRef]
16. Zhao, X.; Xu, Y.; Jia, Y.; Wu, Y.; Xu, R.; Li, J.; Hu, Z.; Wu, H.; Dai, W.; Cai, S. Temperature-Dependent Access Resistances in Large-Signal Modeling of Millimeter-Wave AlGaN/GaN HEMTs. *IEEE Trans. Microw. Theory Tech.* **2017**, *65*, 2271–2278. [CrossRef]
17. Bouslama, M.; Gillet, V.; Chang, C.; Nallatamby, J.-C.; Sommet, R.; Prigent, M.; Quere, R.; Lambert, B. Dynamic Performance and Characterization of Traps Using Different Measurements Techniques for the New AlGaN/GaN HEMT of 0.15-μm Ultrashort Gate Length. *IEEE Trans. Microw. Theory Tech.* **2019**, *67*, 2475–2482. [CrossRef]
18. Alim, M.A.; Hasan, M.A.; Rezazadeh, A.A.; Gaquiere, C.; Crupi, G. Multibias and temperature dependence of the current-gain peak in GaN HEMT. *Int. J. RF Microw. Comput. Aided Eng.* **2019**, *30*, e22129. [CrossRef]
19. Crupi, G.; Raffo, A.; Vadalà, V.; Vannini, G.; Caddemi, A. High-periphery GaN HEMT modeling up to 65 GHz and 200 °C. *Solid-State Electron.* **2019**, *152*, 11–16. [CrossRef]
20. Chen, Y.; Xu, Y.; Kong, Y.; Chen, T.; Zhang, Y.; Yan, B.; Xu, R. Temperature-dependent small signal performance of GaN-on-diamond HEMTs. *Int. J. Numer. Model. Electron. Netw. Dev. Field* **2020**, *33*, e2620. [CrossRef]
21. Gryglewski, D.; Wojtasiak, W.; Kamińska, E.; Piotrowska, A. Characterization of self-heating process in GaN-based HEMTs. *Electronics* **2020**, *9*, 1305. [CrossRef]
22. Majumdar, A.; Chatterjee, S.; Chatterjee, S.; Chaudhari, S.S.; Poddar, D.R. An ambient temperature dependent small signal model of GaN HEMT using method of curve fitting. *Int. J. RF Microw. Comput. Aided Eng.* **2020**, *30*, e22434. [CrossRef]
23. Jarndal, A.; Husain, S.; Hashmi, M. Genetic algorithm initialized artificial neural network based temperature dependent small-signal modeling technique for GaN high electron mobility transistors. *Int. J. RF Microw. Comput. Aided Eng.* **2021**, *31*, e22542. [CrossRef]
24. Luo, H.; Zhong, Z.; Hu, W.; Guo, Y. Analysis and modeling of the temperature-dependent nonlinearity of intrinsic capacitances in AlGaN/GaN HEMTs. *IEEE Microw. Wirel. Comp. Lett.* **2021**, *31*, 373–376. [CrossRef]
25. Colangeli, S.; Ciccognani, W.; Longhi, P.E.; Pace, L.; Poulain, J.; Leblanc, R.; Limiti, E. Linear characterization and modeling of GaN-on-Si HEMT technologies with 100 nm and 60 nm gate lengths. *Electronics* **2021**, *10*, 134. [CrossRef]

26. Jarndal, A.; Alim, M.A.; Raffo, A.; Crupi, G. 2-mm-gate-periphery GaN high electron mobility transistors on SiC and Si substrates: A comparative analysis from a small-signal standpoint. *Int. J. RF Microw. Comput. Aided Eng.* **2021**, *31*, e22642. [CrossRef]
27. Root, D.E.; Hughes, B. Principles of nonlinear active device modeling for circuit simulation. In Proceedings of the 32nd ARFTG Conference Digest, Tempe, AZ, USA, 1–2 December 1988; pp. 1–24.
28. Xu, Y.; Wang, C.; Sun, H.; Wen, Z.; Wu, Y.; Xu, R.; Yu, X.; Ren, C.; Wang, Z.; Zhang, B.; et al. A Scalable Large-Signal Multiharmonic Model of AlGaN/GaN HEMTs and Its Application in C-Band High Power Amplifier MMIC. *IEEE Trans. Microw. Theory Tech.* **2017**, *65*, 2836–2846. [CrossRef]
29. Raffo, A.; Avolio, G.; Vadalà, V.; Schreurs, D.M.M.-P.; Vannini, G. A non-quasi-static FET model extraction procedure using the dynamic-bias technique. *IEEE Microw. Wirel. Comp. Lett.* **2015**, *25*, 841–843. [CrossRef]
30. Lee, S.; Webb, K.J.; Tialk, V.; Eastman, L.F. Intrinsic noise equivalent-circuit parameters for AlGaN/GaN HEMTs. *IEEE Trans Microw. Theory Tech.* **2003**, *51*, 1567–1577.
31. Pace, L.; Colangeli, S.; Ciccognani, W.; Longhi, P.E.; Limiti, E.; Leblanc, R.; Feudale, M.; Vitobello, F. Design and validation of 100 nm GaN-On-Si Ka-band LNA based on custom noise and small signal models. *Electronics* **2020**, *9*, 150. [CrossRef]
32. Jarndal, A.; Hussein, A.; Crupi, G.; Caddemi, A. Reliable noise modeling of GaN HEMTs for designing low-noise amplifiers. *Int. J. Numer. Model. Electron. Netw. Dev. Field* **2020**, *33*, e2585. [CrossRef]
33. Cabral, P.M.; Pedro, J.C.; Carvalho, N.B. Nonlinear device model of microwave power GaN HEMTs for high power-amplifier design. *IEEE Trans. Microw. Theory Tech.* **2004**, *52*, 2585–2592. [CrossRef]
34. Raffo, A.; Scappaviva, F.; Vannini, G. A new approach to microwave power amplifier design based on the experimental characterization of the intrinsic electron-device load line. *IEEE Trans. Microw. Theory Tech.* **2009**, *57*, 1743–1752. [CrossRef]
35. Kim, S.; Lee, M.-P.; Hong, S.-J.; Kim, D.-W. Ku-Band 50 W GaN HEMT power amplifier using asymmetric power combining of transistor cells. *Micromachines* **2018**, *9*, 619. [CrossRef]
36. Crupi, G.; Vadalà, V.; Colantonio, P.; Cipriani, E.; Caddemi, A.; Vannini, G.; Schreurs, D.M.M.-P. Empowering GaN HEMT models: The gateway for power amplifier design. *Int. J. Numer. Model. Electron. Netw. Devices Fields* **2015**, *30*, e2125. [CrossRef]
37. Colangeli, S.; Bentini, A.; Ciccognani, W.; Limiti, E.; Nanni, A. GaN-based robust low-noise amplifiers. *IEEE Trans. Electron Dev.* **2013**, *60*, 3238–3248. [CrossRef]
38. Nalli, A.; Raffo, A.; Crupi, G.; D'Angelo, S.; Resca, D.; Scappaviva, F.; Salvo, G.; Caddemi, A.; Vannini, G. GaN HEMT Noise Model Based on Electromagnetic Simulations. *IEEE Trans. Microw. Theory Tech.* **2015**, *63*, 2498–2508. [CrossRef]
39. Marinković, Z.; Crupi, G.; Caddemi, A.; Markovic, V.; Schreurs, D.M.M.-P. A review on the artificial neural network applications for small-signal modeling of microwave FETs. *Int. J. Numer. Model. Electron. Netw. Dev. Field* **2020**, *33*, e2668. [CrossRef]
40. Jin, J.; Feng, F.; Na, W.C.; Yan, S.X.; Liu, W.Y.; Zhu, L.; Zhang, Q.J. Recent advances in neural network-based inverse modeling techniques for microwave applications. *Int. J. Numer. Model. Electron. Netw. Dev. Field* **2020**, *33*, e2732. [CrossRef]
41. Dambrine, G.; Cappy, A.; Heliodore, F.; Playez, E. A new method for determining the FET small-signal equivalent circuit. *IEEE Trans Microw. Theory Tech.* **1988**, *36*, 1151–1159. [CrossRef]
42. Alt, R.; Marti, D.; Bolognesi, C.R. Transistor modeling: Robust small-signal equivalent circuit extraction in various HEMT technologies. *IEEE Microw. Mag.* **2013**, *14*, 83–101. [CrossRef]
43. Crupi, G.; Caddemi, A.; Schreurs, D.M.M.-P.; Dambrine, G. The large world of FET small-signal equivalent circuits. *Int. J. RF Microw. Comput. Aided Eng.* **2016**, *26*, 749–762. [CrossRef]
44. Chen, C.-H.; Sadler, R.; Wang, D.; Hou, D.; Yang, Y.; Yau, W.; Sutton, W.; Shim, J.; Wang, S.; Duong, A. The causes of GaN HEMT bell-shaped transconductance degradation. *Solid-State Electron.* **2016**, *126*, 115–124. [CrossRef]
45. Ancona, M.G.; Calame, J.P.; Meyer, D.J.; Rajan, S.; Downey, B.P. Compositionally graded III-N HEMTs for improved linearity: A simulation study. *IEEE Trans. Electron Device* **2019**, *66*, 2151–2157. [CrossRef]
46. Jarndal, A.; Kompa, G. A new small signal modeling approach applied to GaN devices. *IEEE Trans. Microw. Theory Tech.* **2005**, *53*, 3440–3448. [CrossRef]
47. Brady, R.G.; Oxley, C.H.; Brazil, T.J. An improved small-signal parameter-extraction algorithm for GaN HEMT devices. *IEEE Trans. Microw. Theory Tech.* **2008**, *56*, 1535–1544. [CrossRef]
48. Al Sabbagh, M.; Yagoub, M.C.E.; Park, J. New small-signal extraction method applied to GaN HEMTs on different substrates. *Int. J. R. F. Microw. Comput. Aided Eng.* **2020**, *30*, e22291. [CrossRef]
49. Chen, Y.; Xu, Y.; Luo, Y.; Wang, C.; Wen, Z.; Yan, B.; Xu, R. A reliable and efficient small-signal parameter extraction method for GaN HEMTs. *Int. J. Numer. Model. Electron. Netw. Dev. Field* **2020**, *33*, e2540. [CrossRef]
50. Du, X.; Helaoui, M.; Cai, J.; Liu, J.; Ghannouchi, F.M. Improved small-signal hybrid parameter-extraction technique for AlGaN/GaN high electron mobility transistors. *Int. J. RF Microw. Comput. Aided Eng.* **2021**, *31*, e22562. [CrossRef]
51. Crupi, G.; Raffo, A.; Marinkovic, Z.; Avolio, G.; Caddemi, A.; Markovic, V.; Vannini, G.; Schreurs, D.M.M.-P. An Extensive Experimental Analysis of the Kink Effects in $S22$ and $h21$ for a GaN HEMT. *IEEE Trans. Microw. Theory Tech.* **2014**, *62*, 513–520. [CrossRef]
52. Ahsan, S.A.; Ghosh, S.; Khandelwal, S.; Chauhan, Y.S. Modeling of kink-effect in RF behaviour of GaN HEMTs using ASM-HEMT model. In Proceedings of the IEEE International Conference Electron Devices Solid-State Circuits, Hong Kong, China, 3–5 August 2016; pp. 426–429. [CrossRef]

53. Alim, M.A.; Hasan, M.A.; Rezazadeh, A.A.; Gaquiere, C.; Crupi, G. Thermal influence on S_{22} kink behavior of a 0.15-μm gate length AlGaN/GaN/SiC HEMT for microwave applications. *Semicond. Sci. Technol.* **2019**, *34*, 1–8. [CrossRef]
54. Crupi, G.; Raffo, A.; Vadalà, V.; Vannini, G.; Caddemi, A. A new study on the temperature and bias dependence of the kink effects in S_{22} and h_{21} for the GaN HEMT technology. *Electronics* **2018**, *7*, 353. [CrossRef]

 *micromachines*

MDPI

Article

Fabrication of All-GaN Integrated MIS-HEMTs with High Threshold Voltage Stability Using Supercritical Technology

Meihua Liu, Yang Yang, Changkuan Chang, Lei Li and Yufeng Jin *

School of Electronic and Computer Engineering, Peking University Shenzhen Graduate School, University Town, Xili, Nanshan, Shenzhen 518055, China; liumh@pku.edu.cn (M.L.); yang1994@pku.edu.cn (Y.Y.); kcchang@pkusz.edu.cn (C.C.); lilei@pkusz.edu.cn (L.L.)
* Correspondence: yfjin@pku.edu.cn

Abstract: In this paper, a novel method to achieve all-GaN integrated MIS-HEMTs in a Si-CMOS platform by self-terminated and self-alignment process is reported. Furthermore, a process of repairing interface defects by supercritical technology is proposed to suppress the threshold voltage shift of all GaN integrated MIS-HEMTs. The threshold voltage characteristics of all-GaN integrated MIS-HEMTs are simulated and analyzed. We found that supercritical NH_3 fluid has the characteristics of both liquid NH_3 and gaseous NH_3 simultaneously, i.e., high penetration and high solubility, which penetrate the packaging of MIS-HEMTs. In addition, NH_2^- produced via the auto coupling ionization of NH_3 has strong nucleophilic ability, and is able to fill nitrogen vacancies near the GaN surface created by high temperature process. The fabricated device delivers a threshold voltage of 2.67 V. After supercritical fluid treatment, the threshold voltage shift is reduced from 0.67 V to 0.13 V. Our demonstration of the supercritical technology to repair defects of wide-bandgap family of semiconductors may bring about great changes in the field of device fabrication.

Keywords: GaN; MIS-HEMTs; fabrication; threshold voltage stability; supercritical technology

Citation: Liu, M.; Yang, Y.; Chang, C.; Li, L.; Jin, Y. Fabrication of All-GaN Integrated MIS-HEMTs with High Threshold Voltage Stability Using Supercritical Technology. *Micromachines* **2021**, *12*, 572. https://doi.org/10.3390/mi12050572

Academic Editor: Giovanni Verzellesi

Received: 19 April 2021
Accepted: 15 May 2021
Published: 18 May 2021

Publisher's Note: MDPI stays neutral with regard to jurisdictional claims in published maps and institutional affiliations.

Copyright: © 2021 by the authors. Licensee MDPI, Basel, Switzerland. This article is an open access article distributed under the terms and conditions of the Creative Commons Attribution (CC BY) license (https://creativecommons.org/licenses/by/4.0/).

1. Introduction

GaN-based high electron mobility transistors (HEMTs) are good candidates for high frequency and high efficiency power switching applications owing to their attractive superiorities of high breakdown electric field and high saturation electron velocity [1]. Normally-off property is strongly required for GaN devices used in the power electronics systems. To date, there are totally four possible ways to realize enhance-mode (E-mode) GaN devices, which are (a) cascode configuration [2], (b) P-GaN gate GaN HEMT, (c) recessed gate GaN MIS-HEMT or MIS-FET and (d) fluoride implanted gate GaN (MIS-) HEMT. Among them, cascode structure is compatible with Si CMOS platform reducing the production cost and complexity. Furthermore, the Miller capacitance is eliminated because of blocking the reverse recovery diode by GaN devices, thus improving the switching speed and reducing the switching loss [3].

However, a few issues have been reported that may negate the speed advantage in the GaN plus Si hybrid cascode devices, such as increased parasitic inductance [4] and mismatch in intrinsic capacitances between the Si and GaN devices [5]. All-GaN integrated cascode device by replacing the Si MOSFET with a low voltage GaN E-mode device achieved using fluoride ion implantation has proven to be able to address the issues mentioned above and improve the switching speed [3]. However, fluoride ion implantation tends to result in V_{th} instability and drain current degradation in HEMTs [6]. In addition, fluoride ion implantation requires high energy, and it is difficult to realize general silicon process lines.

Furthermore, metal-insulator-semiconductor (MIS) gate structure is typically adopted to maintain a relatively large gate swing a low gate leakage current. However, there are several raliability issues realted to GaN MIS-HEMTs. When a positive gate bias is applied, defects located in the gate stack act as charge trapping sites. This induces a shift of the device transfer characteristics toward more positive values [7,8].

Generally, the trapping effect in an MIS gate stack could be related to the traps at/near the interface, named interface states/border traps, or in the bulk of the insulator [9]. There have been several techniques used to suppress threshold voltage shift to date, including pre-fluorination argon treatment [10], sputter-deposited Al_2O_3 [11], in-situ pre-deposition plasma nitridation [12], metal-organic chemical vapor deposition-grown in situ SiN [13,14], hybrid ferroelectric charge trap gate stack [15], etc., to reduce the trapping effect at/near the insulator/semiconductor interface. Due to the low deposition temperature, there also exists large density of traps in the bulk of the gate insulators deposited by PECVD or ALD. Recently, high temperature deposited gate insulator, such as low-pressure chemical vapor deposition (LPCVD) grown SiN_x has been proven to be a robust gate dielectric for both normally-on GaN MIS-HEMTs and normally-off gate recessed hybrid MIS-HEMTs with low bulk trap density. However, the interface quality between LPCVD SiN_x and (Al)GaN is degraded due to the high growth temperature and H erosion. Despite that low temperature deposited insertion layer or N surface plasma treatment have been adopted to improve the interface quality, the drift of V_{th} still exists in those devices. Supercritical fluid technology can effectively bring elements into materials through supercritical CO_2 fluid to reduce trap density because of its penetration and damage-free diffusion ability in the devices [16]. Supercritical technology has been applied in the field of memory [17] and LED [18], but there is no research on the effectiveness of GaN power devices.

In this work, a new way is presented to achieve all-GaN integrated MIS-HEMTs in a Si CMOS platform by replacing the Si MOSFET with a low voltage GaN recessed gate MIS-HEMT. In the process, self-terminated gate open method and quasi-self-alignment technology are adopted allowing the recessed gate was defined and fabricated at the beginning of the process. In addition, we propose the application of supercritical nitridation treatment (SNT) to passivate the defects and mitigate the shift of V_{th} in the all-GaN MIS-HEMTs. After SNT, the interface trap density in LPCVD Si_3N_4/AlGaN layer interface is effectively reduced and near 0.13 V shift of V_{th} in the transfer curve of a GaN power device is observed with a bidirectional gate bias sweep up to 15 V.

2. Device Fabrications

The AlGaN/GaN heterostructure was grown by the metal organic chemical vapor deposition(MOCVD) on a Si(111) substrate, which consists of a 4-μm C-doped GaN buffer layer, a 300-nm unintentionally doped GaN channel layer, a 1-nm AlN insertion layer, a 25-nm $Al_{0.25}Ga_{0.75}N$ barrier layer and a 3-nm GaN cap layer for improving surface morphology. On wafer Hall measurement yields a sheet resistance of 363 Ω/square, a 2DEG density of 1.1×10^{13} cm^{-2}, and an electron mobility of 1547 cm^2/V·s. The reported devices were fabricated in Founder Microelectronics International Corporation, Ltd, a 6-inch Si CMOS platform.

The main process flow is:

(1) Defining (Figure 1(2)) of the mesa isolation by Cl_2/BCl_3 based plasma etching.
(2) Patterning (Figure 1(3)) of the recessed gate by etching the AlGaN layer completely.
(3) Deposition of a 35-nm Si_3N_4 layer using low pressure chemical vapor deposition (LPCVD) (Figure 1(4)). The Si_3N_4 layer acts as a surface passivation layer and a gate insulator. The LPCVD Si_3N_4 exhibits good insulating property and passivation effects.
(4) Deposition of a 500-nm oxide layer over the Si_3N_4 layer by plasma enhanced chemical vapor deposition (PECVD) (Figure 1(5)). The oxide layer acts as the plasma etching sacrificial layer in the follow process patterning source and drain contacts and gate strips, and the gate field plate dielectrics.
(5) Opening (Figure 1(6)) of the source and drain contact windows by etching the oxide layer,the Si_3N_4 layer and partial AlGaN layer .
(6) Deposition of Ti/Al/Ti/TiN multi metal layers by physical vapor deposition (PVD) as ohmic metal and Patterning (Figure 1(7)) of the source and drain electrode.

(7)   Metallization by rapid thermal annealing at 850 °C for 30 s in ambient N_2 (Figure 1(8)).

(8)   Patterning (Figure 1(9)) of the D-mode gate. In this step, the low power SF_6-based inductively coupled plasma (ICP) etching and the buffered HF (BHF) wet etching were adopted sequentially to define the gate stem, realizing a self-terminated dielectric etching (PECVD SiO_2/LPCVD Si_3N_4 etching selectivity is 200:1) on the surface of the LPCVD Si_3N_4 gate dielectric layer. The self-terminated nature guaranteed good performance uniformity along the whole wafer. Meanwhile, quasi-self-alignment is realized, the E-mode GaN HEMT recessed gate can be fabricated at the same time.

(9)   Deposition of TiN/Ti/Al multi metal layers by PVD as gate metal and patterning (Figure 1(10)) of the gate electrode.

(10) After the PAD metal and Final passivation (Figure 1(11),(12)), the devices were annealed at 450 °C for 30 min in ambient H_2.

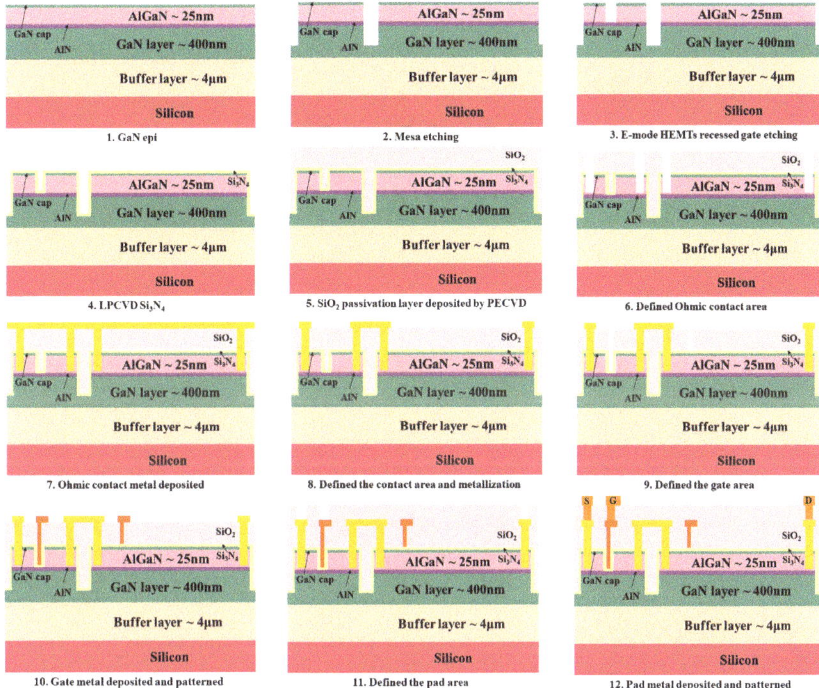

Figure 1. (**1–12**) Main process steps of the all GaN integrated MIS-HEMT in CMOS fab.

Figure 1 shows schematics flow of the all-GaN integrated devices The characterization was performed on the devices with a dimension of $L_G/L_{GS}/L_{GD}/W = 2/4.5/3/24$ μm for recessed gate MIS-FET and a dimension of $L_G/L_{GS}/L_{GD}/W = 1/4.5/8/24$ μm for D-mode MIS-HEMT. Figure 2 shows the schematic view and cross-section TEM image of the fabricated all-GaN integrated MIS-HEMT.

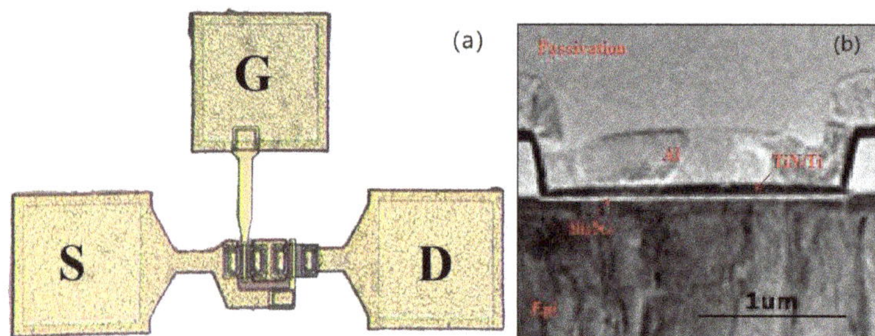

Figure 2. (**a**) Schematic view and (**b**) Cross-section TEM image of the fabricated all-GaN integrated MIS-HEMT.

3. Results and Discussion

All our electrical transport measurements were carried out in an Agilent B1500 semiconductor parameter analyzer and an automated Keithley SCS 4200 system.

3.1. Transport Measurements

Figure 3 shows the DC transfer characteristics and output characteristics of the fabricated all-GaN MIS-HEMTs. The V_{th} of the fresh device is about 2.67 V and the output saturation current is 305 mA/mm at a gate bias of 12 V. The I_{DS}-V_{GS} transfer characteristics of the all-GaN MIS-HEMTs with and without SNT are shown in Figure 4. All the curves are swept in bidirectional mode at V_{DS}=1 V. For the all-GaN MIS-HEMTs without SNT, significant hysteresis was observed during the sweep, suggesting severe trap-induced V_{th} shift.

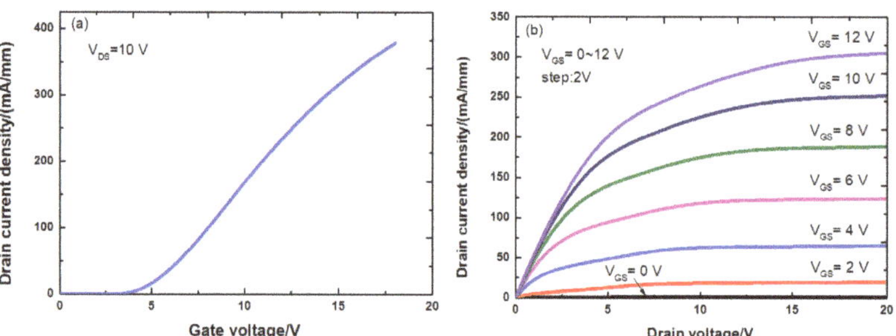

Figure 3. (**a**) DC transfer characteristics and (**b**) output characteristics of the fabricated all-GaN integrated MIS-HEMTs.

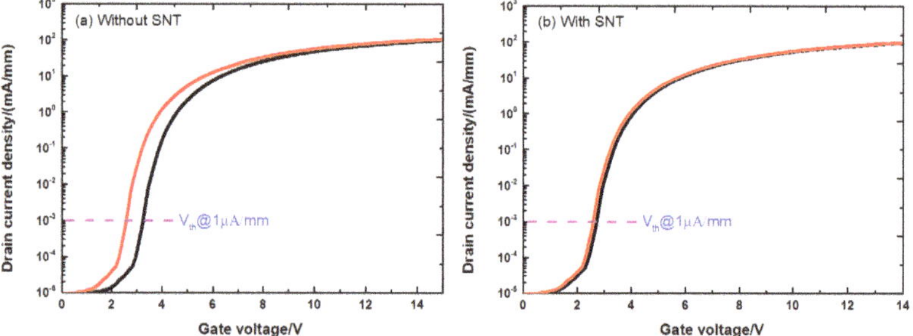

Figure 4. Transfer curves of the all-GaN integrated MIS-HEMTs in didiretional $V_{GS_{sweep}}$ from 0 V to 15 V and then back to 0 V (**a**) without SNT and (**b**) with SNT.

3.2. Threshold Voltage Stability

The temperature dependent V_{th} hysteresis is evaluated by submitting the all-GaN integrated MIS-HEMTs to thermal simulation from 300 K to 370 K with a 10 K step at V_{DS} =1 V. Figure 5 shows the transfer characteristics at various temperatures. When the temperature rises from 300 K to 370 K, the threshold voltage shifts about 1.4 V without SNT. Meanwhile, when the temperature rises from 300 K to 370 K, the threshold voltage shifts about 0.95 V with SNT. Compared with that without SNT, the threshold voltage shifts decrease by 0.45 V.

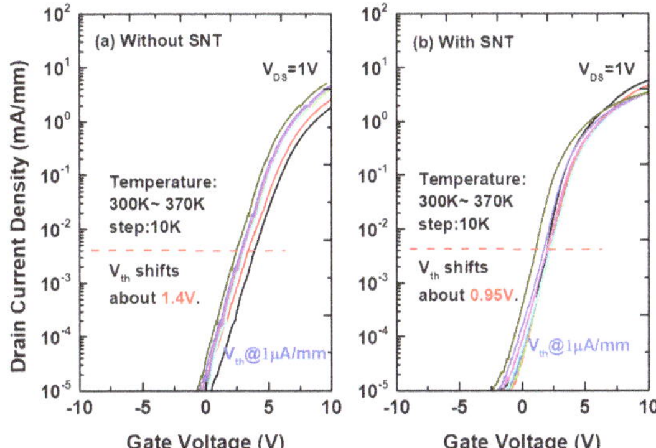

Figure 5. Temperature−dependent transfer characteristics of the all-GaN integrated MIS-HEMTs (**a**) without SNT and (**b**) with SNT.

To further study the effect of SNT on the shift of V_{th} after various time intervals and different gate-bias-induced stresses, time-of-fly gate-bias-induced stress and V_{th} measurement have been performed on the device with both positive and negative gate biases. The shift of V_{th} with bias-induced stress time from 1 ms to 1000 s at different voltages is shown in Figure 6.The shift of V_{th} shows a nearly linear relationship with logarithmic time scale, suggesting broad distribution of the time constant of deep traps in the devices without SNT. On the other hand, the shift of V_{th} in the all-GaN MIS-HEMTs with SNT after the stress is much smaller. The maximum shift of V_{th} at 8 V gate stress for 1000 s is only 0.58 V.

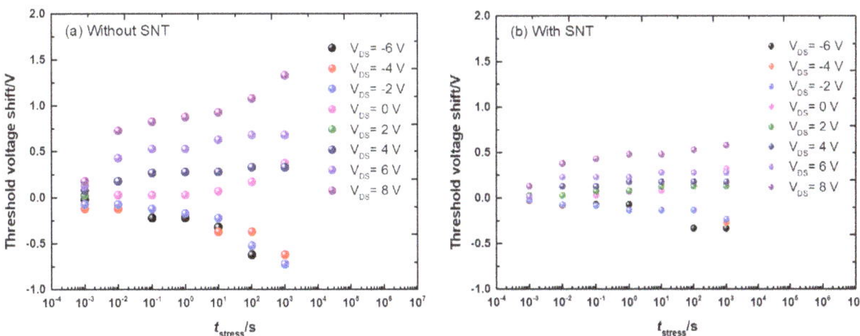

Figure 6. Stress−sequence of the all-GaN integrated MIS-HEMTs (**a**) without SNT and (**b**) with SNT.

3.3. Si₃N₄/AlGaN Interface Trap Characterization

Figure 7 shows the Ga 3d (left), Al 2p (middle) and Si 2p (right) core-level spectra at the Si_3N_4/AlGaN interface in the 8-nm Si_3N_4/AlGaN/GaN samples, which is simulated by Avantage. For the control sample without SNT, the native oxide exhibits a large shoulder (Ga-O bonds) on the high binding energy side of the Ga-N peak [Figure 7a left]. The Ga 3d core-level spectrum in Figure 7a indicates the existence of amorphous native oxide at the Si_3N_4/AlGaN interface without SNT. Ga-O/Al-O/Si-O bonds may also be a kind of the interface state [19], which could lead to the V_{th} shift phenomena in GaN MIS-HEMTs. During the formation of Ga-O/Al-O/Si-O bonds, some of them are natural oxidation and some are O filled with N vacancy. With SNT, a higher intensity of Al-N bonds is observed at the Si_3N_4/AlGaN interface [Figure 7b middle]. The proportion of Ga-N bonds increases from 79.01–89.35%. The proportion of Al-N bonds increases from 69.72–85.01%. The proportion of Si-N bonds increases from 65.63–83.61%. At the same time, the proportion of Ga dangling bonds decreases from 0.08–0.06%. The proportion of Al dangling bonds decreases from 1.16–0.37%. The proportion of Si dangling bonds decreases from 1.25–0.54%. The change of the proportion of chemical bonds indicates that part of the N vacancies is filled, and some O atoms in Ga-O bonds are replaced by N during SNT treatment.

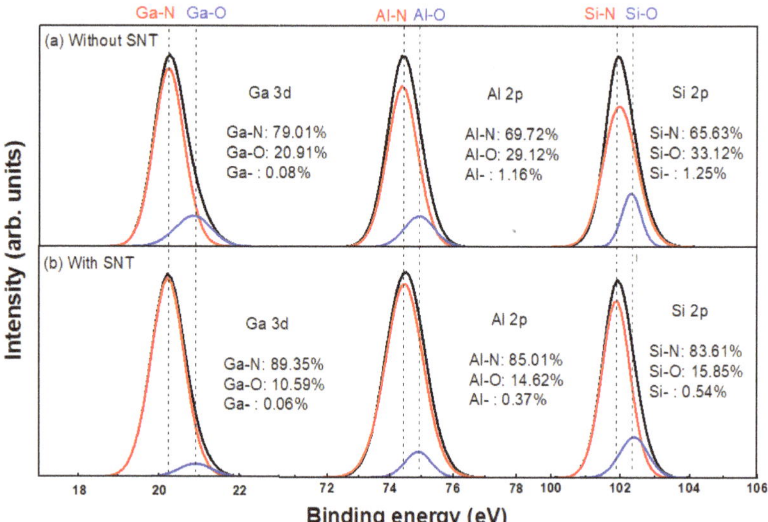

Figure 7. Ga 3d (left), Al 2p (middle) and Si 2p (right) core-level spectra at the Si_3N_4/AlGaN interface in the 8-nm Si_3N_4/AlGaN/GaN samples (**a**) without SNT and (**b**) with SNT. Each measured spectrum (symbol) is resolved into two Gaussian functions that correspond to M in M-N (solid red line) and M-O (solid blue line) bonds. M represents Ga (**left**), Al (**middle**) or Si (**right**). The solid blue red line is a superposition of the two fitting functions.

CO_2 is a double bond structure with large structure with large activation energy and stable chemical properties. It does not participate in the reaction during supercritical nitridation treatment. It acts only as a solvent for supercritical NH_3, avoiding the supercritical NH_3 reaction and eroding the device electrodes. As mentioned above, due to the dry etch process and high growth temperature in LPCVD, the very surface of the AlGaN layer and the interface between Si_3N_4 and AlGaN layer may be relatively defective, which could lead to the trap-induced V_{th} shift phenomena in GaN MIS-HEMTs [20].

Figure 8 shows the supercritical nitridation technology model of the fabricated all-GaN integrated MIS-HEMTs, which is simulated by Nanodcal. In the process of SNT, NH_2^- produced by auto coupling ionization of NH_3 has strong nucleophilic ability and can fill the nitrogen vacancy near the Si_3N_4/AlGaN interface caused by the dry etch and high temperature process. During the nucleophilic reaction [21], NH_2^- can react with Ga-O/Al-O/Si-O to replace O and from the Ga-N/Al-N/Si-N bonds. The exact passivation mechanism during SNT is still under investigation.

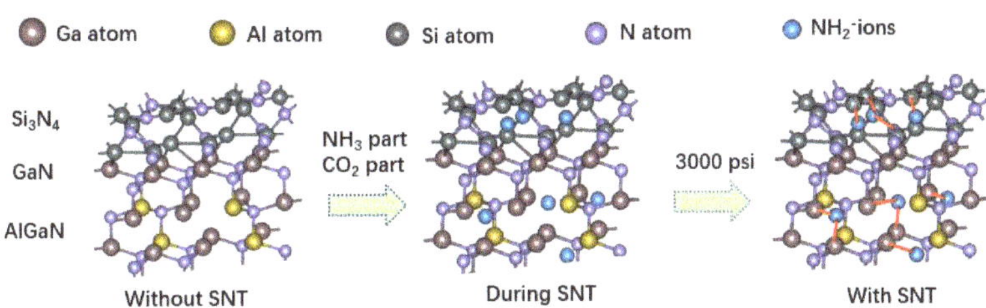

Figure 8. The supercritical nitridation technology model of the fabricated all-GaN integrated MIS-HEMTs.

4. Conclusions

In summary, high performance AlGaN/GaN MIS-HEMTs realized via supercritical nitridation technology has been designed, fabricated, and measured. By comparing the devices without and with SNT, we find that supercritical nitridation technology can effectively repair the defects and suppress the shift of threshold voltage. With SNT, the optimized MIS-HEMTs demonstrate a low V_{th} shift decrease of about 0.54 V at $V_{G_s sweep}$ from 0 V to 15 V. Our demostration of the supercritical nitridation technology to repair defects of wide-bandgap family of semiconductors may bring about great changes in the field of device fabrication.

Author Contributions: Conceptualization, M.L.; methodology, M.L.; software, M.L.; validation, M.L. and Y.J.; investigation, M.L.; resources, M.L.; data curation, M.L.and Y.Y.; writing—original draft preparation, M.L.; writing—review and editing, M.L. and L.L.; visualization, M.L.; supervision, Y.J. and C.C.; project administration, Y.J.; funding acquisition, Y.J. All authors have read and agreed to the published version of the manuscript.

Funding: This research was funded by the Shenzhen Science and Technology Innovation Committee (grant number: ZDSYS201802061805105 and JCYJ20190808155007550). Among them grant ZDSYS201802061805105 is the TSV 3D Integrated Micro/Nanosystem Laboratory program.

Institutional Review Board Statement: Not applicable.

Informed Consent Statement: Not applicable.

Data Availability Statement: All data, models, and code generated or used during the study appear in the submitted article.

Conflicts of Interest: The authors declare no conflict of interest.

References

1. Wu, T.L.; Tang, S.W.; Jiang, H.J. Investigation of Recessed Gate AlGaN/GaN MIS-HEMTs with Double AlGaN Barrier Designs toward an Enhancement-Mode Characteristic. *Micromachines* **2020**, *11*, 163. [CrossRef] [PubMed]
2. Chen, K.J.; Häberlen, O.; Lidow, A.; Tsai, C.L.; Ueda, T.; Uemoto, Y.; Wu, Y. GaN-on-Si Power Technology: Devices and Applications. *IEEE Trans. Electron Devices* **2017**, *64*, 779–795. [CrossRef]
3. Jiang, S.; Lee, K.B.; Guiney, I.; Miaja, P.F.; Zaidi, Z.H.; Qian, H.T.; Wallis, D.J.; Forsyth, A.J.; Humphreys, C.J.; Houston, P.A. All-GaN-Integrated Cascode Heterojunction Field Effect Transistors. *IEEE Trans. Power Electron.* **2017**, *32*, 8743–8750. [CrossRef]
4. Liu, Z.Y.; Huang, X.C.; Lee, F.C.; Li,Q. Package Parasitic Inductance Extraction and Simulation Model Development for the High-Voltage Cascode GaN HEMT. *IEEE Trans. Power Electron.* **2014**, *29*, 1977–1985. [CrossRef]
5. Du, W.J.; Huang, X.C.; Lee, F.C.; Li, Q.; Zhang, W.L. Avoiding divergent oscillation of cascode GaN device under high current turn-off condition. In Proceedings of the 2016 IEEE Applied Power Electronics Conference and Exposition (APEC), Long Beach, CA, USA, 20–24 March 2016; pp. 1002–1009.
6. Sun, W.W.; Zheng, X.F.; Fan, S.; Wang, C.; Du, M.; Zhang, K.; Chen, W.W.; Cao, Y.R.; Mao, W.; Ma, X.H.; et al. Degradation mechanism of enhancement-mode AlGaN/GaN HEMTs using fluorine ion implantation under the on-state gate overdrive stress. *Chin. Phys. B* **2015**, *24*, 017303. [CrossRef]
7. Saito, W.; Nitta, T.; Kakiuchi, Y.; Saito, Y.; Tsuda, K.; Omura, I.; Yamaguchi, M. Suppression of Dynamic On-Resistance Increase and Gate Charge Measurements in High-Voltage GaN-HEMTs With Optimized Field-Plate Structure. *IEEE Trans. Electron Devices* **2007**, *54*, 1825–1830. [CrossRef]

8. Uesugi, T.; Kachi, T. Which are the Future GaN Power Devices for Automotive Applications, Lateral Structures or Vertical Structures. In Proceedings of the CS MANTECH Conference, Palm Springs, CA, USA, 16–19 May 2011; pp. 307–310.

9. Wu, T.L.; Marcon, D.; De Jaeger, B.; Van Hove, M.; Bakeroot, B.; Stoffels, S.; Groeseneken, G.; Decoutere, S.; Roelofs, R. Time Dependent Dielectric Breakdown (TDDB) evaluation of PE-ALD SiN gate dielectrics on AlGaN/GaN recessed gate D-mode MIS-HEMTs and E-mode MIS-FETs. In Proceedings of the 2015 IEEE International Reliability Physics Symposium, Monterey, CA, USA, 19–23 April 2015; pp. 6C.4.1–6C.4.6.

10. Uemoto, Y.; Hikita, M.; Ueno, H.; Matsuo, H.; Ishida, H.; Yanagihara, M.; Ueda, T.; Tanaka, T.; Ueda, D. Gate Injection Transistor (GIT)—A Normally-Off AlGaN/GaN Power Transistor Using Conductivity Modulation. *IEEE Trans. Electron Devices* **2007**, *54*, 3393–3399. [CrossRef]

11. Dutta, G.; Dasgupta, N.; Dasgupta, A. Effect of Sputtered-Al_2O_3 Layer Thickness on the Threshold Voltage of III-Nitride MIS-HEMTs. *IEEE Trans. Electron Devices* **2016**, *63*, 1450–1458. [CrossRef]

12. Zhang, Z.L.; Li, W.Y.; Fu, K.; Yu, G.H.; Zhang, X.Z.; Zhao, X.D.; Zhao, Y.F.; Sun, S.C.; Song, L.; Deng, X.G.; et al. AlGaN/GaN MIS-HEMTs of Very-Low Vth Hysteresis and Current Collapse with In-Situ Pre-Deposition Plasma Nitridation and LPCVD-Si_3N_4 Gate Insulator. *IEEE Electron Device Lett.* **2017**, *38*, 236–239. [CrossRef]

13. Wu, C.H.; Han, P.C.; Liu, S.C.; Hsieh, T.E.; Lumbantoruan, F.J.; Ho, Y.H.; Chen, J.Y.; Yang, K.S.; Wang, H.C.; Lin, Y.K.; et al. High-Performance Normally-OFF GaN MIS-HEMTs Using Hybrid Ferroelectric Charge Trap Gate Stack (FEG-HEMT) for Power Device Applications. *IEEE Electron Device Lett.* **2018**, *39*, 991–994. [CrossRef]

14. Ikeda, N.; Niiyama, Y.; Kambayashi, H.; Sato, Y.; Nomura, T.; Kato, S.; Yoshida, S. GaN Power Transistors on Si Substrates for Switching Applications. *Proc. IEEE.* **2010**, *98*, 1151–1161. [CrossRef]

15. Benoit, D.; Regolini, J.; Morin, P. Hydrogen desorption and diffusion in PECVD silicon nitride. Application to passivation of CMOS active pixel sensors. *Microelectron. Eng.* **2007**, *84*, 2169–2172. [CrossRef]

16. Tsai, C.T.; Chang, K.M.; Liu, P.T.; Kuo, Y.L.; Kin, K.T.; Chang, P.L.; Huang, F.S. Low-temperature method for enhancing sputter-deposited HfO_2 films with complete oxidization. *Appl. Phys. Lett.* **2007**, *91*, 012109. [CrossRef]

17. Chang, K.C.; Pan, C.H.; Chang, T.C.; Tsai, T.M.; Zhang, R.; Lou, J.C.; Young, T.F.; Chen, J.H.; Shih, C.C.; Chu, T.J.; et al. Hopping effect of hydrogen-doped silicon oxide insert RRAM by supercritical CO_2 fluid treatment. *IEEE Electron Device Lett.* **2013**, *34*, 617–619. [CrossRef]

18. Chattopadhyay, P.; Gupta, R.B. Production of griseofulvin nanoparticles using supercritical CO_2 antisolvent with enhanced mass transfer. *Int. J. Pharm.* **2001**, *228*, 19–31. [CrossRef]

19. Robertson, J. Model of interface states at III-V oxide interfaces. *Appl. Phys. Lett.* **2009**, *94*, 214–215. [CrossRef]

20. Hashizume, T.; Hasegawa, H. Effects of nitrogen deficiency on electronic properties of AlGaN surfaces subjected to thermal and plasma processes. *Appl. Surf. Sci.* **2004**, *234*, 387–394. [CrossRef]

21. Yu, J.S.; Liu, Y.L.; Tang, J.; Wang, X.; Zhou, J. Highly efficient "on water" catalyst-free nucleophilic addition reactions using difluoroenoxysilanes: Dramatic fluorine effects. *Angew. Chem.* **2014**, *53*, 9512–9516. [CrossRef] [PubMed]

 *micromachines*

Article

On the Modeling of the Donor/Acceptor Compensation Ratio in Carbon-Doped GaN to Univocally Reproduce Breakdown Voltage and Current Collapse in Lateral GaN Power HEMTs

Nicolò Zagni [1,*], Alessandro Chini [1], Francesco Maria Puglisi [1], Paolo Pavan [1] and Giovanni Verzellesi [2]

[1] Department of Engineering "Enzo Ferrari", University of Modena and Reggio Emilia, via P. Vivarelli 10, 41125 Modena, Italy; alessandro.chini@unimore.it (A.C.); francescomaria.puglisi@unimore.it (F.M.P.); paolo.pavan@unimore.it (P.P.)
[2] Department of Sciences and Methods for Engineering (DISMI) and EN&TECH Center, University of Modena and Reggio Emilia, via G. Amendola, 2, 42122 Reggio Emilia, Italy; giovanni.verzellesi@unimore.it
* Correspondence: nicolo.zagni@unimore.it

Citation: Zagni, N.; Chini, A.; Puglisi, F.M.; Pavan, P.; Verzellesi, G. On the Modeling of the Donor/ Acceptor Compensation Ratio in Carbon-Doped GaN to Univocally Reproduce Breakdown Voltage and Current Collapse in Lateral GaN Power HEMTs. *Micromachines* 2021, 12, 709. https://doi.org/10.3390/mi12060709

Academic Editor: Ali Soltani

Received: 26 May 2021
Accepted: 15 June 2021
Published: 16 June 2021

Publisher's Note: MDPI stays neutral with regard to jurisdictional claims in published maps and institutional affiliations.

Copyright: © 2021 by the authors. Licensee MDPI, Basel, Switzerland. This article is an open access article distributed under the terms and conditions of the Creative Commons Attribution (CC BY) license (https://creativecommons.org/licenses/by/4.0/).

Abstract: The intentional doping of lateral GaN power high electron mobility transistors (HEMTs) with carbon (C) impurities is a common technique to reduce buffer conductivity and increase breakdown voltage. Due to the introduction of trap levels in the GaN bandgap, it is well known that these impurities give rise to dispersion, leading to the so-called "current collapse" as a collateral effect. Moreover, first-principles calculations and experimental evidence point out that C introduces trap levels of both acceptor and donor types. Here, we report on the modeling of the donor/acceptor compensation ratio (CR), that is, the ratio between the density of donors and acceptors associated with C doping, to consistently and univocally reproduce experimental breakdown voltage (V_{BD}) and current-collapse magnitude (ΔI_{CC}). By means of calibrated numerical device simulations, we confirm that ΔI_{CC} is controlled by the effective trap concentration (i.e., the difference between the acceptor and donor densities), but we show that it is the total trap concentration (i.e., the sum of acceptor and donor densities) that determines V_{BD}, such that a significant CR of at least 50% (depending on the technology) must be assumed to explain both phenomena quantitatively. The results presented in this work contribute to clarifying several previous reports, and are helpful to device engineers interested in modeling C-doped lateral GaN power HEMTs.

Keywords: GaN power HEMTs; breakdown voltage; current collapse; compensation ratio; auto-compensation; carbon doping

1. Introduction

Carbon (C) doping is a common technological solution to reduce buffer conductivity and increase breakdown voltage (V_{BD}) in lateral gallium nitride (GaN)-based power transistors [1,2]. However, this comes at the cost of increased dynamic on-resistance and current-collapse effects [1,3,4]. Depending on the growth conditions, C atoms can either substitute N or Ga sites, occupy interstitial locations in the crystal, or form complexes with intrinsic defects [5–8]. In typical undoped GaN layers used as buffer in high electron mobility transistors (HEMTs), the position of the Fermi level is such that both acceptor and donor traps are likely to form. Several works, discussing either simulation or experimental results, indicate the occurrence of partial "auto-compensation" between the dominant deep acceptor traps, generally attributed to C_N levels, and the concomitantly introduced (i.e., non-pre-existing) shallow donors, which reduce the *effective* concentration of acceptor traps well below the level of the introduced C concentration (especially in the case of extrinsic C doping) [4,9–15].

These aspects call for the correct modeling of C-related trap states in GaN transistors when performing device simulations to investigate important performance-limiting effects, such as buffer leakage and related V_{BD} [13,14], dynamic R_{ON} [4,10,16], current collapse [2,12,17], and threshold voltage instabilities [18–21]. In fact, both the concentration of acceptor states ($N_{C,A}$) and donor states ($N_{C,D}$), as well as their *compensation ratio*, defined as CR = $N_{C,D}/N_{C,A}$, need to be properly determined in order to reproduce the features of realistic devices and calibrate device simulation for a given technology.

In this paper, we present calibrated numerical device simulation results that reveal the functional dependence of V_{BD} and ΔI_{CC}, respectively, on the *total* ($N_{C,TOT} = N_{C,A} + N_{C,D}$) and *effective* ($N_{C,EFF} = N_{C,A} - N_{C,D}$) C-related trap concentrations in the buffer at different CR values. Our results (i) confirm the necessity of assuming compensating donor traps in C-doped GaN to correctly model realistic devices, and (ii) provide physical insights into the origin of the observed V_{BD} and ΔI_{CC} dependence on CR.

2. Modeling Framework

Two-dimensional numerical device simulations were carried out with the commercial simulator SDevice^{TM}. The simulated structure is sketched in Figure 1 with indication of device dimensions (not to scale); the device resembles the AlGaN/GaN Schottky-gate HEMT reported in [1].

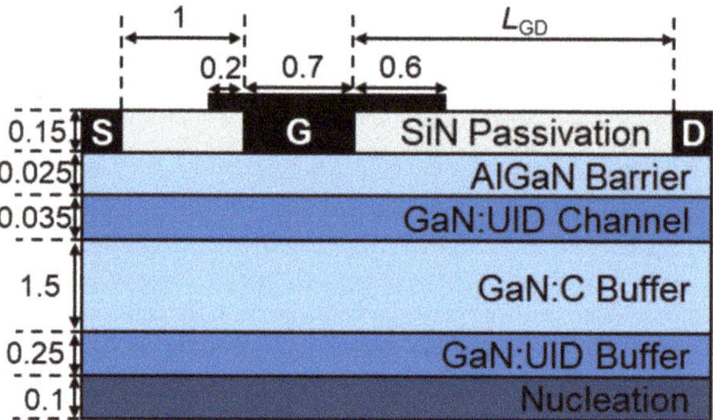

Figure 1. Cross-section of the simulated Schottky-gate HEMT resembling the C-doped device in [1]. Dimensions are in μm (not to scale).

Charge transport was modelled by means of the drift-diffusion model. Piezoelectric polarization was included by using the default strain model of the simulator. Note that at the passivation/barrier interface, the polarization model was deactivated. This approach is equivalent to assuming that the negative polarization charge at this interface is completely compensated by an equal positive surface charge [22]. Therefore, we neglected the possible dynamic effects related to surface traps.

Chynoweth's law was used to model impact ionization; model coefficients for both electrons and holes were set in agreement with Monte Carlo calculations [23].

Gate current was modelled by the thermionic and field emission mechanisms. The field emission component was calculated self-consistently by the simulator through a nonlocal tunnelling model based on the WKB approximation [16].

To account for trap effects, one Shockley-Read–Hall (SRH) trap-balance equation was used for each distinct trap allowing for the dynamics of trap occupation to be described without any quasi-static approximation.

Calibration of the simulation parameters against measurements taken from [1] has already been reported in [14,17]. What makes the measurements reported in [1] instrumental to our scope is the possibility of calibrating our simulation deck against a consistent set of experimental data from devices with several different L_{GD} values.

Key results are shown in Figures 2 and 3, illustrating the agreement achieved in the off-state three-terminal breakdown and current collapse, respectively. Regarding the pulsed

I_D–V_{DS} curves shown in Figure 3, the output curves were obtained by pulsing V_{GS} and V_{DS} from different baselines to either suppress or induce trapping [1]. The current collapse is defined as $\Delta I_{CC} = (I_{D,BL1} - I_{D,BL2})/I_{D,BL1} \times 100$ evaluated at $V_{DS} = 10$ V.

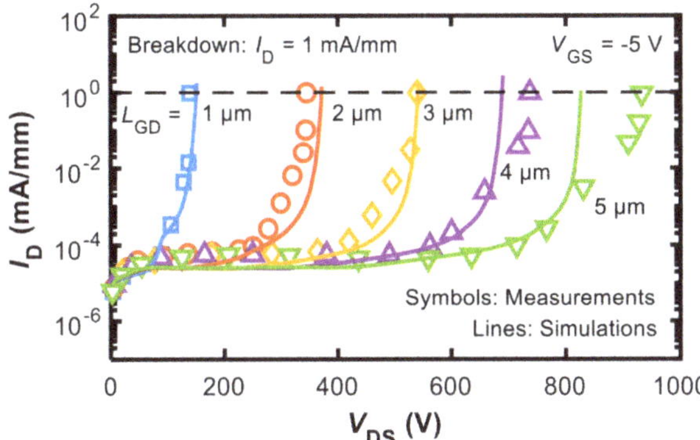

Figure 2. Calibrated simulations (lines) and measurements (symbols) of the off-state I_D–V_{DS} curves. Measured data are taken from [1]. Adapted from [14].

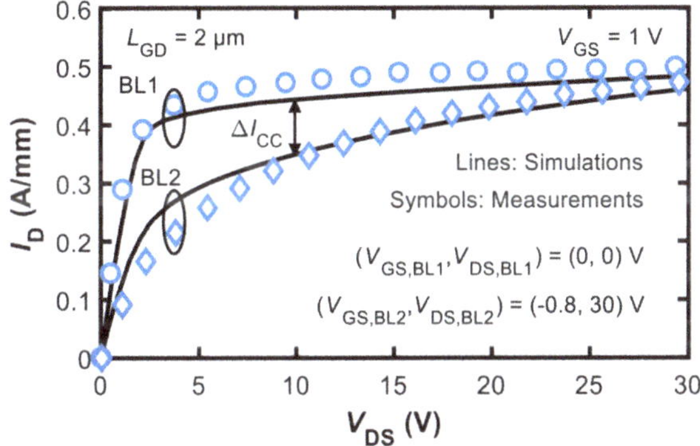

Figure 3. Calibrated pulsed I_D–V_{DS} curve simulations (lines) showing the achieved agreement in current-collapse measurements from [1] (symbols). Adapted from [17].

C doping in the GaN buffer was modelled by considering a dominant deep acceptor trap at E_V + 0.9 eV (generally assumed to correspond to C_N states) and a shallow donor trap at E_C − 0.11 eV (more likely related to C_{Ga} states) as the two major energy states associated with C [24]. When varying trap concentrations and CR, the above energy levels were kept fixed and other possible states related to C doping were neglected. The adopted concentrations for the calibrations shown in Figures 2 and 3 were 8×10^{17} cm^{-3} and 4×10^{17} cm^{-3} for C-related acceptors and donors, respectively.

Although no additional trap levels were considered, in all nitride layers a background doping concentration of 10^{15} cm^{-3} was adopted to account for the unintentional n-type conductivity due to shallow-donor impurities incorporated during growth [2,13]. The C doping model based on discrete point defects adopted here can lose validity for impurity concentrations larger than 10^{19} cm^{-3}, for which a dominant defect band behavior has been proposed to be more appropriate [11]. Therefore, we limited our analysis to cases for which $N_{C,A} < 10^{19}$ cm^{-3}.

As elucidated by the results in Section 3, the key feature of the adopted C doping model is that the dominant deep acceptor-type hole traps are partially compensated by shallow donor-type electron traps. Note that the actual energy position of donor traps, if sufficiently shallow, has little influence on the simulation results. C-related donors could actually be moved even closer to E_C or be modelled as completely ionized dopants (i.e., fixed positive charge) [25], in agreement with recent hybrid-functional density functional theory (DFT) calculations [6,8], without significant changes. This is because as long as the dominant traps are the acceptor states at $E_V + 0.9$ eV, the Fermi level stays well below the shallow energy of donors, thus guaranteeing their complete ionization. The capability of the acceptor–donor model for C doping to reproduce source–drain leakage currents and off-state breakdown is shown in [13,14].

Table 1 lists the main physical mechanisms, along with the respective models and parameters included in the simulations.

Table 1. List of main physical mechanisms, the respective models, and parameters used in the simulations.

Physical Mechanism	Model	Parameters	Value
Impact Ionization	Chynoweth's Law	a (electrons)	2.32×10^6 cm^{-1}
		b (electrons)	1.4×10^7 V/cm
		a (holes)	5.41×10^6 cm^{-1}
		b (holes)	1.89×10^7 V/cm
Carbon Doping (Buffer)	Acceptor Trap Level	Concentration	Variable
		Energy Level	$0.9 + E_V$ eV
	Donor Trap Level	Concentration	Variable
		Energy Level	$E_C - 0.11$ eV
Unintentional Doping (Channel)	Donor Trap Level	Concentration	1×10^{15} cm^{-3}
Schottky Diode (Gate Contact)	Thermionic and Field Emission	Schottky Barrier Height	1 eV
Low-Field Mobility (GaN)		μ_n	1800 cm^2/Vs
High-Field Saturation (GaN)	Canali Model	$v_{n,sat}$	1.5×10^7 cm/s

3. Results

To understand the impact of the total ($N_{C,TOT} = N_{C,A} + N_{C,D}$) and effective ($N_{C,EFF} = N_{C,A} - N_{C,D}$) C-related trap concentration on V_{BD} and ΔI_{CC}, we performed a sensitivity analysis starting from the parameter set, resulting in the calibrated results shown in Figures 2 and 3. L_{GD} was set to 2 µm because only for this case, both V_{BD} and ΔI_{CC} measurement data were available in [1]. Three different CR values were considered for simplicity in the following: 0%, 50%, and 90%. For each CR value, $N_{C,A}$ was set to $\{0.04, 0.08, 0.4, 0.8, 4, 8, 40, 80\} \times 10^{17}$ cm^{-3}, while $N_{C,D}$ was set according to the assumed CR (i.e., 0%, 50%, or 90% of $N_{C,A}$).

3.1. Breakdown Voltage

Figure 4 shows V_{BD} as a function of $N_{C,TOT}$ for the different CR values. As can be noted, for all CR values, V_{BD} first increases and then saturates with $N_{C,TOT}$.

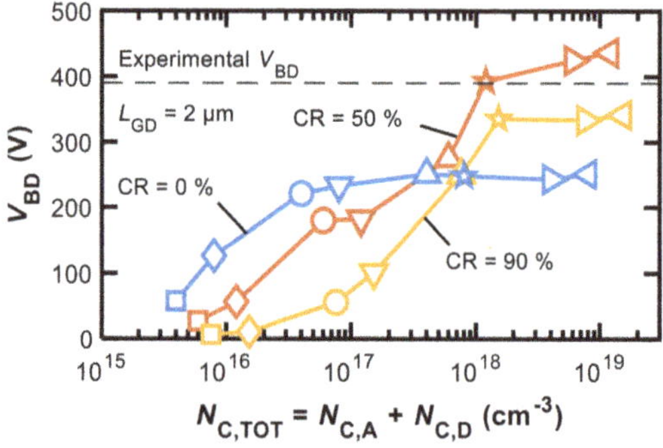

Figure 4. V_{BD} vs. $N_{C,TOT}$ for CR = {0, 50, 90}%. The same symbols correspond to the same $N_{C,A}$ at different $N_{C,D}$ depending on CR.

This behaviour is largely expected, as it is related to the decrease in the electric field peak at the gate edge resulting from the increase in the ionized acceptor density (negative charge). In essence, it is exactly to induce this effect that doping with acceptor impurities (like Fe and C) is adopted in power GaN HEMTs. The V_{BD} saturation is attributable to the fact that once the electric-field peak moves from the gate edge to the drain contact, the beneficial effect of further increasing the acceptor concentration ceases.

In addition to confirming the above behaviour, Figure 4 provides us with two other pieces of information that are key to our purposes: (1) the maximum V_{BD} attainable at large $N_{C,TOT}$ ($V_{BD,max}$) is a non-monotonic function of CR, and (2) a significant CR value of about 50% must be assumed in order to obtain V_{BD} in agreement with experiments. The reasons behind this can be understood with the aid of Figures 5 and 6.

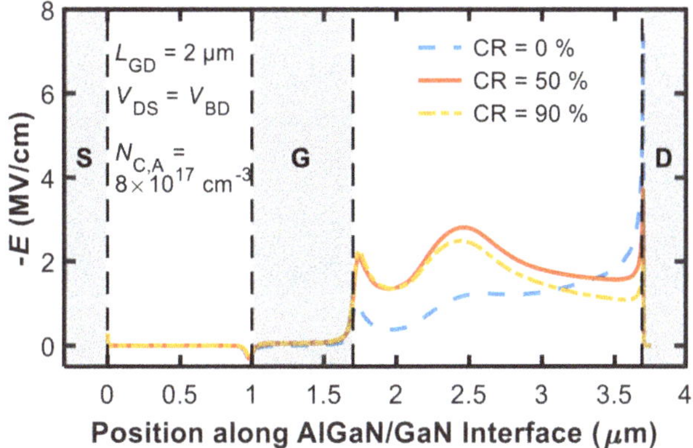

Figure 5. Lateral component of the electric field (E) along the AlGaN/GaN interface at breakdown ($V_{GS} = -5$ V, $V_{DS} = V_{BD}$) for CR = {0, 50, 90}% at the same $N_{C,A}$ value (8×10^{17} cm^{-3}).

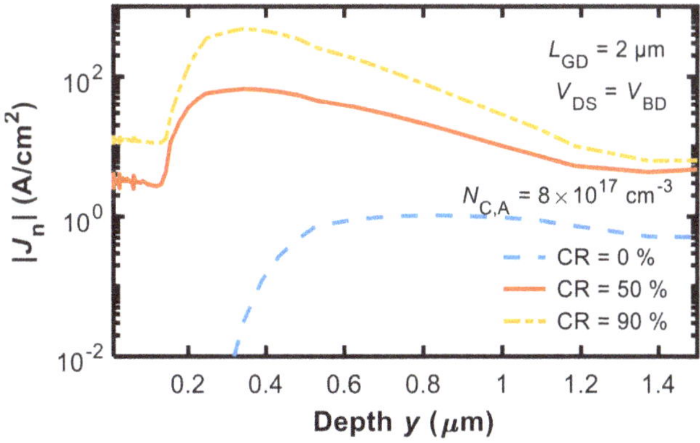

Figure 6. Modulus of the electron current density ($|J_N|$) as a function of the device depth along a cutline taken in the middle of the gate contact for CR = {0, 50, 90}% at the same $N_{C,A}$ value (8×10^{17} cm^{-3}).

Figure 5 shows the lateral component of the electric field as a function of position along a cutline corresponding to the AlGaN/GaN interface at breakdown ($V_{GS} = -5$ V, $V_{DS} = V_{BD}$) for $N_{C,A} = 8 \times 10^{17}$ cm^{-3} in the three cases of CR = 0%, 50%, and 90%. As shown in Figure 5, increasing CR from 0% (i.e., $N_{C,A} = 8 \times 10^{17}$ cm^{-3}, $N_{D,A} = 0$ cm^{-3}) to 50% (i.e., $N_{C,A} = 8 \times 10^{17}$ cm^{-3}, $N_{D,A} = 4 \times 10^{17}$ cm^{-3}) and 90% (i.e., $N_{C,A} = 8 \times 10^{17}$ cm^{-3},

$N_{D,A} = 7.2 \times 10^{17}$ cm^{-3}) effectively modulates the electric field profile, relaxing the peak at the drain contact. Positively charged donors thus contribute to make the electric field profile more uniform (if the electric field peak has already moved to the drain contact). This explains why in Figure 4, $V_{BD,max}$ increases from about 250 V to about 450 V as CR is raised from 0% to 50%.

When further increasing CR to 90%, $V_{BD,max}$ is reduced. This behavior can be explained with the aid of Figure 6, which shows the modulus of the electron current density as a function of the device depth along a vertical cutline taken in the middle of the gate contact. As is clearly shown, the source–drain punch-through (or sub-threshold) current across the buffer increases as CR increases from 0% to 50% and 90%. This is a consequence of the higher $N_{C,D}$ that increases the conductivity of the buffer and thus reduces V_{BD}.

In summary, without significant donor/acceptor compensation resulting in a CR of about 50%, it is not possible, according to our analysis, to explain the state-of-the-art V_{BD} vs. L_{GD} dependence with slopes of 150–200 V/mm [1,26–28], and, specifically in the case considered here, a V_{BD} of about 370 V for a device with an L_{GD} of 2 μm.

3.2. Current Collapse

The results of the sensitivity analysis on ΔI_{CC} are shown in Figure 7, where ΔI_{CC} is plotted against $N_{C,EFF}$ for the same CR values used for Figure 4. As it can be noted, ΔI_{CC} remains small (<10% in the specific devices considered here) regardless of CR when $N_{C,EFF}$ is smaller than 10^{17} cm^{-3}. For higher $N_{C,EFF}$, unless CR is very large (90% in our case), ΔI_{CC} increases steeply with $N_{C,EFF}$, reaching values >60%, which are well above those reported for state-of-the-art C-doped GaN power HEMTs for $N_{C,EFF}$ values >10^{18} cm^{-3}, the latter being instead quite typical for nominal C densities in extrinsically doped devices (i.e., using C precursors). In other words, according to this analysis it is unreasonable that C doping at high concentrations could simply translate to C_N acceptors, as in this case DC-to-dynamic dispersion effects as current collapse and dynamic R_{ON} increase would make the device completely nonfunctional.

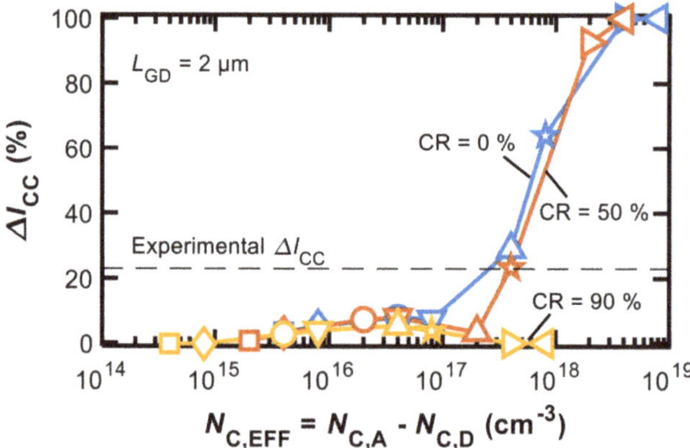

Figure 7. $\Delta I_{D,CC}$ vs. $N_{C,EFF}$ for CR = {0, 50, 90}%. The same symbols correspond to the same $N_{C,A}$ at different $N_{C,D}$ depending on CR.

This is in agreement with previous results that showed how assuming CR = 0% (i.e., acceptors only) with concentrations on the order of the nominal C density (i.e., ~10^{18}–10^{19} cm^{-3}) resulted in large overestimation of current-collapse effects measured in actual devices of different technologies [4,12,18,29].

4. Discussion

By combining the results shown in Figures 4 and 7, we observe that the V_{BD} and ΔI_{CC} values measured in the device under study can be reproduced with a single set of parameters, and specifically, with the same $N_{C,TOT}$ and $N_{C,EFF}$, only when considering a CR of about 50%. More generally, our results point to the necessity that a non-negligible part of incorporated C atoms results in donor-like levels or contribute to donor-like defect-impurity centers, thus compensating to a significant degree the dominant acceptor traps introduced by C doping.

The results presented in this work are relevant for the modeling of any GaN HEMT structure that incorporates C impurities (even unintentionally) in significant concentrations. High unintentional C doping concentrations can likely occur for metal-organic chemical vapor deposition (MOCVD)-grown, intentionally Fe-doped HEMTs for RF applications, where C incorporation comes as an inevitable consequence of the growth processing conditions [30].

5. Conclusions

We reported on the modeling of the compensation ratio (CR) between the donor and acceptor densities due to carbon doping in the buffer of lateral GaN power HEMTs to correctly simulate breakdown voltage (V_{BD}) and current collapse (ΔI_{CC}). We showed that compensating shallow donor traps ($N_{C,D}$) need to be considered in addition to the dominant deep acceptor traps ($N_{C,A}$), in order to reproduce V_{BD} and ΔI_{CC} with a single set of parameters. Furthermore, we identified that the primary dependence of V_{BD} (ΔI_{CC}) on C doping is through the total (effective) concentration of acceptor and donor traps. The results presented here allow device engineers to properly model a given GaN HEMT technology that incorporates C in its structure (even unintentionally) by setting the CR value required to univocally reproduce both V_{BD} and ΔI_{CC} data.

Author Contributions: Conceptualization, N.Z., A.C., and G.V.; formal analysis, N.Z., A.C., F.M.P., P.P., and G.V.; methodology, N.Z., A.C., and G.V.; writing—original draft preparation, N.Z., and G.V.; writing—review and editing, N.Z., A.C., F.M.P., P.P., and G.V.; supervision, A.C., F.M.P., P.P., and G.V. All authors have read and agreed to the published version of the manuscript.

Funding: This research received no external funding.

Conflicts of Interest: The authors declare no conflict of interest.

References

1. Bahat-Treidel, E.; Brunner, F.; Hilt, O.; Cho, E.; Würfl, J.; Trankle, G. AlGaN/GaN/GaN:C back-barrier HFETs with breakdown voltage of over 1 kV and low RON× A. *IEEE Trans. Electron. Devices* **2010**, *57*, 3050–3058. [CrossRef]
2. Uren, M.J.; Moreke, J.; Kuball, M. Buffer design to minimize current collapse in GaN/AlGaN HFETs. *IEEE Trans. Electron. Devices* **2012**, *59*, 3327–3333. [CrossRef]
3. Del Alamo, J.A.; Lee, E.S. Stability and Reliability of Lateral GaN Power Field-Effect Transistors. *IEEE Trans. Electron Devices* **2019**, *66*, 4578–4590. [CrossRef]
4. Chini, A.; Meneghesso, G.; Meneghini, M.; Fantini, F.; Verzellesi, G.; Patti, A.; Iucolano, F. Experimental and Numerical Analysis of Hole Emission Process from Carbon-Related Traps in GaN Buffer Layers. *IEEE Trans. Electron. Devices* **2016**, *63*, 3473–3478. [CrossRef]
5. Remesh, N.; Mohan, N.; Raghavan, S.; Muralidharan, R.; Nath, D.N. Optimum Carbon Concentration in GaN-on-Silicon for Breakdown Enhancement in AlGaN/GaN HEMTs. *IEEE Trans. Electron. Devices* **2020**, *67*, 2311–2317. [CrossRef]
6. Matsubara, M.; Bellotti, E. A first-principles study of carbon-related energy levels in GaN. I. Complexes formed by substitutional/interstitial carbons and gallium/nitrogen vacancies. *J. Appl. Phys.* **2017**, *121*, 195701. [CrossRef]
7. Lyons, J.L.; Wickramaratne, D.; Van de Walle, C.G. A first-principles understanding of point defects and impurities in GaN. *J. Appl. Phys.* **2021**, *129*, 111101. [CrossRef]
8. Lyons, J.L.; Janotti, A.; Van De Walle, C.G. Effects of carbon on the electrical and optical properties of InN, GaN, and AlN. *Phys. Rev. B Condens. Matter Mater. Phys.* **2014**, *89*, 1–8. [CrossRef]
9. Rackauskas, B.; Uren, M.J.; Stoffels, S.; Zhao, M.; Decoutere, S.; Kuball, M. Determination of the self-compensation ratio of carbon in AlGaN for HEMTs. *IEEE Trans. Electron. Devices* **2018**, *65*, 1838–1842. [CrossRef]

10. Uren, M.J.; Karboyan, S.; Chatterjee, I.; Pooth, A.; Moens, P.; Banerjee, A.; Kuball, M. "Leaky Dielectric" Model for the Suppression of Dynamic RON in Carbon-Doped AlGaN/GaN HEMTs. *IEEE Trans. Electron. Devices* **2017**, *64*, 2826–2834. [CrossRef]

11. Koller, C.; Pobegen, G.; Ostermaier, C.; Pogany, D. Effect of Carbon Doping on Charging/Discharging Dynamics and Leakage Behavior of Carbon-Doped GaN. *IEEE Trans. Electron. Devices* **2018**, *65*, 5314–5321. [CrossRef]

12. Verzellesi, G.; Morassi, L.; Meneghesso, G.; Meneghini, M.; Zanoni, E.; Pozzovivo, G.; Lavanga, S.; Detzel, T.; Häberlen, O.; Curatola, G. Influence of buffer carbon doping on pulse and AC behavior of insulated-gate field-plated power AlGaN/GaN HEMTs. *IEEE Electron. Device Lett.* **2014**, *35*, 443–445. [CrossRef]

13. Joshi, V.; Tiwari, S.P.; Shrivastava, M. Part I: Physical Insight Into Carbon-Doping-Induced Delayed Avalanche Action in GaN Buffer in AlGaN/GaN HEMTs. *IEEE Trans. Electron. Devices* **2019**, *66*, 561–569. [CrossRef]

14. Zagni, N.; Puglisi, F.M.; Pavan, P.; Chini, A.; Verzellesi, G. Insights into the off-state breakdown mechanisms in power GaN HEMTs. *Microelectron. Reliab.* **2019**, *100–101*, 113374. [CrossRef]

15. Fariza, A.; Lesnik, A.; Bläsing, J.; Hoffmann, M.P.; Hörich, F.; Veit, P.; Witte, H.; Dadgar, A.; Strittmatter, A. On reduction of current leakage in GaN by carbon-doping. *Appl. Phys. Lett.* **2016**, *109*, 212102. [CrossRef]

16. Zagni, N.; Chini, A.; Puglisi, F.M.; Meneghini, M.; Meneghesso, G.; Zanoni, E.; Pavan, P.; Verzellesi, G. "Hole Redistribution" Model Explaining the Thermally Activated RON Stress/Recovery Transients in Carbon-Doped AlGaN/GaN Power MIS-HEMTs. *IEEE Trans. Electron. Devices* **2021**, *68*, 697–703. [CrossRef]

17. Zagni, N.; Chini, A.; Puglisi, F.M.; Pavan, P.; Verzellesi, G. The Role of Carbon Doping on Breakdown, Current Collapse, and Dynamic On-Resistance Recovery in AlGaN/GaN High Electron Mobility Transistors on Semi-Insulating SiC Substrates. *Phys. Status Solidi* **2019**, *217*, 1900762. [CrossRef]

18. Meneghesso, G.; Silvestri, R.; Meneghini, M.; Cester, A.; Zanoni, E.; Verzellesi, G.; Pozzovivo, G.; Lavanga, S.; Detzel, T.; Haberlen, O.; et al. Threshold voltage instabilities in D-mode GaN HEMTs for power switching applications. In Proceedings of the IEEE International Reliability Physics Symposium (IRPS), Monterey, CA, USA, 14–18 April 2014; pp. 6–10. [CrossRef]

19. Viey, A.G.; Vandendaele, W.; Jaud, M.-A.; Cluzel, J.; Barnes, J.-P.; Martin, S.; Krakovinsky, A.; Gwoziecki, R.; Plissonnier, M.; Gaillard, F.; et al. Investigation of nBTI degradation on GaN-on-Si E-mode MOSc-HEMT. In Proceedings of the IEEE International Electron Devices Meeting (IEDM), San Francisco, CA, USA, 7–11 December 2019; pp. 4.3.1–4.3.4. [CrossRef]

20. Zagni, N.; Chini, A.; Puglisi, F.M.; Pavan, P.; Verzellesi, G. The effects of carbon on the bidirectional threshold voltage instabilities induced by negative gate bias stress in GaN MIS-HEMTs. *J. Comput. Electron.* **2020**, *19*, 1555–1563. [CrossRef]

21. Zagni, N.; Cioni, M.; Chini, A.; Iucolano, F.; Puglisi, F.M.; Pavan, P.; Verzellesi, G. Mechanisms Underlying the Bidirectional V T Shift After Negative-Bias Temperature Instability Stress in Carbon-Doped Fully Recessed AlGaN/GaN MIS-HEMTs. *IEEE Trans. Electron. Devices* **2021**, *68*, 2564–2567. [CrossRef]

22. Ber, E.; Osman, B.; Ritter, D. Measurement of the Variable Surface Charge Concentration in Gallium Nitride and Implications on Device Modeling and Physics. *IEEE Trans. Electron. Devices* **2019**, *66*, 2100–2105. [CrossRef]

23. Bellotti, E.; Bertazzi, F. A numerical study of carrier impact ionization in Al xGa 1-xN. *J. Appl. Phys.* **2012**, *111*, 103711. [CrossRef]

24. Armstrong, A.; Poblenz, C.; Green, D.S.; Mishra, U.K.; Speck, J.S.; Ringel, S.A. Impact of substrate temperature on the incorporation of carbon-related defects and mechanism for semi-insulating behavior in GaN grown by molecular beam epitaxy. *Appl. Phys. Lett.* **2006**, *88*, 1–4. [CrossRef]

25. Joshi, V.; Tiwari, S.P.; Shrivastava, M. Part II: Proposals to Independently Engineer Donor and Acceptor Trap Concentrations in GaN Buffer for Ultrahigh Breakdown AlGaN/GaN HEMTs. *IEEE Trans. Electron. Devices* **2019**, *66*, 570–577. [CrossRef]

26. Moens, P.; Liu, C.; Banerjee, A.; Vanmeerbeek, P.; Coppens, P.; Ziad, H.; Constant, A.; Li, Z.; De Vleeschouwer, H.; Roig-Guitart, J.; et al. An industrial process for 650 V rated GaN-on-Si power devices using in-situ SiN as a gate dielectric. In Proceedings of the IEEE International Symposium on Power Semiconductor Devices & IC's (ISPSD), Waikoloa, HI, USA, 15–19 June 2014; pp. 374–377. [CrossRef]

27. Cornigli, D.; Reggiani, S.; Gnani, E.; Gnudi, A.; Baccarani, G.; Moens, P.; Vanmeerbeek, P.; Banerjee, A.; Meneghesso, G. Numerical Investigation of the Lateral and Vertical Leakage Currents and Breakdown Regimes in GaN-on-Silicon Vertical Structures. In Proceedings of the IEEE International Electron Devices Meeting (IEDM), Washington, DC, USA, 7–9 December 2015; pp. 109–112, ISBN 9781467398947.

28. Hilt, O.; Brunner, F.; Cho, E.; Knauer, A.; Bahat-Treidel, E.; Wurfl, J. Normally-off high-voltage p-GaN gate GaN HFET with carbon-doped buffer. In Proceedings of the IEEE International Symposium on Power Semiconductor Devices and ICs (ISPSD), Virtual, 30 May–3 June 2011; pp. 239–242. [CrossRef]

29. Iucolano, F.; Parisi, A.; Reina, S.; Patti, A.; Coffa, S.; Meneghesso, G.; Verzellesi, G.; Fantini, F.; Chini, A. Correlation between dynamic Rdsou transients and Carbon related buffer traps in AlGaN/GaN HEMTs. In Proceedings of the IEEE International Reliability Physics Symposium (IRPS), Pasadena, CA, USA, 17–21 April 2016; pp. CD21–CD24. [CrossRef]

30. Uren, M.J.; Kuball, M. Current collapse and kink effect in GaN RF HEMTs: The key role of the epitaxial buffer. In Proceedings of the IEEE BiCMOS and Compound Semiconductor Integrated Circuits and Technology Symposium (BCICTS), Monterey, CA, USA, 16–19 November 2020; pp. 1–8. [CrossRef]

 *micromachines*

Review

Development of GaN HEMTs Fabricated on Silicon, Silicon-on-Insulator, and Engineered Substrates and the Heterogeneous Integration

Lung-Hsing Hsu [1,2] , Yung-Yu Lai [3], Po-Tsung Tu [1,2], Catherine Langpoklakpam [1], Ya-Ting Chang [1], Yu-Wen Huang [1], Wen-Chung Lee [1,2], An-Jye Tzou [4], Yuh-Jen Cheng [3] , Chun-Hsiung Lin [5,*], Hao-Chung Kuo [1,6,*] and Edward Yi Chang [5]

[1] Department of Photonics and Institute of Electro-Optical Engineering, College of Electrical and Computer Engineering, National Chiao Tung University, Hsinchu 30010, Taiwan; alger.99g@g2.nctu.edu.tw (L.-H.H.); itriA30378@itri.org.tw (P.-T.T.); cath01.ee09@nycu.edu.tw (C.L.); s922085493@gmail.com (Y.-T.C.); huangwendy227@gmail.com (Y.-W.H.); vincent.lee67@gmail.com (W.-C.L.)

[2] Industrial Technology Research Institute, Hsinchu 31040, Taiwan

[3] Research Center for Applied Sciences, Academia Sinica, Taipei 114699, Taiwan; loveriver031@gmail.com (Y.-Y.L.); yjcheng@sinica.edu.tw (Y.-J.C.)

[4] Taiwan Semiconductor Research Institute, Hsinchu 30078, Taiwan; jerrytzou.ep00g@gmail.com

[5] International College of Semiconductor Technology, National Yang Ming Chiao Tung University, Hsinchu 30010, Taiwan; edc@mail.nctu.edu.tw

[6] Semiconductor Research Center, Hon Hai Research Institute, Taipei 114699, Taiwan

[*] Correspondence: chun_lin@nctu.edu.tw (C.-H.L.); hckuo@faculty.nctu.edu.tw (H.-C.K.); Tel.: +886-3-571-2121 (ext. 31986) (H.-C.K.)

Citation: Hsu, L.-H.; Lai, Y.-Y.; Tu, P.-T.; Langpoklakpam, C.; Chang, Y.-T.; Huang, Y.-W.; Lee, W.-C.; Tzou, A.-J.; Cheng, Y.-J.; Lin, C.-H.; et al. Development of GaN HEMTs Fabricated on Silicon, Silicon-on-Insulator, and Engineered Substrates and the Heterogeneous Integration. *Micromachines* **2021**, *12*, 1159. https://doi.org/10.3390/mi12101159

Academic Editor: Giovanni Verzellesi

Received: 3 August 2021
Accepted: 11 September 2021
Published: 27 September 2021

Publisher's Note: MDPI stays neutral with regard to jurisdictional claims in published maps and institutional affiliations.

Copyright: © 2021 by the authors. Licensee MDPI, Basel, Switzerland. This article is an open access article distributed under the terms and conditions of the Creative Commons Attribution (CC BY) license (https://creativecommons.org/licenses/by/4.0/).

Abstract: GaN HEMT has attracted a lot of attention in recent years owing to its wide applications from the high-frequency power amplifier to the high voltage devices used in power electronic systems. Development of GaN HEMT on Si-based substrate is currently the main focus of the industry to reduce the cost as well as to integrate GaN with Si-based components. However, the direct growth of GaN on Si has the challenge of high defect density that compromises the performance, reliability, and yield. Defects are typically nucleated at the GaN/Si heterointerface due to both lattice and thermal mismatches between GaN and Si. In this article, we will review the current status of GaN on Si in terms of epitaxy and device performances in high frequency and high-power applications. Recently, different substrate structures including silicon-on-insulator (SOI) and engineered poly-AlN (QST®) are introduced to enhance the epitaxy quality by reducing the mismatches. We will discuss the development and potential benefit of these novel substrates. Moreover, SOI may provide a path to enable the integration of GaN with Si CMOS. Finally, the recent development of 3D hetero-integration technology to combine GaN technology and CMOS is also illustrated.

Keywords: gallium nitride; high-electron mobility transistor; heterogeneous integration; SOI; QST

1. Introduction

1.1. History and Applications of GaN HEMT

In the past decades, the wide bandgap GaN semiconductor materials and its alloys (AlGaN and InGaN) are emerging as one of the most promising materials for a variety of applications. Due to its robust thermal stability and electronic properties such as radiative hardness, it is an ideal candidate for working in a harsh and aggressive environment [1,2]. The development of AlGaN/GaN high electron mobility transistors (HEMT) is mainly due to both military and commercial interests in high-temperature, high-frequency, and high-power device applications [2–4]. AlGaN/GaN HEMT has been extensively developed for radio frequency (RF) high power amplifiers with high output power and high efficiency [3,5–7]. GaN's high breakdown electric field (~3.3 MV/cm) and high mobility

(>900 cm^2 Vs) [8] makes GaN-based devices attractive to work in very high powers and microwave frequencies since it can handle high current (~ 10 A) and high voltage (~ 100 V) with high transit speed [9,10]. The first GaN RF device product was presented in 2005 [11] but the insertion of the technology was limited in the military and some high-end RF infra-structure applications due to high cost. Later on, the first commercial product of GaN HEMT for power electronic application was presented in 2009 [12]. Owing to its superior performance and manufacturability on Si, GaN HEMT enters the stage of rapid growth. More recently, GaN HEMT has been rapidly developed for high-performance RF power amplifiers (PAs), such as mmW PA for 5G and beyond applications, owing to the requirement of wide-bandwidth and data rate for future mobile communication systems [13]. The roadmap in GaN HEMTs applications is illustrated in Figure 1.

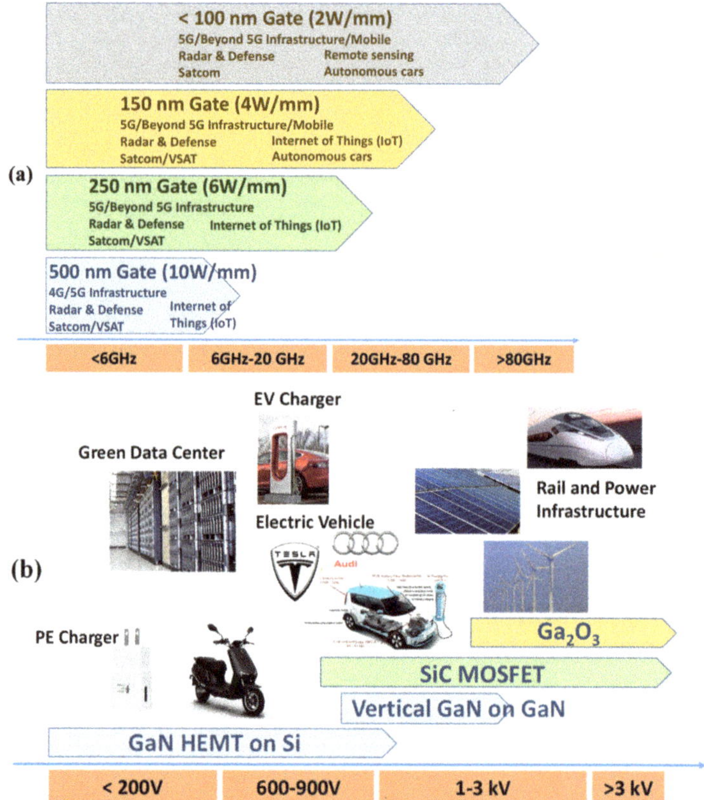

Figure 1. (a) A roadmap of RF GaN HEMTs technology. (b) A roadmap of Power GaN HEMTs technology.

1.2. The Influence of Different Substrates on GaN HEMT Device

In the earlier stage of its development, the progress of GaN/AlGaN HEMT was focused on improving the quality of epitaxial material for RF applications, the major efforts include the selection of the best substrate as well as developing unique processes [14]. GaN HEMTs can be grown on different substrates, including sapphire, silicon (Si) silicon carbide (SiC), or diamond due to the lack of GaN bulk substrate [15,16]. The basic structure of GaN HEMT is shown in Figure 2. For RF GaN HEMT, it was expected to provide a very high output RF power for a single die up to several hundred watts over a wide frequency range [17,18]. The high-power density requires efficient power dissipation on transistors as well as on substrate [19], thus, the performance and reliability of the high-power RF transistor can be seriously affected by the thermal conductivity of the substrate. The thermal properties of the different substrates are shown in Table 1. By using a substrate with high thermal conductivity, like SiC, the high heat generated can be effectively dissipated. AlGaN/GaN HEMTs on SiC substrates with very high RF output power densities of about 40 W/mm have been reported [20]. However, the full advantages of GaN on the high thermal conductivity SiC cannot be obtained as the thermal performance is degraded due to the epitaxial layer defects at the interface between GaN and SiC [21], which we will discuss further in a later paragraph. Cree has been developing GaN on SiC for commercial sale since 2006 [22]. The gradual scale-up of high purity semi-insulating 4H-SiC substrates from 2-inch to 4-inch substrates has greatly enhanced the economic viability of wide band-gap microwave devices.

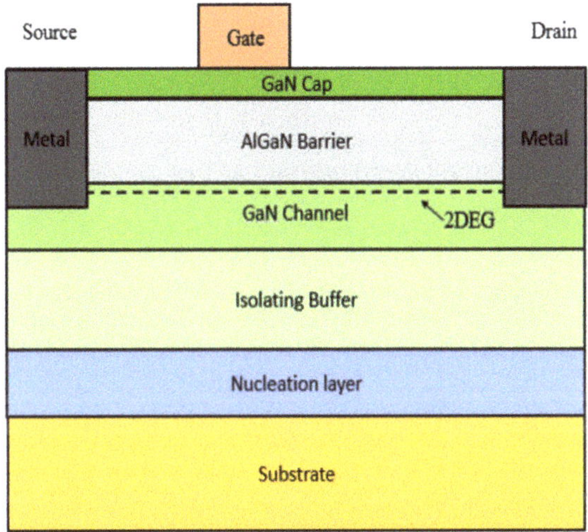

Figure 2. The basic structure of GaN HEMT.

In the past, SiC is mainly used as a substrate for RF GaN HEMT and demonstrated satisfactory performance. However, Si substrate is desired due to its low cost, good thermal conductivity, and availability of a large area. The growth of GaN on Si is more difficult than the growth of GaN on SiC since the growth of GaN on Si tends to result in higher dislocation density or microcracks due to higher thermal expansion coefficient and lattice mismatches [23]. With the progress of material engineering, the low cost and high performance. GaN on Si device has been achieved on a high-quality GaN layer grown on a large area Si substrate (8-inch wafer) [24]. One of the main issues of fabricating RF GaN HEMT on Si is the increased RF losses due to the parasitic conduction channel introduced at the III-nitride–Si interface and lower resistivity of Si substrates [25–28]. High resistive (HR) Si substrates are often adopted to solve the issue. However, the growth of high strain GaN on HR Si substrate shows a worse bowing problem. Moreover, the cost of HR Si is higher than that of low resistive (LR) Si substrate. These factors limit the production of the device on a larger silicon substrate. Substrate removing technique is reported to allow the fabrication of GaN HEMT on LR Si substrate with minimized substrate parasitic to improve RF/MW characteristic [29]. Though high-performance RF GaN HEMT on Si is still in development, GaN HEMT on Si has become the mainstream for power electronics applications. Currently, the power GaN HEMT with breakdown voltage higher than 1200 V has been demonstrated. In the later paragraphs, we will discuss and review the status of GaN on Si devices for both RF and power applications in details.

Among all the potential substrate materials, single-crystal diamond shows the highest thermal conductivity [30]. A high cut-off frequency (85 GHz) of GaN HEMT on the diamond was fabricated using wafer bonding between GaN HEMT structure grown on Si wafer and polycrystalline diamond wafer [31,32]. AlGaN/GaN HEMT structure grown by molecular beam epitaxy (MBE) on diamond (111) was reported with RF small-signal characteristic [31,32]. A single-crystal AlN and AlGaN/GaN HEMT were grown on a diamond substrate by using (111) surface orientation with metal-organic vapor phase epitaxy (MOVPE) [15,33,34]. AlGaN/GaN HEMTs grown on diamond (111) substrate for RF power operation with an output density of 2.13 W/mm were achieved [31]. Although the device performance for GaN on diamond looks promising, the lack of a large diamond substrate limits the development of the technology. Commercially available sapphire is also one of the most prominent substrates for the epitaxy of GaN due to the limited native GaN availability [35]. However, the use of sapphire is limited as the stability of the GaN on sapphire is less than that of GaN on Si due to weaker scattering phenomena which is related to its thermal conductivity at high drain bias [36].

Along with the thermal conductivity of the substrate, Thermal Boundary Resistance (TBR) is an important parameter to affect the overall rise in temperature of a device [37]. The thermal effects of the device are boosted by TBR which is resulted from the interface for the growth of dissimilar materials. TBR is a measure of the resistance imposed by the interface to heat flow due to the different dynamics of the phonons and the poor quality of the crystal near the boundary. It is defined as the maximum temperature increase by the maximum power dissipation of the device. The TBR in commercial GaN HEMTs on SiC can reach levels greater than 6×10^{-4} cm^{2}K/W which can increase the maximum temperature of the device by up to 40–50% [38,39]. The increase in TBR causes an offset to the advantages offered by the substrate with high thermal conductivity due to the discontinuity of the temperature gradient in TBR layers located at the GaN substrate interface [40]. Thermal conductivity for GaN and different substrate materials at 300 K and TBR values of GaN/ substrate interface simulated in [41] are listed in Table 1. The lower TBR values of GaN on Si substrate than that of SiC substrate is mainly due to the different thermal expansion coefficient [42], the roughness of the substrate materials as well as the defect related to the growth techniques. The high value of TBR at the interface of high thermal conductivity substrate and GaN could be reduced and has additional advantage if it has a better thermal coupling and fewer interface defects while low thermal conductivity like sapphire has a less significant influence of TBR on transporting heat [43].

Table 1. Temperature-dependent thermal conductivity of GaN/different substrate materials, TBR values for GaN/substrate interfaces used in simulations are listed [41,44].

Material	Thermal Conductivity κ (W/m·K)	TBR (m^{2}K/GW)	Thermal Expansion Coefficient, $(10^{-6}/K)$	Lattice Mismatch with GaN (%)
GaN	$160\,(300/T)^{1.4}$	–	–	–
Sapphire	$35\,(300/T)^{1}$	10–40	39	16
Si	$150\,(300/T)^{1.3}$	10–40	54	17
SiC	$420\,(300/T)^{1.3}$	30–60	3.2	4
Diamond	$1200\,(300/T)^{1}$	20–50	62.5	12

1.3. Evolution of GaN HEMT

1.3.1. Different Gate Structure Designs

The HEMT structure was based on T. Minura et al. (1975) [45] and M.A. khan et al. (1994) [5]. At the interface of AlGaN and GaN, there exists 2-dimensional electron gas (2DEG) with high electron mobility owing to the difference in spontaneous polarization and piezoelectric polarization [46]. Hence the device operates as a normally-on device naturally. To deplete the 2DEG channel, a gate electrode on top of the AlGaN layer has a negative gate voltage concerning drain and source electrode applied. This type of device is known as depletion-mode (D-mode) HEMT. Moreover, there are two types of D-mode HEMT, namely with a Schottky gate electrode or with an insulating gate [47]. The first d-mode HEMT introduced had a Schottky gate electrode in which the metal gate electrode is directly deposited on top of AlGaN. Ni–Au or Pt metals were used to form the Schottky barrier [23,48,49]. In insulate gated d-mode HEMT, an insulating layer is placed in between the gate electrode and AlGaN similar to that of MOSFET to block the gate current [50]. Schottky gate and insulated gate d-mode HEMTs are shown in Figure 3.

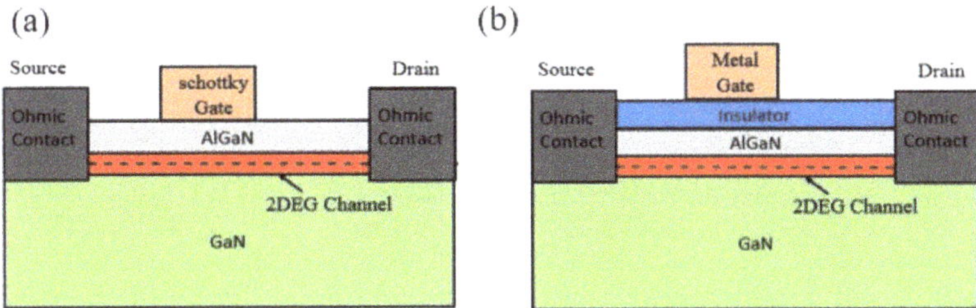

Figure 3. (**a**) Schottky gate D-mode HEMT (**b**) Insulated gate D-mode HEMT.

D-mode HEMT is not preferred in system applications as it requires a negative bias to be applied. On the other hand, there is also a concern in fail-safe operation. Therefore, Enhancement-mode (E-mode, normally-off) device is favored and has become one of current focuses of technology development. To create e-mode devices, there are five popular structures: recessed gate, implanted gate, pGaN gate, direct-drive hybrid, and cascode hybrid [47]. For GaN HEMT in RF applications, the most common fabrication technique used for modifying the threshold voltage is "gate-recess". This process reduces the barrier thickness under the gate metal. The basic structure of gate recess HEMT is shown in Figure 4a.

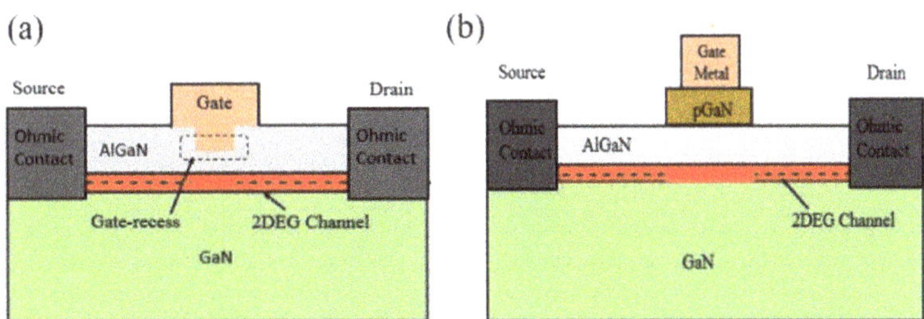

Figure 4. (**a**) Basic structure of Gate-recess HEMT. (**b**) Schematic structure of p-GaN Gate HEMT.

There are also approaches to obtain E-mode devices by heterostructure or gate stack designs. Ohmaki et al. proposed a double-barrier-layer AlGaN/GaN HFET in 2006. The use of double-barrier-layer sustains the device at a BV of 435 V, Ron,sp of 1.9 mΩ cm^2, with a threshold voltage of -0.1 V [51]. Inter University Microelectronics Centre (IMEC) presented another E-mode AlN/GaN/AlGaN HFET structure in 2010 with a double heterostructure-barrier layer. The double structure comprises of a high concentration 2DEG at the surface of the hetero structure and an ultrathin AlN barrier layer grown on silicon substrate; and the structure maintains a BV of 580 V, with an Ron,sp of 1.25 mΩ cm^2, and a threshold voltage of about 0 V [52]. Mizutani et al. in 2007, proposed an E-mode AlGaN/GaN HFET with a thin InGaN cap layer. The proposed structure shifts the threshold voltage towards the positive direction by rising the conduction band of the AlGaN/GaN via the use of a polarization-induced field in the InGaN cap layer [53]. Ostermaier et al. in 2009, proposed an ultrathin InAlN/AlN barrier HEMT that have high performance under normally off operation. They selectively etched the n++ GaN cap layer structure, which in-turn controls the width of the ultrathin barrier layer [54]. The proposed structure operates with a threshold voltage of 0.7 V, a maximum transconductance of 400 mS/mm and a maximum output current of 800 mA/mm. Hughes Research Laboratories, in collaboration with the Next Generation of Nitrides Electronics Project from the US Defense Advanced Research Projects Agency (DARPA), proposed an E-mode HFET with a double-barrier layer via selective-area molecular beam epitaxy (MBE) [55]. For the integrated E-mode and D-mode AlN/GaN/AlGaN double-heterojunction field-effect transistors (DHFETs) on a single SiC substrate, an E-mode channel was achieved without additional procedures or compromise in the electrical characteristics. The idea of developing a monolithic integrated E-mode and D-mode device can be the groundwork for direct-coupled field-effect transistor (FET) logic circuits. Guowang Li et al. proposed an E-mode AlGaN/AlN/GaN HFET having 70% Al composition comprising of a thick AlGaN layer (17 nm), an AlN layer (0.6 nm) and an Al_2O_3 (4 nm) layer. They used Ni/Au in the Schottky gate metal instead of Al/Au thereby enhancing the threshold voltage from -1.0 V to -0.13 V [56]. Chiu et al. proposed an E-mode HFET comprising of a composite dielectric layer via N_2O plasma oxidation technology. The composite dielectric layer (Al_2O_3/Ga_2O_3) after the oxidation in the N_2O for the AlGaN barrier layer improves the threshold voltage from -3.6 V to 0.17 V [57]. Based on this structure, Chiu et al. presented an E-mode HFET with a high-*k* composite dielectric layer in 2012. The AlGaN barrier layer was oxidized by nitric oxide (NO) gas prior to the Schottky metal gate deposition. After the oxidized AlGaN barrier layer stack (Al_2O_3/Ga_2O_3) is formed, a dielectric film of gadolinium oxide (Gd_2O_3) was deposited to complete the structure thereby, enhancing the threshold voltage from -3.15 V to 0.6 V [58]. Massachusetts Institute of Technology in collaboration with Harvard university presented one E-mode HFET with a scandium oxide (Sc_2O_3) high-*k* dielectric. In the study, Wang et al. introduced a Sc_2O_3 high k-dielectric layer in the HFET to reduce the invert leakage current thereby having the switch-current ratio (I_{ON}/I_{OFF}) reach 10s [59].

By reducing AlGaN barrier thickness, it can result in a reduction of polarization induced 2DEG and the threshold voltage is shifted positively with the help of the work-function of the metal gate. A positive threshold voltage can be achieved with a deep enough gate recess etching thus forming an E-mode HEMT [60]. A chloride-based dry inductively coupled plasma reactive ion etching (ICP-RIE) for gate recess etching has been employed by several groups [61–66] which can effectively change the threshold voltage to the positive direction of AlGaN/GaN HEMT. Damage in the subsurface was also reported from ICP dry etching due to low etch selectivity between materials which increases the gate-leakage current [67]. The damages were found to be repaired after post-etching rapid thermal annealing (RTA) at 700 °C [65,66]. For the RF HEMT structure, recessed gate enhances the device performance by providing a better gate control capability of the channel carriers. On the other hand, due to the repaired damage with proper recessed gate process, flicker noise characteristics were reduced, which is beneficial for RF circuit applications such as voltage-controlled oscillators (VCOs) and mixers [68,69]. On the other hand, p-GaN gate technology is more popular for the fabricator of e-mode power GaN HEMT device. The schematic structure of p-GaN gate HEMT is shown in Figure 4b. The conduction band of the AlGaN is risen above the Fermi level when compared with a standard Schottky gate normally-on HEMT due to the presence of p-GaN cap and results in the depletion of the 2DEG channel. Uemoto et al. first proposed a p-AlGaN gate normally-off HEMT [70]. The thickness of the AlGaN barrier as well as Al-concentration needs to be defined appropriately to achieve an efficient depletion region in the channel [71–74]. A high doping concentration of Mg, a p-type dopant, ($>10^{18}$ cm^{-3}) is necessary for the efficient depletion at the interface of the metal gate/p-gate [75]. However, the hole concentration can be reduced at a high temperature above

500 °C due to the formation of Mg-H complexes [76]. Hence, proper care is necessary during device processing like annealing of Ohmic contacts so that Mg concentration does not reduce. Another important feature is the choice of the metal gate as the threshold voltage of the device depends on metal/p-GaN Schottky barrier height. In this aspect, much researches have been done on metal gate work-function influence on p-GaN electrical behavior [77–80]. Schottky metal gates have shown improvement in lowering the leakage and increasing the threshold voltage than Ohmic gate [77,78]. Presently, the TiN gate is one of the good solutions due to its thermal and chemical stability along with the processing compatibility [81–84]. A p-GaN gate HEMT with a "self-aligned" process using Mo-based gate was demonstrated which employed "gate first" process [85]. Mo gate sustained the high-temperature annealing process of source-drain ohmic contact without the barrier degradation. Although p-GaN gate HEMT has reached commercialization, their reliability issues [80] need intensive researches.

1.3.2. N Polar vs. Ga Polar

Typically group III-Nitride devices are fabricated using the Ga-polar (0001) orientation. However, the inverted N-polar polarity possesses a numerous advantage over Ga-polar counterparts. The absence of inversion symmetry in wurtzite group III-Nitride results in opposite polarization of N-polar crystal and Ga-polar crystal. Hence, the polarization induced electric fields of Ga-polar heterostructures is opposite to that of N-polar counterparts which results in formation of 2DEG of N-polar heterostructures above the wide-bandgap barrier layer instead of below [86]. The advantages offered by N-polar GaN HEMTs over Ga-polar HEMTs are as follows: (i) N-polar heterostructures has a strong back-barrier due to its inherent wide-bandgap Al(Ga)N back-barrier for electron confinement which reduces the effects due to short-channel [87], (ii) N-polar HEMTs has low-resistivity Ohmic contact as the channel layer with lower surface barrier to electrons and a narrower bandgap can contact 2DEG of N-polar HEMTs rather than contacting through wide-bandgap Al(Ga)N barrier [88,89] which results in the possibility of lowering the contact resistance by using selective regrowth of ohmic area in N-polar structure, (iii) N-polar heterostructure has improved scalability. The effective gate-channel distance is reduced by the quantum displacement of the 2DEG in N-polar HEMT which is opposite to that of Ga-polar HEMTs owing to the reduction in effective gate-channel capacitance due to quantum capacitance [90]. Such enhancement results in higher N-polar transconductance than that of Ga-polar with same gate-channel thickness. The aspect ratio as well as the charge density under the gate of N-polar HEMTs can be controlled independently where the enhancement of charge depletion due to the scaling of channel thickness are compensated by increasing the thickness and the polarization of charge-inducing back barrier. Whereas, gate aspect ratio in Ga-polar HEMT depends on the barrier thickness and there is a trade-off between the charge density under the gate and the barrier thickness [91]. N-polar heterostructure does not required intentional n-type doping for the formation of 2DEG [89,92,93] due to the presence of high-density unintentional bulk and surface donors [94]. A N-polar GaN HEMTs epitaxy grown by MOCVD on 4° off-cut 2-inch diameter sapphire substrate at UCSB for X-band power performance transceiver systems was reported with 2 tone results at 10 GHz [95]. The reported HEMTs structure is similar with that of structure reported by X. Zheng et al. [96]. The reported HEMT device shows very little 3rd order distortion in intermodulation and 65% of single tone high power added efficiency (PAE) with 3 W/mm power density which can be scaled favorably with 15 V drain voltage. A deep recess W-band power N-polar GaN HEMTs utilizing a new atomic layer deposition (ALD) ruthenium (Ru) gate metallization process with PAE of 33.8% and 6.2 W/mm high power density was demonstrated [97]. The demonstrated HEMTs has an outstanding control over the DC-RF dispersion due to presence of deep recess structure in conjunction with SiN thin passivation layer and has a high gain as the narrow 48-nm gate trench was filled by ALD Ru metal.

1.4. CMOS Compatible Process for GaN HEMT

In wireless communication, to respond to the growing demand for high data rates, there is a push to a higher operating frequency, switching to millimeter-wave from the congested sub 6 GHz band. Beyond operational speed of power amplifiers, output power (P_{out}) and power added efficiency (PAE) in RF Front End Modules are critical for next-generation portable devices and small cells. As opposed to CMOS, the high-power handling capabilities of GaN are advantageous for mm-wave operations [98] as the more energy-efficient system can be achieved owing to its capability in the high output power with high efficiency at high-frequency. However, the integration of GaN HEMTs remains one of the main concerns. The current limitation of GaN RF technology is due to the use of expensive older generation Au-based processing as well as non-Si substrates [99]. Hence to make GaN devices for RF and MM-wave applications, migrating to 200 nm Si platform and using standardized CMOS fabrication tools for manufacturing the device become a crucial step to improve yield and reduce the cost. Similarly, the CMOS compatible technology can also benefit the development of GaN devices for power electronics. A CMOS-compatible 110 V/650 V e-mode GaN HEMT with excellent power converter switching performance with high robustness was fabricated on 6-inch GaN on a Si wafer [100]. Figure 5 shows an example fabrication process flow of GaN devices. GaN devices with low RF loss, low buffer dispersion as well as good leakage blocking capability have been demonstrated by integrating the device on the Si platform based on the Au-free, Si CMOS compatible process [98,101–103]. To enhance the functionality as well as the performance of the RF modules, various approaches to integrate CMOS and GaN devices have been developed [104,105].

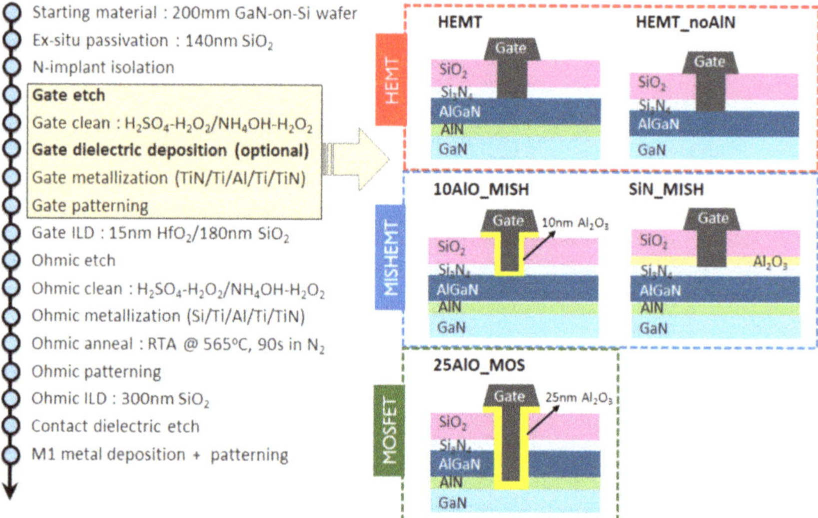

Figure 5. Gate-first process flow for the fabrication of GaN devices. Five device splits were realized based on differences in gate processing [98].

1.5. GaN-Based CMOS Technology

A lot of demonstration has been done on power integrated circuit based on GaN [106–108] which depends on the integration of E-mode and D-mode n- type HEMTs. However, these E/D mode circuits have reduced output voltage swing and also suffers from static power dissipation. Therefore, a CMOS-like circuit technology is required to increase the efficiency of ICs based on GaN as this circuit technology has negligible static power consumption, has higher noise immunity and less circuit complexity [109].

However, the challenges of the monolithic integration of the p-type GaN FETs with n-type GaN FETs along with lack of high-performance of p-type GaN FETs are the major obstacle towards achieving high efficiency GaN-based ICs. A various epitaxial structures of p-type GaN FETs have been demonstrated [110–117]. A GaN complementary inverter circuit comprising of both E-mode n-type GaN FET and p-type GaN FET monolithically integrated on Si Substrate without regrowth technology was also demonstrated [118]. The probe station with thermal chuck were used to characterize the fabricated inverter under high operation temperature. The fabricated circuit shows an outstanding transfer characteristics up to 300 °C with maximum recorded voltage gain of about 27 V/V at 0.59 V input voltage with 5 V as V_{DD} supply. A very high density of 2D hole gas (2DHG) induced by the polarization at the interface of GaN/AlN was discovered [119] which led to the development of p-channel heterostructure field effect transistors (HFETs) that reach the linear current density of 100 mA/mm [112]. A p-channel MISFET with a recessed-gate was grown by metalorganic chemical vapor deposition (MOCVD) using p-GaN/AlGaN/GaN hetrostructure on Si substrate [113,118]. The fabricated structure contains both 2-dimensional electron gas (2DEG) and 2-dimensional hole gas (2DHG) without regrowth technology which is suitable for implementing GaN- based complementary circuit. The fabricated long channel p-type device when compared with p-FET GaN/AlGaN on sapphire substrate exhibits state of the art on-off ratio performance. A p-channel 2DHG GaN/AlN transistors which can break the barrier of GHz speed was demonstrated [120]. The fabricated transistor exhibits an on-current density of 428 mA/mm and a cut-off frequency of 20 GHz. A wide-bandgap CMOS platform formed using these fabricated p-channel HFETs along with excellent performance n-channel HFETs [121] was expected to achieve a new domain in the RF and power electronics applications [122].

In the later paragraphs, we will review in more detail the progress and performance of GaN on Si devices. Moreover, the progress in hetero-integration will also be described.

2. GaN HEMT on Si

2.1. GaN Epitaxial Growth on Silicon

Silicon substrate is a general and commercial materials in semiconductor technology. Actually, a good epitaxy must be fine-matched in lattice and thermal expansion coefficient between GaN and heterogenous substrate, like Si, Sapphire, SiC ... etc. The related summary in physical parameters is listed in Table 2 [123].

Table 2. The lattice and thermal mismatch of Si, SiC, Sapphire, AlN, and GaN.

Mismatch	Si	SiC	Sapphire	AlN	GaN
Crystal Structure	FCC	HCP	HCP	HCP	HCP
Lattice Constant (Å)	5.43	3.08	4.758	3.112	3.189
Lattice Mismatch (%)	−16.9	3.5	16.08	2.4	–
Thermal Expansion (10^{-6} K)	3.59	4.3	7.3	4.15	5.59
Thermal Mismatch (%)	55	30	−23	34	–

Due to the mismatch of lattice constant and thermal expansion coefficient between Si and GaN, it is more difficult to achieve high quality GaN structure growth on Si [124–126]. In general, a lot of defects exist, and the gallium nitride layer may crack when cooling process. In order to solve these problems, gallium nitride epitaxial layers with high uniformity, high quality and no cracks can be grown on silicon substrate by growing buffer layer as shown in Figure 6 [12,127]. The buffer layer can be designed in the following two ways: (1) AlN/GaN superlattice growth through the step gradient AlGaN layer [128], and (2) AlGaN/GaN superlattice [129]. The bending of the wafer will increase with the increase of the thickness. A multilayer-buffer structure reduces the tensile stress caused by the huge difference in thermal expansion coefficient between GaN and Si. It's useful for inducing compressive strain in the growth process, as to counteract tensile strain introduced in the cooling process, prevent cracking and produce a flat wafer.

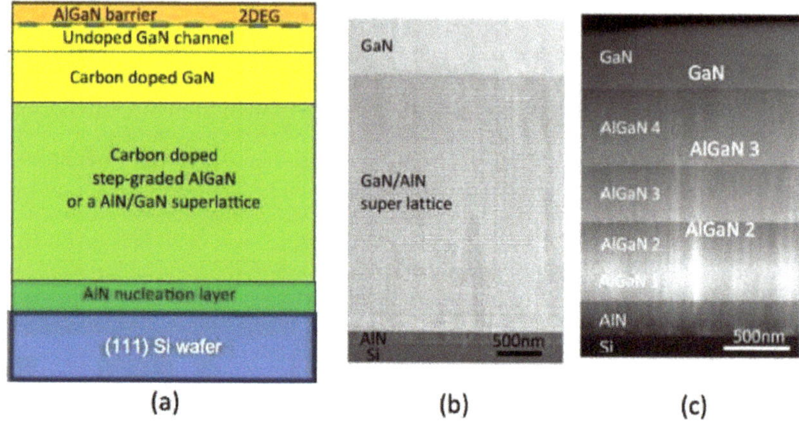

Figure 6. (**a**) Schematic cross-section of the typical epitaxial layer structure used for the manufacture of GaN-on-Si HEMTs. (**b**) TEM image of a GaN/AlN superlattice buffer layer and (**c**) a step graded AlGaN buffer layer, both on Si substrates [12].

Furthermore, the breakdown voltage in GaN HEMT is affected by the quality and resistivity of GaN-based templates. In order to operate an efficient GaN channel, it also needs higher resistivity buffer layer to prevent the DC leakage current and AC coupling. The characteristics of GaN high-frequency power amplifiers will change with the resistance of the underlying buffer layer, which is mainly due to the signal coupling effect. However, several groups have used different methods to improve the buffer layer resistivity, which employed p-type dopant (Mg) to enhance GaN buffer structure (n-type intrinsically). Another method is using Carbon dopant which plays a more attractive role in buffer layer. Compared to Mg case, the storage effect of Carbon-doped method isn't strong. According to doped buffer layer, the carrier concentrations and electrical properties (breakdown voltage) are adjusted by varied epitaxial conditions [130].

On the other hand, the top barrier layer includes AlGaN or InAlN, which results in 2DEG of HEMT through polarization charges in nitride-based materials are critical. While the thin AlGaN layer is grown on GaN channel layer, the Al content and thickness would be limited due to the lattice mismatch of GaN. The interface charge could be adjusted by varied barrier thickness and Al compositions. Compared to AlGaN, InAlN could reduce epitaxial defects as a result of thicker critical thickness. It shows that a good lattice-matched InAlN (18% In) exhibits a stronger spontaneous polarization to generate a higher channel charge density [131].

2.2. Power GaN Performance Si Substrate

As described in Section 1, the traditional power HEMT structure uses a Schottky metal to modulate the 2DEG in the channel. Generally, the metal stack Ni/Au is used for the HEMT. However, in order to efficiently control the gate leakage current, high-k gate dielectric layer was employed to form metal–insulator–semiconductor (MIS) gate, [132] for commercial power GaN HEMT on Si. The passivation layer provides additional protection and reduces the current loss in the surface state of devices. The breakdown voltage of GaN power components on Si in the low- and medium-power fields are predicted above 900 V, and the GaN shows very high potential in power applications due to the benefits of a low switching loss and lower cost [133].

Furthermore, the power electronics application represents one major market in the developments for GaN power devices like p-GaN HEMTs for enhancement-mode (E-mode) operation. The breakdown voltages for p-GaN HEMTs have exceeded 1000 V (Ron,sp of 2 mΩ cm^2). In power switching applications, a normally-off (enhancement-mode) GaN HEMT is desirable due to the safe-operation formation and efficient gate control to switch on/off. A variety of e-mode GaN HEMTs are fabricated by using p-GaN gate [74], gate recess [134], or plasma treatment techniques. The p-GaN gate HEMT showed a good performance, reliability, and commercialization. Figure 7 shows the band structures of normally-on AlGaN/GaN HEMTs and normally-off p-GaN/AlGaN/GaN HEMTs. The 2DEG channel is depleted at a zero-bias condition as the conduction band energy of AlGaN is lifted due to the p-GaN region. The electrical characteristics of the p-GaN gate HEMT shows V_{TH}, the V_{GS} limitation, and the gate leakage current (I_{GSS}) depend on the structure of the gate stack by using normally on and off system [135]. In order to deplete the 2DEG channel at $V_G = 0$, the general AlGaN thickness is 10~15 nm, and the thickness of the p-GaN gate is 50~100 nm. A doping Mg concentration of the p-GaN (or p-AlGaN) gate is around 10^{18}~10^{19} cm^{-3}.

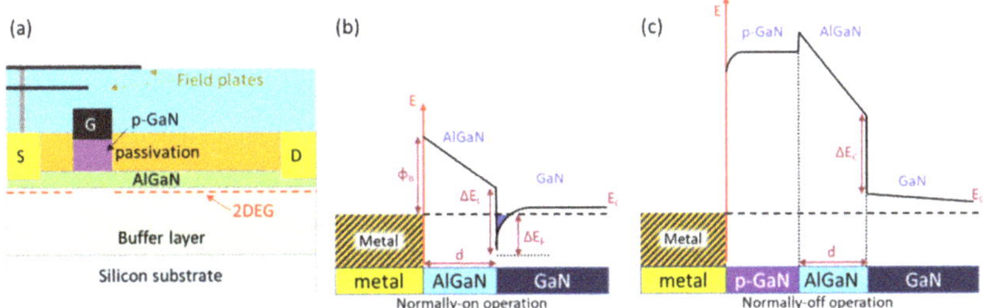

Figure 7. (**a**) Cross-sectional schematic of p-GaN gate HEMT [135] and (**b**) schematic of the operation principle of the normally on HEMT and (**c**) normally off HEMT [74].

2.3. RF GaN Performance Si Substrate

GaN technology that can offer high output power and efficiency at high frequencies is regarded as the most critical technology to reduce the complexity in designing upcoming mm-wave band communication system for 5G or beyond applications. In particular, GaN on Si technology that can greatly reduce production costs has attracted more attention. However, due to the challenges in obtaining higher-quality epitaxy on Si, RF GaN HEMT was fabricated primarily on SiC for high-frequency applications until 2014 when MACOM announced the mass production of 4″ GaN on Si technology. Due to the continuous improvement of epitaxial technology, almost all major semiconductor foundry starts to invest heavily on the development of GaN on Si technologies.

In order to develop RF GaN HEMT with superior high-frequency characteristics, we can refer to the equivalent circuit [136] shown in Figure 8, the small signal model of a GaN HEMT includes gate parasitic capacitance and resistances. According to Eq (1)(2) [137], we must decrease capacitance [138], ohmic contact resistance [139], gate resistance [140] and increase transconductance [141] in order to maximum the f_T and fmax.

$$F_T = \frac{g_m}{2\pi(C_{GS}+C_{GD})[1+(R_S+R_D)G_{DS}+g_m \times C_{GD}(R_S+R_D)]} \tag{1}$$

$$F_{MAX} = \frac{F_T}{2\sqrt{(R_i+R_S+R_G) \times G_{DS}+2\pi F_T R_G C_{GD}}} \tag{2}$$

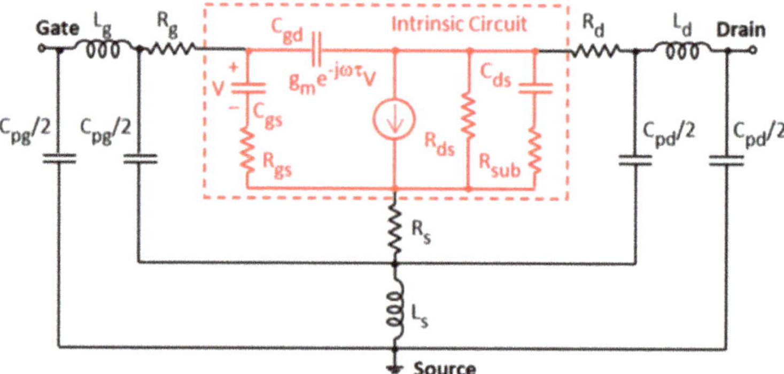

Figure 8. Small-signal equivalent circuit for the tested MOSFETs.

To operate at high frequency, the gate length must be minimum to reduce gate capacitance, as shown in Figure 9. But, the parasitic resistance will be increased and degrade the high frequency performance [142], the T-shaped gate becomes a key element to reduce the gate parasitic resistance. The height and width of the T-shaped gate should be optimized for both capacitance and resistance values. Keisuke Shinohara et al. [142] demonstrated the most suitable T-gate shape through simulation based on gate capacitance as show in Figures 10 and 11. Benchmark of cut-off frequency versus L_G [143–148] illustrates the importance of gate length shrinking to increase f_T. Both NTU [149] and Intel [143] demonstrated 40 nm gate length GaN HEMT on Si with f_T/fmax higher than 300 GHz, though the current record f_T/fmax values of 450 GHz was achieved by HRL [144] with the 20 nm Gate GaN on SiC technology.

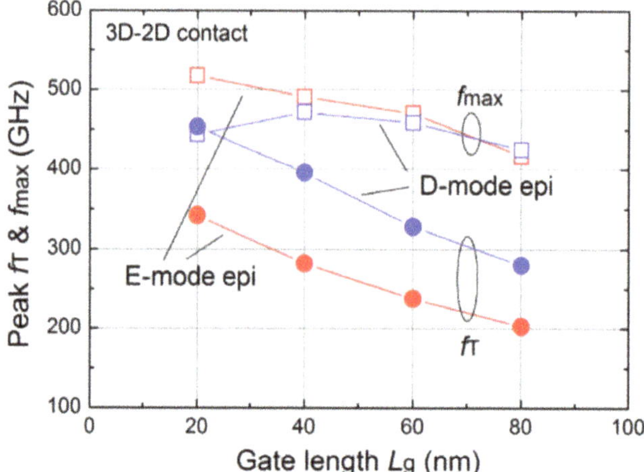

Figure 9. F_T and F_{MAX} versus Lg [142].

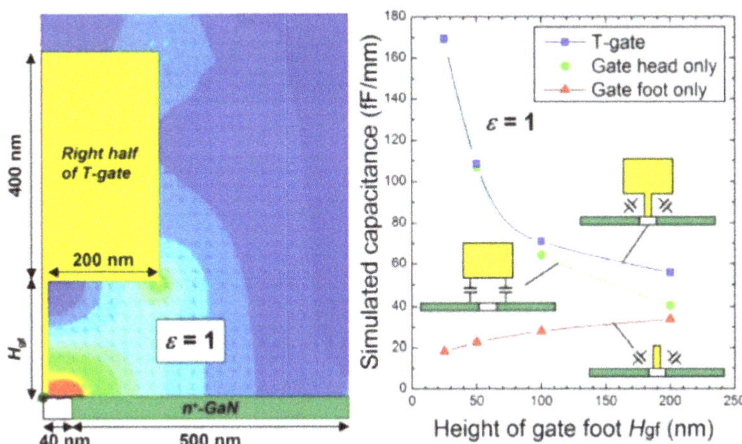

Figure 10. Simulation for T-gate capacitance [142].

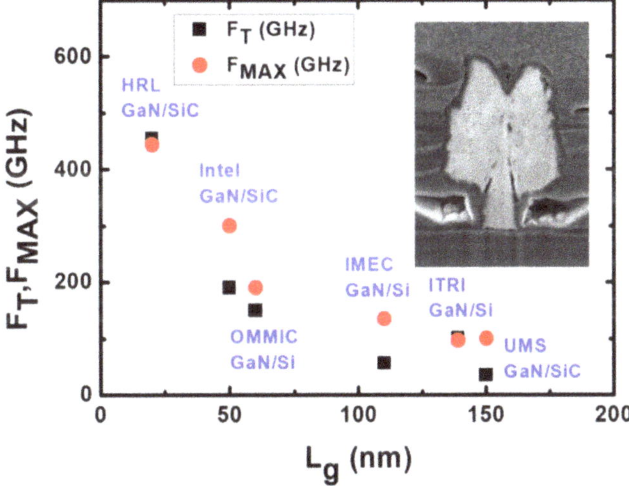

Figure 11. Benchmark for frequency versus Lg.

There are many device technologies have been developed based on GaN HEMT on SiC substrates. Most of those technologies can be applied on GaN HEMT on Si as well. We will describe some of the typical examples below. Hiroyuki Ichikawa et al. [150] designed 150 nm gate length InAlN/GaN and AlGaN/GaN HEMT, the AlInN/GaN HEMT showed high G_M and exhibited a f_T/fmax of 70/150 GHz. Michael L. Schuette et al. [141] demonstrated a peak f_T/fmax of 348/340 GHz for 27 nm gate length on InAlN/GaN HEMT with gate recess. Ezgi Dogmus et al. [151] used ultra-thin AlN (4 nm) barrier to replace AlGaN and in-situ SiN demonstrated a f_T/fmax of 55/235 GHz. Jeong-Sun Moon et al. [152] designed 50 nm gate length AlGaN/GaN HEMT with graded AlGaN barrier and n++regrowth, and the HEMT exhibited a ft/fmax of 156/308 GHz. Lei Li et al. [153] demonstrated f_T/fmax of 250/204 GHz using n++regrowth for InAlN/GaN HEMT on Si substrate. Figure 12 benchmark f_T versus f_{MAX} [141,150–155].

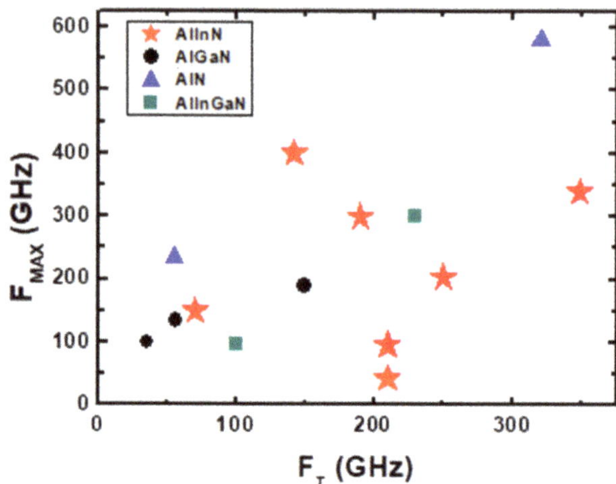

Figure 12. Comparison of the measured f_T and f_{MAX} in GaN-based HEMTs from literature.

GaN HEMT on Si also have high load pull result comparable to SiC substrate [155], nevertheless, GaN HEMT on Si substrate shows high potential. D.C. Dumka et al. [156] demonstrated 13.1 dB linear gain, maximum P_{OUT} = 34.5 dBm, output power density 7 W/mm and PAE 65.6% at 10 GHz in X band. Diego Marti et al. [157] showed 6 dB linear gain, output power density 1.35 W/mm, and PAE 12% at 94 GHz in W band.

For used in 5G mm-Wave communications, higher data rates require more complex communication systems such as Multi-input Multi-output (MIMO). MIMO employs complex frequency and phase division. So high linearity devices are required to avoid interaction each complex frequency bands. J. Vidkjær et al. [158] summarized some solution for linearity include geometrical, layout and epitaxial design. Weichuan Xing et al. [159] designed 150*150 nm nanostrip gate hole structure by BCl_3/Cl_2 and Al_2O_3 insulator which have good linearity. Jeong-sun Moon et al. [160] used AlGaN/GaN graded channel which have good PAE and linearity. Bin Hou et al. [161] used barrier layer of sandwich structure and AlGaN back barrier which show good power performance and linearity. Kai Zhang et al. [162] used Fin-FET HEMT that have good linearity compared to the planar HEMT.

3. GaN HEMT on Silicon-on-Insulator (SOI) Substrates

3.1. GaN Epitaxial Growth on Silicon-on-Insulator (SOI) Substrates

In epitaxial issues, the bowing effect always exists on the hetero-interface due to the lattice mismatch and the thermal expansion differences. T. Egawa from Nagoya Institute of Technology has reported a relative function of the wafer bowing and epitaxial thickness of AlGaN/GaN HEMT on Si [163], as shown in Figure 13. A thicker GaN/AlN superlattice structure exhibits higher bowing value in these experiments. It is a wafer bowing reference in HEMT epitaxial developments on Si. However, based on the outstanding electrical isotropic and mechanical features of the SOI substrate, it is expected to be a significant contender as a technological platform for mass production of GaN HEMT in the near future. However, the SOI substrate still suffer from a bowing effect, which may result in broken wafers or difficulties in subsequent fabrication processing steps, as well as a lower temperature tolerance during wafer process. Recently, from our study, we demonstrate the growth of AlGaN/GaN heterostructure on a 150-mm SOI substrate with different boron doping concentration in handle wafer, as shown in Figure 14. By heavily doping Boron in silicon handle wafer of SOI substrate, we can effectively reduce the bowing effect, increase the thickness of the epitaxial layer, and further improve the device performance. Heavily doped handle wafer causes a reduction in wafer bowing by >97%, as shown in

Table 3. Moreover, it can be seen from Figure 15 that the issue of edge cracks for the heavily doped SOI substrate (sample B) are great improved, and there is no peeling phenomenon, which means that the heavily doped SOI substrate have better ability to resist the stress generated during GaN epitaxy. Figure 16 is the comparison of the half-width values of the GaN epitaxial layer (102) measured by X-ray diffraction analyzer. It can be observed that the BOW value is related to the epitaxial quality. That is, the more severe the bowing effect, the worse the epitaxial quality, which has also been demonstrated in GaN on SiC and bulk GaN substrate [164,165].

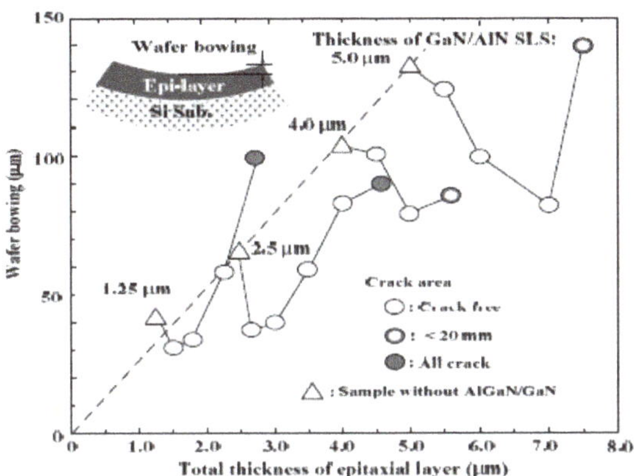

Figure 13. Wafer bowing as a function of total epitaxial layer thickness [163].

GaN Cap 2 nm
AlGaN barrier 20 nm
AlN spacer 1 nm
GaN Channel 800 nm
AlGaN Buffer 1.5 μm
Device layer (111)
Buried oxide
Handle wafer (100)

Figure 14. Schematic cross-section of AlGaN/GaN HEMT on SOI substrate.

Table 3. Device characteristic of different Boron doped level in handle wafer.

Sample	A	B
Dope-level	Light doped	Heavy doped
Doping concentration (atoms-cm^{-3})	1.35×10^{15}~1.49×10^{16}	5.95×10^{19}~1.26×10^{20}
Bowing (μm)	−258	6.7
FWHM (arcsec) (1 0 2)	1484	1188

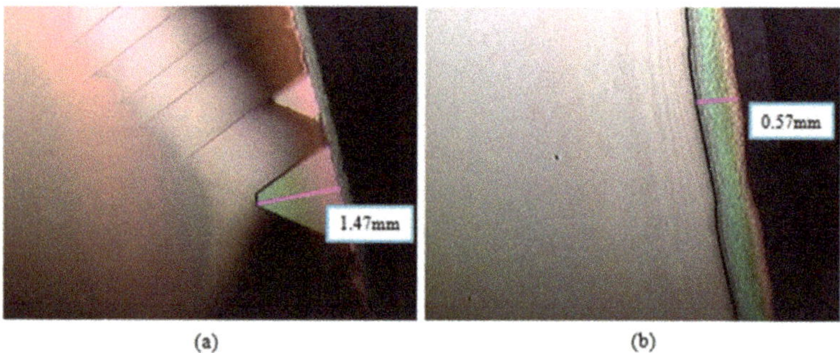

Figure 15. Top section of OM image (**a**) Sample A. (**b**) Sample B.

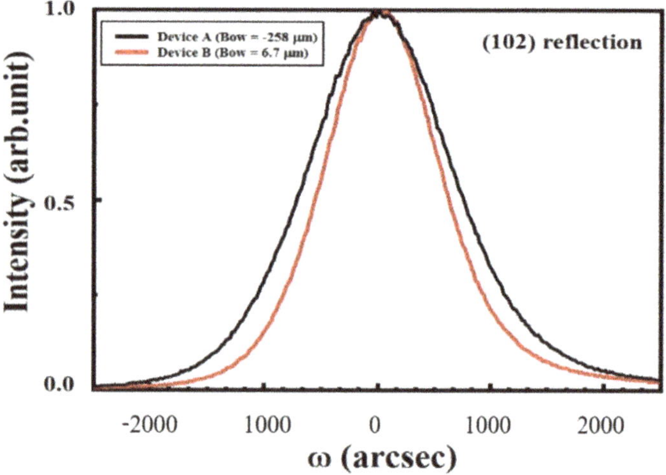

Figure 16. Asymmetric (102) XRD ω-scan rocking curves of substrate with big BOW value (black) and small BOW value (red) measured from surface.

3.2. Power GaN HEMT on SOI

In the past several years, enhancement mode (E-mode) AlGaN/GaN HEMTs have been demonstrated to be the potential devices for next generation high efficiency power switches and converters application [2,166]. Currently, SiC [19] and Si [167] are the most popular substrates for GaN HEMT. However, GaN-on-SOI has been considered as a highly potential option which may provide better performance in high frequency and high-power system, owing to its capability in defect reduction of epitaxial layer, as described in Section 3.1.

According to our studies shown in Section 3.1, the BOW value of GaN on SOI substrate is close related to threading dislocation density (TDD) and epitaxial quality. The HEMT devices on SOI substrate with lower BOW value exhibit 1 order smaller off-state leakage and 8.4% smaller specific on resistance, also 68.8% improvement is observed in 3-terminal off-state breakdown voltage (BV_{GD}), shown as Figure 17. Moreover, the dynamic R_{ON} degradation can be reduced. This implies that by heavily doping in handle wafer not only reduce the bowing effect, but also improve the quality of substrate and the performance of high-power device.

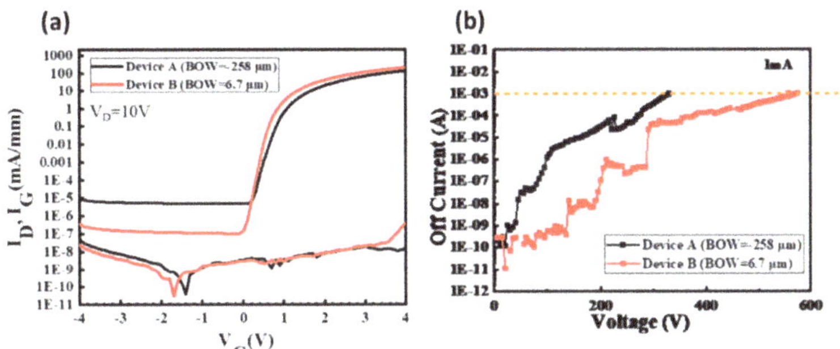

Figure 17. (**a**) I_{DS}-V_{GS} characteristics of HEMTs with big BOW value (Device A, black line) and small BOW value (Device B, red line) (V_{DS} = 10 V). (**b**) Three-terminal off-state characteristic of the E-mode HEMTs at V_{GS} = −5 V.

GaN-on-SOI substrate exhibits a capability to improve the power device performance and also have been proven by many research teams [168–172] including smaller reverse recovery leakage [169], higher breakdown voltage [168,169] and smaller vertical leakage [172]. Kevin J. Chen et al. [169] reported the SOI substrate with a reduced stress in the GaN epilayers (shown as Figure 18a) and an excellent E-mode HEMTs on SOI produced by fluorine plasma implantation method with a high ON/OFF current ratio (10^8–10^9), large breakdown voltage (1471 V with floating substrate), and also a smaller vertical leakage at reverse bias, as shown in Figure 18b.

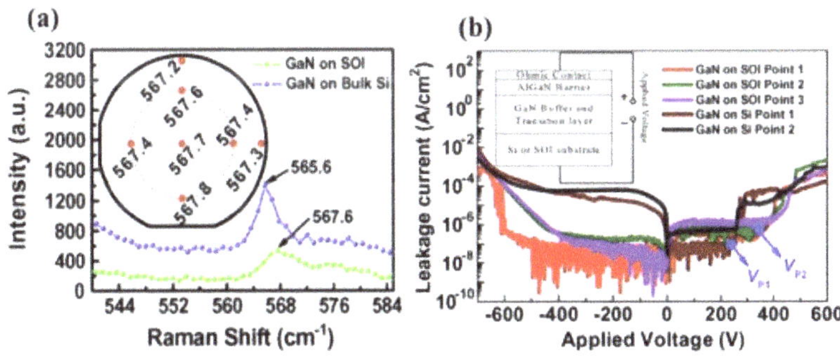

Figure 18. (**a**) GaN-on-SOI and GaN-on-Si (bulk) wafers under Micro-Raman spectroscopy. The E2 peak on the GaN-on-SOI wafer was mapped (Inset) (**b**) Characteristics of vertical leakage on GaN-on-SOI and GaN-on-Si (bulk) platforms [169].

Moreover, most GaN power switching systems are currently produced using a multichip approach, resulting in significant complexity and expense [173–175]. Monolithic integration of GaN-based power devices has steadily gained interest for GaN high-power systems. The benefits of monolithic integration of GaN power systems on a single chip include minimizing parasitic inductance, reducing die size, and enhancing design flexibility [107,176]. To prevent mutual influence between the devices in monolithic GaN power integrated circuit, it suggests that the low and high side HEMT transistors must be fully isolated for a half bridge, as shown in Figure 19 [172,177]. However, it is challenging to accomplish on GaN-on-Si substrates because those HEMTs share a common conductive Si substrate. Figure 20a shows the transfer characteristics of a GaN monolithic half bridge with a common Si substrate biased from −200 to 200 V at 150 °C. Significant variations in threshold voltage (Vth) and drive current are seen when the Si substrate is biased negatively. Nevertheless, using GaN-on-SOI and a trench isolation method, this issue could be overcome [172,177,178]. The transfer characteristics of GaN-on-SOI HEMT, as shown in Figure 20b, illustrate the benefits of device isolation. When the substrate of a neighboring device is biased between −200 V and 200 V, transfer characteristics vary relatively little, which is in sharp contrast to characteristics on a silicon substrate.

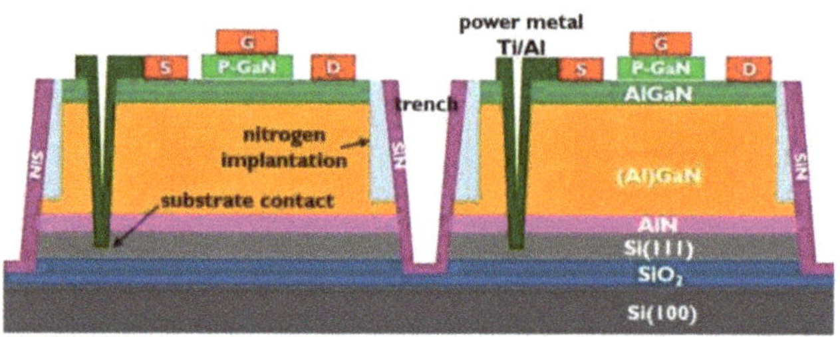

Figure 19. Schematic cross-section of the isolated e-mode p-GaN HEMT [177].

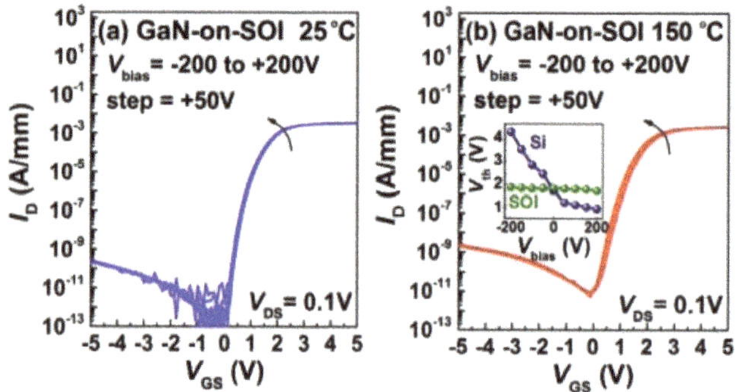

Figure 20. Transfer characteristics while simultaneously biasing the silicon substrate from −200 V to 200 V (**a**) with a common silicon substrate (**b**) with SOI substrate [177].

3.3. RF GaN HEMT on SOI

In addition to the power devices, AlGaN/GaN HEMTs have also been demonstrated to be the potential devices for RF applications [179–182]. Moreover, GaN-based MMIC (Microwave Monolithic Integrated Circuit) has steadily gained interest as compared to a system in package or a multi-chip module, since monolithic integration of GaN RF systems allows for smaller, cheaper, and less complicated circuitry [181,182].

SOI substrate is outstanding for its better vertical isolation performance and a lower substrate loss [183]. Besides, when compared to GaN on Si substrate devices, GaN on SOI substrate devices demonstrated better DC, breakdown voltage, and RF properties [170,184,185]. It was demonstrated that GaN-on-SOI substrates perform better in terms of tensile stress relaxation and surface flatness than Si substrates. It can result in a reduction of defect density, which is further supported by pulse and low-frequency noise measurements. The SOI substrate capacitances extracted from the small signal model are lower than the HR-Si substrate, as illustrated in Figure 21, owing to the series connection of device layer, buried oxide layer, and handle wafer of SOI substrate. As a result of the small substrate capacitances of SOI substrate shunt to the C_{DS}, the effective C_{DS} items that dominated the feedback capacitance were decreased. Furthermore, the utilization of SOI substrate can increase the device's bandwidth and linearity was proven at the same time, as shown in Figure 21b [185].

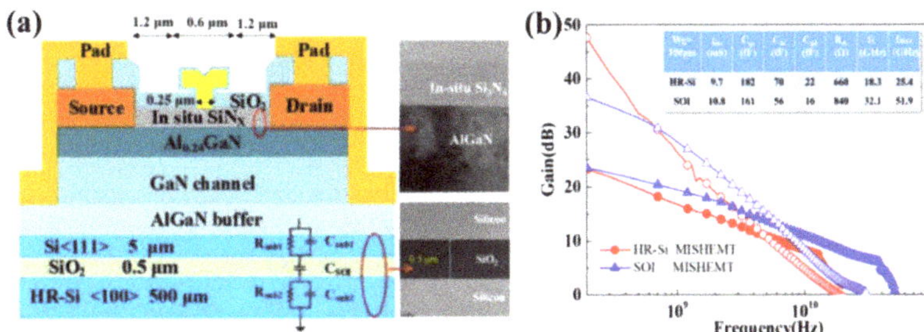

Figure 21. (**a**) Cross-sectional structure of MISHEMT on SOI substrate (**b**) high frequency parameters for AlGaN/GaN MISHEMT on SOI substrate(blue) and AlGaN/GaN MISHEMT on HR-Si substrate (red). [185].

4. GaN HEMTs on QST Substrates

Recently, a new engineered substrate consists of polycrystalline core and single crystalline surface layer, which exhibits a good thermal expansion and crystalline match to GaN, are presented. It provides a good matching in coefficient of thermal expansion (CTE) characteristics with Gallium Nitride (GaN). The template enables growth of thick and high quality GaN semiconductor layers on 8- and even 12-inch wafers and support a lower cost for GaN devices in power supplies and RF transmitters commercial markets. According to Qromis Inc., Qromis Substrate Technology (QST) promises a thicker GaN epitaxy to expands the GaN HEMT's limitation in breakdown voltage roadmaps (up to 1200 V) in power devices in vertical electron paths. Compatible with conventional GaN growth platforms, as the substrates are thermally matched to GaN, it offers low defect density, high crystal quality, and low wafer bow. As previous sections described, the high quality GaN power devices enabled are potential for a higher switching speed, simpler and smaller form, and higher-temperature operation.

The Naval Research Laboratory (NRL), Kyma and Qromis Technology reported some material characterization studies. It's interesting that GaN device layers up to 15μm were demonstrated with a wafer bow of 1 μm for growth on 150-mm-diameter substrates [186]. The development to manufacture 200 mm freestanding GaN from 300 mm QST® [187] also looks promising. According to several reports, the thermal conductivity and CTE relationship for different substrates are depicted in Figure 22. It also shows a high quality AlGaN/GaN buffers grown on substrates with a less mismatch in coefficient of thermal expansion (CTE). Figure 23 shows an illustration of evaluating the vertical buffer leakage currents in both reverse and forward bias mode, and it exhibits a maximum of reverse current of 1 μA/mm at 25 °C and 10 μA/mm at 150 °C as the reverse voltage exceeds 700 V. The leakage current increases by ~3 orders from 25 to 150 °C [188]. According to a MIT group, a GaN vertical power FinFETs on engineered substrate was demonstrated [189]. Figure 24 shows the schematic of the quasi-vertical device architecture, which consists of 132 fins with 100 nm, 700 nm and 21 um widths, spacing and length respectively. It exhibits a current density of J_{DS}=3.8 kA/cm^2 at V_{GS}= 1.5 V and V_{DS}= 4 V, and a maximum g_m = 2 kS/cm^2 at V_{DS} = 4 V. The current density in each fin is higher than 30 kA/cm^2 at the same bias condition. They also benchmark vertical and quasi-vertical MOSFETs on non-GaN substrates, as shown in Figure 25.

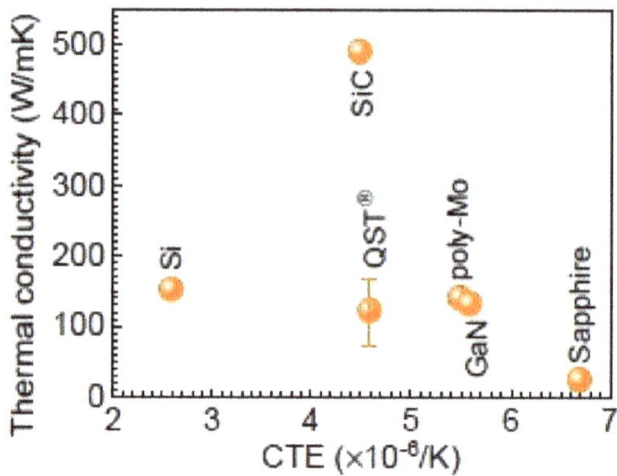

Figure 22. Different substrate for coefficient of thermal expansion (CTE) and thermal conductivity [188].

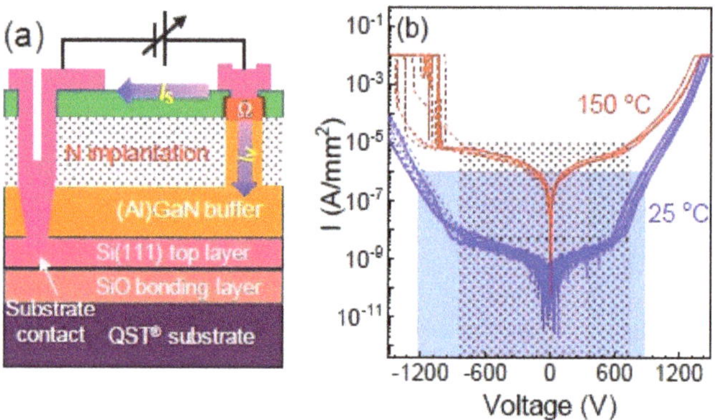

Figure 23. (**a**) The illustration of vertical buffer leakage measurements (**b**) The leakage current density under voltage bias for a 5.6 mm-thick buffer grown on 200 mm GaN-on-QST® substrate [188].

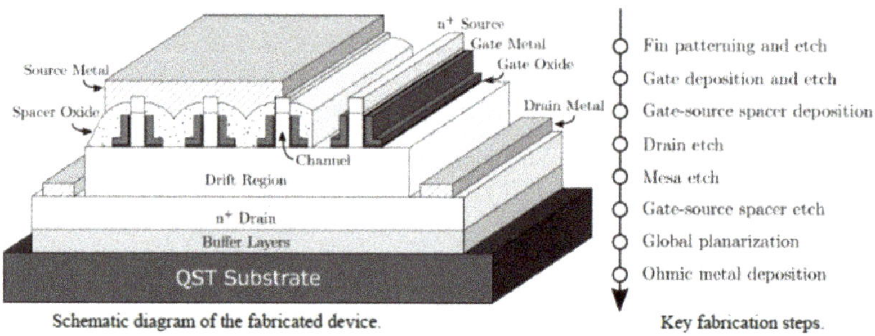

Figure 24. A schematic diagram of the vertical GaN FinFET devices and fabrication steps [189].

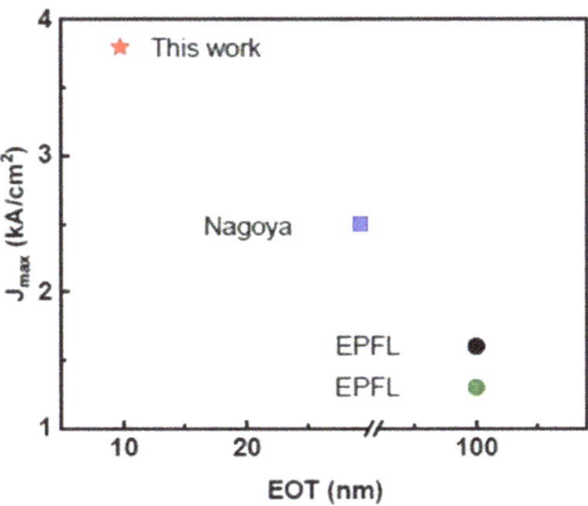

Figure 25. Benchmarking maximum current density in vertical GaN-on-Silicon transistors as a function of equivalent oxide thickness [189].

5. Heterogeneous Integration of GaN HEMT

The conventional silicon-based RF devices fabricated on 32 nm Si complementary metal-oxide-semiconductor (CMOS) with a cut-off frequency of 445 GHz were exhibited and the cut-off frequency are expected to further scale [190]. However, despite having impressive cut-off frequency, Si CMOS is not well suited with high voltage or high-power density due to lower breakdown voltage whereas GaN devices are more suitable for these types of applications. Hence, co-integrating Si with GaN on a single chip may help in achieving high power and high-performance application. The main motivation for integrating GaN and CMOS is due to the superior GaN performance in fast power switching and the high functionality of CMOS logic, reduction in interconnect distance as well as losses, smaller form factor, reduction in power consumption, lower cost, and lower assembling complexity [191,192]. There are two types of GaN and CMOS integration variants on wafer-level namely Monolithic Integration and Heterogeneous Integration (HI). In the past few years, the key technology development and production is heterogeneous integration. For high-frequency applications in space and defense and the 5G application in the commercial world, heterogeneous integration for RF has become an essential task. Typically, hetero-integration of RF devices is done with different semiconductor materials not only CMOS to address the required specific performance like Diverse Accessible Heterogeneous Integration (DAHI) technology [193]. HI methods can be wafer to wafer (WTW), chip to wafer (CTW), or Chip to Chip (CTC), etc [194]. CTW and CTC are usually used for integrating dissimilar heterogeneous materials as it minimizes coefficient of Thermal Expansion (CTE) as well as wafer warpage issues which enable known good die (KGD) and pretty good die (PGD) that helps in achieving good yield. In WTW integration, CTE issues are challenging, and the yield is compromised. High frequency 3D HI faces challenges such as implementing effective 3D screening method as in some cases chiplets don't have test pads and there is insufficient availability of process design kit (PDK) with RF functionality, co-simulation capability, and 3D parasitic extraction [194]. One heterogeneous integration method that makes the GaN power system compatible with CMOS fabrication using SOI substrate was stated by the IBM research division [195] as shown in Figure 26.

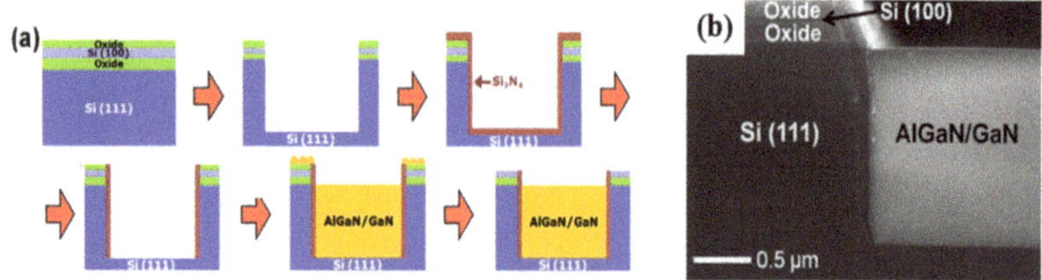

Figure 26. (**a**) Heterogeneous integration of SOI substrate with top Si (100) and bottom Si (111) and GaN epitaxial (**b**) SEM cross-section [195].

A DC-DC boost converter was designed using GaN power transistors integrated with bipolar-CMOS-DMOS (BCD) which combines both the advantages of high-voltage low-loss GaN devices and high-integration BCD circuits [196]. The designed GaN2BCD technology is a promising power converter application platform. Deeply scaled E/D-mode GaN HEMTs integrated with monolithically integrable GaN Schottky diodes were able to offer advantages in MMIC applications [142]. GaN-on-Si monolithic microwave integrated circuits (MMICs) were fabricated on 200-mm-diameter using a fully CMOS-compatible fabrication process which enables integration of wafer-level 3D GaN MMICs with Si CMOS circuits for performance and functionality enhancement while the size, weight, power, and cost is reduced [197]. As shown in Figure 27, a team from Raytheon [198] in the United States successfully demonstrated fabricating GaN HEMTs in windows on SOI wafers containing Si CMOS transistors, with DC and RF performance comparable to GaN HEMTs on SiC substrate, as well as a first GaN–Si CMOS heterogeneously integrated MMIC: GaN amplifier with CMOS gate bias control circuitry (a current mirror) and heterogeneous interconnects, as shown in Figure 28.

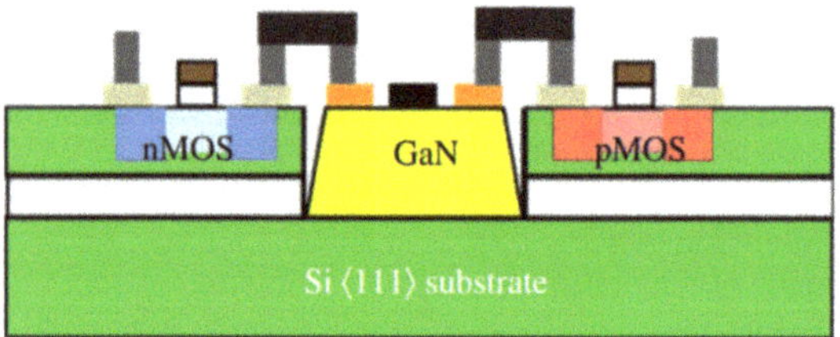

Figure 27. GaN HEMT and Si CMOS are heterogeneously combined on a modified SOI wafer [198].

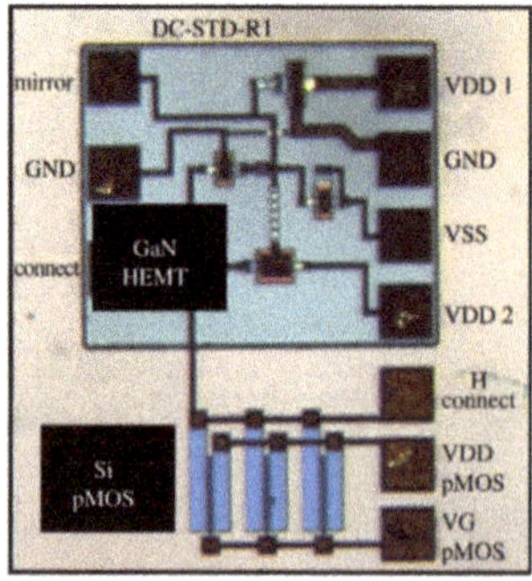

Figure 28. GaN–Si CMOS heterogeneously integrated MMIC [198].

For heterogeneous integration, wafer bonding is one of the most promising integration approach for integrating group III-V materials and CMOS on Si [191]. Monolithic structure can be done by direct wafer bonding [199] or heteroepitaxy [200]. Direct wafer bonding can be used for integrating non-lattice matched semiconductors and also for integrating different crystal structures. Moreover, no additional intermediate layers are required and can also integrate two or more wafers. However, it requires a very flat, smooth, and particle-free surface along with fitting wafer diameter and chip sizes.

A new 3D integrated circuit (3DIC) solution, System on Integrated Chips (SoICTM), was developed by Taiwan Semiconductor Manufacturing Company (TSMC) [201] to integrate active and passive chips into a new integrated SoC system. Comparing the typical 3DIC stacking with SoIC, the latter offers higher I/O density bonding density, lower energy consumption/ bit data, lower electrical parasites, and lower thermal resistance [190] which might help in unleashing the boundary of IC designer on heterogeneous integrations in future 5G, AI, mobile, and HPC applications. IMEC developed NaNO-TSV (Through Silicon Vias) connection for heterogeneous integration as 3D system-on-chip (3D-SoC) integration technology which possess a wafer-to-wafer bonding approach combined with via-last TSV connection [202]. Finally, Wafer-level packaging (WLP) of the heterogeneous integrated devices is required to be protected from the environment [203]. WLP also eliminates assembly equipment, reduces package cost, and minimizes the chip size as well as provides high yield and high reliability.

6. Conclusions

In the previous GaN HEMTs development roadmap, the heterogeneous epitaxy has been one of the issues affecting devices performance. The wide application of compound semiconductors has attracted wide attention and become matured gradually, including and ranging from RF power amplifiers to electronic systems. The demand tendency for power devices is increasing, especially in electric vehicles and the fast-charging applications. As high-frequency communications keep developing, the GaN HEMT technology will be very critical. Traditional silicon substrates will no longer be only GaN HEMTs template but will be replaced by other smooth and friendly substrate for some applications. These advanced substrate technologies efficiently improve device characteristics, performance, and reliability. It will bring thicker GaN buffer layer, high thermal conductivity, and high resistance substrate in the future high-power high-frequency components. In addition, the heterogeneous integration of GaN HEMTs and CMOS structure have become a new direction. In this article, we provide a brief and comprehensive overview of these important technology developments.

Author Contributions: Data curation, Y.-Y.L., C.L., Y.-T.C., Y.-W.H., P.-T.T. and L.-H.H.; project administration, Y.-Y.L., W.-C.L., A.-J.T. and L.-H.H.; supervision, Y.-J.C., C.-H.L., H.-C.K. and E.Y.C.; writing—original draft, Y.-Y.L., C.L., Y.-T.C., P.-T.T. and L.-H.H.; writing—review and editing, L.-H.H. and Y.-Y.L. All authors have read and agreed to the published version of the manuscript.

Funding: Ministry of Science and Technology, Taiwan (107-2221-E-009-113-MY3, 108-2221-E-009-113-MY3, 109-2222-E-009-001-MY3, 110-2218-E-182-001-MY3).

Acknowledgments: The authors would like to thank Ministry of Science and Technology, Industrial Technology Research Institute, Taiwan Semiconductor Research Institute, and the Semiconductor Research Center, Hon Hai Research Institute, for the helpful discussion.

Conflicts of Interest: The authors declare no conflict of interest.

References

1. Chuang, R.W.; Chang, S.; Chang, S.-J.; Chiou, Y.; Lu, C.; Lin, T.; Lin, Y.; Kuo, C.; Chang, H.-M. Gallium nitride metal-semiconductor-metal photodetectors prepared on silicon substrates. *J. Appl. Phys.* **2007**, *102*, 073110. [CrossRef]
2. Mishra, U.K.; Parikh, P.; Wu, Y.-F. AlGaN/GaN HEMTs-an overview of device operation and applications. *Proc. IEEE* **2002**, *90*, 1022–1031. [CrossRef]

3. Wu, Y.-F.; Kapolnek, D.; Ibbetson, J.P.; Parikh, P.; Keller, B.P.; Mishra, U.K. Very-high power density AlGaN/GaN HEMTs. *IEEE Trans. Electron Devices* **2001**, *48*, 586–590.
4. Wu, Y.-F.; Saxler, A.; Moore, M.; Smith, R.; Sheppard, S.; Chavarkar, P.; Wisleder, T.; Mishra, U.; Parikh, P. 30-W/mm GaN HEMTs by field plate optimization. *IEEE Electron Device Lett.* **2004**, *25*, 117–119. [CrossRef]
5. Asif Khan, M.; Bhattarai, A.; Kuznia, J.; Olson, D. High electron mobility transistor based on a GaN-Al$_x$Ga$_{1-x}$N heterojunction. *Appl. Phys. Lett.* **1993**, *63*, 1214–1215. [CrossRef]
6. Dumka, D.; Lee, C.; Tserng, H.; Saunier, P.; Kumar, M. AlGaN/GaN HEMTs on Si substrate with 7 W/mm output power density at 10 GHz. *Electron. Lett.* **2004**, *40*, 1023–1024. [CrossRef]
7. Eastman, L.F.; Tilak, V.; Smart, J.; Green, B.M.; Chumbes, E.M.; Dimitrov, R.; Kim, H.; Ambacher, O.S.; Weimann, N.; Prunty, T. Undoped AlGaN/GaN HEMTs for microwave power amplification. *IEEE Trans. Electron Devices* **2001**, *48*, 479–485. [CrossRef]
8. Kemerley, R.T.; Wallace, H.B.; Yoder, M.N. Impact of wide bandgap microwave devices on DoD systems. *Proc. IEEE* **2002**, *90*, 1059–1064. [CrossRef]
9. Duboz, J.Y. GaN as seen by the industry. *Phys. Status Solidi (A)* **1999**, *176*, 5–14. [CrossRef]
10. Dunleavy, L.; Baylis, C.; Curtice, W.; Connick, R. Modeling GaN: Powerful but challenging. *IEEE Microw. Mag.* **2010**, *11*, 82–96. [CrossRef]
11. Mimura, T.; Hiyamizu, S.; Fujii, T.; Nanbu, K. A new field-effect transistor with selectively doped GaAs/n-AlxGa1-xAs heterojunctions. *Jpn. J. Appl. Phys.* **1980**, *19*, L225. [CrossRef]
12. Amano, H.; Baines, Y.; Beam, E.; Borga, M.; Bouchet, T.; Chalker, P.R.; Charles, M.; Chen, K.J.; Chowdhury, N.; Chu, R. The 2018 GaN power electronics roadmap. *J. Phys. D: Appl. Phys.* **2018**, *51*, 163001. [CrossRef]
13. Sano, H.; Ui, N.; Sano, S. A 40W GaN HEMT Doherty power amplifier with 48% efficiency for WiMAX applications. In *Proceedings of the IEEE Compound Semiconductor Integrated Circuits Symposium, Portland, OR, USA, 14–17 October 2007*; IEEE: Piscataway, NJ, USA, 2007; pp. 1–4.
14. Gaska, R.; Yang, J.; Osinsky, A.; Chen, Q.; Khan, M.A.; Orlov, A.; Snider, G.; Shur, M. Electron transport in AlGaN–GaN heterostructures grown on 6H–SiC substrates. *Appl. Phys. Lett.* **1998**, *72*, 707–709. [CrossRef]
15. Hirama, K.; Taniyasu, Y.; Kasu, M. AlGaN/GaN high-electron mobility transistors with low thermal resistance grown on single-crystal diamond (111) substrates by metalorganic vapor-phase epitaxy. *Appl. Phys. Lett.* **2011**, *98*, 162112. [CrossRef]
16. Tang, X.; Rousseau, M.; Defrance, N.; Hoel, V.; Soltani, A.; Langer, R.; De Jaeger, J.C. Thermal behavior analysis of GaN based epi-material on different substrates by means of a physical–thermal model. *Phys. Status Solidi (A)* **2010**, *207*, 1820–1826. [CrossRef]
17. Yamaki, F.; Inoue, K.; Ui, N.; Kawano, A.; Sano, S. A 65% drain efficiency GaN HEMT with 200 W peak power for 20 V to 65 V envelope tracking base station amplifier. In *Proceedings of the 2011 IEEE MTT-S International Microwave Symposium, Baltimore, MD, USA, 5–10 June 2011*; IEEE: Piscataway, NJ, USA, 2011; pp. 1–4.
18. Mitani, E.; Aojima, M.; Maekawa, A.; Sano, S. An 800-W AlGaN/GaN HEMT for S-band high-power application. *CSMantech Line Dig.* **2007**, *0.2*, 1–3.
19. Pengelly, R.S.; Wood, S.M.; Milligan, J.W.; Sheppard, S.T.; Pribble, W.L. A review of GaN on SiC high electron-mobility power transistors and MMICs. *IEEE Trans. Microw. Theory Tech.* **2012**, *60*, 1764–1783. [CrossRef]
20. Wu, Y.-F.; Moore, M.; Saxler, A.; Wisleder, T.; Parikh, P. 40-W/mm double field-plated GaN HEMTs. In *Proceedings of the 2006 64th device research conference, State College, PA, USA, 26–28 June 2006*; IEEE: Piscataway, NJ, USA, 2006; pp. 151–152.
21. Dumka, D.; Chou, T.; Jimenez, J.; Fanning, D.; Francis, D.; Faili, F.; Ejeckam, F.; Bernardoni, M.; Pomeroy, J.; Kuball, M. Electrical and thermal performance of AlGaN/GaN HEMTs on diamond substrate for RF applications. In *Proceedings of the 2013 IEEE Compound Semiconductor Integrated Circuit Symposium (CSICS), Monterey, CA, USA, 13–16 October 2013*; IEEE: Piscataway, NJ, USA, 2013; pp. 1–4.
22. Smith, R.P.; Sheppard, S.; Wu, Y.-F.; Heikman, S.; Wood, S.; Pribble, W.; Milligan, J. AlGaN/GaN-on-SiC HEMT technology status. In *Proceedings of the 2008 IEEE Compound Semiconductor Integrated Circuits Symposium, Monterey, CA, USA, 12–15 October 2008*; IEEE: Piscataway, NJ, USA, 2008; pp. 1–4.
23. Javorka, P.; Alam, A.; Wolter, M.; Fox, A.; Marso, M.; Heuken, M.; Luth, H.; Kordos, P. AlGaN/GaN HEMTs on (111) silicon substrates. *IEEE Electron Device Lett.* **2002**, *23*, 4–6. [CrossRef]
24. Lenci, S.; De Jaeger, B.; Carbonell, L.; Hu, J.; Mannaert, G.; Wellekens, D.; You, S.; Bakeroot, B.; Decoutere, S. Au-free AlGaN/GaN power diode on 8-in Si substrate with gated edge termination. *IEEE Electron Device Lett.* **2013**, *34*, 1035–1037. [CrossRef]
25. Yacoub, H.; Fahle, D.; Finken, M.; Hahn, H.; Blumberg, C.; Prost, W.; Kalisch, H.; Heuken, M.; Vescan, A. The effect of the inversion channel at the AlN/Si interface on the vertical breakdown characteristics of GaN-based devices. *Semicond. Sci. Technol.* **2014**, *29*, 115012. [CrossRef]
26. Sahoo, A.; Subramani, N.; Nallatamby, J.-C.; Sylvain, L.; Loyez, C.; Quéré, R.; Medjdoub, F. Small signal modeling of high electron mobility transistors on silicon and silicon carbide substrate with consideration of substrate loss mechanism. *Solid State Electron.* **2016**, *115*, 12–16. [CrossRef]
27. Luong, T.T.; Lumbantoruan, F.; Chen, Y.Y.; Ho, Y.T.; Weng, Y.C.; Lin, Y.C.; Chang, S.; Chang, E.Y. RF loss mechanisms in GaN-based high-electron-mobility-transistor on silicon: Role of an inversion channel at the AlN/Si interface. *Phys. Status Solidi (A)* **2017**, *214*, 1600944. [CrossRef]
28. Chandrasekar, H.; Bhat, K.; Rangarajan, M.; Raghavan, S.; Bhat, N. Thickness dependent parasitic channel formation at AlN/Si interfaces. *Sci. Rep.* **2017**, *7*, 1–10. [CrossRef]

29. Hsu, S.S.; Tsou, C.-W.; Lian, Y.-W.; Lin, Y.-S. GaN-on-silicon devices and technologies for RF and microwave applications. In *Proceedings of the 2016 IEEE International Symposium on Radio-Frequency Integration Technology (RFIT), Taipei, Taiwan, 24–26 August 2016*; IEEE: Piscataway, NJ, USA, 2016; pp. 1–3.

30. Pan, L.S.; Kania, D.R. *Diamond: Electronic Properties and Applications*; Springer Science & Business Media: Boston, MA, USA, 2013.

31. Hirama, K.; Kasu, M.; Taniyasu, Y. RF high-power operation of AlGaN/GaN HEMTs epitaxially grown on diamond. *IEEE Electron Device Lett.* **2012**, *33*, 513–515. [CrossRef]

32. Diduck, Q.; Felbinger, J.; Eastman, L.; Francis, D.; Wasserbauer, J.; Faili, F.; Babic, D.; Ejeckam, F. Frequency performance enhancement of AlGaN/GaN HEMTs on diamond. *Electron. Lett.* **2009**, *45*, 758–759. [CrossRef]

33. Hirama, K.; Taniyasu, Y.; Kasu, M. Heterostructure growth of a single-crystal hexagonal AlN (0001) layer on cubic diamond (111) surface. *J. Appl. Phys.* **2010**, *108*, 013528. [CrossRef]

34. Taniyasu, Y.; Kasu, M. MOVPE growth of single-crystal hexagonal AlN on cubic diamond. *J. Cryst. Growth* **2009**, *311*, 2825–2830. [CrossRef]

35. Yagi, S.; Shimizu, M.; Inada, M.; Yamamoto, Y.; Piao, G.; Okumura, H.; Yano, Y.; Akutsu, N.; Ohashi, H. High breakdown voltage AlGaN/GaN MIS–HEMT with SiN and TiO2 gate insulator. *Solid-State Electron.* **2006**, *50*, 1057–1061. [CrossRef]

36. Mukhopadhyay, P.; Bag, A.; Gomes, U.; Banerjee, U.; Ghosh, S.; Kabi, S.; Chang, E.Y.; Dabiran, A.; Chow, P.; Biswas, D. Comparative DC characteristic analysis of AlGaN/GaN HEMTs grown on Si (111) and sapphire substrates by MBE. *J. Electron. Mater.* **2014**, *43*, 1263–1270. [CrossRef]

37. Douglas, E.; Ren, F.; Pearton, S. Finite-element simulations of the effect of device design on channel temperature for AlGaN/GaN high electron mobility transistors. *J. Vac. Sci. Technol. B Nanotechnol. Microelectron. Mater. Process. Meas. Phenom.* **2011**, *29*, 020603. [CrossRef]

38. Manoi, A.; Pomeroy, J.W.; Killat, N.; Kuball, M. Benchmarking of thermal boundary resistance in AlGaN/GaN HEMTs on SiC substrates: Implications of the nucleation layer microstructure. *IEEE Electron Device Lett.* **2010**, *31*, 1395–1397. [CrossRef]

39. Riedel, G.J.; Pomeroy, J.W.; Hilton, K.P.; Maclean, J.O.; Wallis, D.J.; Uren, M.J.; Martin, T.; Forsberg, U.; Lundskog, A.; Kakanakova-Georgieva, A. Reducing thermal resistance of AlGaN/GaN electronic devices using novel nucleation layers. *IEEE Electron Device Lett.* **2008**, *30*, 103–106. [CrossRef]

40. Nochetto, H.C.; Jankowski, N.R.; Bar-Cohen, A. The impact of GaN/substrate thermal boundary resistance on a HEMT device. In Proceedings of the ASME International Mechanical Engineering Congress and Exposition, Denver, CO, USA, 11–17 November 2011; pp. 241–249.

41. Park, K.; Bayram, C. Thermal resistance optimization of GaN/substrate stacks considering thermal boundary resistance and temperature-dependent thermal conductivity. *Appl. Phys. Lett.* **2016**, *109*, 151904. [CrossRef]

42. Hull, R.; Osgood, R.; Sakaki, H. *Nitride Semiconductors and Devices*; Springer: Berlin/Heidelberg, Germany, 1999.

43. Kuzmík, J.; Bychikhin, S.; Pogány, D.; Gaquière, C.; Pichonat, E.; Morvan, E. Investigation of the thermal boundary resistance at the III-nitride/substrate interface using optical methods. *J. Appl. Phys.* **2007**, *101*, 054508. [CrossRef]

44. Fletcher, A.A.; Nirmal, D.; Ajayan, J.; Arivazhagan, L. An intensive study on assorted substrates suitable for high JFOM AlGaN/GaN HEMT. *Silicon* **2021**, *13*, 1591–1598. [CrossRef]

45. Mimura, T.; Yokoyama, N.; Kusakawa, H.; Suyama, K.; Fukuta, M. MP-A4 GaAs MOSFET for low-power high-speed logic applications. *IEEE Trans. Electron Devices* **1979**, *26*, 1828. [CrossRef]

46. Schwierz, F.; Ambacher, O. Recent advances in GaN HEMT development. In *Proceedings of the 11th IEEE International Symposium on Electron Devices for Microwave and Optoelectronic Applications, 2003, EDMO 2003, Orlando, FL, USA, 18 November 2003*; IEEE: Piscataway, NJ, USA, 2003; pp. 204–209.

47. Lidow, A.; De Rooij, M.; Strydom, J.; Reusch, D.; Glaser, J. *GaN Transistors for Efficient Power Conversion*; John Wiley & Sons: Hoboken, NJ, USA, 2019.

48. Liu, Q.; Lau, S. A review of the metal–GaN contact technology. *Solid State Electron.* **1998**, *42*, 677–691. [CrossRef]

49. Liu, Q.; Yu, L.; Lau, S.; Redwing, J.; Perkins, N.; Kuech, T. Thermally stable PtSi Schottky contact on n-GaN. *Appl. Phys. Lett.* **1997**, *70*, 1275–1277. [CrossRef]

50. Kordoš, P.; Heidelberger, G.; Bernát, J.; Fox, A.; Marso, M.; Lüth, H. High-power SiO2/AlGaN/GaN metal-oxide-semiconductor heterostructure field-effect transistors. *Appl. Phys. Lett.* **2005**, *87*, 143501. [CrossRef]

51. Ohmaki, Y.; Tanimoto, M.; Akamatsu, S.; Mukai, T. Enhancement-Mode AlGaN/AlN/GaN High Electron Mobility Transistor with Low On-State Resistance and High Breakdown Voltage. *Jpn. J. Appl. Phys.* **2006**, *45*, L1168–L1170. [CrossRef]

52. Medjdoub, F.; Derluyn, J.; Cheng, K.; Leys, M.; Degroote, S.; Marcon, D.; Visalli, D.; van Hove, M.; Germain, M.; Germain, G. Low On-Resistance High-Breakdown Normally Off AlN/GaN/AlGaN DHFET on Si Substrate. *IEEE Electron Device Lett.* **2010**, *31*, 111–113. [CrossRef]

53. Mizutani, T.; Ito, M.; Kishimoto, S.; Nakamura, F. AlGaN/GaN HEMTs With Thin InGaN Cap Layer for Normally Off Operation. *IEEE Electron Device Lett.* **2007**, *28*, 549–551. [CrossRef]

54. Ostermaier, C.; Pozzovivo, G.; Carlin, J.F.; Basnar, B.; Schrenk, W.; Douvry, Y.; Gaquiere, C.; DeJaeger, J.-C.; Cico, K.; Frohlich, K.; et al. Ultrathin InAlN/AlN Barrier HEMT With High Performance in Normally Off Operation. *IEEE Electron Device Lett.* **2009**, *30*, 1030–1032. [CrossRef]

55. Brown, D.F.; Shinohara, K.; Williams, A.; Milosavljevic, I.; Grabar, R.; Hashimoto, P.; Willadsen, P.J.; Schmitz, A.; Corrion, A.L.; Kim, S. Monolithic Integration of Enhancement- and Depletion-Mode AlN/GaN/AlGaN DHFETs by Selective MBE Regrowth. *IEEE Trans. Electron Devices* **2011**, *58*, 1063–1067. [CrossRef]

56. Guowang, L.; Zimmermann, T.; Cao, Y.; Chuanxin, L.; Xiu, X.; Ronghua, W. Threshold Voltage Control in Al0.72Ga0.28N/AlN/GaN HEMTs by Work-Function Engineering. *IEEE Electron Device Lett.* **2010**, *31*, 954–956.

57. Chiu, H.-C.; Yang, C.-W.; Chen, C.-H.; Fu, J.S.; Chien, F.-T. Characterization of enhancement-mode AlGaN/GaN high electron mobility transistor using N2O plasma oxidation technology. *Appl. Phys. Lett.* **2011**, *99*, 153508. [CrossRef]

58. Chiu, H.; Wu, J.; Yang, C.; Huang, F.; Kao, H. Low-Frequency Noise in Enhancement-Mode GaN MOS-HEMTs by Using Stacked Al2O3/Ga2O3/Gd2O3 Gate Dielectric. *IEEE Electron Device Lett.* **2012**, *33*, 958–960. [CrossRef]

59. Wang, X.; Saadat, O.I.; Xi, B.; Lou, X.; Molnar, R.J.; Palacios, T.; Gordon, R.G. Atomic layer deposition of Sc2O3 for passivating AlGaN/GaN high electron mobility transistor devices. *Appl. Phys. Lett.* **2012**, *101*, 232109. [CrossRef]

60. Cai, Y.; Zhou, Y.; Lau, K.M.; Chen, K.J. Control of threshold voltage of AlGaN/GaN HEMTs by fluoride-based plasma treatment: From depletion mode to enhancement mode. *IEEE Trans. Electron Devices* **2006**, *53*, 2207–2215. [CrossRef]

61. Moon, J.; Wu, S.; Wong, D.; Milosavljevic, I.; Conway, A.; Hashimoto, P.; Hu, M.; Antcliffe, M.; Micovic, M. Gate-recessed AlGaN-GaN HEMTs for high-performance millimeter-wave applications. *IEEE Electron Device Lett.* **2005**, *26*, 348–350. [CrossRef]

62. Moon, J.S.; Wong, D.; Hussain, T.; Micovic, M.; Deelman, P.; Hu, M.; Antcliffe, M.; Ngo, C.; Hashimoto, P.; McCray, L. Submicron enhancement-mode AlGaN/GaN HEMTs. In *Proceedings of the 60th DRC. Conference Digest Device Research Conference, Santa Barbara, CA, USA, 24–26 June 2002*; IEEE: Piscataway, NJ, USA, 2002; pp. 23–24.

63. Okita, H.; Kaifu, K.; Mita, J.; Yamada, T.; Sano, Y.; Ishikawa, H.; Egawa, T.; Jimbo, T. High transconductance AlGaN/GaN-HEMT with recessed gate on sapphire substrate. *Phys. Status Solidi (A)* **2003**, *200*, 187–190. [CrossRef]

64. Wang, W.-K.; Li, Y.-J.; Lin, C.-K.; Chan, Y.-J.; Chen, G.-T.; Chyi, J.-I. Low damage, Cl 2-based gate recess etching for 0.3-μm gate-length AlGaN/GaN HEMT fabrication. *IEEE Electron Device Lett.* **2004**, *25*, 52–54. [CrossRef]

65. Kumar, V.; Kuliev, A.; Tanaka, T.; Otoki, Y.; Adesida, I. High transconductance enhancement-mode AlGaN/GaN HEMTs on SiC substrate. *Electron. Lett.* **2003**, *39*, 1758–1760. [CrossRef]

66. Lanford, W.; Tanaka, T.; Otoki, Y.; Adesida, I. Recessed-gate enhancement-mode GaN HEMT with high threshold voltage. *Electron. Lett.* **2005**, *41*, 449–450. [CrossRef]

67. Zhuang, D.; Edgar, J. Wet etching of GaN, AlN, and SiC: A review. *Mater. Sci. Eng. R: Rep.* **2005**, *48*, 1–46. [CrossRef]

68. Andreani, P.; Sjoland, H. Tail current noise suppression in RF CMOS VCOs. *IEEE J. Solid-State Circuits* **2002**, *37*, 342–348. [CrossRef]

69. Sugimoto, M.; Ueda, H.; Uesugi, T.; Kachi, T. Wide-bandgap semiconductor devices for automotive applications. *Int. J. High Speed Electron. Syst.* **2007**, *17*, 3–9. [CrossRef]

70. Uemoto, Y.; Hikita, M.; Ueno, H.; Matsuo, H.; Ishida, H.; Yanagihara, M.; Ueda, T.; Tanaka, T.; Ueda, D. Gate injection transistor (GIT)—A normally-off AlGaN/GaN power transistor using conductivity modulation. *IEEE Trans. Electron Devices* **2007**, *54*, 3393–3399. [CrossRef]

71. Fujii, T.; Tsuyukuchi, N.; Hirose, Y.; Iwaya, M.; Kamiyama, S.; Amano, H.; Akasaki, I. Fabrication of enhancement-mode AlxGa1–xN/GaN junction heterostructure field-effect transistors with p-type GaN gate contact. *Phys. Status Solidi C* **2007**, *4*, 2708–2711. [CrossRef]

72. Hilt, O.; Knauer, A.; Brunner, F.; Bahat-Treidel, E.; Würfl, J. Normally-off AlGaN/GaN HFET with p-type Ga Gate and AlGaN buffer. In *Proceedings of the 2010 22nd International Symposium on Power Semiconductor Devices & IC's (ISPSD), Hiroshima, Japan, 6–10 June 2010*; IEEE: Piscataway, NJ, USA, 2010; pp. 347–350.

73. Hilt, O.; Brunner, F.; Cho, E.; Knauer, A.; Bahat-Treidel, E.; Würfl, J. Normally-off high-voltage p-GaN gate GaN HFET with carbon-doped buffer. In *Proceedings of the 2011 IEEE 23rd International Symposium on Power Semiconductor Devices and ICs, Hiroshima, Japan, 6–10 June 2010*; IEEE: Piscataway, NJ, USA, 2011; pp. 239–242.

74. Greco, G.; Iucolano, F.; Roccaforte, F. Review of technology for normally-off HEMTs with p-GaN gate. *Mater. Sci. Semicond. Process.* **2018**, *78*, 96–106. [CrossRef]

75. Efthymiou, L.; Longobardi, G.; Camuso, G.; Chien, T.; Chen, M.; Udrea, F. On the physical operation and optimization of the p-GaN gate in normally-off GaN HEMT devices. *Appl. Phys. Lett.* **2017**, *110*, 123502. [CrossRef]

76. Götz, W.; Johnson, N.; Walker, J.; Bour, D.; Amano, H.; Akasaki, I. Hydrogen passivation of Mg acceptors in GaN grown by metalorganic chemical vapor deposition. *Appl. Phys. Lett.* **1995**, *67*, 2666–2668. [CrossRef]

77. Hwang, I.; Kim, J.; Choi, H.S.; Choi, H.; Lee, J.; Kim, K.Y.; Park, J.-B.; Lee, J.C.; Ha, J.; Oh, J. p-GaN gate HEMTs with tungsten gate metal for high threshold voltage and low gate current. *IEEE Electron Device Lett.* **2013**, *34*, 202–204. [CrossRef]

78. Lee, F.; Su, L.-Y.; Wang, C.-H.; Wu, Y.-R.; Huang, J. Impact of gate metal on the performance of p-GaN/AlGaN/GaN high electron mobility transistors. *IEEE Electron Device Lett.* **2015**, *36*, 232–234. [CrossRef]

79. Greco, G.; Iucolano, F.; Di Franco, S.; Bongiorno, C.; Patti, A.; Roccaforte, F. Effects of annealing treatments on the properties of Al/Ti/p-GaN interfaces for normally OFF p-GaN HEMTs. *IEEE Trans. Electron Devices* **2016**, *63*, 2735–2741. [CrossRef]

80. Meneghini, M.; Hilt, O.; Wuerfl, J.; Meneghesso, G. Technology and reliability of normally-off GaN HEMTs with p-type gate. *Energies* **2017**, *10*, 153. [CrossRef]

81. Posthuma, N.; You, S.; Liang, H.; Ronchi, N.; Kang, X.; Wellekens, D.; Saripalli, Y.; Decoutere, S. Impact of Mg out-diffusion and activation on the p-GaN gate HEMT device performance. In *Proceedings of the 2016 28th International Symposium on Power Semiconductor Devices and ICs (ISPSD), Prague, Czech Republic, 12–16 June 2016*; IEEE: Piscataway, NJ, USA, 2016; pp. 95–98.

82. Wu, T.-L.; Marcon, D.; You, S.; Posthuma, N.; Bakeroot, B.; Stoffels, S.; Van Hove, M.; Groeseneken, G.; Decoutere, S. Forward bias gate breakdown mechanism in enhancement-mode p-GaN gate AlGaN/GaN high-electron mobility transistors. *IEEE Electron Device Lett.* **2015**, *36*, 1001–1003. [CrossRef]

83. Greco, G.; Iucolano, F.; Giannazzo, F.; Di Franco, S.; Corso, D.; Smecca, E.; Alberti, A.; Patti, A.; Roccaforte, F. Metal/p-GaN contacts on AlGaN/GaN heterostructures for normally-off HEMTs. In Proceedings of the Materials Science Forum; Trans Tech Publication Ltd.: Zurich, Switzerland, 2016; pp. 1170–1173.

84. Sayadi, L.; Iannaccone, G.; Sicre, S.; Häberlen, O.; Curatola, G. Threshold voltage instability in p-GaN gate AlGaN/GaN HFETs. *IEEE Trans. Electron Devices* **2018**, *65*, 2454–2460. [CrossRef]

85. Lükens, G.; Hahn, H.; Kalisch, H.; Vescan, A. Self-aligned process for selectively etched p-GaN-gated AlGaN/GaN-on-Si HFETs. *IEEE Trans. Electron Devices* **2018**, *65*, 3732–3738. [CrossRef]

86. Crespo, A.; Bellot, M.; Chabak, K.; Gillespie, J.; Jessen, G.; Miller, V.; Trejo, M.; Via, G.; Walker, D.; Winningham, B. High-power Ka-band performance of AlInN/GaN HEMT with 9.8-nm-thin barrier. *IEEE Electron Device Lett.* **2009**, *31*, 2–4. [CrossRef]

87. Rajan, S.; Chini, A.; Wong, M.H.; Speck, J.S.; Mishra, U.K. N-polar GaN/AlGaN/GaN high electron mobility transistors. *J. Appl. Phys.* **2007**, *102*, 044501. [CrossRef]

88. Wong, M.H.; Pei, Y.; Palacios, T.; Shen, L.; Chakraborty, A.; McCarthy, L.S.; Keller, S.; DenBaars, S.P.; Speck, J.S.; Mishra, U.K. Low nonalloyed Ohmic contact resistance to nitride high electron mobility transistors using N-face growth. *Appl. Phys. Lett.* **2007**, *91*, 232103. [CrossRef]

89. Meyer, D.; Katzer, D.; Bass, R.; Garces, N.; Ancona, M.; Deen, D.; Storm, D.; Binari, S. N-polar n+ GaN cap development for low ohmic contact resistance to inverted HEMTs. *Phys. Status Solidi C* **2012**, *9*, 894–897. [CrossRef]

90. Mahboob, I.; Veal, T.Á.; McConville, C.F.; Lu, H.; Schaff, W.Á. Intrinsic electron accumulation at clean InN surfaces. *Phys. Rev. Lett.* **2004**, *92*, 036804. [CrossRef]

91. Nidhi; Rajan, S.; Keller, S.; Wu, F.; DenBaars, S.P.; Speck, J.S.; Mishra, U.K. Study of interface barrier of SiNx/GaN interface for nitrogen-polar GaN based high electron mobility transistors. *J. Appl. Phys.* **2008**, *103*, 124508. [CrossRef]

92. Meyer, D.; Katzer, D.; Deen, D.; Storm, D.; Binari, S.; Gougousi, T. HfO2-insulated gate N-polar GaN HEMTs with high breakdown voltage. *Phys. Status Solidi (A)* **2011**, *208*, 1630–1633. [CrossRef]

93. Wong, M.H.; Singisetti, U.; Lu, J.; Speck, J.S.; Mishra, U.K. Anomalous output conductance in N-polar GaN high electron mobility transistors. *IEEE Trans. Electron Devices* **2012**, *59*, 2988–2995. [CrossRef]

94. Ibbetson, J.P.; Fini, P.; Ness, K.; DenBaars, S.; Speck, J.; Mishra, U. Polarization effects, surface states, and the source of electrons in AlGaN/GaN heterostructure field effect transistors. *Appl. Phys. Lett.* **2000**, *77*, 250–252. [CrossRef]

95. Arias, A.; Rowell, P.; Bergman, J.; Urteaga, M.; Shinohara, K.; Zheng, X.; Li, H.; Romanczyk, B.; Guidry, M.; Wienecke, S. High performance N-polar GaN HEMTs with OIP3/P dc~12 dB at 10 GHz. In *Proceedings of the 2017 IEEE Compound Semiconductor Integrated Circuit Symposium (CSICS), Miami, FL, USA, 22–25 October 2017*; IEEE: Piscataway, NJ, USA, 2017; pp. 1–3.

96. Zheng, X.; Li, H.; Ahmadi, E.; Hestroffer, K.; Guidry, M.; Romanczyk, B.; Wienecke, S.; Keller, S.; Mishra, U.K. High frequency N-polar GaN planar MIS-HEMTs on sapphire with high breakdown and low dispersion. In *Proceedings of the 2016 Lester Eastman Conference (LEC), Bethlehem, PA, USA, 2–4 August 2016*; IEEE: Piscataway, NJ, USA, 2016; pp. 42–45.

97. Liu, W.; Romanczyk, B.; Guidry, M.; Hatui, N.; Wurm, C.; Li, W.; Shrestha, P.; Zheng, X.; Keller, S.; Mishra, U.K. 6.2 W/mm and Record 33.8% PAE at 94 GHz from N-polar GaN Deep Recess MIS-HEMTs with ALD Ru Gates. *IEEE Microw. Wirel. Compon. Lett.* **2021**, *31*, 748–751. [CrossRef]

98. Peralagu, U.; Alian, A.; Putcha, V.; Khaled, A.; Rodriguez, R.; Sibaja-Hernandez, A.; Chang, S.; Simoen, E.; Zhao, S.; De Jaeger, B. CMOS-compatible GaN-based devices on 200mm-Si for RF applications: Integration and Performance. In *Proceedings of the 2019 IEEE International Electron Devices Meeting (IEDM), San Francisco, CA, USA, 7–11 December 2019*; IEEE: Piscataway, NJ, USA, 2019; pp. 17.2. 1–17.2. 4.

99. Yuk, K.; Branner, G.; Cui, C. Future directions for GaN in 5G and satellite communications. In *Proceedings of the 2017 IEEE 60th International Midwest Symposium on Circuits and Systems (MWSCAS), Boston, MA, USA, 6–9 August 2017*; IEEE: Piscataway, NJ, USA, 2017; pp. 803–806.

100. Wong, K.-Y.R.; Kwan, M.-H.; Yao, F.-W.; Tsai, M.-W.; Lin, Y.-S.; Chang, Y.-C.; Chen, P.-C.; Su, R.-Y.; Yu, J.-L.; Yang, F.-J. A next generation CMOS-compatible GaN-on-Si transistors for high efficiency energy systems. In *Proceedings of the 2015 IEEE International Electron Devices Meeting (IEDM), Washington, DC, USA, 7–9 December 2015*; IEEE: Piscataway, NJ, USA, 2015; pp. 9.5. 1–9.5. 4.

101. Xie, H.; Liu, Z.; Gao, Y.; Ranjan, K.; Lee, K.E.; Ng, G.I. CMOS-compatible GaN-on-Si HEMTs with cut-off frequency of 210 GHz and high Johnson's figure-of-merit of 8.8 THz V. *Appl. Phys. Express* **2020**, *13*, 026503. [CrossRef]

102. Yeh, P.-C.; Tu, P.-T.; Liu, H.-H.; Hsu, C.-H.; Yang, H.-Y.; Fu, Y.-K.; Lee, L.-H.; Tzeng, P.-J.; Wu, Y.-R.; Sheu, S.-S. CMOS-compatible GaN HEMT on 200mm Si-substrate for RF application. In *Proceedings of the 2021 International Symposium on VLSI Technology, Systems and Applications (VLSI-TSA), Hsinchu, Taiwan, 19–22 April 2021*; IEEE: Piscataway, NJ, USA, 2021; pp. 1–2.

103. Then, H.W.; Dasgupta, S.; Radosavljevic, M.; Agababov, P.; Ban, I.; Bristol, R.; Chandhok, M.; Chouksey, S.; Holybee, B.; Huang, C. 3D heterogeneous integration of high performance high-K metal gate GaN NMOS and Si PMOS transistors on 300 mm high-resistivity Si substrate for energy-efficient and compact power delivery, RF (5G and beyond) and SoC applications. In *Proceedings of the 2019 IEEE International Electron Devices Meeting (IEDM), San Francisco, CA, USA, 7–11 Decmber 2019v*; IEEE: Piscataway, NJ, USA, 2019; pp. 17.3. 1–17.3. 4.

104. Chen, C.; Yost, D.-R.; Knecht, J.; Wey, J.; Chapman, D.; Oakley, D.; Soares, A.; Mahoney, L.; Donnelly, J.; Chen, C. Wafer-scale 3D integration of InGaAs photodiode arrays with Si readout circuits by oxide bonding and through-oxide vias. *Microelectron. Eng.* **2011**, *88*, 131–134. [CrossRef]
105. Green, D.S.; Dohrman, C.L.; Demmin, J.; Chang, T.-H. Path to 3D heterogeneous integration. In *Proceedings of the 2015 International 3D Systems Integration Conference (3DIC), Sendai, Japan, 31 August–2 September 2015*; IEEE: Piscataway, NJ, USA, 2015; pp. FS7. 1–FS7. 3.
106. Li, X.; Geens, K.; Guo, W.; You, S.; Zhao, M.; Fahle, D.; Odnoblyudov, V.; Groeseneken, G.; Decoutere, S. Demonstration of GaN integrated half-bridge with on-chip drivers on 200-mm engineered substrates. *IEEE Electron Device Lett.* **2019**, *40*, 1499–1502. [CrossRef]
107. Reusch, D.; Strydom, J.; Glaser, J. Improving high frequency DC-DC converter performance with monolithic half bridge GaN ICs. In *Proceedings of the 2015 IEEE Energy Conversion Congress and Exposition (ECCE), Montreal, QC, Canada, 20–24 September 2015*; IEEE: Piscataway, NJ, USA, 2015; pp. 381–387.
108. Jiang, Q.; Tang, Z.; Zhou, C.; Yang, S.; Chen, K.J. Substrate-coupled cross-talk effects on an AlGaN/GaN-on-Si smart power IC platform. *IEEE Trans. Electron Devices* **2014**, *61*, 3808–3813. [CrossRef]
109. Kang, S.; Choi, B.; Kim, B. Linearity analysis of CMOS for RF application. *IEEE Trans. Microw. Theory Tech.* **2003**, *51*, 972–977. [CrossRef]
110. Chowdhury, N. *p-Channel Gallium Nitride Transistor on Si Substrate*; Massachusetts Institute of Technology: Boston, MA, USA, 2018.
111. Reuters, B.; Hahn, H.; Pooth, A.; Holländer, B.; Breuer, U.; Heuken, M.; Kalisch, H.; Vescan, A. Fabrication of p-channel heterostructure field effect transistors with polarization-induced two-dimensional hole gases at metal–polar GaN/AlInGaN interfaces. *J. Phys. D: Appl. Phys.* **2014**, *47*, 175103. [CrossRef]
112. Bader, S.; Chaudhuri, R.; Hickman, A.; Nomoto, K.; Bharadwaj, S.; Then, H.; Xing, H.; Jena, D. GaN/AlN Schottky-gate p-channel HFETs with InGaN contacts and 100 mA/mm on-current. In *Proceedings of the 2019 IEEE International Electron Devices Meeting (IEDM), San Francisco, CA, USA, 7–11 December 2019*; IEEE: Piscataway, NJ, USA, 2019; pp. 4.5. 1–4.5. 4.
113. Chowdhury, N.; Lemettinen, J.; Xie, Q.; Zhang, Y.; Rajput, N.S.; Xiang, P.; Cheng, K.; Suihkonen, S.; Then, H.W.; Palacios, T. P-channel GaN transistor based on p-GaN/AlGaN/GaN on Si. *IEEE Electron Device Lett.* **2019**, *40*, 1036–1039. [CrossRef]
114. Zheng, Z.; Song, W.; Zhang, L.; Yang, S.; Wei, J.; Chen, K.J. High I_{ON} and I_{ON}/I_{OFF} Ratio Enhancement-Mode Buried *p*-Channel GaN MOSFETs on *p*-GaN Gate Power HEMT Platform. *IEEE Electron Device Lett.* **2019**, *41*, 26–29. [CrossRef]
115. Raj, A.; Krishna, A.; Hatui, N.; Gupta, C.; Jang, R.; Keller, S.; Mishra, U.K. Demonstration of a GaN/AlGaN superlattice-based p-channel FinFET with high ON-current. *IEEE Electron Device Lett.* **2020**, *41*, 220–223. [CrossRef]
116. Chowdhury, N.; Xie, Q.; Yuan, M.; Rajput, N.S.; Xiang, P.; Cheng, K.; Then, H.W.; Palacios, T. First demonstration of a self-aligned GaN p-FET. In *Proceedings of the 2019 IEEE International Electron Devices Meeting (IEDM), San Francisco, CA, USA, 7–11 December 2019*; IEEE: Piscataway, NJ, USA, 2019; pp. 4.6. 1–4.6. 4.
117. Chen, F.; Hao, R.; Yu, G.; Zhang, X.; Song, L.; Wang, J.; Cai, Y.; Zhang, B. Enhancement-mode n-GaN gate p-channel heterostructure field effect transistors based on GaN/AlGaN 2D hole gas. *Appl. Phys. Lett.* **2019**, *115*, 112103. [CrossRef]
118. Chowdhury, N.; Xie, Q.; Yuan, M.; Cheng, K.; Then, H.W.; Palacios, T. Regrowth-free GaN-based complementary logic on a Si substrate. *IEEE Electron Device Lett.* **2020**, *41*, 820–823. [CrossRef]
119. Chaudhuri, R.; Bader, S.J.; Chen, Z.; Muller, D.A.; Xing, H.G.; Jena, D. A polarization-induced 2D hole gas in undoped gallium nitride quantum wells. *Science* **2019**, *365*, 1454–1457. [CrossRef]
120. Nomoto, K.; Chaudhuri, R.; Bader, S.; Li, L.; Hickman, A.; Huang, S.; Lee, H.; Maeda, T.; Then, H.; Radosavljevic, M. GaN/AlN p-channel HFETs with I max> 420 mA/mm and~ 20 GHz f T/f MAX. In *Proceedings of the 2020 IEEE International Electron Devices Meeting (IEDM), San Francisco, CA, USA, 12–18 December 2020*; IEEE: Piscataway, NJ, USA, 2020; pp. 8.3. 1–8.3. 4.
121. Hickman, A.; Chaudhuri, R.; Bader, S.J.; Nomoto, K.; Lee, K.; Xing, H.G.; Jena, D. High breakdown voltage in RF AlN/GaN/AlN quantum well HEMTs. *IEEE Electron Device Lett.* **2019**, *40*, 1293–1296. [CrossRef]
122. Bader, S.J.; Lee, H.; Chaudhuri, R.; Huang, S.; Hickman, A.; Molnar, A.; Xing, H.G.; Jena, D.; Then, H.W.; Chowdhury, N. Prospects for wide bandgap and ultrawide bandgap CMOS devices. *IEEE Trans. Electron Devices* **2020**, *67*, 4010–4020. [CrossRef]
123. Zhao, D.; Xu, S.; Xie, M.; Tong, S.; Yang, H. Stress and its effect on optical properties of GaN epilayers grown on Si (111), 6H-SiC (0001), and c-plane sapphire. *Appl. Phys. Lett.* **2003**, *83*, 677–679. [CrossRef]
124. Nakamura, S. GaN growth using GaN buffer layer. *Jpn. J. Appl. Phys.* **1991**, *30*, L1705. [CrossRef]
125. Selvaraj, S.L.; Watanabe, A.; Wakejima, A.; Egawa, T. 1.4-kV breakdown voltage for AlGaN/GaN high-electron-mobility transistors on silicon substrate. *IEEE Electron Device Lett.* **2012**, *33*, 1375–1377. [CrossRef]
126. Rowena, I.B.; Selvaraj, S.L.; Egawa, T. Buffer thickness contribution to suppress vertical leakage current with high breakdown field (2.3 MV/cm) for GaN on Si. *IEEE Electron Device Lett.* **2011**, *32*, 1534–1536. [CrossRef]
127. Ito, S.; Nakagita, T.; Sawaki, N.; Ahn, H.S.; Irie, M.; Hikosaka, T.; Honda, Y.; Yamaguchi, M.; Amano, H. Nature of yellow luminescence band in GaN grown on Si substrate. *Jpn. J. Appl. Phys.* **2014**, *53*, 11RC02. [CrossRef]
128. Weeks, T.W., Jr.; Piner, E.L.; Gehrke, T.; Linthicum, K.J. Gallium Nitride Materials and Methods. US6617060B2, 18 November 2003.
129. Ikeda, N.; Niiyama, Y.; Kambayashi, H.; Sato, Y.; Nomura, T.; Kato, S.; Yoshida, S. GaN power transistors on Si substrates for switching applications. *Proc. IEEE* **2010**, *98*, 1151–1161. [CrossRef]

130. Kato, S.; Satoh, Y.; Sasaki, H.; Masayuki, I.; Yoshida, S. C-doped GaN buffer layers with high breakdown voltages for high-power operation AlGaN/GaN HFETs on 4-in Si substrates by MOVPE. *J. Cryst. Growth* **2007**, *298*, 831–834. [CrossRef]
131. Kuzmík, J. Power electronics on InAlN/(In) GaN: Prospect for a record performance. *IEEE Electron Device Lett.* **2001**, *22*, 510–512. [CrossRef]
132. Liu, S.-C.; Chen, B.-Y.; Lin, Y.-C.; Hsieh, T.-E.; Wang, H.-C.; Chang, E.Y. GaN MIS-HEMTs with nitrogen passivation for power device applications. *IEEE Electron Device Lett.* **2014**, *35*, 1001–1003.
133. Roccaforte, F.; Fiorenza, P.; Greco, G.; Nigro, R.L.; Giannazzo, F.; Patti, A.; Saggio, M. Challenges for energy efficient wide band gap semiconductor power devices. *Phys. Status Solidi (A)* **2014**, *211*, 2063–2071. [CrossRef]
134. Miyamoto, H.; Okamoto, Y.; Kawaguchi, H.; Miura, Y.; Nakamura, M.; Nakayama, T.; Masumoto, I.; Miyake, S.; Hirai, T.; Fujita, M. Enhancement-mode GaN-on-Si MOS-FET using Au-free Si process and its operation in PFC system with high-efficiency. In *Proceedings of the 2015 IEEE 27th International Symposium on Power Semiconductor Devices & IC's (ISPSD), Hong Kong, China, 10–14 May 2015*; IEEE: Piscataway, NJ, USA, 2015; pp. 209–212.
135. Wang, H.; Wei, J.; Xie, R.; Liu, C.; Tang, G.; Chen, K.J. Maximizing the performance of 650-V p-GaN gate HEMTs: Dynamic RON characterization and circuit design considerations. *IEEE Trans. Power Electron.* **2016**, *32*, 5539–5549. [CrossRef]
136. Crupi, G.; Schreurs, D.M.-P.; Caddemi, A. Effects of gate-length scaling on microwave MOSFET performance. *Electronics* **2017**, *6*, 62. [CrossRef]
137. Chung, J.W. Millimeter-Wave GaN High Electron Mobility Transistors and Their Integration with Silicon Electronics. Ph.D. Thesis, Massachusetts Institute of Technology, Cambridge, MA, USA, 2011.
138. Sun, H.; Alt, A.R.; Tirelli, S.; Marti, D.; Benedickter, H.; Piner, E.; Bolognesi, C.R. Nanometric AlGaN/GaN HEMT performance with implant or mesa isolation. *IEEE Electron Device Lett.* **2011**, *32*, 1056–1058. [CrossRef]
139. Lan, Y.-L.; Lin, H.-C.; Liu, H.-H.; Lee, G.-Y.; Ren, F.; Pearton, S.J.; Chang, M.-N.; Chyi, J.-I. Low-resistance smooth-surface Ti/Al/Cr/Mo/Au n-type ohmic contact to AlGaN/GaN heterostructures. *Appl. Phys. Lett.* **2009**, *94*, 243502. [CrossRef]
140. Ture, E. *GaN-Based Tri-Gate High Electron Mobility Transistors*; BoD–Books on Demand: Freiburg, Germany, 2018.
141. Schuette, M.L.; Ketterson, A.; Song, B.; Beam, E.; Chou, T.-M.; Pilla, M.; Tserng, H.-Q.; Gao, X.; Guo, S.; Fay, P.J. Gate-recessed integrated E/D GaN HEMT technology with f_T/f_{max} > 300 GHz. *IEEE Electron Device Lett.* **2013**, *34*, 741–743. [CrossRef]
142. Shinohara, K.; Regan, D.C.; Tang, Y.; Corrion, A.L.; Brown, D.F.; Wong, J.C.; Robinson, J.F.; Fung, H.H.; Schmitz, A.; Oh, T.C. Scaling of GaN HEMTs and Schottky diodes for submillimeter-wave MMIC applications. *IEEE Trans. Electron Devices* **2013**, *60*, 2982–2996. [CrossRef]
143. Then, H.W.; Radosavljevic, M.; Agababov, P.; Ban, I.; Bristol, R.; Chandhok, M.; Chouksey, S.; Holybee, B.; Huang, C.; Krist, B. GaN and Si Transistors on 300mm Si (111) enabled by 3D Monolithic Heterogeneous Integration. In *Proceedings of the 2020 IEEE Symposium on VLSI Technology, Honolulu, HI, USA, 16–19 June 2020*; IEEE: Piscataway, NJ, USA, 2020; pp. 1–2.
144. Tang, Y.; Shinohara, K.; Regan, D.; Corrion, A.; Brown, D.; Wong, J.; Schmitz, A.; Fung, H.; Kim, S.; Micovic, M. Ultrahigh-speed GaN high-electron-mobility transistors with f_T/f_{max} of 454/444 GHz. *IEEE Electron Device Lett.* **2015**, *36*, 549–551. [CrossRef]
145. Tu, P.-T.; Sanyal, I.; Yeh, P.-C.; Lee, H.-Y.; Lee, L.-H.; Wu, C.-I.; Chyi, J.-I. Quaternary Barrier AlInGaN/GaN-on-Si High Electron Mobility Transistor with Record F T-L g Product of 13.9 GHz-μm. In *Proceedings of the 2020 International Symposium on VLSI Technology, Systems and Applications (VLSI-TSA), Hsinchu, Taiwan, 10–13 August 2020*; IEEE: Piscataway, NJ, USA, 2020; pp. 130–131.
146. Parvais, B.; Alian, A.; Peralagu, U.; Rodriguez, R.; Yadav, S.; Khaled, A.; ElKashlan, R.; Putcha, V.; Sibaja-Hernandez, A.; Zhao, M. GaN-on-Si mm-wave RF Devices Integrated in a 200mm CMOS Compatible 3-Level Cu BEOL. In *Proceedings of the 2020 IEEE International Electron Devices Meeting (IEDM), San Francisco, CA, USA, 12–18 December 2020*; IEEE: Piscataway, NJ, USA, 2020; pp. 8.1. 1–8.1. 4.
147. Ciccognani, W.; Colangeli, S.; Serino, A.; Pace, L.; Fenu, S.; Longhi, P.; Limiti, E.; Poulain, J.; Leblanc, R. Comparative noise investigation of high-performance GaAs and GaN millimeter-wave monolithic technologies. In *Proceedings of the 2019 14th European Microwave Integrated Circuits Conference (EuMIC), Paris, France, 30 September–1 October 2019*; IEEE: Piscataway, NJ, USA, 2019; pp. 192–195.
148. Di Giacomo-Brunel, V.; Byk, E.; Chang, C.; Grünenpütt, J.; Lambert, B.; Mouginot, G.; Sommer, D.; Jung, H.; Camiade, M.; Fellon, P. Industrial 0.15-μm AlGaN/GaN on SiC technology for applications up to Ka band. In *Proceedings of the 2018 13th European Microwave Integrated Circuits Conference (EuMIC), Madrid, Spain, 23–25 September 2018*; IEEE: Piscataway, NJ, USA, 2018; pp. 1–4.
149. Xie, H.; Liu, Z.; Gao, Y.; Ranjan, K.; Lee, K.E.; Ng, G.I. Deeply-scaled GaN-on-Si high electron mobility transistors with record cut-off frequency f T of 310 GHz. *Appl. Phys. Express* **2019**, *12*, 126506. [CrossRef]
150. Ichikawa, H.; Mizue, C.; Makabe, I.; Tateno, Y.; Nakata, K.; Inoue, K. AlGaN/GaN HEMTs versus InAlN/GaN HEMTs fabricated by150-nm Y-gate process. In *Proceedings of the 2014 Asia-Pacific Microwave Conference, Sendai, Japan, 4–7 November 2014*; IEEE: Piscataway, NJ, USA, 2014; pp. 780–782.
151. Dogmus, E.; Kabouche, R.; Linge, A.; Okada, E.; Zegaoui, M.; Medjdoub, F. High power, high PAE Q-band sub-10 nm barrier thickness AlN/GaN HEMTs. *Phys. Status Solidi (A)* **2017**, *214*, 1600797. [CrossRef]
152. Moon, J.-S.; Grabar, B.; Wong, J.; Chuong, D.; Arkun, E.; Morales, D.V.; Chen, P.; Malek, C.; Fanning, D.; Venkatesan, N. Power Scaling of Graded-Channel GaN HEMTs With Mini-Field-Plate T-gate and 156 GHz f T. *IEEE Electron Device Lett.* **2021**, *42*, 796–799. [CrossRef]

153. Li, L.; Nomoto, K.; Pan, M.; Li, W.; Hickman, A.; Miller, J.; Lee, K.; Hu, Z.; Bader, S.J.; Lee, S.M. GaN HEMTs on Si with regrown contacts and cutoff/maximum oscillation frequencies of 250/204 GHz. *IEEE Electron Device Lett.* **2020**, *41*, 689–692. [CrossRef]

154. Denninghoff, D.; Lu, J.; Laurent, M.; Ahmadi, E.; Keller, S.; Mishra, U. N-polar GaN/InAlN MIS-HEMT with 400-GHz f max. In *Proceedings of the 70th Device Research Conference, University Park, PA, USA, 18–20 June 2012*; IEEE: Piscataway, NJ, USA, 2012; pp. 151–152.

155. Alomari, M.; Medjdoub, F.; Carlin, J.-F.; Feltin, E.; Grandjean, N.; Chuvilin, A.; Kaiser, U.; Gaquière, C.; Kohn, E. InAlN/GaN MOSHEMT with self-aligned thermally generated oxide recess. *IEEE Electron Device Lett.* **2009**, *30*, 1131–1133. [CrossRef]

156. Dumka, D.; Saunier, P. GaN on Si HEMT with 65% power added efficiency at 10 GHz. *Electron. Lett.* **2010**, *46*, 946–947. [CrossRef]

157. Marti, D.; Tirelli, S.; Teppati, V.; Lugani, L.; Carlin, J.-F.; Malinverni, M.; Grandjean, N.; Bolognesi, C.R. 94-GHz large-signal operation of AlInN/GaN high-electron-mobility transistors on silicon with regrown ohmic contacts. *IEEE Electron Device Lett.* **2014**, *36*, 17–19. [CrossRef]

158. Vidkjær, J.; Shevchenko, S.; Würfl, J.; Uren, M.; Hirche, K.; Jost, R. Linearity Assessment of GaN Technology. *Abstract ESA/ESTEC contract 20456*, (07).

159. Xing, W.; Liu, Z.; Ranjan, K.; Ng, G.I.; Palacios, T. Planar nanostrip-channel Al_2O_3/InAlN/GaN MISHEMTs on Si with improved linearity. *IEEE Electron Device Lett.* **2018**, *39*, 947–950. [CrossRef]

160. Moon, J.-S.; Grabar, B.; Antcliffe, M.; Wong, J.; Dao, C.; Chen, P.; Arkun, E.; Khalaf, I.; Corrion, A.; Chappell, J. High-speed graded-channel GaN HEMTs with linearity and efficiency. In *Proceedings of the 2020 IEEE/MTT-S International Microwave Symposium (IMS), Los Angeles, CA, USA, 4–6 August 2020*; IEEE: Piscataway, NJ, USA, 2020; pp. 573–575.

161. Hou, B.; Yang, L.; Mi, M.; Zhang, M.; Yi, C.; Wu, M.; Zhu, Q.; Lu, Y.; Zhu, J.; Zhou, X. High linearity and high power performance with barrier layer of sandwich structure and Al0. 05GaN back barrier for X-band application. *J. Phys. D: Appl. Phys.* **2020**, *53*, 145102. [CrossRef]

162. Zhang, K.; Kong, Y.; Zhu, G.; Zhou, J.; Yu, X.; Kong, C.; Li, Z.; Chen, T. High-linearity AlGaN/GaN FinFETs for microwave power applications. *IEEE Electron Device Lett.* **2017**, *38*, 615–618. [CrossRef]

163. Egawa, T. Heteroepitaxial growth and power electronics using AlGaN/GaN HEMT on Si. In *Proceedings of the 2012 International Electron Devices Meeting, San Francisco, CA, USA, 10–13 December 2012*; IEEE: Piscataway, NJ, USA, 2012; pp. 27.1. 1–27.1. 4.

164. Cho, E.; Mogilatenko, A.; Brunner, F.; Richter, E.; Weyers, M. Impact of AlN nucleation layer on strain in GaN grown on 4H-SiC substrates. *J. Cryst. Growth* **2013**, *371*, 45–49. [CrossRef]

165. Foronda, H.M.; Romanov, A.E.; Young, E.C.; Robertson, C.A.; Beltz, G.E.; Speck, J.S. Curvature and bow of bulk GaN substrates. *J. Appl. Phys.* **2016**, *120*, 035104. [CrossRef]

166. Ma, C.-T.; Gu, Z.-H. Review of GaN HEMT applications in power converters over 500 W. *Electronics* **2019**, *8*, 1401. [CrossRef]

167. Chen, K.J.; Häberlen, O.; Lidow, A.; lin Tsai, C.; Ueda, T.; Uemoto, Y.; Wu, Y. GaN-on-Si power technology: Devices and applications. *IEEE Trans. Electron Devices* **2017**, *64*, 779–795. [CrossRef]

168. Arulkumaran, S.; Lin, V.; Dolmanan, S.; Ng, G.; Vicknesh, S.; Tan, J.; Teo, S.; Kumar, M.K.; Tripathy, S. Improved OFF-state breakdown voltage in AlGaN/GaN HEMTs grown on 150-mm diameter silicon-on-insulator (SOI) substrate. In *Proceedings of the 70th Device Research Conference, University Park, PA, USA, 18–20 June 2012*; IEEE: Piscataway, NJ, USA, 2012; pp. 59–60.

169. Jiang, Q.; Liu, C.; Lu, Y.; Chen, K.J. 1.4-kV AlGaN/GaN HEMTs on a GaN-on-SOI platform. *IEEE Electron Device Lett.* **2013**, *34*, 357–359. [CrossRef]

170. Chiu, H.-C.; Peng, L.-Y.; Wang, H.-Y.; Wang, H.-C.; Kao, H.-L.; Lee, G.-Y.; Chyi, J.-I. The Characterization of InAlN/AlN/GaN HEMTs Using Silicon-on-Insulator (SOI) Substrate Technology. *J. Electrochem. Soc.* **2015**, *163*, H110. [CrossRef]

171. Tham, W.H.; Ang, D.S.; Bera, L.K.; Dolmanan, S.B.; Bhat, T.N.; Lin, V.K.; Tripathy, S. Comparison of the Al x Ga 1–x N/GaN heterostructures grown on silicon-on-insulator and bulk-silicon substrates. *IEEE Trans. Electron Devices* **2015**, *63*, 345–352. [CrossRef]

172. Li, X.; Amirifar, N.; Geens, K.; Zhao, M.; Guo, W.; Liang, H.; You, S.; Posthuma, N.; De Jaeger, B.; Stoffels, S. GaN-on-SOI: Monolithically integrated all-GaN ICs for power conversion. In *Proceedings of the 2019 IEEE International Electron Devices Meeting (IEDM), San Francisco, CA, USA, 7–11 December 2019*; IEEE: Piscataway, NJ, USA, 2019; pp. 4.4. 1–4.4. 4.

173. Nomura, T.; Masuda, M.; Ikeda, N.; Yoshida, S. Switching characteristics of GaN HFETs in a half bridge package for high temperature applications. *IEEE Trans. Power Electron.* **2008**, *23*, 692–697. [CrossRef]

174. Morita, T.; Tamura, S.; Anda, Y.; Ishida, M.; Uemoto, Y.; Ueda, T.; Tanaka, T.; Ueda, D. 99.3% efficiency of three-phase inverter for motor drive using GaN-based gate injection transistors. In *Proceedings of the 2011 Twenty-Sixth Annual IEEE Applied Power Electronics Conference and Exposition (APEC), Fort Worth, TX, USA, 6–11 March 2011*; IEEE: Piscataway, NJ, USA, 2011; pp. 481–484.

175. Ishida, M.; Ueda, T.; Tanaka, T.; Ueda, D. GaN on Si technologies for power switching devices. *IEEE Trans. Electron Devices* **2013**, *60*, 3053–3059. [CrossRef]

176. Chen, H.-Y.; Kao, Y.-Y.; Zhang, Z.-Q.; Liao, C.-H.; Yang, H.-Y.; Hsu, M.-S.; Chen, K.-H.; Lin, Y.-H.; Lin, S.-R.; Tsai, T.-Y. A Fully Integrated GaN-on-Silicon Gate Driver and GaN Switch with Temperature-compensated Fast Turn-on Technique for Improving Reliability. In *Proceedings of the 2021 IEEE International Solid-State Circuits Conference (ISSCC), San Francisco, CA, USA, 13–22 February 2021*; IEEE: Piscataway, NJ, USA, 2021; pp. 460–462.

177. Li, X.; Van Hove, M.; Zhao, M.; Geens, K.; Lempinen, V.-P.; Sormunen, J.; Groeseneken, G.; Decoutere, S. 200 V enhancement-mode p-GaN HEMTs fabricated on 200 mm GaN-on-SOI with trench isolation for monolithic integration. *IEEE Electron Device Lett.* **2017**, *38*, 918–921. [CrossRef]

178. Yamashita, Y.; Stoffels, S.; Posthuma, N.; Decoutere, S.; Kobayashi, K. Monolithically integrated E-mode GaN-on-SOI gate driver with power GaN-HEMT for MHz-switching. In Proceedings of the 2018 IEEE 6th Workshop on Wide Bandgap Power Devices and Applications (WiPDA), Atlanta, GA, USA, 31 October–2 November 2018; IEEE: Piscataway, NJ, USA, 2018; pp. 231–236.

179. Giofrè, R.; Colantonio, P.; Giannini, F. GaN broadband power amplifiers for terrestrial and space transmitters. In *Proceedings of the 2012 19th International Conference on Microwaves, Radar & Wireless Communications, Warsaw, Poland, 21–23 May 2012*; IEEE: Piscataway, NJ, USA, 2012; pp. 605–609.

180. Ciccognani, W.; De Dominicis, M.; Ferrari, M.; Limiti, E.; Peroni, M.; Romanini, P. High-power monolithic AlGaN/GaN HEMT switch for X-band applications. *Electron. Lett.* **2008**, *44*, 911–912. [CrossRef]

181. Ciccognani, W.; Limiti, E.; Longhi, P.; Mitrano, C.; Nanni, A.; Peroni, M. An ultra-broadband robust LNA for defence applications in AlGaN/GaN technology. In *Proceedings of the 2010 IEEE MTT-S International Microwave Symposium, Anaheim, CA, USA, 23–28 May 2010*; IEEE: Piscataway, NJ, USA, 2010; pp. 493–496.

182. Flack, T.J.; Pushpakaran, B.N.; Bayne, S.B. GaN technology for power electronic applications: A review. *J. Electron. Mater.* **2016**, *45*, 2673–2682. [CrossRef]

183. Rack, M.; Raskin, J.-P. SOI Technologies for RF and Millimeter Wave Applications. *ECS Trans.* **2019**, *92*, 79. [CrossRef]

184. Hao-Yu-Wang, L.-Y.P.; Cheng, Y.-H.; Chiu, H.-C. The Demonstration and Characterization of In-situ SiNx/AlGaN/GaN HEMT on 6-inch Silicon on Insulator (SOI) Substrate. In Proceedings of the CS ManTec Conference, Miami, FL, USA, 16–19 May 2016.

185. Chiu, H.-C.; Wang, H.-Y.; Peng, L.-Y.; Wang, H.-C.; Kao, H.-L.; Hu, C.-W.; Xuan, R. RF Performance of In Situ SiN x Gate Dielectric AlGaN/GaN MISHEMT on 6-in Silicon-on-Insulator Substrate. *IEEE Trans. Electron Devices* **2017**, *64*, 4065–4070. [CrossRef]

186. Anderson, T.J.; Koehler, A.D.; Tadjer, M.J.; Hite, J.K.; Nath, A.; Mahadik, N.A.; Aktas, O.; Odnoblyudov, V.; Basceri, C.; Hobart, K.D. Electrothermal evaluation of thick GaN epitaxial layers and AlGaN/GaN high-electron-mobility transistors on large-area engineered substrates. *Appl. Phys. Express* **2017**, *10*, 126501. [CrossRef]

187. Leach, J.; Udwary, K.; Quayle, P.; Odnoblyudov, V.; Basceri, C.; Aktas, O.; Splawn, H.; Evans, K. Towards manufacturing large area GaN substrates from QST® seeds. *Proceeding Compd. Semicond. Manuf. Technol.* **2018**, 014.17.

188. Geens, K.; Li, X.; Zhao, M.; Guo, W.; Wellekens, D.; Posthuma, N.; Fahle, D.; Aktas, O.; Odnoblyudov, V.; Decoutere, S. 650 V p-GaN gate power HEMTs on 200 mm engineered substrates. In *Proceedings of the 2019 IEEE 7th Workshop on Wide Bandgap Power Devices and Applications (WiPDA), Raleigh, NC, USA, 29–31 October 2019*; IEEE: Piscataway, NJ, USA, 2019; pp. 292–296.

189. Zubair, A.; Perozek, J.; Niroula, J.; Aktas, O.; Odnoblyudov, V.; Palacios, T. First demonstration of GaN vertical power FinFETs on engineered substrate. In *Proceedings of the 2020 Device Research Conference (DRC), Columbus, OH, USA, 21–24 June 2020*; IEEE: Piscataway, NJ, USA, 2020; pp. 1–2.

190. Jan, C.-H.; Agostinelli, M.; Deshpande, H.; El-Tanani, M.; Hafez, W.; Jalan, U.; Janbay, L.; Kang, M.; Lakdawala, H.; Lin, J. RF CMOS technology scaling in high-k/metal gate era for RF SoC (system-on-chip) applications. In *Proceedings of the 2010 International Electron Devices Meeting, San Francisco, CA, USA, 6–8 December 2010*; IEEE: Piscataway, NJ, USA, 2010; pp. 27.2. 1–27.2. 4.

191. Lee, K.H.; Bao, S.; Kohen, D.; Huang, C.C.; Lee, K.E.K.; Fitzgerald, E.; Tan, C.S. Monolithic integration of III–V HEMT and Si-CMOS through TSV-less 3D wafer stacking. In *Proceedings of the 2015 IEEE 65th Electronic Components and Technology Conference (ECTC), San Diego, CA, USA, 26–29 May 2015*; IEEE: Piscataway, NJ, USA, 2015; pp. 560–565.

192. Zhu, M.; Matioli, E. Monolithic integration of GaN-based NMOS digital logic gate circuits with E-mode power GaN MOSHEMTs. In *Proceedings of the 2018 IEEE 30th International Symposium on Power Semiconductor Devices and ICs (ISPSD), Chicago, IL, USA, 13–17 May 2018*; IEEE: Piscataway, NJ, USA, 2018; pp. 236–239.

193. Gutierrez-Aitken, A.; Wu, B.Y.-C.; Scott, D.; Sato, K.; Poust, B.; Watanabe, M.; Monier, C.; Lin, N.; Zeng, X.; Nakamura, E. A meeting of materials: Integrating diverse semiconductor technologies for improved performance at lower cost. *IEEE Microw. Mag.* **2017**, *18*, 60–73. [CrossRef]

194. Gutierrez-Aitken, A. Heterogeneous Integration for High Frequency RF Applications. In *Proceedings of the 2020 IEEE International Electron Devices Meeting (IEDM), San Francisco, CA, USA, 12–18 December 2020*; IEEE: Piscataway, NJ, USA, 2020; pp. 34.5. 1–34.5. 4.

195. Lee, K.-T.; Bayram, C.; Piedra, D.; Sprogis, E.; Deligianni, H.; Krishnan, B.; Papasouliotis, G.; Paranjpe, A.; Aklimi, E.; Shepard, K. GaN devices on a 200 mm Si platform targeting heterogeneous integration. *IEEE Electron Device Lett.* **2017**, *38*, 1094–1096. [CrossRef]

196. Meng, F.; Disney, D.; Liu, B.; Volkan, Y.B.; Zhou, A.; Liang, Z.; Yi, X.; Selvaraj, S.L.; Peng, L.; Ma, K. Heterogeneous integration of GaN and BCD technologies and its applications to high conversion-ratio DC–DC boost converter IC. *IEEE Trans. Power Electron.* **2018**, *34*, 1993–1996. [CrossRef]

197. Warnock, S.; Chen, C.-L.; Knechtl, J.; Molnar, R.; Yost, D.-R.; Cook, M.; Stull, C.; Johnson, R.; Galbraith, C.; Daulton, J. InAlN/GaN-on-Si HEMT with 4.5 W/mm in a 200-mm CMOS-Compatible MMIC Process for 3D Integration. In *Proceedings of the 2020 IEEE/MTT-S International Microwave Symposium (IMS), Los Angeles, CA, USA, 4–6 August 2020*; IEEE: Piscataway, NJ, USA, 2020; pp. 289–292.

198. Kazior, T.; Chelakara, R.; Hoke, W.; Bettencourt, J.; Palacios, T.; Lee, H. High performance mixed signal and RF circuits enabled by the direct monolithic heterogeneous integration of GaN HEMTs and Si CMOS on a silicon substrate. In *Proceedings of the 2011 IEEE Compound Semiconductor Integrated Circuit Symposium (CSICS), Waikoloa, HI, USA, 16–19 October 2011*; IEEE: Piscataway, NJ, USA, 2011; pp. 1–4.

199. Chung, J.W.; Lee, J.-k.; Piner, E.L.; Palacios, T. Seamless on-wafer integration of Si (100) MOSFETs and GaN HEMTs. *IEEE Electron Device Lett.* **2009**, *30*, 1015–1017. [CrossRef]
200. Chyurlia, P.; Semond, F.; Lester, T.; Bardwell, J.; Rolfe, S.; Tang, H.; Tarr, N. Monolithic integration of AlGaN/GaN HFET with MOS on silicon (111) substrates. *Electron. Lett.* **2010**, *46*, 253–254. [CrossRef]
201. Chen, M.-F.; Chen, F.-C.; Chiou, W.-C.; Doug, C. System on integrated chips (SoIC (TM) for 3D heterogeneous integration. In *Proceedings of the 2019 IEEE 69th Electronic Components and Technology Conference (ECTC), Las Vegas, NV, USA, 28–31 May 2019*; IEEE: Piscataway, NJ, USA, 2019; pp. 594–599.
202. Jourdain, A.; Schleicher, F.; De Vos, J.; Stucchi, M.; Chery, E.; Miller, A.; Beyer, G.; Van der Plas, G.; Walsby, E.; Roberts, K. Extreme wafer thinning and nano-TSV processing for 3D heterogeneous integration. In *Proceedings of the 2020 IEEE 70th Electronic Components and Technology Conference (ECTC), Orlando, FL, USA, 3–30 June 2020*; IEEE: Piscataway, NJ, USA, 2020; pp. 42–48.
203. Esashi, M. Wafer level packaging of MEMS. *J. Micromech. Microeng.* **2008**, *18*, 073001. [CrossRef]

 *micromachines*

Article

Design and Implementation of a SiC-Based VRFB Power Conditioning System

Chao-Tsung Ma * and Yi-Hung Tian

Department of Electrical Engineering, CEECS, National United University, Miaoli 36063, Taiwan; jakk226601735@yahoo.com.tw
* Correspondence: ctma@nuu.edu.tw; Tel.: +886-37-382-482; Fax: +886-37-382-488

Received: 21 November 2020; Accepted: 11 December 2020; Published: 12 December 2020

Abstract: An energy storage system using secondary batteries combined with advanced power control schemes is considered the key technology for the sustainable development of renewable energy-based power generation and smart micro-grids. The performance of energy storage systems in practical application mainly depends on their power conditioning systems. This paper proposes a silicon carbide-based multifunctional power conditioning system for the vanadium redox flow battery. The proposed system is a two-stage circuit topology, including a three-phase grid-tie inverter that can perform four-quadrant control of active and reactive power and a bi-directional multi-channel direct current converter that is responsible for the fast charging and discharging control of the battery. To achieve the design objectives, i.e., high reliability, high efficiency, and high operational flexibility, silicon carbide-based switching devices, and advanced digital control schemes are used in the construction of a power conditioning system for the vanadium redox flow battery. This paper first describes the proposed system topologies and controller configurations and the design methods of controllers for each converter in detail, and then results from both simulation analyses and experimental tests on a 5 kVA hardware prototype are presented to verify the feasibility and effectiveness of the proposed system and the designed controllers.

Keywords: energy storage system; power conditioning system; silicon carbide; vanadium redox flow batteries

1. Introduction

In recent years, the development of secure, low-carbon, and renewable energy sources and various smart micro-grid systems [1], power converter-based system compensating devices [2,3], advanced power converters using state-of-the-art wide-bandgap (WBG) switching devices, and digital-integrated intelligent control schemes [4,5] have become very popular research topics in the field of electric power and energy engineering. To best facilitate the above-mentioned technologies, various types of power converters are normally required [6–11], whose main components are semiconductor power switches and various system control units. To further enhance and optimize the performance of power converters, advanced semiconductor switches based on WBG materials, also known as third-generation semiconductor materials, such as gallium nitride (GaN) and silicon carbide (SiC), are emerging as very promising solutions [12,13]. It is well known that WBG materials offer superior characteristics over silicon in terms of band gap, electron mobility, electric breakdown field, saturated electron velocity, and thermal conductivity, which make WBG devices much desired for switching applications with high-voltage, -power, -temperature, and -frequency requirements. In particular, SiC devices with high-frequency switching capability and superior thermal conductivity are suitable for high-voltage and -power applications, while GaN has the highest bandgap, electron mobility, electric breakdown field, and saturated electron velocity, normally used in low- to mid-power systems [14]. In [15], the performance of a digitally controlled 2 kVA three-phase shunt-active power filter using GaN high electron mobility transistors was demonstrated for the first time. In addition, since decarbonization and green energy are two of the modern trends, renewable energy-based distributed power generation and on-line energy management systems have been intensively researched over the past decade. How to improve the performance and optimize the application of grid-level energy storage systems (ESSs) has also become one of the necessary technologies to promote the sustainable development of renewable power generation, active power distribution systems, and micro-grids. At present, practical electric energy storage technologies include pumped hydro system, compressed air energy storage, battery energy storage systems (BESS), flow battery, superconducting magnetic energy storage, flywheel, supercapacitor, etc. [16,17]. Among the above-mentioned grid-level ESSs, the vanadium redox flow battery (VRFB) has the advantages of independent and flexible design of output power and energy storage capacity, high energy conversion efficiency, safety, and low maintenance costs, which make it very suitable for a wide range of applications, such as distributed power generation optimization, energy management and integrated power quality control technology related applications [18,19]. In general, the performance of ESSs in practical application mainly depends on their power conditioning systems (PCS). The PCS topology required by the general grid-connected BESS can be divided into two categories: single-stage [20–23] and two-stage [24–27] according to the circuit architecture. The single-stage system is more suitable for high-voltage, high-capacity battery packs, while the two-stage circuit architecture usually includes a single-phase or three-phase direct current to alternative current (DC/AC) converter and a bi-directional direct current to direct current (DC/DC) power converter for matching with a wider range of battery pack voltage specifications, and enabling the realization of different charging and discharging strategies. In fact, various battery-based ESSs have been developed for a long time; however, most BESSs reported in the literature are based on some specific operation and control functions required by the system concerned and the system operating functions in this kind of BESS are quite limited and cannot be universal leading a very high system cost, long payback period, and serious lack of application flexibility. To improve the above-mentioned shortcomings and to achieve an advanced and versatile ESS, this paper proposes a SiC-based multifunctional PCS for the VRFB.

2. The VRFB System and the Proposed PCS Topology

The system architecture of a VRFB is shown in Figure 1. The two electrolytes, positive (V^{4+}/V^{5+}) and negative (V^{2+}/V^{3+}) electrolytes, in a VRFB are stored in different electrolyte storage tanks. During charging or discharging, the two electrolytes are separated by an isolation membrane, but selected ions are allowed to pass through the membrane forming a current path. The concentration and amount of electrolyte determine the system capacity of VRFB, the design specifications of electrodes determine the rated power of VRFB, and the number of single cells in series in the battery stack determines the maximum working voltage of VRFB. It is important to note that to achieve a cost-effective and high-efficiency design, the number of cells in series cannot be too high. This has resulted in a preferable lower system voltage. Considering this condition, a two-stage circuit topology is proposed for the VRFB PCS in this paper. In operation, both the DC/AC power converter and the interleaved multi-channel DC/DC converter are activated at the same time according to the due operating mode and system conditions. The detailed circuit architecture of the VRFB PCS proposed in this paper is shown in Figure 2, where the main function of the interleaved buck-boost converter, consisting of six SiC power semiconductor switches and inductors L_b, is fast charging/discharging current command tracking. L_b is used to filter out ripple components in the current caused by the switching of the semiconductor switch. As can be seen in Figure 2, the architecture consists of three parallel synchronous buck-boost converters, where the output switching signal of each converter is 120 degrees apart from another, offsetting each other's ripples and reducing total output ripple. The left side of Figure 2 shows the grid-tied 3-phase inverter, whose main functions are DC bus voltage regulation via active power balancing control and system reactive power compensation via bi-directional reactive power tracking control. To provide a clear picture of the above-mentioned control functions, Figure 3 shows the possible active and reactive power flows in the proposed VRFB PCS.

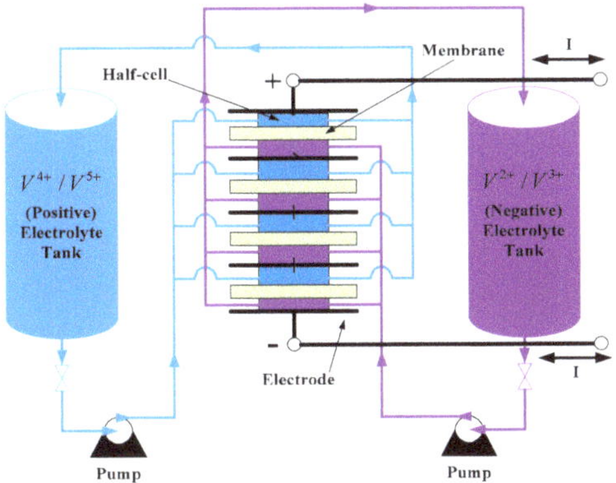

Figure 1. The system architecture of a vanadium redox flow battery (VRFB) [19].

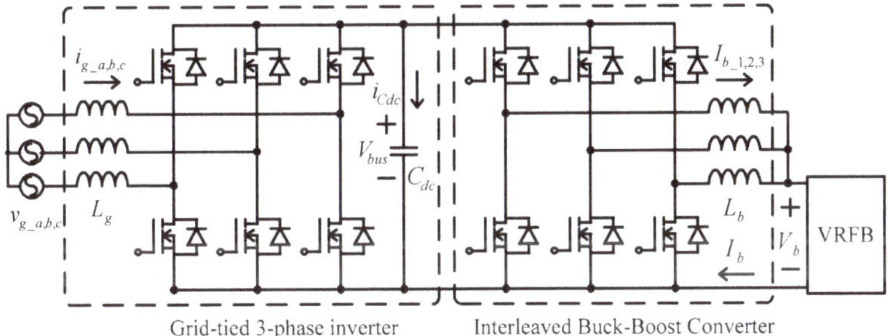

Figure 2. VRFB power conditioning system (PCS) circuit topology.

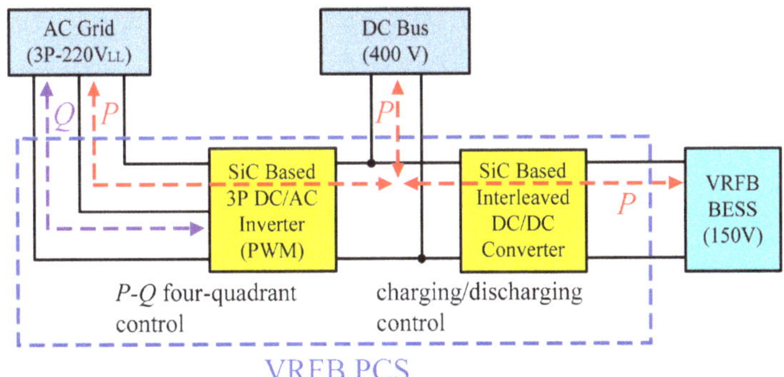

Figure 3. Possible active and reactive power flows in the proposed VRFB PCS.

3. Controller Design of VRFB PCS

The relevant system parameters and hardware specifications of the proposed VRFB PCS are shown in Table 1. Following in this section, the required mathematical model derivation and controller design will be carried out according to the specifications given in Table 1.

Table 1. Specifications of the proposed system.

Component	Item	Value
Grid	Three-phase line voltage	220 Vrms, V_{LL}
	Voltage frequency	60 Hz
Interleaved buck-boost converter	Rated power	5 kW
	Number of channels	3
	VRFB pack voltage	136–153.6 V (48 cells)
	Switching frequency	100 kHz
	Switching device	SiC MOSFET
	Filter inductor	383 μH (20%)
	Current sensing factor	0.05 V/A
Grid-tie inverter	Rated power	5 kVA
	DC bus voltage	400 V
	Switching frequency	100 kHz
	Carrier voltage	5 V
	LPF	1st order (270 μH)
	DC bus capacitor	600 V/1620 μF
	DC voltage sensing factor	0.006 V/V
	AC voltage sensing factor	0.0031 V/V
	Current sensing factor	0.05 V/A
Controller	DSP	TI TMS320F28335

3.1. Grid-Tie Inverter Modeling and Design of Controllers

To achieve a reliable control scheme, the grid-tie inverter adopts a dual-loop control architecture, where the inner loop controls inductor currents, and the outer loop controls DC bus voltage and AC-side reactive power. The overall control architecture is shown in Figure 4.

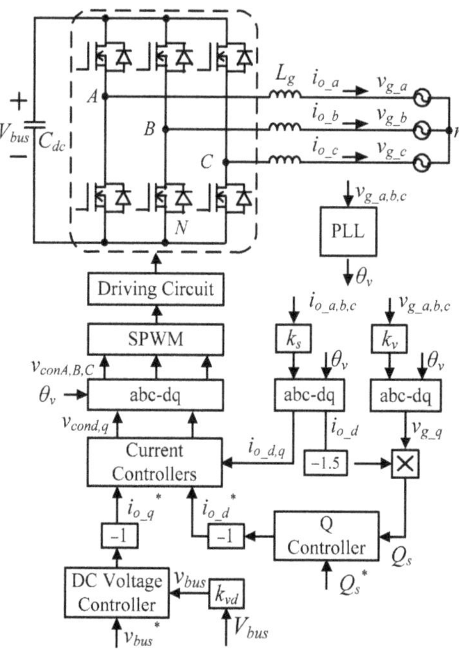

Figure 4. Overall control architecture of grid-tie inverter (* indicates commands).

3.1.1. Design of Inductor Current Controllers

The mathematical model of inverter's inductor current in synchronous reference frame can be derived according to Figure 3:

$$
\begin{bmatrix} L_g\frac{dI_{o\_d}}{dt} \\ L_g\frac{dI_{o\_q}}{dt} \\ L_g\frac{dI_{o\_0}}{dt} \end{bmatrix} = K_{pwm}\begin{bmatrix} 1 & 0 & 0 \\ 0 & 1 & 0 \\ 0 & 0 & 1 \end{bmatrix}\begin{bmatrix} v_{cond} \\ v_{conq} \\ v_{con0} \end{bmatrix} - \begin{bmatrix} 1 & 0 & 0 \\ 0 & 1 & 0 \\ 0 & 0 & 1 \end{bmatrix}\begin{bmatrix} V_{g\_d} \\ V_{g\_q} \\ V_{g\_0} \end{bmatrix} - \begin{bmatrix} 0 & \omega L_g & 0 \\ -\omega L_g & 0 & 0 \\ 0 & 0 & 0 \end{bmatrix}\begin{bmatrix} I_{o\_d} \\ I_{o\_q} \\ I_{o\_0} \end{bmatrix}
\tag{1}
$$

In this paper, the Type II controller is used to control the inductor currents. Using (1) and the mathematical form of Type II controller, dq-axis inner inductor current control loops can be obtained, as shown in Figure 5.

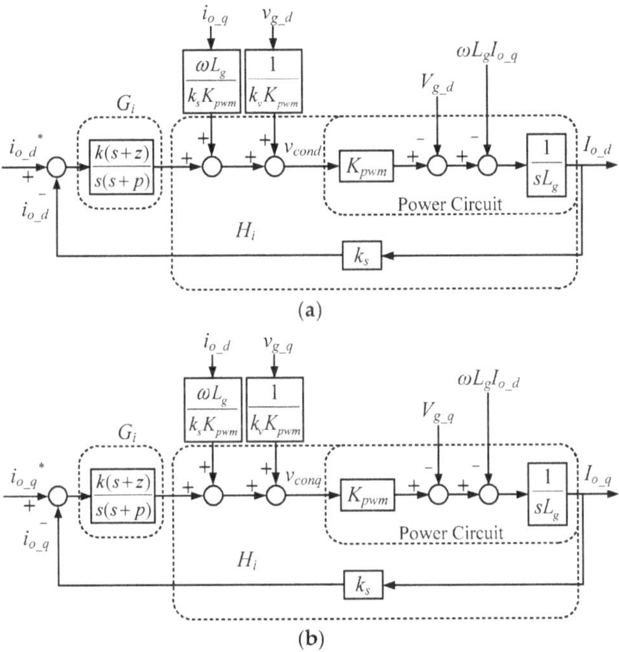

Figure 5. Inner inductor current control loops (* indicates commands): (**a**) d-axis; (**b**) q-axis.

The quantification design of the inner loop inductor current controller is as follows: choosing the crossover frequency ω_i = 41,888 rad/s; zero = 8377.5 rad/s and pole = 136,282.2 rad/s, yielding the required Type II controller as follows:

$$
G_i(s) = \frac{3.953 \times 10^5(s + 8377.5)}{s(s + 1.363 \times 10^5)}
\tag{2}
$$

Figure 6 shows the Bode plot of inner inductor current control loop. The phase margin is 62°.

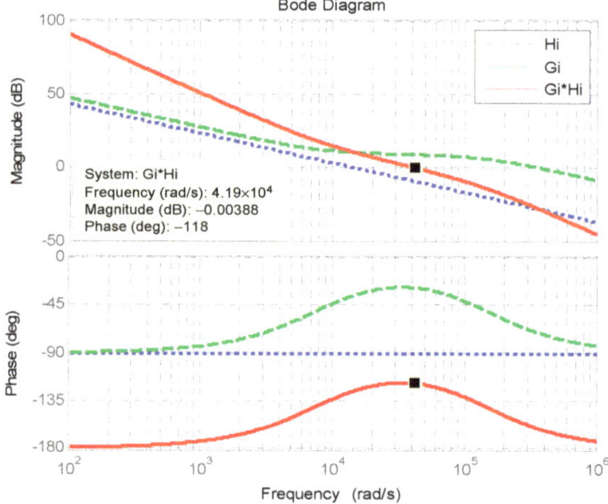

Figure 6. Bode plot of inner inductor current control loop.

3.1.2. Design of the DC Bus Voltage Controller

Considering the steady-state operating point, the equivalent circuit of the DC bus voltage loop is as shown in Figure 7.

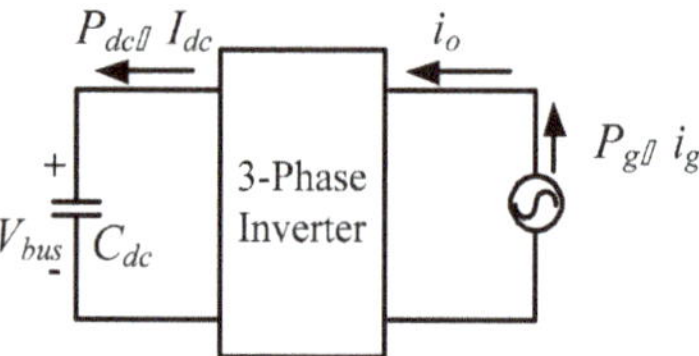

Figure 7. Equivalent circuit of direct current (DC) bus voltage loop.

The mathematical model of DC bus voltage can be derived according to Figure 7:

$$\frac{V_{bus}}{I_{o\_q}} = \frac{-k_{dc}}{sC_{dc}}, k_{dc} = 1.5\frac{V_{g\_q}}{V_{bus}} \tag{3}$$

In this paper, the Type II controller is used to control the DC bus voltage, and thus outer DC bus voltage control loop can be obtained, as shown in Figure 8.

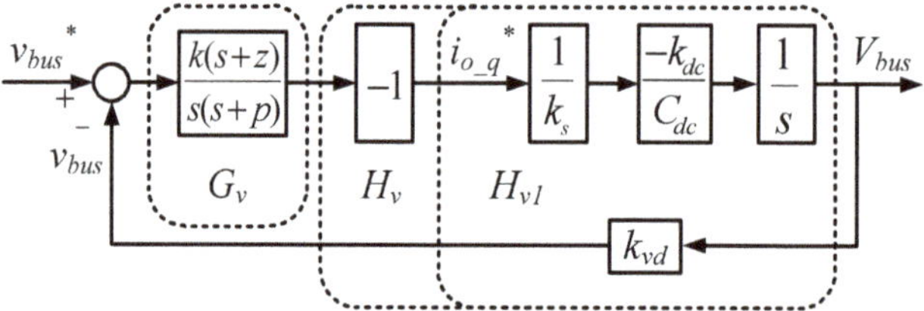

Figure 8. Outer DC bus voltage control loop (* indicates commands).

The quantification design of the DC bus voltage controller is as follows: choosing the crossover frequency ω_v = 5235.9877 rad/s; zero = 523.598 rad/s; pole = 68,141.144 rad/s, yielding the required Type II controller as follows:

$$G_v(s) = \frac{1.071 \times 10^7 (s + 523.3426)}{s(s + 6.814 \times 10^4)} \tag{4}$$

Figure 9 shows the Bode plot of outer DC bus voltage control loop. The phase margin is 80°.

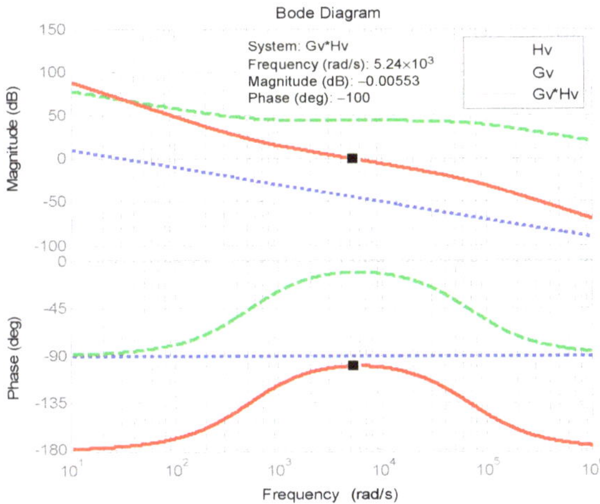

Figure 9. Bode plot of outer DC bus voltage control loop.

3.1.3. Design of Reactive Power Controller

The derivation of the AC-side reactive power controller in this paper takes the grid-side current flowing into the converter as positive:

$$Q_g = -1.5 \times V_{g\_q} \times I_{o\_d} \tag{5}$$

In this case, the Type II controller is used to control the reactive power, and thus outer reactive power control loop can be obtained, as shown in Figure 10.

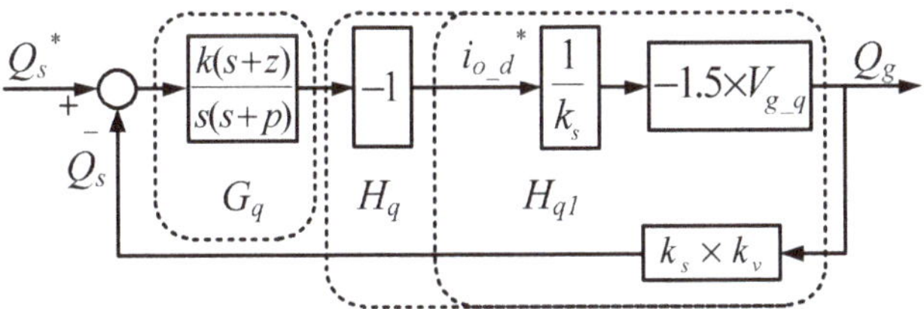

Figure 10. Outer reactive power control loop (* indicates commands).

The quantification design of the reactive power controller is as follows: choosing the crossover frequency ω_q = 3490.6585 rad/s; zero = 1745.329 rad/s; pole = 6981.317 rad/s, yielding the required Type II controller as follows:

$$G_v(s) = \frac{8359(s + 1745.4241)}{s(s + 6981)} \tag{6}$$

Figure 11 shows the Bode plot of outer reactive power control loop. The phase margin is 127°.

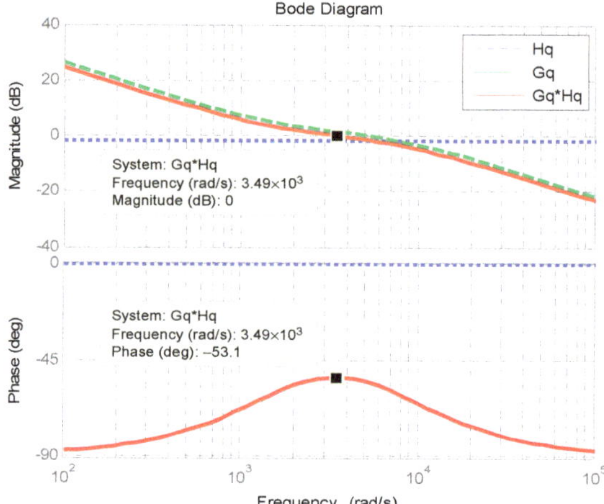

Figure 11. Bode plot of outer reactive power control loop.

3.2. Interleaved Buck-Boost Converter Controllers

The interleaved buck-boost converter adopts a single-loop inductor current controller, and each channel is individually controlled and uses a different phase shift angle. The overall control architecture is shown in Figure 12.

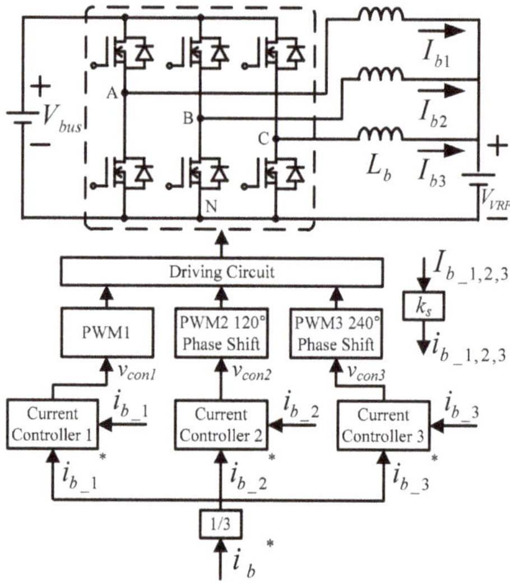

Figure 12. Overall control architecture of interleaved buck-boost converter (* indicates commands).

Design of Inductor Current Controller

The proposed interleaved buck-boost converter is composed of multiple buck-boost converters, and its operating principle is the same as that of a single buck-boost converter. Therefore, only the controller design of a single buck-boost converter is illustrated. Taking leg A as an example, the mathematical model of the inductor current is as follows:

$$L_{b1}\frac{di_{Lb1}}{dt} = v_{con1}K_{pwm} - V_b, \; K_{pwm} = \frac{V_{bus}}{v_{tri}} \tag{7}$$

In this control case, the Type II controller is again used to control the inductor current, and thus inductor current control loop can be obtained, as shown in Figure 13.

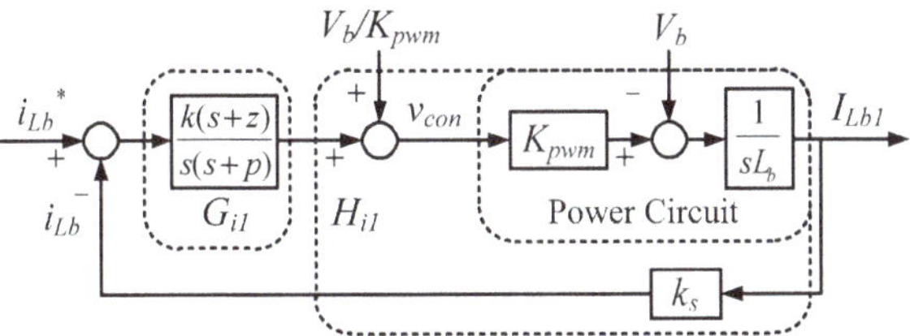

Figure 13. Buck-boost converter inductor current control loop (* indicates command).

According to Figure 13, the transfer function of inductor current loop is as follows:

$$H_{i1}(s) = K_{pwm} \times \frac{1}{sL_b} \times k_s \tag{8}$$

The quantification design of the inductor current controller is as follows: choosing the crossover frequency ω_i = 39,270 rad/s; zero = 5167.1 rad/s; pole = 298,280 rad/s, yielding the required Type II controller as follows:

$$G_i(s) = \frac{1.122 \times 10^6(s + 5167.1)}{s(s + 2.9828 \times 10^5)} \tag{9}$$

Figure 14 shows the Bode plot of the designed inductor current control loop. The phase margin is 75°.

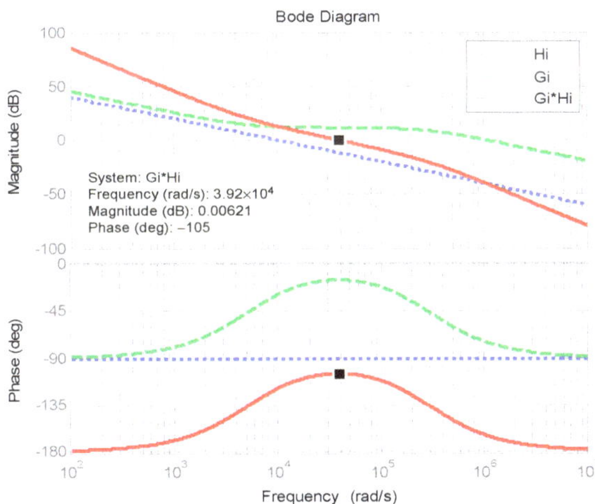

Figure 14. Bode plot of buck-boost converter inductor current control loop.

4. Cases Simulation

4.1. Ramp-Up Procedure

To verify the correctness of the designed PCS controllers presented in the previous section, a software model of the proposed VRFB PCS is developed with power simulation software as shown in Figure 15. Two typical simulation cases, the ramp-up procedure and charging/discharging with four-quadrant P-Q control of the grid-tie inverter, are carried out in this study. Figure 16 shows the result of simulating ramp-up procedure of the system. This is to verify that the designed PCS can securely establish the required DC bus voltage of 400 V. As can be seen in Figure 16, after the grid-tie converter confirms the status of synchronization with the grid, the circuit starts to charge the DC bus capacitor slowly, and the rated DC bus voltage of PCS is boosted from 360 V and finally controlled at the target value of 400 V to complete the preparation of the system for various functional operations.

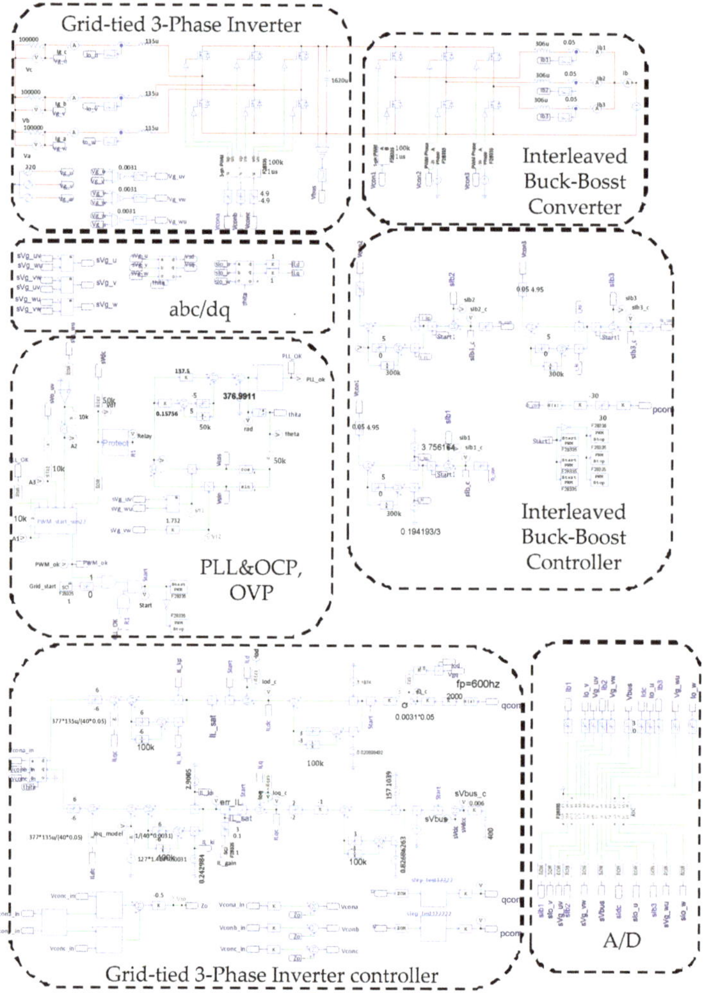

Figure 15. The power simulation software model of the proposed VRFB PCS.

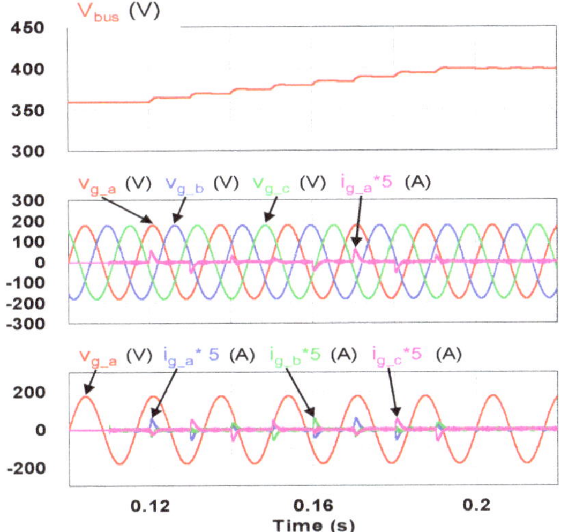

Figure 16. Waveforms of DC bus voltage (**top**), grid three-phase voltages and a-phase current (**mid**), grid a-phase voltage and three-phase currents (**bottom**).

4.2. Charging/Discharging with P-Q Four-Quadrant Control of the Grid-Tie Inverter

This case verifies simultaneous operation of the charging/discharging of VRFB and the function of reactive power regulation. In this operation mode, the charging/discharging current of the VRFB respectively corresponds to the positive and negative active power of the grid-tied inverter. With the independent control function of positive and negative reactive power regulation, a four-quadrant P-Q control is achieved by the grid-tied inverter. In this simulation case, the battery voltage = 150 V, a charging and discharging current command of ±30 A (equivalent to ±4.5 kW) and a ±2 kVAR reactive power command is arranged. Figure 17 shows the schematic diagram of PCS operating in the 1st and 3rd quadrants. Figures 18–22 show a set of complete simulation results. As shown in Figure 18a, the three interleaved inductor currents are regulated evenly while the DC bus voltage is stably controlled at its rated value of 400 V. It can be clearly seen from Figures 19 and 20, with the proposed direct current control scheme, the cross interference between active and reactive power of the grid-tied inverter is negligible. Figures 21 and 22 show the tracking performance of the designed reactive power, charging and discharging controllers.

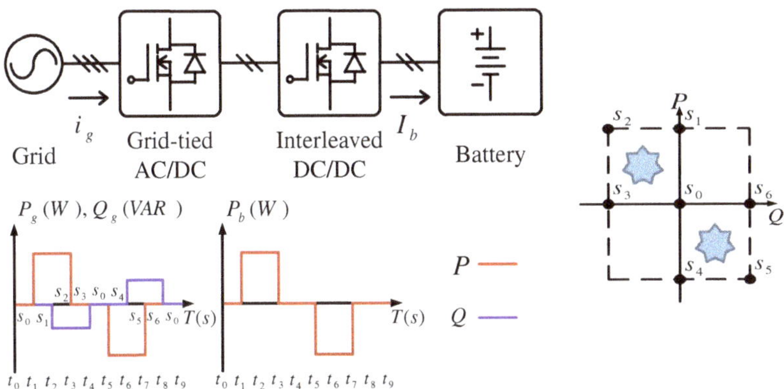

Figure 17. The schematic diagram of PCS operating in the 1st and 3rd quadrants.

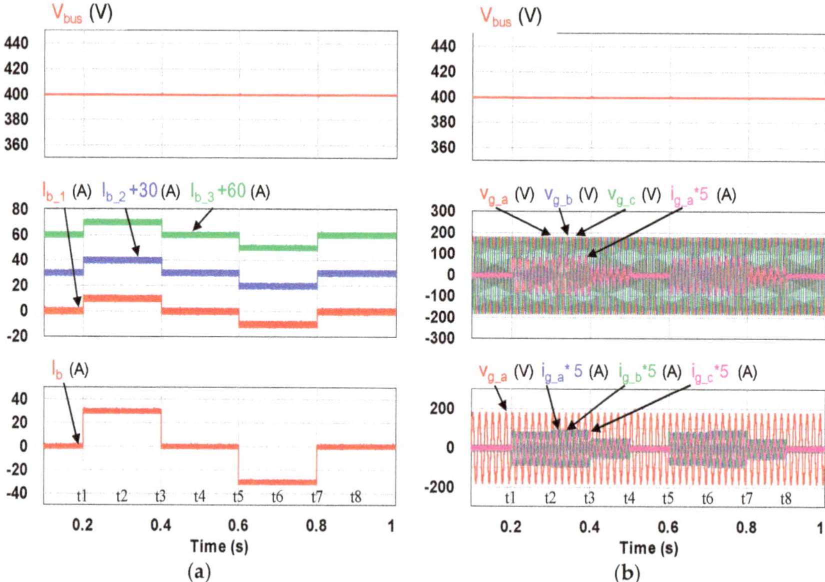

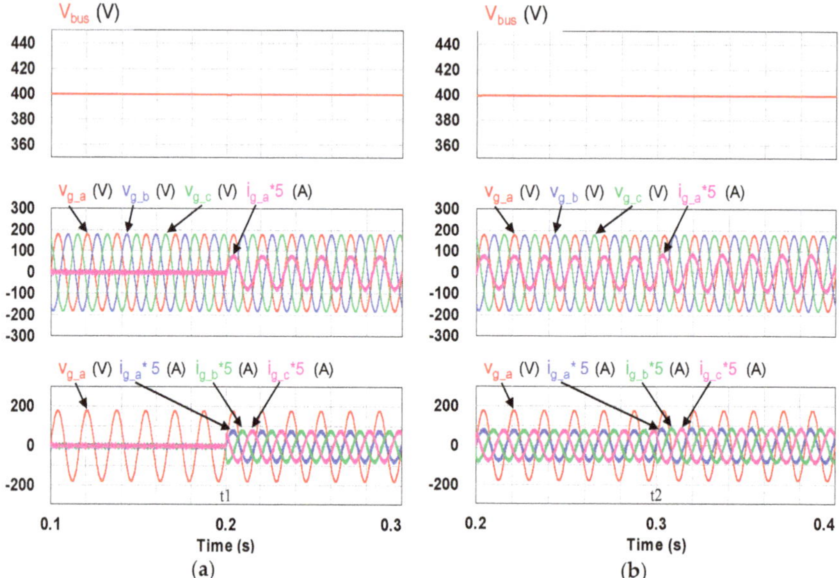

Figure 18. Waveforms of (**a**) interleaved buck-boost converter: DC bus voltage (**top**), interleaved inductor currents (**mid**), battery current (**bottom**) and (**b**) grid-tied inverter: DC bus voltage (**top**), grid three-phase voltages and a-phase current (**mid**), grid a-phase voltage and three-phase currents (**bottom**).

Figure 19. The detailed view of Figure 18b: (**a**) near t1; (**b**) near t2.

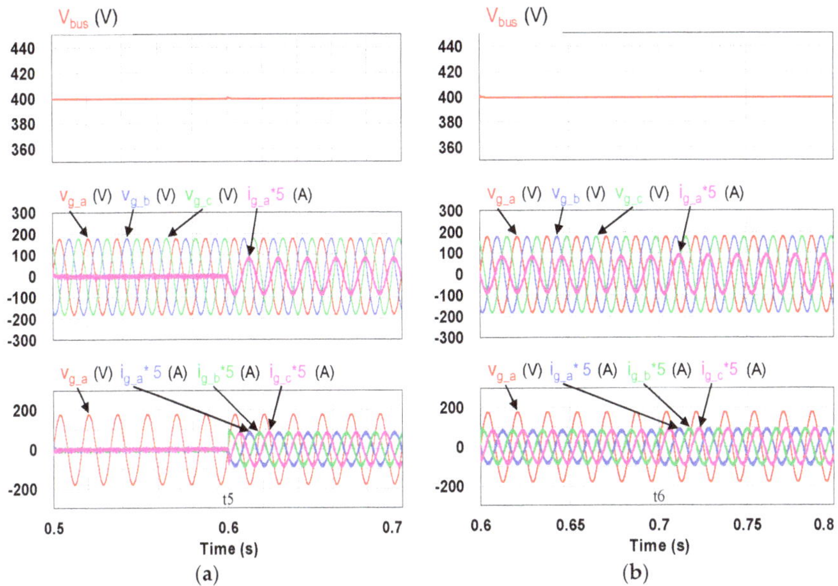

Figure 20. The detailed view of Figure 18b: (**a**) near t5; (**b**) near t6.

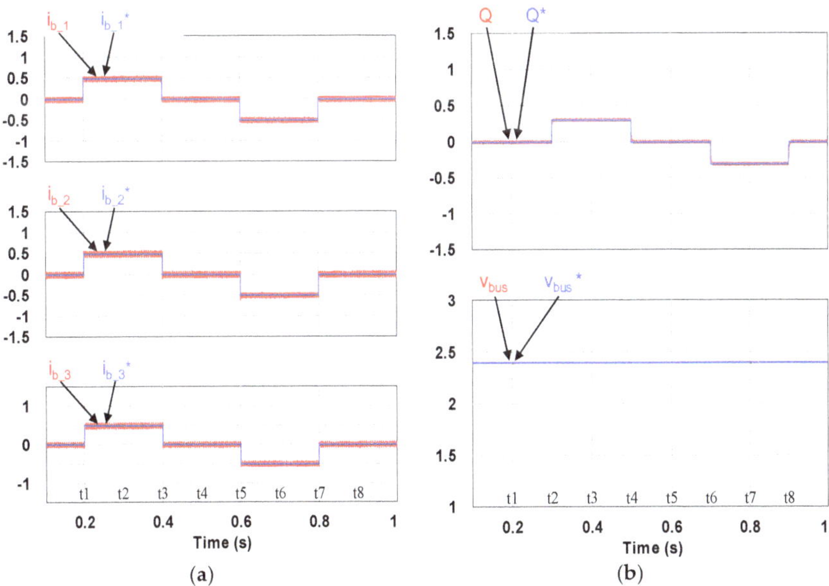

Figure 21. Waveforms of control commands and feedbacks: (**a**) three interleaved inductor currents; (**b**) reactive power (top), DC bus voltage (bottom).

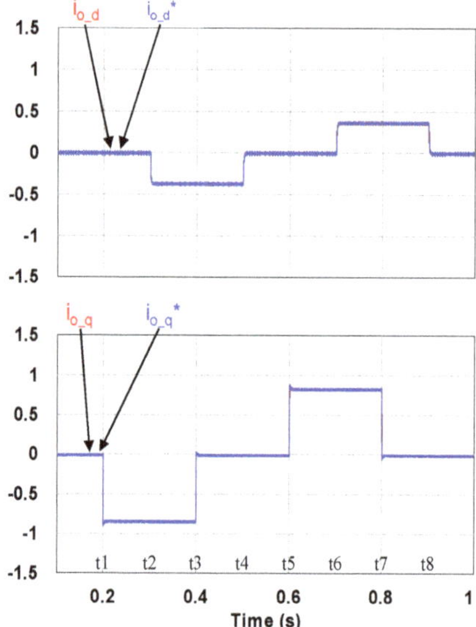

Figure 22. Waveforms of control commands and feedbacks: qd-axis inductor currents.

With the same operating condition as described previously, Figure 23 depicts a schematic diagram of PCS operating in the 2nd and 4th P-Q quadrants. Figures 24–28 show a set of complete simulation results. In this case, with the same charging/discharging command, the current waveforms of interleaved buck-boost converter are identical to those shown in Figure 18a, so they are not shown in this case. As shown in Figure 24, during the charging and discharging operation of the battery the DC bus voltage is stably controlled at its rated value of 400 V with the designed voltage controller. It can be clearly seen from Figures 25 and 26, with the proposed control scheme, the cross interference between active and reactive power of the grid-tied inverter is negligible. Figures 27 and 28 verify the tracking performance of the designed reactive power, charging and discharging controllers working at different operating points.

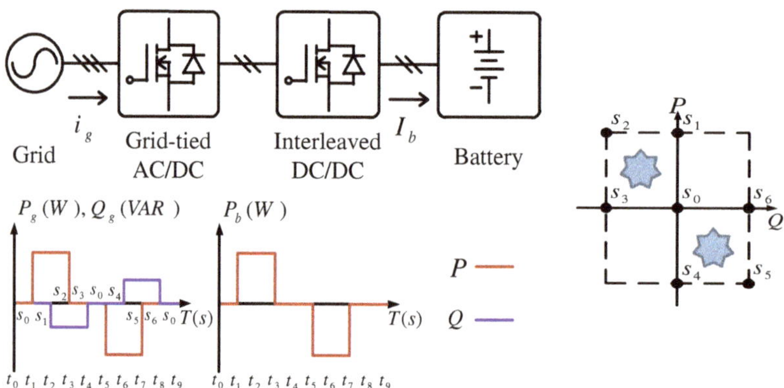

Figure 23. PCS operating in the 2nd and 4th quadrants.

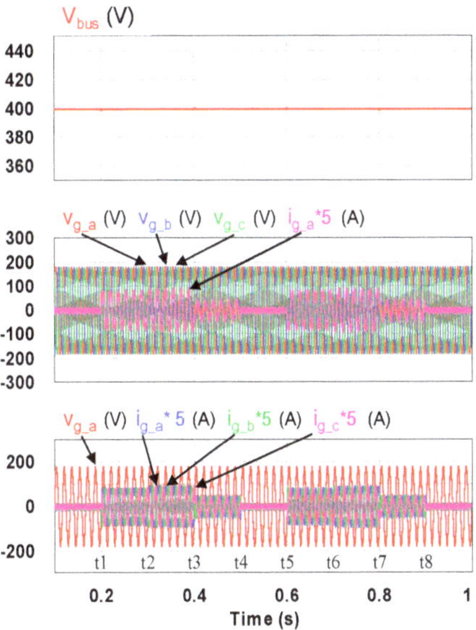

Figure 24. Waveforms of DC bus voltage (**top**), grid three-phase voltages and a-phase current (**mid**), grid a-phase voltage and three-phase currents (**bottom**).

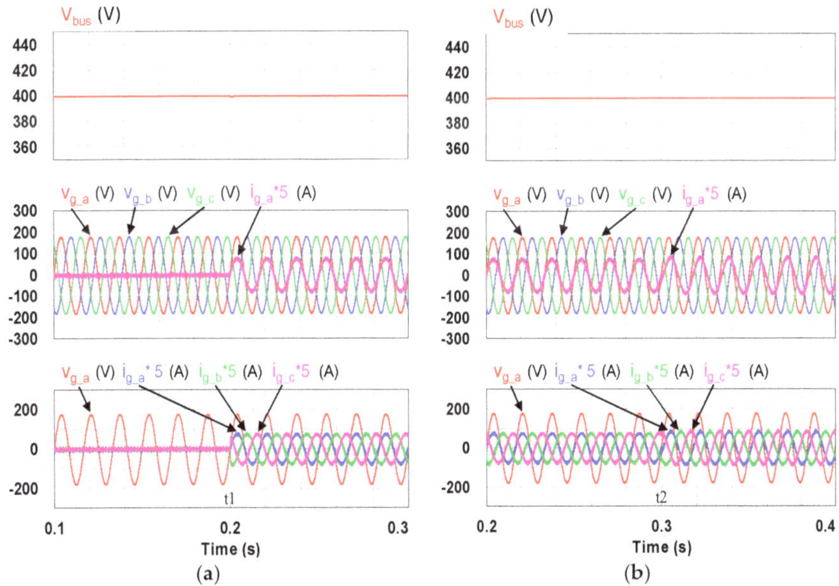

Figure 25. The detailed view of Figure 23: (**a**) near t1; (**b**) near t2.

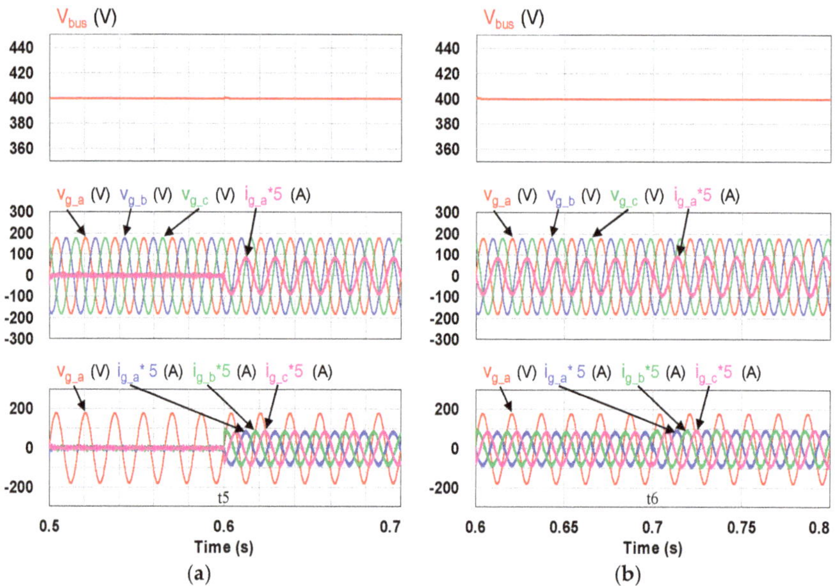

Figure 26. The detailed view of Figure 23: (**a**) near t5; (**b**) near t6.

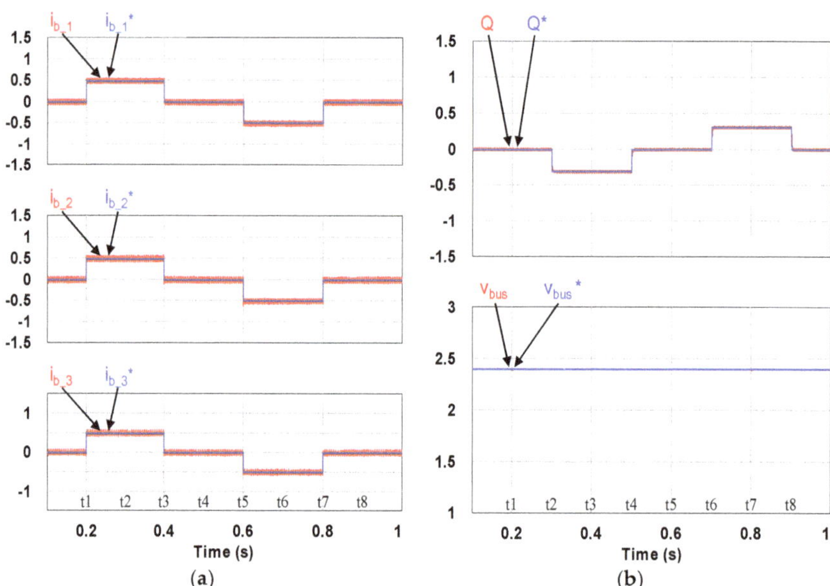

Figure 27. Waveforms of control commands and feedbacks: (**a**) three interleaved inductor currents; (**b**) reactive power (**top**), DC bus voltage (**bottom**).

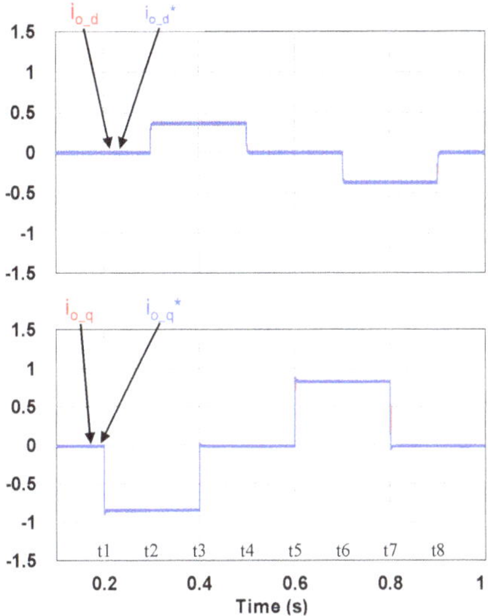

Figure 28. Waveforms of control commands and feedbacks: qd-axis inductor currents.

5. Hardware Implementation and Test Results

To further verify the performance of the proposed VRFB PCS, a 5 kVA hardware experimental platform using SiC MOSFET is built according to the system specifications listed in Table 1 and the operating scenarios of the test cases are identical to that used in the simulated cases presented in the previous section. Figure 29 shows a photo of the constructed SiC-based VRFB PCS hardware system and the experimental platform, including (1) auxiliary power, (2) oscilloscope, (3) SiC-based grid-tie three-phase inverter, (4) SiC-based interleaved DC-DC buck-boost converter, (5) current probe, and (6) voltage probes. Figure 30 shows the test result of ramp-up procedure of the PCS hardware system. Figures 31–34 show a set of experimental results of the proposed PCS operating in the 1st and 3rd quadrants. As can be seen in Figures 31–34, the measured waveforms are very close to those obtained from simulation studies presented in the previous section. This has verified the feasibility and effectiveness of the proposed control schemes.

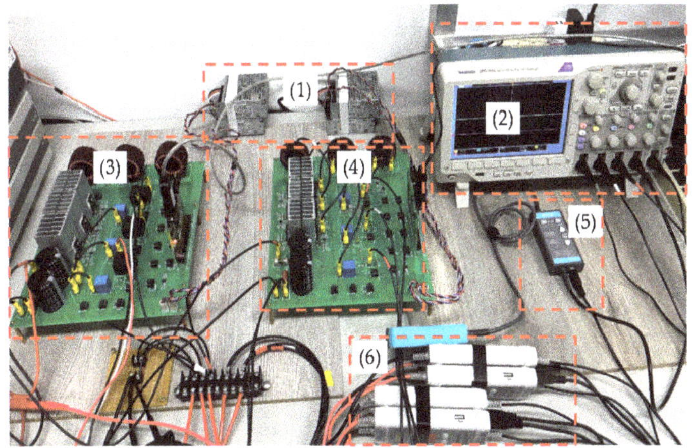

Figure 29. Photo of the constructed VRFB PCS hardware and the experimental platform.

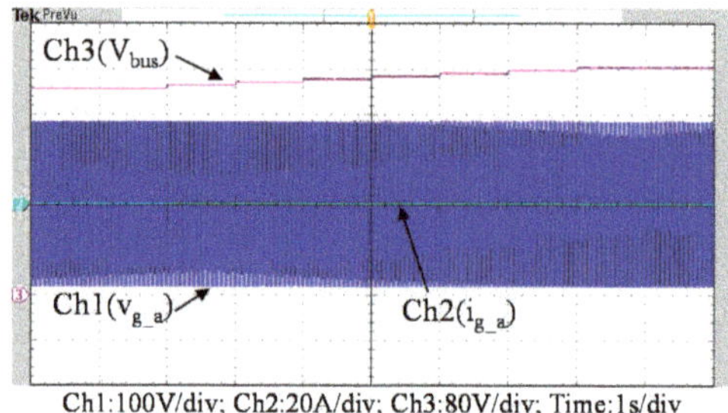

Figure 30. Waveforms of DC bus voltage (V_{bus}), phase-a grid voltage (V_{g_a}) and a-phase grid current (i_{g_a}).

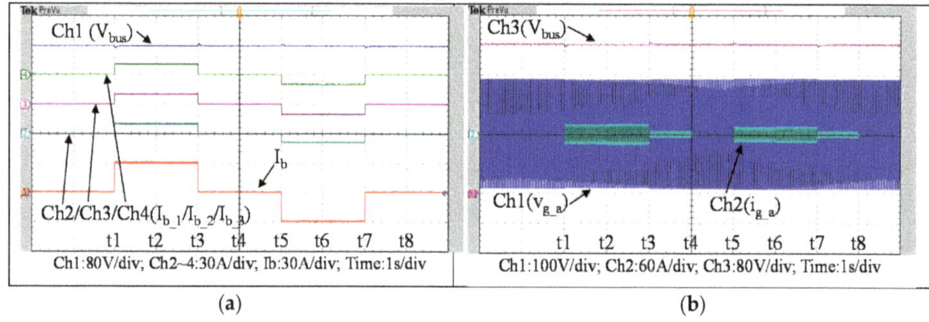

(**a**)   (**b**)

Figure 31. Waveforms of (**a**) interleaved buck-boost converter: DC bus voltage, interleaved inductor currents, and battery current, (**b**) DC bus voltage and the grid phase-a voltage and current.

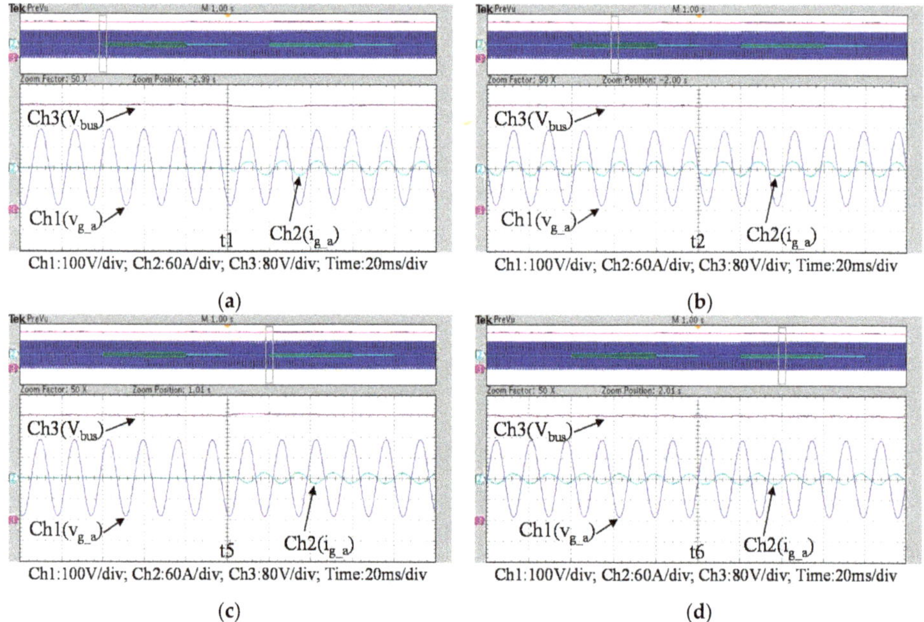

(**a**)   (**b**)

(**c**)   (**d**)

Figure 32. Detailed view of Figure 31b: (**a**) near t1; (**b**) near t2; (**c**) near t5; (**d**) near t6.

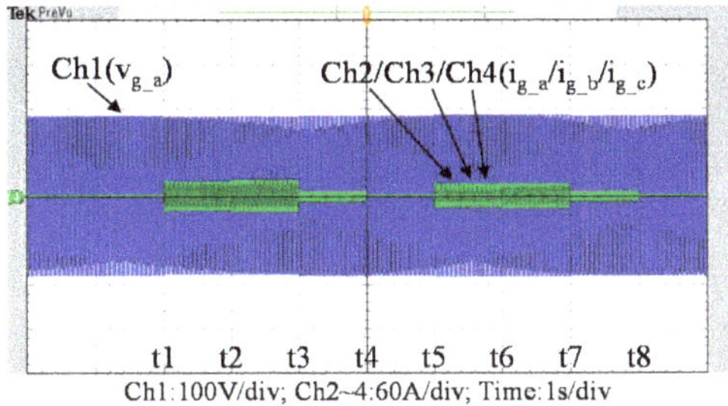

Figure 33. Waveforms of grid phase-a voltage and three-phase currents.

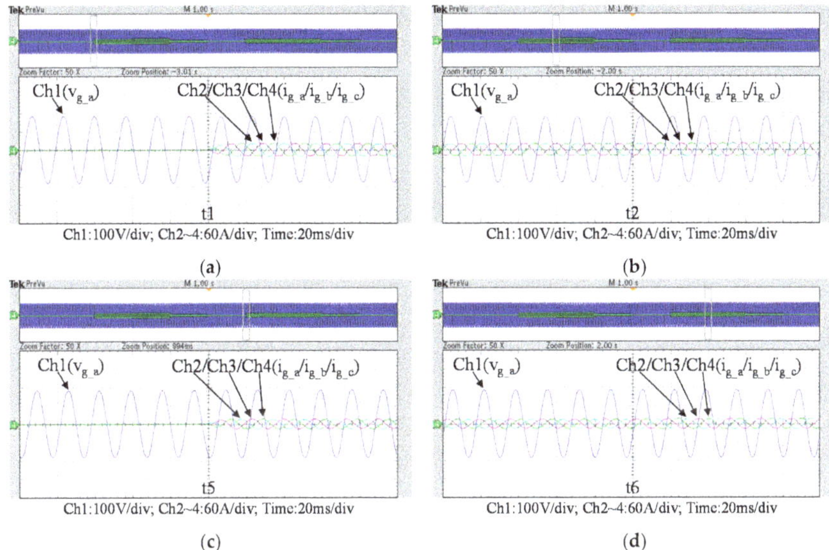

Figure 34. Detailed view of Figure 33: (**a**) near t1; (**b**) near t2; (**c**) near t5; (**d**) near t6.

To fully verify the performance of the proposed SiC-based hardware system and control scheme, Figure 35 show a second set of experimental results, in which the proposed PCS is operating in the 2nd and 4th quadrants. As can be seen in Figures 35–37, satisfactory performances of the proposed PCS grid-tied inverter and the interleaved buck-boost converter are achieved.

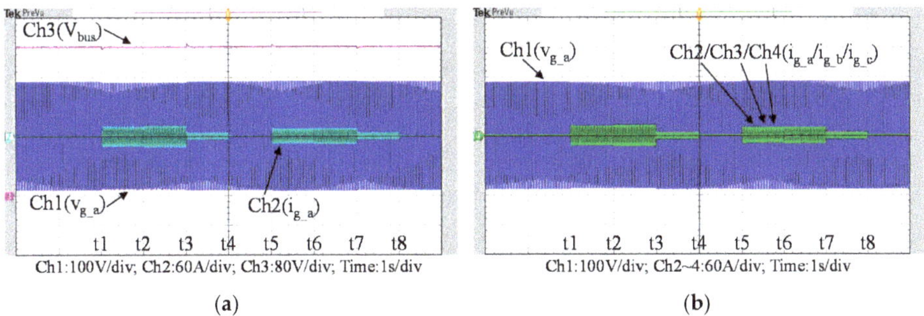

Figure 35. Waveforms of (**a**) DC bus and grid phase-a voltage and current; (**b**) three-phase currents.

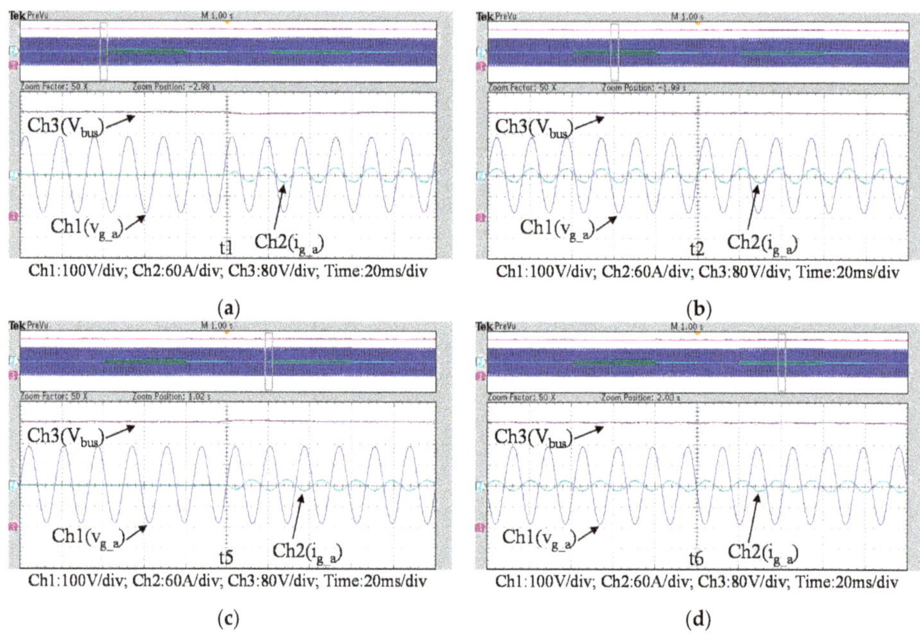

Figure 36. Detailed view of Figure 35a: (a) near t1; (b) near t2; (c) near t5; (d) near t6.

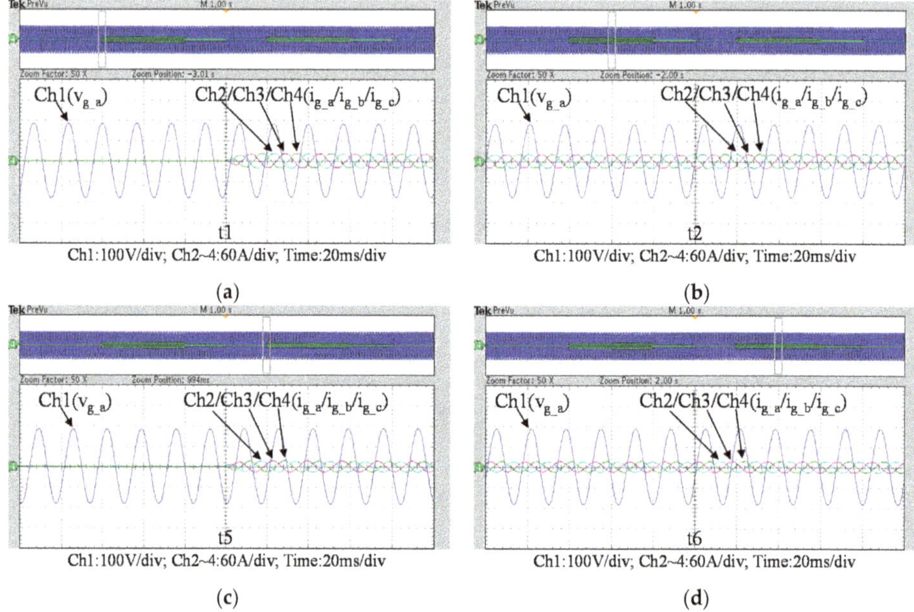

Figure 37. Detailed view of Figure 35b: (a) near t1; (b) near t2; (c) near t5; (d) near t6.

The efficiency test results of the proposed 5 kVA, SiC-based PCS's grid-connected three-phase inverter and the 5 kW, 3-channel interleaved buck-boost converter are shown in Figure 38a,b respectively. The highest efficiency of the grid-tied three-phase inverter and 3-channel interleaved buck-boost converter system is measured as 94.1% at 80% system rated power and 96.3% at 60% system rated power, respectively.

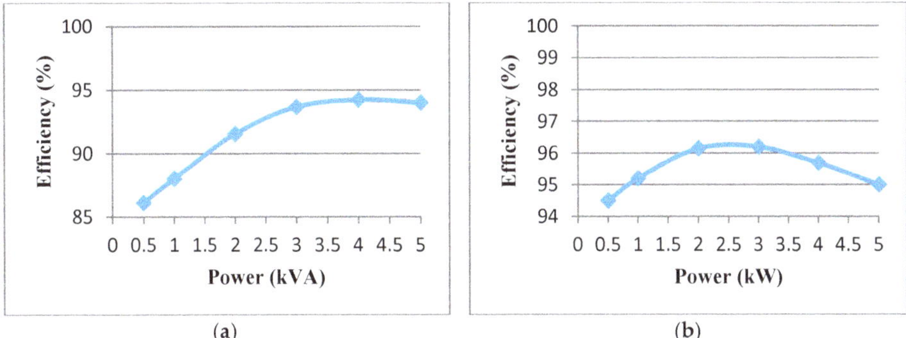

Figure 38. The efficiency analysis of the PCS: (**a**) grid-tied inverter, (**b**) interleaved buck-boost converter.

6. Conclusions

It has been well accepted that the economic benefits of distributed power generation and micro-grids are multifaceted. For power users, the economic benefits lie in efficient use of energy, environmental protection, and reliable customized electrical energy services, while optimizing resource allocation and providing highly efficient energy management with operational flexibility are the main factors for achieving the economic benefits of micro-grids. However, with the addition of renewable power generations and various types of micro-grids in the power systems the complexity in system control and operation is significantly increased and certain compensating devices, e.g., ESSs integrated with advanced PCSs are urgently needed to be proposed and verified for feasibility. In this regard, this paper has proposed a SiC-based multifunctional PCS for the VRFB. In this study, it has been found that SiC switching devices with their excellent thermal and voltage capability can meet the requirement of a cost-effective design of grid-tied inverter system, in which the pulse width modulation technique can be used to reduce the hardware cost of PCS while improving system reliability. In this paper, the consideration of circuit topology and the detailed design steps of related controllers of the proposed SiC-based VRFB PCS have been fully addressed. The highest efficiency of the constructed SiC-based grid-tied three-phase inverter and 3-channel interleaved buck-boost converter system is measured to be 94.1% and 96.3% respectively. With the proposed PCS control schemes, four-quadrant control of active and reactive power and fast charging and discharging control of the VRFB have been achieved. Both simulation studies and experimental tests on a 5 kVA hardware prototype have verified the feasibility and overall performance of the proposed SiC-based VEFB PCS. It is worth noting that with the decoupled active and reactive power control capability and fast current command tracking feature the proposed VRFB PCS is expected to perform multiple system compensating functions, e.g., real-time support for renewable power generation, voltage and frequency support for micro-grids, and power quality improvement for power distribution systems.

Author Contributions: The corresponding author, C.-T.M. conducted the research work, proposed the VRFB PCS concept and design methods, verified the results, wrote the manuscript draft and polished the final manuscript. Y.-H.T. a postgraduate student in the department of EE, CEECS, National United University, Taiwan assisted hardware tests and organized results. All authors have read and agreed to the published version of the manuscript.

Funding: This research was funded by MOST, Taiwan, with the grant number, MOST 109-3116-F-006-020-CC1.

Acknowledgments: The author would like to thank the Ministry of Science and Technology (MOST) of Taiwan for financially support the energy-related research regarding key technologies in micro-grids and the design of advanced power and energy systems.

Conflicts of Interest: The authors declare no conflict of interest.

References

1. Energy Information Administration. *International Energy Outlook 2019*; Office of Energy Analysis: Washington, DC, USA, 2019.
2. She, X.; Wang, F.; Burgos, R.; Huang, A.Q. Solid state transformer interfaced wind energy system with integrated active power transfer, reactive power compensation and voltage conversion functions. In Proceedings of the IEEE Energy Conversion Congress and Exposition (ECCE) 2012, Raleigh, NC, USA, 15–20 September 2012; pp. 3140–3147.
3. Herrera, R.S.; Salmeron, P. Instantaneous Reactive Power Theory: A Reference in the Nonlinear Loads Compensation. *IEEE Trans. Ind. Electron.* **2009**, *56*, 2015–2022. [CrossRef]
4. Sheng, H.; Chen, Z.; Wang, F.; Millner, A. Investigation of 1.2 kV SiC MOSFET for high frequency high power applications. In Proceedings of the Twenty-Fifth Annual IEEE Applied Power Electronics Conference and Exposition (APEC), Palm Springs, CA, USA, 21–25 February 2010; pp. 1572–1577.
5. Jones, E.A.; Wang, F.F.; Costinett, D. Review of Commercial GaN Power Devices and GaN-Based Converter Design Challenges. *IEEE J. Emerg. Sel. Top. Power Electron.* **2016**, *4*, 707–719. [CrossRef]
6. Figueiredo, J.P.M.; Tofoli, F.L.; Silva, B.L.A. A review of single-phase PFC topologies based on the boost converter. In Proceedings of the 2010 9th IEEE/IAS International Conference on Industry Applications-INDUSCON 2010, Sao Paulo, Brazil, 8–10 November 2010; pp. 1–6.
7. Bellar, M.D.; Wu, T.S.; Tchamdjou, A.; Mahdavi, J.; Ehsani, M. A review of soft-switched DC-AC converters. *IEEE Trans. Ind. Appl.* **1998**, *34*, 847–860. [CrossRef]
8. Mohammad Noor, S.Z.; Omar, A.M.; Mahzan, N.N.; Ibrahim, I.R. A review of single-phase single stage inverter topologies for photovoltaic system. In Proceedings of the IEEE 4th Control and System Graduate Research Colloquium 2013, Shah Alam, Malaysia, 19–20 August 2013; pp. 69–74.
9. Aryal, A.; Hellany, A.; Nagrial, M.; Rizk, J. A Critical Review of Single Phase Inverter Grid Synchronization Topologies. In Proceedings of the 2016 International Conference on Power, Energy Engineering and Management (PEEM 2016), Bangkok, Thailand, 24–25 January 2016.
10. Mechouma, R.; Azoui, B.; Chaabane, M. Three-phase grid connected inverter for photovoltaic systems, a review. In Proceedings of the 2012 First International Conference on Renewable Energies and Vehicular Technology, Hammamet, Tunisia, 26–28 March 2012; pp. 37–42.
11. Lee, H.; Smet, V.; Tummala, R. A Review of SiC Power Module Packaging Technologies: Challenges, Advances, and Emerging Issues. *IEEE J. Emerg. Sel. Top. Power Electron.* **2020**, *8*, 239–255. [CrossRef]
12. Spaziani, L.; Lu, L. Silicon, GaN and SiC: There's room for all: An application space overview of device considerations. In Proceedings of the IEEE 30th International Symposium on Power Semiconductor Devices and ICs (ISPSD) 2018, Chicago, IL, USA, 13–17 May 2018; pp. 8–11.
13. Viswan, V. A review of silicon carbide and gallium nitride power semiconductor devices. *Int. J. Res. Eng. Sci. Manag.* **2018**, *1*, 224–225.
14. Ma, C.-T.; Gu, Z.-H. Review of GaN HEMT Applications in Power Converters over 500 W. *Electronics* **2019**, *8*, 1401. [CrossRef]
15. Ma, C.-T.; Gu, Z.-H. Design and Implementation of a GaN-Based Three-Phase Active Power Filter. *Micromachines* **2020**, *11*, 134. [CrossRef] [PubMed]
16. Eghtedarpour, N.; Farjah, E. Distributed charge/discharge control of energy storages in a renewable-energy-based DC micro-grid. *IET Renew. Power Gener.* **2014**, *8*, 45–57. [CrossRef]
17. Zhang, K.; Mao, C.; Xie, J.; Lu, J.; Wang, D.; Zeng, J.; Chen, X.; Zhang, J. Determination of characteristic parameters of battery energy storage system for wind farm. *IET Renew. Power Gener.* **2014**, *8*, 22–32. [CrossRef]
18. Sarkar, A.; Bhattacharjee, H.; Samanta, K.; Bhattacharya, K.; Saha, H. Optimal design and implementation of solar PV-wind-biogas-VRFB storage integrated smart hybrid microgrid for ensuring zero loss of power supply probability. *Energy Convers. Manag.* **2019**, *191*, 102–118. [CrossRef]
19. Ma, C.-T. A Novel State of Charge Estimating Scheme Based on an Air-Gap Fiber Interferometer Sensor for the Vanadium Redox Flow Battery. *Energies* **2020**, *13*, 291. [CrossRef]

20. Drobnič, K.; Grandi, G.; Hammami, M.; Mandrioli, R.; Viatkin, A.; Vujacic, M. A Ripple-Free DC Output Current Fast Charger for Electric Vehicles Based on Grid-Tied Modular Three-Phase Interleaved Converters. In Proceedings of the 2018 International Symposium on Industrial Electronics (INDEL), Banja Luka, Bosnia and Herzegovina, 1–3 November 2018.

21. Veneri, O.; Capasso, C.; Iannuzzi, D. Experimental evaluation of DC charging architecture for fully-electrified low-power two-wheeler. *Appl. Energy* **2016**, *162*, 1428–1438. [CrossRef]

22. Kim, B.; Kim, H.; Choi, S. Three-Phase On-board Charger with Three Modules of Single-stage Interleaved Soft-switching AC-DC Converter. In Proceedings of the 2018 IEEE Applied Power Electronics Conference and Exposition (APEC), San Antonio, TX, USA, 4–8 March 2018.

23. Monteiro, V.; Exposto, B.; Pinto, J.G.; Sepúlveda, M.J.; Meléndez, A.A.N.; Yue, M. Three-Phase Three-Level Current-Source Converter for EVs Fast Battery Charging Systems. In Proceedings of the 2015 IEEE International Conference on Industrial Technology (ICIT), Seville, Spain, 17–19 March 2015.

24. Monteiro, V.; Ferreira, J.C.; Meléndez, A.A.N.; Couto, C.; Afonso, J.L. Experimental Validation of a Novel Architecture Based on a Dual-Stage Converter for Off-Board Fast Battery Chargers of Electric Vehicles. *IEEE Trans. Veh. Technol.* **2018**, *67*, 1000–1011. [CrossRef]

25. Mortezaei, A.; Abdul-Hak, M.; Simoes, M.G. A Bidirectional NPC-based Level 3 EV Charging System with Added Active Filter Functionality in Smart Grid Applications. In Proceedings of the 2018 IEEE Transportation Electrification Conference and Expo (ITEC), Long Beach, CA, USA, 13–15 June 2018.

26. Kamble, A.S.; Swami, P.S. On-Board Integreted Charger for Electric Vehicle Based on Split Three Phase Insuction Motor. In Proceedings of the 2018 International Conference on Emerging Trends and Innovations In Engineering And Technological Research (ICETIETR), Ernakulam, India, 11–13 July 2018.

27. Yong, J.Y.; Fazeli, S.M.; Ramachandaramurthy, V.K.; Tan, K.M. Design and development of a three-phase off-board electric vehicle charger prototype for power grid voltage regulation. *Energy* **2017**, *133*, 128–141. [CrossRef]

Publisher's Note: MDPI stays neutral with regard to jurisdictional claims in published maps and institutional affiliations.

 © 2020 by the authors. Licensee MDPI, Basel, Switzerland. This article is an open access article distributed under the terms and conditions of the Creative Commons Attribution (CC BY) license (http://creativecommons.org/licenses/by/4.0/).

micromachines

MDPI

Review

Review on Driving Circuits for Wide-Bandgap Semiconductor Switching Devices for Mid- to High-Power Applications

Chao-Tsung Ma * and Zhen-Huang Gu

Department of Electrical Engineering, CEECS, National United University, Miaoli 36063, Taiwan; M0621002@smail.nuu.edu.tw
* Correspondence: ctma@nuu.edu.tw; Tel.: +886-37-382-482; Fax: +886-37-382-488

Abstract: Wide-bandgap (WBG) material-based switching devices such as gallium nitride (GaN) high electron mobility transistors (HEMTs) and silicon carbide (SiC) metal-oxide-semiconductor field-effect transistors (MOSFETs) are considered very promising candidates for replacing conventional silicon (Si) MOSFETs for various advanced power conversion applications, mainly because of their capabilities of higher switching frequencies with less switching and conduction losses. However, to make the most of their advantages, it is crucial to understand the intrinsic differences between WBG- and Si-based switching devices and investigate effective means to safely, efficiently, and reliably utilize the WBG devices. This paper aims to provide engineers in the power engineering field a comprehensive understanding of WBG switching devices' driving requirements, especially for mid- to high-power applications. First, the characteristics and operating principles of WBG switching devices and their commercial products within specific voltage ranges are explored. Next, considerations regarding the design of driving circuits for WBG switching devices are addressed, and commercial drivers designed for WBG switching devices are explored. Lastly, a review on typical papers concerning driving technologies for WBG switching devices in mid- to high-power applications is presented.

Keywords: wide-bandgap (WBG); gallium nitride (GaN); silicon carbide (SiC); high electron mobility transistor (HEMT); metal-oxide-semiconductor field effect transistor (MOSFET); driving technology

Citation: Ma, C.-T.; Gu, Z.-H. Review on Driving Circuits for Wide-Bandgap Semiconductor Switching Devices for Mid- to High-Power Applications. *Micromachines* **2021**, *12*, 65. https://doi.org/10.3390/mi12010065

Received: 25 December 2020
Accepted: 6 January 2021
Published: 8 January 2021

Publisher's Note: MDPI stays neutral with regard to jurisdictional claims in published maps and institutional affiliations.

Copyright: © 2021 by the authors. Licensee MDPI, Basel, Switzerland. This article is an open access article distributed under the terms and conditions of the Creative Commons Attribution (CC BY) license (https://creativecommons.org/licenses/by/4.0/).

1. Introduction

In modern industries, requirements for the performance of various power electronic-based converters are becoming stricter in terms of capacity, voltage level, efficiency, and size (switching frequency related issues). In order to enhance the performance of existing power converters, replacing conventional Si switching devices with wide-bandgap (WBG) switching devices such as gallium nitride (GaN) high electron mobility transistors (HEMTs) and silicon carbide (SiC) metal-oxide-semiconductor field-effect transistors (MOSFETs) is currently a popularly adopted method. WBG semiconductor materials offer superior characteristics to those of Si, as shown in Figure 1. The respective merits of GaN and SiC lead to the advantageous adoption of GaN HEMTs for low (<1 kW) to mid (<10 kW) power applications and SiC MOSFETs for mid (<10 kW) to high (>10 kW) power applications in practical design scenarios. The superiority of GaN HEMTs is yet to be fully utilized because they feature some form of heterogeneous integration with dissimilar substrate. This leads to large thermal boundary resistance between GaN and substrate, causing the self-heat issue [1], which may cause the switching device to overheat. However, GaN HEMTs offer the highest efficiency and switching speed, and SiC MOSFETs provide the highest voltage, current, and temperature capabilities. The main challenge of using the WBG semiconductor switching devices is overcoming potential difficulties introduced from their high slew rates, which could worsen electromagnetic interference (EMI) level and may cause voltage oscillation and instability [2–7].

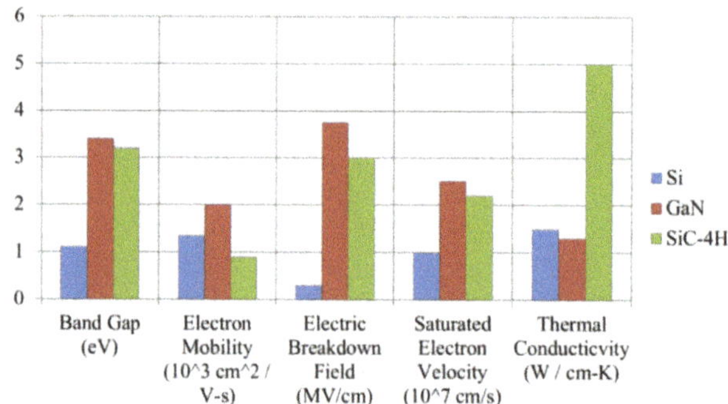

Figure 1. Comparison of Si, gallium nitride (GaN), and silicon carbide (SiC).

The GaN HEMT is designed with a unique aluminum gallium nitride (AlGaN)/GaN heterojunction structure where two-dimensional electron gas (2DEG) is formed. The 2DEG allows large bidirectional current and yields extremely low on resistance. GaN HEMTs are currently divided into three types: depletion mode (D-mode), enhancement mode (E-mode), and cascode devices. The D-mode GaN HEMT, as shown in Figure 2, is naturally on because of the 2DEG and can be turned off with negative gate-source voltage. The E-mode GaN HEMT, as shown in Figure 3, is normally off because the 2DEG has been depleted by an additional P-doped layer of GaN or AlGaN on the gate, and it can be turned on with appropriate gate-source voltage. The cascode GaN HEMT, as shown in Figure 4, is also normally off because it consists of a D-mode GaN HEMT and an additional high-speed low-voltage Si MOSFET, and it can be turned on with appropriate gate-source voltage applied on the Si MOSFET. E-mode and cascode GaN HEMTs possess different characteristics mainly because of the additional Si MOSFET in the cascode device: the E-mode device offers lower on resistance, higher operating temperature, and no body diode, while the cascode device offers less strict driving requirements, as shown in Table 1 [4–7].

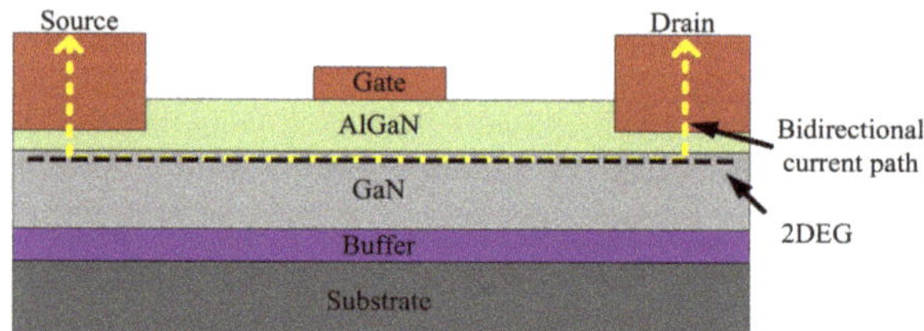

Figure 2. D-mode GaN high electron mobility transistors (HEMT).

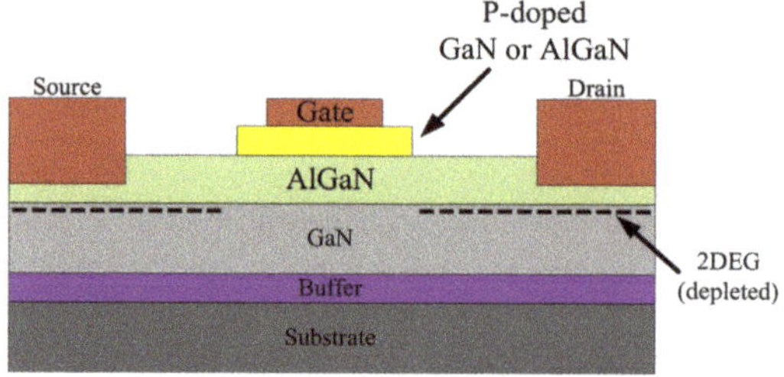

Figure 3. E-mode GaN HEMT.

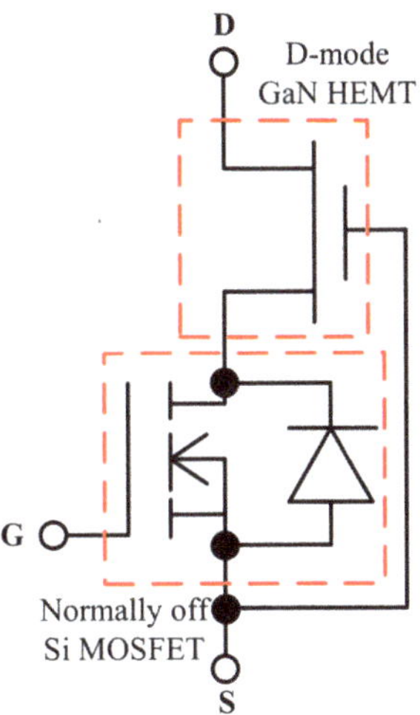

Figure 4. Cascode GaN HEMT.

Table 1. General comparison of normally off GaN high electron mobility transistors (HEMTs).

Device	Driving Voltage Threshold	Driving Voltage Range	Operating Temperature	On Resistance	Body Diode
E-mode	<2 V	−10 V~7 V	Higher	Lower	X
Cascode	~4 V	±20 V	Lower	Higher	O

The SiC MOSFET has a similar structure to that of Si MOSFET, as shown in Figure 5, but the thickness can be made an order smaller because of SiC's higher voltage capability. This leads to much smaller on resistance (although not as small as that of the GaN HEMT). Additionally, the SiC MOSFET offers the highest power capability. The operation of the SiC MOSFET is the same as that of the Si MOSFET: with appropriate gate-source voltage, the device can be turned on, and the body diode is used for reverse conduction during off state [8,9]. A general comparison of Si MOSFET, normally off GaN HEMTs, and SiC MOSFET is shown in Table 2.

Table 2. General comparison of Si metal-oxide-semiconductor field-effect transistors (MOSFET), normally off GaN HEMTs, and SiC MOSFET.

Device	Driving Voltage Strictness	Power Rating	Switching Speed	On Resistance	Operating Temperature	Body Diode
Si MOSFET	4th	2nd	4th	4th	3rd	O
E-GaN	Highest	3rd	Fastest	Lowest	2nd	X
Cascode-GaN	2nd	4th	2nd	2nd	3rd	O
SiC MOSFET	3rd	Highest	3rd	3rd	Highest	O

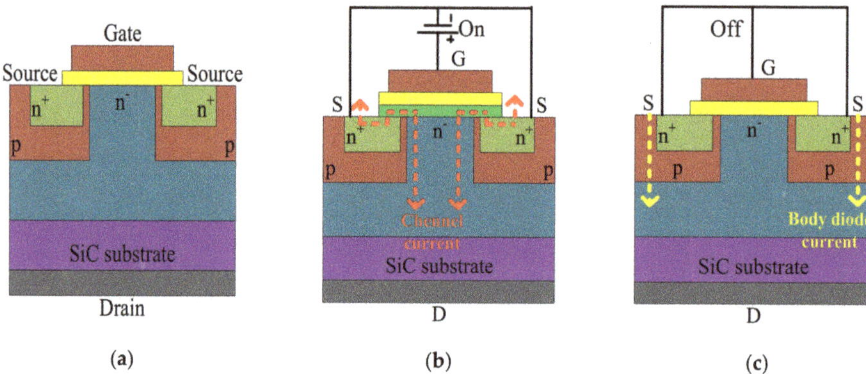

Figure 5. Schematic diagrams of a SiC MOSFET, (**a**) basic structure, (**b**) the gate-source voltage and current path in on-state, (**c**) the gate-source voltage and current path in off-state.

In recent years, because of the high slew rate of the WBG semiconductor switching devices, the philosophy of replacing conventional Si devices with WBG devices is an ongoing research trend. This paper aims to review issues concerning the driving technologies of WBG semiconductor switching devices. First, a general introduction of the GaN HEMT and the SiC MOSFET is given in the first section. In the second section, a survey of commercial GaN HEMTs above 600 V and SiC MOSFETs between 600 and 1200 V is presented. In the third section, the challenges and solutions of driving GaN HEMTs and SiC MOSFETs are addressed. A survey of commercial drivers designed for WBG switching devices is then provided in the fourth section. The fifth and sixth sections cover the literature review on driving circuits for GaN HEMTs and SiC MOSFETs, respectively. Lastly, this paper is concluded in the seventh section.

2. Commercial Wide-Bandgap (WBG) Switching Devices

2.1. Discrete Commercial GaN High Electron Mobility Transistors (HEMTs)

According to two famous electronic device providers, Digi-Key [10] and Mouser [11], GaN HEMT products can be purchased from several manufacturers, including EPC (15~200 V) [12], Infineon Technologies (400 and 600 V) [13], GaN Systems (100 and 650 V) [14], Panasonic (600 and 650 V) [15], Nexperia (650 V) [16], and Transphorm (650 and 900 V) [17]. Currently, the two highest voltage ratings of commercial GaN HEMTs are 900 and 650 V, respectively. The 900 V GaN HEMTs are produced by Transphorm, and 650 V GaN HEMTs are produced by GaN Systems, Panasonic, Nexperia (formerly Standard Products business unit of NXP Semiconductors), and Transphorm. Table 3 presents the device specifications of commercial GaN HEMTs above 600 V, where MFR stands for manufacturer, V_{ds} denotes drain-source voltage, I_{ds} denotes drain-source current, V_{TH} denotes threshold voltage, V_{gs} denotes gate-source voltage, $R_{ds(on)}$ denotes on resistance, and C_{iss} denotes input capacitance. Some specifications of the latest large-current devices from GaN Systems are not published. The maximum V_{gs} values of the products from Infineon Technologies and Panasonic are not specified because these devices are current-controlled, which offers good robustness but leads to higher gate losses.

Table 3. Commercial GaN HEMTs above 600 V.

Manufacturer	Type	V_{ds} (V)	I_{ds} (A)	V_{TH} (V)	V_{gs} (V)	$R_{ds(on)}$ (mΩ)	C_{iss} (pF)
Infineon Technologies		600	10	1.2	−10~5	140	157
			12.5				
			15			55	380
			31				
GaN Systems	E-mode	650	3.5	1.4	−10~7	500	30
			7.5	1.3		200	65
			8	1.4		225	52
			11	1.5		150	74
			15	1.3		100	130
			22			67	195
			30	1.7		50	260
			60	1.3		25	520
			80	-	-	18	-
			150			10	
Panasonic		600	15	3.5	−10~5	140	160
			31			56	405
		650	9.4			270	80
Nexperia		650	34.5	3.9	±20	50	1000
			47.2			35	1500
Transphorm	Cascode	650	6.5	4	±18	240	760
			15		±20	150	576
			16	2.1	±18		720
			20		±18	130	760
			25	4	±20	72	600
			28	2.6	±18		1130
			34	4	±20	50	1000
						60	
			36			50	
			46.5			35	1500
			47				
		900	15	2.1	±18	205	780
			34	3.9	±20	50	980

2.2. Discrete Commercial SiC Metal-Oxide-Semiconductor Field-Effect Transistors (MOSFETs)

Since SiC MOSFETs have been developed for a longer time, there is a much larger variety of companies that produce SiC MOSFETs: ON Semiconductor (900 and 1200 V) [18], Littelfuse (600~1700 V) [19], Infineon Technologies (650~1700 V) [13], Cree (650~1700 V) [20], Rohm Semiconductor (650~1700 V) [21], STMicroelectronics (650~1700 V) [22], United Silicon Carbide (650~1700 V) [23], Microchip (700~1700 V) [24], and GeneSiC Semiconductor (1200~3300 V) [25]. Tables 4–12 present the device specifications of commercial SiC MOSFETs with voltages ratings between 600 and 1200 V. When comparing the listed commercial devices, we can see that it is common for SiC MOSFETs to possess much higher current capabilities than those of GaN HEMTs, which makes SiC devices more suitable for high-power applications such as high-speed railway, power transmission, industrial drives, smart grid, and wind power generation. On the other hand, GaN devices offer smaller on resistances and input capacitances, which indicates that GaN devices have the potential to yield lower conduction losses and faster switching with such ratings. Therefore, GaN HEMTs are currently applied to improve the efficiencies of mid-voltage and mid-power applications such as switching power supply, solar PV, AC/DC adapter, medical equipment, electric vehicle (EV), and uninterruptible power supply. Figure 6 shows the application fields of Si, SiC, and GaN switching devices [13].

Table 4. Commercial 600~1200 V SiC MOSFETs by On Semiconductor.

V_{ds} (V)	I_{ds} (A)	V_{TH} (V)	V_{gs} (V)	$R_{ds(on)}$ (mΩ)	C_{iss} (pF)
900	44	2.7	−10~19	60	1800
	46	2.7	−10~20	60	1770
	112	2.6	−10~19	20	4415
	118	2.7	−10~19	20	4415
1200	17	3.1	−15~25	160	665
	17.3	3.1	−15~25	160	665
	19.5	3	−15~25	160	678
	29	2.75	−15~25	80	1112
	30	3	−15~25	80	1154
	31	2.7	−15~25	80	1112
	58	3	−15~25	80	1762
	60	3	−15~25	40	1789
	60	2.97	−15~25	40	1781
	98	2.7	−15~25	20	2943
	102	2.7	−15~25	20	2943
	103	2.7	−15~25	20	2890

Table 5. Commercial 600~1200 V SiC MOSFETs by Littelfuse.

V_{ds} (V)	I_{ds} (A)	V_{TH} (V)	V_{gs} (V)	$R_{ds(on)}$ (mΩ)	C_{iss} (pF)
600	15	3	±20	150	2000
1200	22	2.8	−10~25	160	870
	27	2.8	−10~25	120	1125
	39	2.8	−10~25	80	1825
	47	2.6	−10~25	40	1900
	48	2.8	−10~25	40	1895
	68	2.6	−10~25	25	2790
	90	2.6	−10~25	25	2790

Table 6. Commercial 600~1200 V SiC MOSFETs by Infineon Technologies.

V_{ds} (V)	I_{ds} (A)	V_{TH} (V)	V_{gs} (V)	$R_{ds(on)}$ (mΩ)	C_{iss} (pF)
650	20	4.5	−5~23	107	496
	26			72	744
	28				
	39			48	1118
	47			27	2131
	59				
1200	4.7		−7~23	350	182
	13			220	289
	19			140	454
	26			90	707
	36			60	1060
	52		−7~20	45	2130
			−10~20		1900
	56	4.5	−7~23	30	2120

Table 7. Commercial600~1200 V SiC MOSFETs by Cree.

V_{ds} (V)	I_{ds} (A)	V_{TH} (V)	V_{gs} (V)	$R_{ds(on)}$ (mΩ)	C_{iss} (pF)
650	36	2.3	−8~19	60	1020
	37				
	120			15	5011
900	11.5	2.1	−8~18	280	150
	22		−8~19	120	350
	23		−8~18		
	35		−8~19	65	760
					660
	36		−8~18		760
	63	2.4	−8~19	30	1747
1000	22	2.1	−8~19	120	350
	35			65	660

Table 7. *Cont.*

V_{ds} (V)	I_{ds} (A)	V_{TH} (V)	V_{gs} (V)	$R_{ds(on)}$ (mΩ)	C_{iss} (pF)
	7.2	2.5	−8~19	350	345
	7.6				
	10	2.6	−10~25	280	259
	17	2.8	−8~19	160	632
	18	2.9	−10~25		606
1200	30	2.5	−8~19	75	1350
					1390
	36	2.9	−10~25	80	1130
	60	2.6		40	1893
	63	2.5	−8~19	32	3357
	90	2.6	−10~25	25	2788
	100	2.5	−8~19	21	4818
	115			16	6085

Table 8. Commercial 600~1200 V SiC MOSFETs by Rohm Semiconductor.

V_{ds} (V)	I_{ds} (A)	V_{TH} (V)	V_{gs} (V)	$R_{ds(on)}$ (mΩ)	C_{iss} (pF)
	21	2.7	−4~26	120	460
	29	2.8	−10~26	120	1200
	30			80	571
650	39			60	852
	70	2.7	−4~26	30	1526
	93			22	2208
	118			17	2884

Table 9. Commercial 600~1200 V SiC MOSFETs by STMicroelectronics.

V_{ds} (V)	I_{ds} (A)	V_{TH} (V)	V_{gs} (V)	$R_{ds(on)}$ (mΩ)	C_{iss} (pF)
				45	
	45	3.2		55	1370
				75	
650	90		−10~22	18	3300
	95	3.1		20	3315
	100				
	116	3.2		15	3380
	119			18	

Table 9. *Cont.*

V_{ds} (V)	I_{ds} (A)	V_{TH} (V)	V_{gs} (V)	$R_{ds(on)}$ (mΩ)	C_{iss} (pF)
1200	12	3.5	−10~25	520	290
					300
	20			189	650
	33	3.2	−10~22	75	1230
	45	3.5	−10~25	90	1700
	52	3.1	−10~22	45	2086
	65	3	−10~25	59	1900
	75	3.1	−10~22	30	3400
	91	3.45		21	3540

Table 10. Commercial 600~1200 V SiC MOSFETs by United Silicon Carbide.

V_{ds} (V)	I_{ds} (A)	V_{TH} (V)	V_{gs} (V)	$R_{ds(on)}$ (mΩ)	C_{iss} (pF)
650	18	5	±25	34	1500
				45	
	25			80	
				111	
	31			80	
	41			42	
	54				
	65			27	
	85				
	120	4.7	±20	6.7	8360
1200	7.6	4.7	±25	410	740
	18.4	4.4		150	738
	33			80	1500
	34.5	5		70	
	65			35	
	107	4.7	±20	16	7824
	120			8.6	8512

Table 11. Commercial 600~1200 V SiC MOSFETs by Microchip.

V_{ds} (V)	I_{ds} (A)	V_{TH} (V)	V_{gs} (V)	$R_{ds(on)}$ (mΩ)	C_{iss} (pF)
700	25	2.4	−10~23	90	785
	28			86	
	37			60	1175
	39				
	65	2.7		35	2010
	77				
	126	2.4		15	4500
	140		−10~25		
1200	35	2.8	−10~23	80	838
	37				
	53		−10~25	40	1990
	64	2.6	−10~23	40	1990
	66	2.7			
	77	2.8	−10~25	25	3020
	89		−10~23		
	103				

Table 12. Commercial 600~1200 V SiC MOSFETs by GeneSiC.

V_{ds} (V)	I_{ds} (A)	V_{TH} (V)	V_{gs} (V)	$R_{ds(on)}$ (mΩ)	C_{iss} (pF)
1200	8	3	−10~25	350	225
	16			160	493
	32			75	1053
	33				
	57			40	1974
	59				
	74			30	2633
	78				
	95			20	3949
	107				

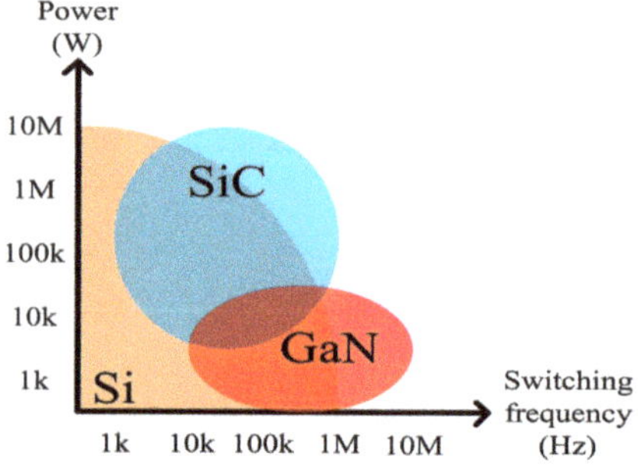

Figure 6. Applications fields of Si, SiC, and GaN switching devices [13].

2.3. Modular WBG Switching Devices

Some commercial switching modules based on GaN HEMTs can also be found from EPC [12], as numerated in Table 13, where HB stands for half bridge. EPC produces up to 3 kW modules. Consequently, these products are not yet matured for high-power applications.

Table 13. Commercial GaN HEMT modules by EPC.

Configuration	Voltage Rating (V)	Current Rating (A)
HB	30	10/40
		16
	60	10/40
		30
	80	10/40
		30
	100	1.7
		30
HB + bootstrap	60	1.7/0.5
	100	
Dual common source	120	3.4

Unlike GaN modules, there are several manufacturers with commercial SiC modules available for purchase: Infineon Technologies [13], Cree [20], Rohm Semiconductor [21], Microchip [24], Powerex [26], and SemiQ [27], as listed in Tables 14–19, where FB stands for full bridge. As can be imagined, the voltage and current ratings of SiC modules easily exceed those of GaN modules. Among SiC modules manufactured by the above-mentioned companies, HB is the most common configuration. Particularly, the configuration of DF23MR12W1M1P_B11, DF23MR12W1M1_B11, DF11MR12W1M1P_B11, and DF11MR12W1M1_B11 by Infineon Technologies is presented in Figure 7. The suitable applications of them are specified as solar applications.

Table 14. Commercial SiC MOSFET modules by Infineon Technologies.

Configuration	Voltage Rating (V)	Current Rating (A)
HB	1200	25
		50
		100
		150
		200
		250
		375
		500
FB		50
HB (3-arm)		25
Figure 7		25
		50
Vienna rectifier phase leg		75
		100

Table 15. Commercial SiC MOSFET modules by Cree.

Configuration	Voltage Rating (V)	Current Rating (A)
HB	1200	20
		50
		120
		225
		300
		325
		400
		425
		450
Three-phase		20
		50

Table 16. Commercial SiC MOSFET modules by Rohm Semiconductor.

Configuration	Voltage Rating (V)	Current Rating (A)
HB	1200	80
		134
		180
		204
		250
		300
		358
		397
		576
Chopper		134
		180
		204
		300
		358
		576

Table 17. Commercial SiC MOSFET modules by Microchip.

Configuration	Voltage Rating (V)	Current Rating (A)
HB	700	124
		241
		353
	1200	55
		89
		173
		254
		337
		495
		733
		805
		947
	1700	50
		100
		280
FB	700	98
	1200	55
		89
		173
HB (3-arm)	700	98
		189
		278
	1200	89
		171
		251
Chopper	700	98
	1200	55
		89
		173
		254
Vienna rectifier phase leg	700	124
		238

Table 18. Commercial SiC MOSFET modules by Powerex.

Configuration	Voltage Rating (V)	Current Rating (A)
Split dual SiC MOSFET	1200	100
Dual MOSFET	1700	540

Table 19. Commercial SiC MOSFET modules by SemiQ.

Configuration	Voltage Rating (V)	Current Rating (A)
HB	1200	160
		200
		240
		320
HB (2-arm)		40
		80
FB		20
HB (3-arm)		20

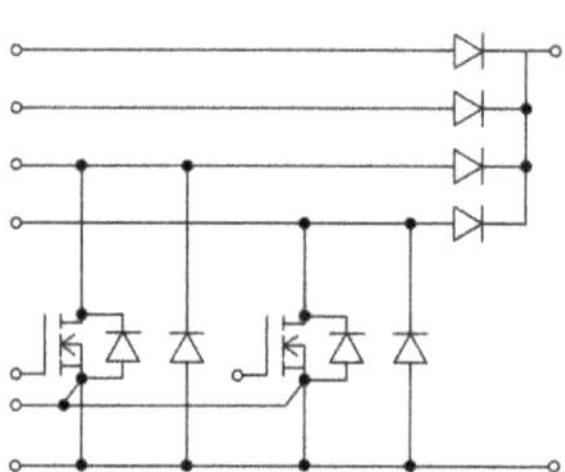

Figure 7. Configurations of Infineon Technologies SiC switching modules DF23MR12W1M1P_B11, DF23MR12W1M1_B11, DF11MR12W1M1P_B11, and DF11MR12W1M1_B11 [13].

3. Considerations for the Design of Driving Circuits for WBG Switching Devices

It has been well accepted that the key factor of realizing WBG switching devices' full potential is their driving circuits. The main difference in the driving characteristics of WBG and Si switching devices is due to WBG devices' much faster transient. The fast switching and high switching frequency require shorter driver rise and fall times and propagation delay. Additionally, the slew rate of the WBG devices can reach up to 100 times that of conventional Si devices, which can severely worsen EMI-related problems such as gate ringing and measurement. In order to deal with the fast transient, the driving circuit design and printed circuit board (PCB) layout must be optimized [28,29].

In general, the fundamental rules of driving high-power GaN HEMTs and SiC MOS-FETs are to apply high drive strength, provide enough isolation between driving and power circuits, prevent voltage oscillation, limit gate voltage spikes, and optimize dead time. Isolation can be provided by various types of isolators or isolated drivers suitable for WBG switching devices. In general, it is also possible to use high-speed MOSFET and IGBT drivers (similar to Figure 8) to drive WBG devices, but the complexity of the driving circuits and cost may be increased. However, another special characteristic of WBG switching devices is that they do not always use symmetrical driving voltages (such as ±18 and ±20 V), so there is often a need for asymmetric driving voltage design. Next, it is recommended to minimize the parasitic inductance and capacitance by minimizing the length of the driving loop and the overlapping between circuits and using devices with short or no wire bond [30–32]. In particular, GaN HEMTs (especially E-mode) have faster transients and more narrow driving voltage ranges, and SiC MOSFETs have higher power ratings. Consequently, the EMI issues are more dominant when driving GaN HEMTs, and higher driving strength is required when driving high-power SiC MOSFETs.

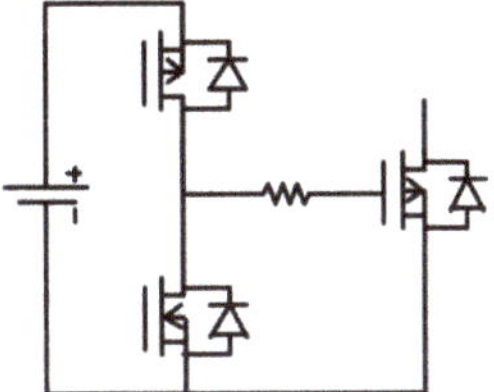

Figure 8. Conventional totem-pole gate driver.

For the turn-on period, the sum of external resistance and driver's output resistance should be designed to be much larger than the internal resistance of the power switching device in order to reduce the influence of internal resistance on the switching speed and damp voltage overshoot. If there is a need to damp gate ringing of certain frequencies, ferrite beads can be used as gate impedance as well. However, a low-impedance turn-off path is also required to ensure fast turn-offs and thus prevent shoot-through. As a result, it is usually recommended to design separate turn-on and turn-off paths, where drivers with separate high and low outputs can provide more flexibility, as shown in Figure 9. Moreover, active Miller clamps can be used to directly limit gate-source voltage range; negative turn-off voltage can increase the turn-off speed even more; and Kelvin source connection can separate the driving loop and power loop, so that the influence of parasitic inductance on the driving loop can be minimized, as shown in Figure 10 [31–33]. Particularly, active gate drive can be used to control the slew rate directly [34,35].

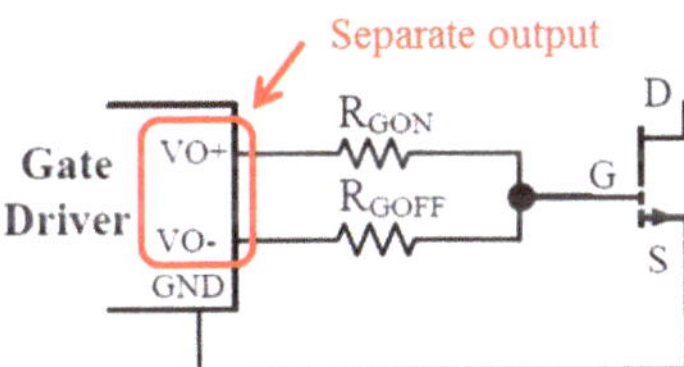

Figure 9. Gate driver with separate high and low outputs [31].

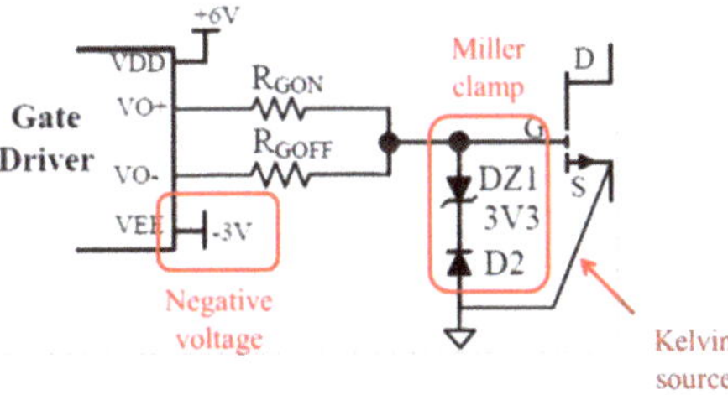

Figure 10. Driving loop with active Miller clamp, Kelvin source connection, and negative turn-off voltage [31].

4. Commercial Drivers for WBG Switching Devices

4.1. GaN HEMT Drivers

According to Digi-Key [10] and Mouser [11], commercial GaN HEMT drivers are currently available from several companies: Infineon Technologies [13], On Semiconductor [21], Maxim Integrated [36], pSemi [37], Silicon Laboratories [38], and Texas Instruments [39], as given in Tables 20–25. It is desirable to use drivers integrated with multiple functions such as digital control and signal detection in order to reduce the number of external devices required. Some companies also produce GaN power modules that integrate GaN

HEMTs with designed drivers: EPC [12], Texas Instruments [39], and Navitas Semiconductor [40], as presented in Tables 26–28.

Table 20. Commercial GaN HEMT drivers by Infineon Technologies.

Num. of Channels	Peak Source Current (A)	Peak Sink Current (A)	Supply Voltage (V)	Rise Time (ns)	Fall Time (ns)	Prop. Delay (ns)
1	4	8				41
2	1	2	3~20	6.5	4.5	37
	4	8				

Table 21. Commercial GaN HEMT driver by On Semiconductor.

Num. of Channels	Peak Source Current (A)	Peak Sink Current (A)	Supply Voltage (V)	Rise Time (ns)	Fall Time (ns)	Prop. Delay (ns)
2	1	1.3	9~17	1	1	25

Table 22. Commercial GaN HEMT drivers by Maxim Integrated.

Num. of Channels	Peak Source Current (A)	Peak Sink Current (A)	Supply Voltage (V)	Rise Time (ns)	Fall Time (ns)	Prop. Delay (ns)
1	3	7	4~14	4~37	4~18	8
	4	2.85	3~36	3.6	2.5	35
		5.7	−16~36		1.8	53

Table 23. Commercial GaN HEMT drivers by pSemi.

Num. of Channels	Peak Source Current (A)	Peak Sink Current (A)	Supply Voltage (V)	Rise Time (ns)	Fall Time (ns)	Prop. Delay (ns)
2	2	4	4~6.5	1	1	11
			4~6	0.9	0.9	9.1

Table 24. Commercial GaN HEMT drivers by Silicon Laboratories.

Num. of Channels	Peak Source Current (A)	Peak Sink Current (A)	Supply Voltage (V)	Rise Time (ns)	Fall Time (ns)	Prop. Delay (ns)
1	0.3	0.5	6.5~24	20	20	60
	1.5	2.5				
	2.8	3.4	2.8~30	5.5	8.5	40
2	0.25	0.5	3~30	20	20	30
	0.4	0.6	6.5~30	5.5	8.5	40
	1.8		2.5~30	10.5	13.3	45
	2	4	2.5~24	12	12	30
			3~30			

Table 25. Commercial GaN HEMT drivers by Texas Instruments.

Num. of Channels	Peak Source Current (A)	Peak Sink Current (A)	Supply Voltage (V)	Rise Time (ns)	Fall Time (ns)	Prop. Delay (ns)
1	1.3	7.6	4~12.6	12	3	12
	4	4	4.5~18	9	7	13
		4	4.5~18	8	7	17
		6	4~18	9	4	14
		6	3~18	5	6	27
		8	4.5~18	9	7	13
	5		3~33	10	10	65
	7	5	4.75~5.25	0.4	0.4	2.5
				0.65	0.85	2.9
2	1.2	5	4.5~5.5	7	3.5	30
						35
	1.5	2.5	3~18	8	9	28
		3	3.8~18	0.5	0.5	10
	4	6	3~18	5	6	25
						28
			3~25	6	8	19
					7	
	5	5	4.5~18	7	6	13
				9	6	17

Table 26. Commercial switch-driver-integrated module by EPC.

Configuration	Voltage Rating (V)	Current Rating (A)	Supply Voltage (V)
HB	70	12.5	11~13

Table 27. Commercial switch-driver-integrated module by Texas Instruments.

Configuration	Voltage Rating (V)	Current Rating (A)	Supply Voltage (V)
Single switch	600	17	9.5~18
		40	9.5~18
		34	9.5~18
HB	80	10	4.57~5.25

Table 28. Commercial switch-driver-integrated module by Navitas Semiconductor.

Configuration	Voltage Rating (V)	Current Rating (A)	Supply Voltage (V)
Single switch	650	5	5.5~24
		8	
		12	

4.2. SiC MOSFET Drivers

Since SiC devices have been developed for a longer time, a larger variety of SiC MOSFET drivers than that of GaN HEMT drivers have been developed by many manufacturers, including Infineon Technologies [13], On Semiconductor [21], Microchip [24],

Maxim Integrated [36], Silicon Laboratories [38], Texas Instruments [39], Analog Devices [41], Tamura [42], Rohm Semiconductor [21], Littelfuse [19], Diodes Incorporated [43], NXP Semiconductors [44], and Power Integrations [45], as listed in Tables 29–37. Because SiC MOSFETs are suitable for and often used in high-power applications, the peak output current ratings of SiC MOSFET drivers are generally larger than those of GaN HEMT drivers. Particularly, Tamura's drivers offer as large as 43 A peak driving current.

Table 29. Commercial SiC MOSFET drivers by Infineon Technologies.

Num. of Channels	Peak Source Current (A)	Peak Sink Current (A)	Supply Voltage (V)	Rise Time (ns)	Fall Time (ns)	Prop. Delay (ns)
1	2	2	−12~28	34	50	170
	4	3.5	3.1~35	10	9	125
	4.4	4.1	3.1~18	9	6	300
	10	9.4	3.1~35	10	9	125
2	2	2	−12~28	30	50	170
	4	8	3~20	6.5	4.5	37

Table 30. Commercial SiC MOSFET drivers by On Semiconductor.

Num. of Channels	Peak Source Current (A)	Peak Sink Current (A)	Supply Voltage (V)	Rise Time (ns)	Fall Time (ns)	Prop. Delay (ns)
1	6	6	−8~22	8	8	25
	7.8	7.1	−10~24	10	15	66
2	1.9	2.3	0~20	13	8	90

Table 31. Commercial SiC MOSFET drivers by Microchip.

Num. of Channels	Peak Source Current (A)	Peak Sink Current (A)	Supply Voltage (V)	Rise Time (ns)	Fall Time (ns)	Prop. Delay (ns)
2	10	10	14~16	80	90	250
	20	20				

Table 32. Commercial SiC MOSFET drivers by Maxim Integrated.

Num. of Channels	Peak Source Current (A)	Peak Sink Current (A)	Supply Voltage (V)	Rise Time (ns)	Fall Time (ns)	Prop. Delay (ns)
1	4	2.85	3~36	3.6	2.5	35
		5.7	−16~36		1.8	53

Table 33. Commercial SiC MOSFET drivers by Silicon Laboratories.

Num. of Channels	Peak Source Current (A)	Peak Sink Current (A)	Supply Voltage (V)	Rise Time (ns)	Fall Time (ns)	Prop. Delay (ns)
1	2.8	3.4	0~30	5.5	8.5	40
	4	4	3~30	12	12	19
2	1.8	4	2.5~30	10.5	13.3	45
	4		3~30	12	12	19
						89
						39

Table 34. Commercial SiC MOSFET drivers by Texas Instruments.

Num. of Channels	Peak Source Current (A)	Peak Sink Current (A)	Supply Voltage (V)	Rise Time (ns)	Fall Time (ns)	Prop. Delay (ns)
1	1.5	2	−13~33	28	25	70
	2.5	5	−15~30 / 0~30	18	20	76
			−5~32	15	7	17
	4.5	5.3	−13~33	28	25	70
	8.5		−16~33	10	10	65
		10	−5~15	33	27	90
	10		−5~15	28	24	90
			−16~33 / 3~33	10	10	65
	15	15	−12~30	150	150	150
	17	17	−16~33	10	10	65
2	4	6	3~25	6	7	19

Table 35. Commercial SiC MOSFET drivers by Analog Devices.

Num. of Channels	Peak Source Current (A)	Peak Sink Current (A)	Supply Voltage (V)	Rise Time (ns)	Fall Time (ns)	Prop. Delay (ns)
1	0.2	0.2	4.5~17	15	15	60
	2	2	2.5~35	18	18	38
			3.3~35	17	17	30
	2.3	2.3	2.5~35	18	18	43
	4	4	3~18	12	12	46
			3~18	22	22	53
			−15~30 / −15~35	16	16	55
	6	6	4.5~25	-	-	107
			6~25	-	-	100
2	0.1	0.1	4.5~18	25	25	124
		0.3	4.5~18.5	25	10	100
	4	4	3~18	12	12	47
			3~18	12	12	46
			4.5~18	14	14	160

Table 36. Commercial SiC MOSFET drivers by Tamura.

Num. of Channels	Peak Source Current (A)	Peak Sink Current (A)	Supply Voltage (V)	Rise Time (ns)	Fall Time (ns)	Prop. Delay (ns)
2	1.8	1.8	13.5~26.4	-	-	90
	2.5	3.5		-	-	
	3	3				
	4	4		-	-	
	4.5	4.5		-	-	
	6	6		-	-	
	7	7		-	-	
	18	18		-	-	80
	43	43		-	-	100

Table 37. Commercial SiC MOSFET drivers by Rohm Semiconductor, Littelfuse, Diodes Incorporated, NXP Semiconductors, and Power Integrations.

MFR	Num. of Channels	Peak Source Current (A)	Peak Sink Current (A)	Supply Voltage (V)	Rise Time (ns)	Fall Time (ns)	Prop. Delay (ns)
Rohm		>4 (self-limited)	>4 (self-limited)	4.5~20	15	15	65
Littelfuse		9	9	−10~25	10	10	75
Diodes	1	10	10	40	48	35	10
NXP		15	15	−12~40	-	-	-
Power Int.		8	8	4.75~28	113	105	270

5. Review on GaN HEMT Driving Circuits

5.1. Single-Channel GaN HEMT Driving Circuits

To provide readers with direct design references, typical papers presenting the design of GaN HEMT driving circuits are reviewed in this subsection with examples of single-channel drive. Gurpinar and Castellazzi [46] conducted a benchmark of Si-, SiC-, and GaN-based switching devices at a 600 V class in 3.5 kW, 700 V/230 V, 16~160 kHz single-phase T-type inverter. Evaluated items included gate driver requirements, switching performance, inverter efficiency performance, heat sink volume, output filter volume, and dead-time effect for each technology. A Broadcom gate drive optocoupler ACPL-P346 was selected as the isolated driver for Panasonic PGA26A10DS, and an XP Power isolated DC/DC converter IH0512S-H was used to provide +12 V supply. The design offered small footprint, but the drive strength was limited at 3 A. The series capacitor C_s in the proposed driving circuit was designed at 2.82 nF in order to provide −4.5 V during turn-off and speed up turn-on transient. In [47], a low-inductance switching power cell was designed for a three-level ANPC inverter based on GaN System GS66508T. Texas Instruments UCC27511 provided separate turn-on and turn-off outputs. PWM signal was generated using a fiber optic link, and an inverting Schmitt trigger were used to transfer the PWM signal and avoid any false turn-on or turn-off. The signal was then isolated using Silicon Laboratories Si861x. A 7 V power supply was provided using an isolated DC/DC converter and a low-dropout (LDO) regulator. A Schottky diode was used for voltage clamp. The four-layer PCB and surface-mount components significantly reduced the loop inductance using flux cancellation, where the layout was required to eliminate common-mode current circulation. In [48], a 1.5 kW HB bidirectional DC/DC converter was proposed based on GaN System GS66508T driven by Silicon Laboratories Si8271 and −3~6 V driving voltage. A CUI isolated DC/DC converter PES1-S5-S9-M-TR was used to provide power supply. Four-layer PCB layout was adopted in this paper, where flux cancellation was adopted in order to minimize the loop inductance. Advanced Thermal Solutions heat sink ATS-FPX060060013-112-C2-R1 was chosen to match the small thermal pad of GS66508T, and a copper bar was placed between the switching device and the heat sink for enhanced thermal performance.

A 3 kW bidirectional GaN-HEMT DC/DC converter was proposed in [49]. The self-designed driving circuit was designed with an additional NPN bipolar transistor that acted as a voltage clamp and showed no impact on the switching speed. Zero voltage turn-on and negative turn-off voltage were utilized to reach >99% efficiency. A fast GaN HEMT driving circuit was designed for Panasonic PGA26E19BA with a voltage clamped to achieve optimized switching performance, freewheeling conduction, and short-circuit robustness [50]. The manufacturer-recommended driving circuit design was modified by adding a diode-resistor network that helped the capacitor on the driving path quickly discharge as well as provided more flexibility in gate resistor design.

5.2. Dual-Channel GaN HEMT Driving Circuits

Dual-channel GaN HEMT driving circuits are generally more complex than single-channel driving circuits. However, there are some commercial dual-channel GaN HEMT drivers ICs that greatly reduce the design complexity. Zhao et al. [51] designed a 2 kVA GaN-based single-phase inverter. The driving circuit for GaN System GS66508T consisted of Analog Devices ADuM7223 and separate turn-on and turn-off paths. The driver IC selection was limited in the writing of this paper (2016), so the selected driver failed to meet the common-mode transient immunity (CMTI) requirement of over 150 V/ns, but no failure occurred during testing. In [52], a 2 kW inductive power transfer system was designed based on a 600 V GaN gate injection transistor (GIT). The slew rate was lowered so that the Analog Devices ADuM3223, which had state of the art CMTI capability of 50 V/ns back in 2016, was compatible. In [53], a high-power density single-phase FB inverter was designed based on GaN Systems GS66502B. Silicon Laboratories Si8274 was used to drive the GaN HEMTs. Ways to minimize parasitic inductance and optimize heat dissipation were discussed, including Kelvin source connection, short and wide PCB trails (preferably copper), minimized overlap between driving loop and power loop, flux cancellation, sufficient area and thickness of copper, small thermal vias but large in number, direct copper and thermal via placement, increased number of layers, and reduced thickness of PCBs.

A totem-pole bridgeless power factor correction (PFC) converter based on GaN Systems GS66508B was proposed in [54]. The Silicon Laboratories Si8273 was used to drive the GaN HEMTs, and Kelvin source connection, voltage clamp, discharging resistor, and separate turn-on and turn-off paths were utilized. Elrajoubi et al. [55] designed an AC/DC converter for battery charging application based on GaN Systems GS66504B. The Silicon Laboratories Si8273 was used to drive the switches, and the Texas Instruments TMS320F28335 was used to generate PWM signals. Kelvin source connection, four-layer PCB, optimized layout, and voltage clamp were taken into consideration.

In [56], a bidirectional buck–boost converter for EV charging application was designed based on 1.3 kV series-stacked GS66508T switching modules. The modules were each designed with two GaN devices, proper gate impedances, and voltage clamps and required only one driving signal to function. The design yielded almost equally shared DC bus voltage.

5.3. Brief Summary of Reviewed GaN HEMT Driving Circuits

Table 38 categorizes the reviewed papers according to whether they adopted commercial drivers or self-designed drivers and whether the designed circuits are single-channel or dual-channel circuits, and adopted deriver ICs (if used) are also listed. Sun et al. [57] compared resonant gate drivers for both Si MOSFETs and GaN HEMTs applications. It was concluded that the resonant driver reduced switching losses because of fast charging/discharging capability.

Table 38. Reviewed GaN HEMT driving circuits.

Ref	Channel per Driver	f_{sw} (kHz)	System P_{out}	System Efficiency	GaN HEMT	Commercial Driver
[46]		160	2.5 kW	97.3%	PGA26A10DS	ACPL-P346
[47]		10	1 kW	unrevealed	GS66508T	UCC27511
[48]	1	100	1.5 kW	>97%	GS66508T	Si8271
[49]		120~200	3 kW	>99%	GS66508T	N/A
[50]		200	N/A	N/A	PGA26E19BA	
[51]		100	2 kVA	97.4%	GS66508T	ADuM7223
[52]		100~250	2 kW	95%	unrevealed	ADuM3223
[53]	2	160	500 W	96.2%	GS66502B	Si8274
[54]		-	800 W	98%	GS66508B	Si8273
[55]		65~100	1.5 kW	90%	GS66504B	Si8273
[56]		25	unrevealed	unrevealed	GS66508T	N/A

5.4. GaN HEMT Power Stages with Integrated Driving Circuits

Switching module design is also a research focus in the field of GaN HEMT driving. A high-power density, high-efficiency half-bridge module based on insulated metal substrate was proposed in [58] for >3 kW applications. In both high side and low side, two GaN System GS66508Bs were paralleled. The 6-layer module consists of four high-side and four low-side switches. The most crucial work was to minimize and balance parasitic inductances of the GaN switching devices. Brothers and Beechner [59] proposed a three-phase 100 V/270 A (per phase) GaN module consisting of six GaN System GS61008Ts. The module layout was then designed based on references from commercial modules and modules in literature. The proposed module successfully hard switched up to 375A.

6. Review on SiC MOSFET Driving Circuits

6.1. Single-Channel SiC MOSFET Driving Circuits

Since the SiC MOSFET has been developed longer than the GaN HEMT, a lot more papers regarding its driving circuit can be found in open literature. Pirc et al. [60] improved the performance of a nanosecond pulse electroporator by adopting Cree C2M0025120D. RECOM DC/DC converters RP-0512D and RP-0505S were used to provide −5 and 24 V isolated power supply, the Broadcom optocoupler HCPL-0723 was used for signal isolation, and the Littelfuse ultrafast MOSFET driver IXDD609SI was used to drive the SiC MOSFETs. In [61], an isolated smart self-driving multilevel SiC MOSFET driver for fast switching and crosstalk suppression was proposed using variable gate voltage generated with an auxiliary circuit that acted differently during turn-on and turn-off periods. The Isolated Analog Devices single-channel driver ADuM4135 was used to drive Cree C2M0040120D. The designed driving circuit adopted a two-state turn-off scheme. At first, the turn off was ensured using negative voltage, and the voltage was then switched to zero to avoid negative voltage breakdown. The circuit was simple, suitable for integration, highly efficient, compact, and cost-effective. A SiC-based 4MHz 10 kW single-phase zero-voltage-switching inverter for high-density plasma generators was proposed in [62]. The Littelfuse single-channel driver IXRFD630 was used to drive two parallel-connected Cree C2M0080120Ds. The required negative voltage was generated with an auxiliary circuit consisting of a resistor, a capacitor, and a zener diode.

Kim et al. [63] proposed a MHz SiC MOSFET driving circuit using parallel connected FBs based on EPC E-mode device EPC2016. The Broadcom ACPL-346 was used to drive the GaN FBs. The more FBs connected in parallel, the higher the output frequency of pulse width modulation (PWM) could be. The circuit was successfully tested at 2 MHz SiC MOSFET switching frequency using two FBs for a 600 W DC–DC converter. The switching frequency of 5 MHz was also verified achievable using two FBs (2.5 MHz * 2). In [64], a high-speed gate driver was designed focusing on high-temperature capability (180 °C) with low cost, and the circuit was integrated with overcurrent and undervoltage lockout protection. High-temperature transistors (from On Semiconductor), diodes, zener diodes, and pulse transformer were used to realize the design. Even higher operating temperature could be further realized by using polyimide- and hydrocarbon-based PCBs. The cost was successfully reduced from United States dollar (USD) 2250 to 100 compared with using Cissoid EVK-HADES1210. The tested SiC MOSFET was Fuji Electric MT5F31003. Qi et al. [65] developed a 30 kVA three-phase inverter based on Cree SiC HB power module CAS300M12BM2 in order to investigate how to achieve cost-effective outstanding high-temperature performance (targeted at 180 °C ambient temperature). The designed driving circuit was based on Central Semiconductor high-temperature transistors in metal can packages (rated at 200 °C) and consisted of signal isolation, gate drive, saturation detection, undervoltage detection, and protection logic circuits. It achieved 90% cost reduction compared with driving circuits using commercial silicon-on-insulator (SOI) ICs (from USD 2250 to 50 for active components).

A low-cost analog active driver was proposed in [34] for a higher parasitic environment. The designed driver consisted of a current amplifier stage and turn-on and turn-off switching controllers and outputted continuous analog current. Cree C2M0080120D was used to verify that the designed driver reduced losses compared with a hard switched gate driver and that good dynamic was achieved with much larger parasitic inductance and capacitance. In [66], a driving circuit with minimum propagation delay was designed to drive Rohm SCH2080KE for high-temperature applications. The design consisted of signal isolation and level shifting circuits and a two-level isolated auxiliary power supply. The isolation circuit achieved very small propagation delay by using non-delay RC differential circuits and a set-rest flip-flop. The auxiliary power supply was compatible with wide input voltage and operating temperature ranges because of the exclusion of low-temperature linear optocouplers. Li et al. [67] focused on crosstalk elimination. High off-state gate impedance was employed to eliminate the voltage drop on the common-source inductance, while the potential fault turn-on was prevented by utilizing the pre-charged voltage in the gate-source capacitance. Cree C2M0025120D was used to verify the driving circuit, and the design was compatible with most of the commercial SiC MOSFETs. Zhao et al. [68] proposed an intelligent and versatile active gate driver with three turn-on speeds and two turn-off speeds using an adjustable voltage regulator, a voltage selector, and a current sinking circuit. Cree C2M0080120D was used to verify the design.

6.2. Dual-Channel SiC MOSFET Driving Circuits

A 1 kW interleaved high-conversion ratio bidirectional DC-DC converter based on four Rohm SiC MOSFET SCTMU001F and four Si MOSFETs was proposed in [69] for distributed energy storage systems. Each driving circuit drove one SiC MOSFET and one Si MOSFET and consisted of Texas Instruments UCC27531, a transformer with two secondary windings, and a voltage clamp. In [70], two series-connected Cree C2M0160120Ds were driven by a low-cost, simple, and reliable driving circuit based on the Broadcom ACPL-344JT, coupling circuits, dv/dt limiting circuit, and a voltage limit circuit. Good voltage balancing and reliable switching were obtained. Lower switching frequency and smaller DC bus voltage were also compatible for the designed circuit. Wang et al. [71] proposed an enhanced gate driver consisting of the STMicroelectronics galvanically isolated MOS-FET/IGBT driver STGAP1AS, a bipolar junction transistor-based multi-cell current booster, a high-bandwidth and high-accuracy nonintrusive Rogowski switch-current sensor, and a noise-free isolation architecture. The designed driver was verified with Cree HB module CAS300M17BM2 and compatible with almost all the SiC MOSFET modules. Yang et al. [72] proposed a driving circuit with dynamic voltage balancing for series-connected SiC MOS-FETs. Only one external driving IC was required to drive both switches. An overdrive control method helped adapt to DC-bus voltage variation. Switched capacitors could be utilized to further widen the control range. The Cree C2M1000170D was used to verify the design, which was suitable for various high-voltage applications.

A single gate driver was designed to drive four cascaded series-connected SiC MOSFETs for medium voltage applications [73]. This was realized using an auxiliary circuits consisting of diodes, zener diodes, resistors, and capacitors. A 2400 V 10 kHz synchronous boost converter was demonstrated using the designed driving circuit.

6.3. Brief Summary of Reviewed SiC MOSFET Driving Circuits

As can be imagined, many similarities related to safety and loss reduction can be observed in GaN HEMT and SiC MOSFET driving circuit designs. A difference is that SiC MOSFETs require higher driving strength and high temperature capability when used in high-power applications. Table 39 categorizes the reviewed papers according to whether they adopted commercial drivers or self-designed drivers and whether the designed circuits are single-channel or multichannel driving circuits, and adopted deriver ICs are also listed. Sakib et al. [74] compared various gate drivers for SiC MOSFETs (2017). Covered aspects included passively triggered gate drive, negative spike mitigation, crosstalk prevention, and resonance and clamping. Another review was conducted on SiC MOSFET devices and individual SiC MOSFET gate drivers (2018) [75]. Covered items included the adjustment of switching speed, voltage, and power level and other special functions. In [76], the status and applications of SiC-based power converters, challenges regarding high-switching frequency gate driver design, and problems related to commercial drivers were reviewed (2018). It was pointed out that the commercial drivers back then were far from universal, which was due to very specific driving requirements of various SiC switching devices. Liu and Yang [77] reviewed the characteristics of SiC MOSFETs and different driving circuits (2019). It was suggested that, for >150 °C applications, discrete components or Si-on-insulator-based gate drivers should be used rather than conventional Si MOSFET drivers. In terms of crosstalk suppressing, it was recommended that a combination of additional capacitors, variable voltage/resistance driver, and auxiliary discharging path could be used. In [78], slew rate control methods for SiC MOSFET active gate drivers were reviewed (2020). Reviewed aspects included the principle of slew rate control, factors that influenced slew rate, and issues induced by high slew rate. Slew rate control methods included variable gate resistance, input capacitance, gate current, and gate voltage. Control strategies included open-loop, measurement-based, estimation-based, and timing-based controls. The advantages and disadvantages of each control strategy were also listed. Next, the conventional and emerging applications of active gate drive were also reviewed, including EMI noise mitigation, dead-time adaption, motor drive, reliability enhancement of SiC MOSFET, and parallel SiC MOSFET connection.

Table 39. Reviewed SiC MOSFET driving circuits.

Ref	Channel per Driver	f_{sw}	System P_{out}	System Efficiency	SiC MOSFET	Commercial Driver
[60]		N/A	unrevealed	N/A	C2M0025120D	IXDD609SI
[61]		1 MHz	51 kW	unrevealed	C2M0040120D	ADuM4135
[62]		4 MHz	10 kW	>97.5%	C2M0080120D	IXRFD630
[34]		20 kHz	5.9 kW	unrevealed	C2M0080120D	
[63]	1	2 MHz	600 W	unrevealed	C2M0080120D	N/A
[64]		150 kHz	4 kW	>99%	MT5F31003	

Table 39. *Cont.*

Ref	Channel per Driver	f_{sw}	System P_{out}	System Efficiency	SiC MOSFET	Commercial Driver
[65]		10 kHz	30 kW	99%	CAS300M12BM2	
[66]		100 kHz	2 kW	82%	SCH2080KE	
[67]		unrevealed	4.5 kW	unrevealed	C2M0025120D	
[68]		unrevealed	9 kW	unrevealed	SCH2080KE	
[69]		200 kHz	1 kW	96%	SCTMU001F	UCC27531
[70]		25 kHz	2.5 kW	N/A	C2M0160120D	ACPL-344JT
[71]	2	100 kHz	252 kW	99.4%	CAS300M17BM2	STGAP1AS
[72]		100 kHz	1.8 kW	unrevealed	C2M1000170D	unrevealed
[73]		10 kHz	unrevealed	unrevealed	unrevealed	N/A

6.4. SiC MOSFET Power Stages with Integrated Driving Circuits

Some examples of modular SiC MOSFET power stages can be found in [79–81]. Jørgensen et al. [79] proposed a 10 kV single-switch module adopting −5~20 V hard-switched Littlefuse IXRFD630, Kelvin connection, no external gate resistance for the fastest switching speed possible, and low inductance design for better heat dissipation. In [80], a 1200 V/120 A HB module was designed based on a direct bonding copper-stacked hybrid packaging structure for minimized thermal resistance and commutation power loop inductance. The designed module was tested as a 5.5 kW single-phase inverter, yielding 97.7% efficiency, and the power loss was 28.3% less than Cree HB module CAS120M12BM2. A module with adjustable drive strength based on hybrid combination of logics and high temperature capability was proposed in [81]. The Cree bare die SiC MOSFET CPM3-0900-0065B was successfully switched with less than 75 ns rise/fall time from room temperature to over 500 °C. Overshoot and dv/dt were successfully and dynamically controlled.

7. Conclusions

The desire of replacing conventional Si-based switching devices with WBG material-based switching devices for higher switching frequency and efficiency has led to intensive research on the driving technologies of GaN HEMT and SiC MOSFET. This paper has addressed the characteristics and operating principles of GaN HEMT and SiC MOSFET. Commercially available products of WBG switching devices with V_{ds} ranging from 600V to 1200 V were explored. GaN HEMTs are currently suitable for low- to mid-power and high-frequency applications because of their ultrafast switching speed and ultralow conduction losses, and SiC MOSFETs are especially suitable for high-power applications because of their high thermal capability. In this paper, the driving requirements of WBG switching devices have been explained, where overcoming high slew rate is the biggest challenge. Commercial drivers designed for WBG switching devices were surveyed. It has been observed that drivers for GaN HEMTs and SiC MOSFETs are normally designed based on their specific system requirements. Finally, typical papers discussing the driving circuits of GaN HEMT and SiC MOSFET, previously published review papers, and some papers focusing on modular design of WBG switching devices integrated with driving circuits have been reviewed with brief discussions.

Author Contributions: The corresponding author, C.-T.M. conducted the review work, verified the content of technical issues, and polished the final manuscript. Z.-H.G., a postgraduate student in the department of EE, CEECS, National United University, Taiwan, assisted in paper searching and organized results. All authors have read and agreed to the published version of the manuscript.

Funding: This research was funded by MOST, Taiwan, with grant numbers MOST 109-2221-E-239-007.

Institutional Review Board Statement: Not applicable.

Informed Consent Statement: Not applicable.

Data Availability Statement: No new data were created or analyzed in this study. Data sharing is not applicable to this article.

Acknowledgments: The author would like to thank the Ministry of Science and Technology (MOST) of Taiwan for financially support the energy-related research regarding key technologies in microgrids and the design of advanced power and energy systems.

Conflicts of Interest: The authors declare no conflict of interest.

References

1. Yuan, C.; Pomeroy, J.W.; Kuball, M. Above bandgap thermoreflectance for non-invasive thermal characterization of GaN-based wafers. *Appl. Phys. Lett.* **2018**, *113*, 1–5. [CrossRef]
2. Lee, H.; Smet, V.; Tummala, R. A review of SiC power module packaging technologies: Challenges, advances, and emerging issues. *IEEE J. Emerg. Sel. Top. Power Electron.* **2020**, *8*, 239–255. [CrossRef]
3. Sheng, H.; Chen, Z.; Wang, F.; Millner, A. Investigation of 1.2 kV SiC MOSFET for high frequency high power applications . In Proceedings of the 2010 Twenty-Fifth Annual IEEE Applied Power Electronics Conference and Exposition (APEC), Palm Springs, CA, USA, 21–25 February 2010.
4. Jones, E.A.; Wang, F.F.; Costinett, D. Review of commercial GaN power devices and GaN-based converter design challenges. *IEEE J. Emerg. Sel. Top. Power Electron.* **2016**, *4*, 707–719. [CrossRef]
5. Spaziani, L.; Lu, L. Silicon, GaN and SiC: There's room for all: An application space overview of device considerations. In Proceedings of the 2018 IEEE 30th International Symposium on Power Semiconductor Devices and ICs (ISPSD), Chicago, IL, USA, 13–17 May 2018.
6. Viswan, V. A review of silicon carbide and gallium nitride power semiconductor devices. *IJRESM* **2018**, *1*, 224–225.
7. Bosshard, T.R.; Kolar, J.W. Performance comparison of a GaN GIT and a Si IGBT for high-speed drive applications. In Proceedings of the 2014 International Power Electronics Conference (IPEC-Hiroshima 2014—ECCE ASIA), Hiroshima, Japan, 18–21 May 2014.
8. High-performance SiC MOSFET Technology for Power Electronics Design. Available online: https://www.eeweb.com/profile/eeweb/articles/high-performance-coolsic-mosfet-technology-for-power-electronics-design (accessed on 13 April 2020).
9. Adappa, R.; Suryanarayana, K.; Swathi Hatwar, H.; Ravikiran Rao, M. Review of SiC based Power Semiconductor Devices and their Applications. In Proceedings of the 2019 2nd International Conference on Intelligent Computing, Instrumentation and Control Technologies (ICICICT), Kannur, Kerala, India, 5–6 July 2019.
10. Digi-Key. Available online: https://www.digikey.com (accessed on 16 July 2020).
11. Mouser. Available online: https://www.mouser.com (accessed on 16 July 2020).
12. EPC. Available online: https://epc-co.com/epc (accessed on 16 July 2020).
13. Infienon Technologies. Available online: https://www.infineon.com (accessed on 16 July 2020).
14. GaN Systems. Available online: https://gansystems.com (accessed on 16 July 2020).
15. Panasonic. Available online: https://industrial.panasonic.com/ww (accessed on 16 July 2020).
16. Nexperia. Available online: https://www.nexperia.com (accessed on 16 July 2020).
17. Transphorm. Available online: https://www.transphormusa.com (accessed on 16 July 2020).
18. On Semiconductor. Available online: https://www.onsemi.com (accessed on 16 July 2020).
19. Littelfuse. Available online: https://www.littelfuse.com (accessed on 16 July 2020).
20. Cree. Available online: https://www.cree.com (accessed on 16 July 2020).
21. ROHM Semiconductor. Available online: https://www.rohm.com (accessed on 16 July 2020).
22. STMicroelectronics. Available online: https://www.st.com (accessed on 16 July 2020).
23. United Silicon Carbide. Available online: https://unitedsic.com (accessed on 16 July 2020).
24. Microchip. Available online: https://www.microchip.com (accessed on 16 July 2020).
25. GeneSiC Semiconductor. Available online: https://www.genesicsemi.com (accessed on 16 July 2020).
26. Powerex. Available online: https://www.pwrx.com/Home.aspx (accessed on 16 July 2020).
27. SemiQ. Available online: https://www.semiq.com (accessed on 16 July 2020).
28. Papers, Articles and Presentations. Available online: https://gansystems.com/design-center/papers (accessed on 16 July 2020).
29. Observations about the Impact of GaN Technology. Available online: https://www.powerelectronicsnews.com (accessed on 31 March 2020).
30. Use SiC and GaN Power Components to Address EV Design Requirements. Available online: https://www.digikey.tw/en/articles/use-sic-and-gan-power-components-ev-design-requirements (accessed on 13 April 2020).
31. Application Notes. Available online: https://gansystems.com/design-center/application-notes/ (accessed on 14 April 2020).
32. IGBT & SiC Gate Driver Fundamentals. Texas Instruments, IGBT & SiC Gate Driver Fundamentals, Texas Instruments. 2019. Available online: https://www.ti.com/lit/wp/slyy169/slyy169.pdf?ts=1594784799613&ref_url=https%253A%252F%252Fwww.google.com%252F (accessed on 22 April 2020).

33. Pajnić, M.; Pejović, P.; Despotović, Ž.; Lazić, M.; Skender, M. Characterization and gate drive design of high voltage cascode GaN HEMT. In Proceedings of the 2017 International Symposium on Power Electronics (Ee), Novi Sad, Serbia, 19–21 October 2017.
34. Krishna Miryala, V.; Hatua, K. Low-cost analogue active gate driver for SiC MOSFET to enable operation in higher parasitic environment. *IET Power Electron.* **2020**, *13*, 463–474. [CrossRef]
35. Soldati, A.; Imamovic, E.; Concari, C. Bidirectional bootstrapped gate driver for high-density SiC-based automotive DC/DC converters. *IEEE J. Em. Sel. Top. Power Electron.* **2020**, *8*, 475–485. [CrossRef]
36. Maxim Integrated. Available online: https://www.maximintegrated.com (accessed on 16 July 2020).
37. pSemi. Available online: https://www.psemi.com (accessed on 16 July 2020).
38. Silicon Laboratories. Available online: https://www.silabs.com (accessed on 16 July 2020).
39. Texas Instruments. Available online: https://www.ti.com (accessed on 16 July 2020).
40. Navitas Semiconductor. Available online: https://www.navitassemi.com (accessed on 16 July 2020).
41. Analog Devices. Available online: https://www.analog.com/en/index.html (accessed on 16 July 2020).
42. Tamura. Available online: https://www.tamuracorp.com/global/index.html (accessed on 16 July 2020).
43. Diodes Incorporated. Available online: https://www.diodes.com (accessed on 16 July 2020).
44. NXP. Available online: https://www.nxp.com (accessed on 16 July 2020).
45. Power Integrations. Available online: https://www.power.com (accessed on 16 July 2020).
46. Gurpinar, E.; Castellazzi, A. Single-phase T-type inverter performance benchmark using Si IGBTs, SiC MOSFETs, and GaN HEMTs. *IEEE Trans. Power Electron.* **2016**, *31*, 7148–7160. [CrossRef]
47. Gurpinar, E.; Iannuzzo, F.; Yang, Y.; Castellazzi, A.; Blaabjerg, F. Design of low-inductance switching power cell for GaN HEMT based inverter. *IEEE Trans. Ind. Appl.* **2018**, *54*, 1592–1601. [CrossRef]
48. Gurudiwan, S.; Roy, S.K.; Basu, K. Design and implementation of 1.5 kW half bridge bidirectional DC-DC converter based on gallium nitride devices. In Proceedings of the 2019 National Power Electronics Conference (NPEC), Tiruchirappalli, India, 13–15 December 2019.
49. Ebli, M.; Wattenberg, M.; Pfost, M. A high-efficiency bidirectional GaN-HEMT DC/DC converter. In Proceedings of the PCIM Europe 2016: International Exhibition and Conference for Power Electronics, Intelligent Motion, Renewable Energy and Energy Management, Nuremberg, Germany, 10–12 May 2016.
50. Wu, H.; Fayyaz, A.; Castellazzi, A. P-gate GaN HEMT gate-driver design for joint optimization of switching performance, freewheeling conduction and short-circuit robustness. In Proceedings of the 2018 IEEE 30th International Symposium on Power Semiconductor Devices and ICs (ISPSD), Chicago, IL, USA, 13–17 May 2018.
51. Zhao, C.; Trento, B.; Jiang, L.; Jones, E.A.; Liu, B.; Zhang, Z.; Costinett, D.; Wang, F.F.; Tolbert, L.M.; Jansen, J.F.; et al. Design and implementation of a GaN-Based, 100-kHz, 102-W/in3 single-phase inverter. *IEEE J. Em. Sel. Top. Power Electron.* **2016**, *4*, 824–840. [CrossRef]
52. Cai, A.Q.; Siek, L. A 2-kW, 95% efficiency inductive power transfer system using gallium nitride gate injection transistors. *IEEE J. Emerg. Sel. Top. Power Electron.* **2017**, *5*, 458–468. [CrossRef]
53. Meng, W.; Zhang, F.; Fu, Z. A high power density inverter design based on GaN power devices. In Proceedings of the 2018 8th International Conference on Power and Energy Systems (ICPES), Colombo, Sri Lanka, 21–22 December 2018.
54. Li, B.; Zhang, R.; Zhao, N.; Wang, G.; Huo, J.; Zhu, L.; Xu, D. GaN HEMT driving scheme of totem-pole bridgeless PFC converter. In Proceedings of the 2018 IEEE International Power Electronics and Application Conference and Exposition (PEAC), Shenzhen, China, 4–7 November 2018.
55. Elrajoubi, A.M.; Ang, S.S.; George, K. Design and analysis of a new GaN-Based AC/DC converter for battery charging application. *IEEE Trans. Ind. Appl.* **2019**, *55*, 4044–4052. [CrossRef]
56. Shojaie, M.; Elsayad, N.; Tabarestani, S.; Mohammed, O.A. A bidirectional buck-boost converter using 1.3 kV series-stacked GaN E-HEMT modules for electric vehicle charging application. In Proceedings of the 2018 IEEE 6th Workshop on Wide Bandgap Power Devices and Applications (WiPDA), Atlanta, GA, USA, 31 October–2 November 2018.
57. Sun, B.; Zhang, Z.; Andersen, M.A.E. A comparison review of the resonant gate driver in the silicon MOSFET and the GaN transistor application. *IEEE Trans. Ind. Appl.* **2019**, *55*, 7776–7786. [CrossRef]
58. Lu, J.L.; Chen, D.; Yushyna, L. A high power-density and high efficiency insulated metal substrate based GaN HEMT power module. In Proceedings of the 2017 IEEE Energy Conversion Congress and Exposition (ECCE), Cincinnati, OH, USA, 1–5 October 2017.
59. Brothers, J.A.; Beechner, T. GaN module design recommendations based on the analysis of a commercial 3-phase GaN module. In Proceedings of the 2019 IEEE Energy Conversion Congress and Exposition (ECCE), Baltimore, MD, USA, 29 Septmber–3 October 2019.
60. Pirc, E.; Miklavčič, D.; Reberšek, M. Nanosecond pulse electroporator with silicon carbide mosfets: Development and evaluation. *IEEE Trans. Biomed. Eng.* **2019**, *66*, 3526–3533. [CrossRef] [PubMed]
61. Liu, C.; Zhang, Z.; Liu, Y.; Si, Y.; Lei, Q. Smart self-driving multilevel gate driver for fast switching and crosstalk suppression of SiC MOSFETs. *IEEE J. Emerg. Sel. Top. Power Electron.* **2020**, *8*, 442–453. [CrossRef]
62. Park, S.; Sohn, Y.; Cho, G. SiC-based 4 MHz 10 kW ZVS inverter with fast resonance frequency tracking control for high-density plasma generators. *IEEE Trans. Power Electron.* **2020**, *35*, 3266–3275. [CrossRef]
63. Kim, T.; Jang, M.; Agelidis, V.G. Ultra-fast MHz range driving circuit for SiC MOSFET using frequency multiplier with eGaN FET. *IET Power Electron.* **2016**, *9*, 2085–2094. [CrossRef]

64. Nayak, P.; Pramanick, S.K.; Rajashekara, K. A high-temperature gate driver for silicon carbide mosfet. *IEEE Trans. Ind. Electron.* **2018**, *65*, 1955–1964. [CrossRef]

65. Qi, F.; Wang, M.; Xu, L. Investigation and review of challenges in a high-temperature 30-kVA three-phase inverter using SiC MOSFETs. *IEEE Trans. Ind. Appl.* **2018**, *54*, 2483–2491. [CrossRef]

66. Zhang, Z.; Yao, K.; Ke, G.; Zhang, K.; Gao, Z.; Wang, Y.; Ren, X.; Chen, Q. SiC MOSFETs gate driver with minimum propagation delay time and auxiliary power supply with wide input voltage range for high-temperature applications. *IEEE J. Em. Sel. Top. Power Electron.* **2020**, *8*, 417–428. [CrossRef]

67. Li, C.; Lu, Z.; Chen, Y.; Li, C.; Luo, H.; Li, W.; He, X. High off-state impedance gate driver of SiC MOSFETs for crosstalk voltage elimination considering common-source inductance. *IEEE Trans. Power Electron.* **2020**, *35*, 2999–3011. [CrossRef]

68. Zhao, S.; Zhao, X.; Dearien, A.; Wu, Y.; Zhao, Y.; Mantooth, H.A. An intelligent versatile model-based trajectory-optimized active gate driver for silicon carbide devices. *IEEE J. Emerg. Sel. Top. Power Electron.* **2020**, *8*, 429–441. [CrossRef]

69. Wang, Y.; Xue, L.; Wang, C.; Wang, P.; Li, W. High off-state impedance gate driver of SiC MOSFETs for crosstalk voltage elimination considering common-source inductance. *IEEE Trans. Power Electron.* **2016**, *31*, 5547–5561. [CrossRef]

70. Wang, R.; Liang, L.; Chen, Y.; Kang, Y. A single voltage-balancing gate driver combined with limiting snubber circuits for series-connected SiC MOSFETs. *IEEE J. Em. Sel. Top. Power Electron.* **2020**, *8*, 465–474. [CrossRef]

71. Wang, J.; Mocevic, S.; Burgos, R.; Boroyevich, D. High-scalability enhanced gate drivers for SiC MOSFET modules with transient immunity beyond 100 V/ns. *IEEE Trans. Power Electron.* **2020**, *35*, 10180–10199. [CrossRef]

72. Yang, C.; Pei, Y.; Xu, Y.; Zhang, F.; Wang, L.; Zhu, M.; Yu, L. A gate drive circuit and dynamic voltage balancing control method suitable for series-connected SiC mosfets. *IEEE Trans. Power Electron.* **2020**, *35*, 6625–6635. [CrossRef]

73. Jørgensen, A.B.; Sønderskov, S.H.; Beczkowski, S.; Bidoggia, B.; Munk-Nielsen, S. Analysis of cascaded silicon carbide MOSFETs using a single gate driver for medium voltage applications. *IET Power Electron.* **2020**, *13*, 413–419. [CrossRef]

74. Sakib, N.; Manjrekar, M.; Ebong, A. An overview of advances in high reliability gate driving mechanisms for SiC MOSFETs. In Proceedings of the 2017 IEEE 5th Workshop on Wide Bandgap Power Devices and Applications (WiPDA), Albuquerque, NM, USA, 30 October–1 November 2017.

75. Alves, L.F.S.; Lefranc, P.; Jeannin, P.; Sarrazin, B. Review on SiC-MOSFET devices and associated gate drivers. In Proceedings of the 2018 IEEE International Conference on Industrial Technology (ICIT), Lyon, France, 20–22 February 2018.

76. Choudhury, A. Present status of SiC based power converters and gate drivers—A review. In Proceedings of the 2018 International Power Electronics Conference (IPEC-Niigata 2018 -ECCE Asia), Niigata, Japan, 20–24 May 2018.

77. Liu, Y.; Yang, Y. Review of SiC MOSFET drive circuit. In Proceedings of the 2019 IEEE International Conference on Electron Devices and Solid-State Circuits (EDSSC), Xi'an, China, 12–14 June 2019.

78. Zhao, S.; Zhao, X.; Wei, Y.; Zhao, Y.; Mantooth, H.A. A review on switching slew rate control for silicon carbide devices using active gate drivers. *IEEE J. Emerg. Sel. Top. Power Electron.* **2020**. early access. [CrossRef]

79. Jørgensen, A.B.; Aunsborg, T.S.; Beczkowski, S.; Uhrenfeldt, C.; Munk-Nielsen, S. High-frequency resonant operation of an integrated medium-voltage SiC MOSFET power module. *IET Power Electron.* **2020**, *13*, 475–482. [CrossRef]

80. Huang, Z.; Chen, C.; Xie, Y.; Yan, Y.; Kang, Y.; Luo, F. A High-Performance Embedded SiC Power Module Based on a DBC-Stacked Hybrid Packaging Structure. *IEEE J. Emerg. Sel. Top. Power Electron.* **2020**, *8*, 351–366. [CrossRef]

81. Barlow, M.; Ahmed, S.; Francis, A.M.; Mantooth, H.A. An integrated SiC CMOS gate driver for power module integration. *IEEE Trans. Power Electron.* **2019**, *34*, 11191–11198. [CrossRef]

 *micromachines*

Article

Design and Implementation of a Flexible Photovoltaic Emulator Using a GaN-Based Synchronous Buck Converter

Chao-Tsung Ma [1,*], Zhen-Yu Tsai [1], Hung-Hsien Ku [2] and Chin-Lung Hsieh [2]

[1] Applied Power Electronics Systems Research Group, Department EE, CEECS, National United University, Miaoli City 36063, Taiwan; M0721013@o365.nuu.edu.tw

[2] Chemistry Division, Institute of Nuclear Energy Research (INER), AEC, Executive Yuan, Taoyuan 325207, Taiwan; HHKu@iner.gov.tw (H.-H.K.); clhsieh@iner.gov.tw (C.-L.H.)

* Correspondence: ctma@nuu.edu.tw

Abstract: In order to efficiently facilitate various research works related to power converter design and testing for solar photovoltaic (PV) generation systems, it is a great merit to use advanced power-converter-based and digitally controlled PV emulators in place of actual PV modules to reduce the space, cost, and time to obtain the required scenarios of solar irradiances for various functional tests. This paper presents a flexible PV emulator based on gallium nitride (GaN), a wide-bandgap (WBG) semiconductor, and a based synchronous buck converter and controlled with a digital signal processor (DSP). With the help of GaN-based switching devices, the proposed emulator can accurately mimic the dynamic voltage-current characteristics of any PV module under normal irradiance and partial shading conditions. With the proposed PV emulator, it is possible to closely emulate any PV module characteristic both theoretically, based on manufacturer's datasheets, and experimentally, based on measured data from practical PV modules. A curve fitting algorithm is used to handle the real-time generation of control signals for the digital controller. Both simulation with computer software and implementation on 1 kW GaN-based experimental hardware using Texas Instruments DSP as the controller have been carried out. Results show that the proposed emulator achieves efficiency as high as 99.05% and exhibits multifaceted application features in tracking various PV voltage and current parameters, demonstrating the feasibility and excellent performance of the proposed PV emulator.

Keywords: wide bandgap semiconductor; gallium nitride; photovoltaic module; digital signal processor; synchronous buck converter

Citation: Ma, C.-T.; Tsai, Z.-Y.; Ku, H.-H.; Hsieh, C.-L. Design and Implementation of a Flexible Photovoltaic Emulator Using a GaN-Based Synchronous Buck Converter. *Micromachines* **2021**, *12*, 1587. https://doi.org/10.3390/mi12121587

Academic Editor: Giovanni Verzellesi

Received: 5 November 2021
Accepted: 17 December 2021
Published: 20 December 2021

Publisher's Note: MDPI stays neutral with regard to jurisdictional claims in published maps and institutional affiliations.

Copyright: © 2021 by the authors. Licensee MDPI, Basel, Switzerland. This article is an open access article distributed under the terms and conditions of the Creative Commons Attribution (CC BY) license (https://creativecommons.org/licenses/by/4.0/).

1. Introduction

It has been predicted that the world's energy consumption will increase by roughly 28% between 2015 and 2040, with residential and commercial buildings' electricity usage being the major contributor to the growth [1]. Conventional electricity generation technologies tend to produce carbon dioxide emission, which is believed to be associated with global warming and some other environmental issues. This has led to an urgent need to increase the penetration of renewable energy (RE)-based distributed power generation (DG). In the past decade, the penetration of typical RE-based DG technologies, such as wind power generation and solar photovoltaic (PV) generation, has grown drastically worldwide [2]. Because of the inherently unpredictable and nonlinear characteristics of these RE-based DG technologies, improvement in energy harvesting and conversion technologies is much desired. Among currently used RE sources, solar PV offers many advantages: it is inexhaustible and pollution-free, and it exhibits the highest energy density, with an average of 170 W/m^2. In addition, solar PV generation systems require relatively low maintenance costs. Increasing the usage of distributed PV systems to a certain level can substantially reduce transmission losses, thus increasing energy utilization. Based on the data reported in [3], until 2017, the total share of global RE-based generation was only roughly 4%, but it has been expected that the solar PV generation level will catch up in the share of future RE-based generation.

The PV cells in a PV module produce DC electricity out of sunlight. As a result, a proper power converter interface is required for a grid-connected or standalone solar PV generation system when supplying electricity to AC loads. When power converters are tested for practical PV generation applications, environmental conditions such as solar irradiance and ambience temperature have to be considered. Testing the converters under actual and various environmental conditions is a valid yet time-consuming method, and the associated cost may be very high. Moreover, it is also challenging to obtain the current and voltage (I-V) characteristics of a PV module in a laboratory with limited space. A better approach is to design PV emulators using power converters, which require much less space and are easily programmable. In the literature, many PV emulation systems have been reported. In [4], a buck converter-based PV emulator was controlled using a fast-convergence resistance-feedback strategy that required a current-resistance model of a PV module. A dynamic field programmable analog array (FPAA)-based PV emulator was presented in [5] with the following advantages: (1) it did not require numerical interpolation or large amount of data in memory; (2) it was easily characterized, compared with a field programmable gate array (FPGA)- or digital signal processor (DSP)-based implementations. In [6], a portable, inexpensive and light-weight plug-and-play PV emulator was developed using a DC/DC converter. Great promise for PV module emulation with a minimum amount of output error was shown. In [7], S. M. Azharuddin et al. proposed a near-accurate PV emulator using a simple power converter and a dSPACE controller for real-time control. The proposed device enabled fast converter response with a high bandwidth current regulator. A low-cost, portable, embedded design for implementing a real-time, high-accuracy PV emulator using new processors was proposed in [8].

For industrial applications, a PV emulator with a higher power rating is required. In [9], a dynamic 10-kVA PV emulator based on high-power inverters and an active LCL filter was developed and implemented. The maximum power point tracking (MPPT) algorithms of a real PV module were tested using the proposed hardware. The minimized harmonic current content and fast adjustment for power factor (PF) were achieved with the selected power stage. A. Koran et al. [10] proposed a high-efficiency PV source simulator with the advantages of using analog and digital-based controllers. A three-phase AC-DC dual boost rectifier cascaded with a three-phase DC-DC interleaved buck converter was used in this system. A low-cost PV module emulation system developed using LabVIEW software was proposed in [11]. Several analytical models of PV cells and the estimation of power and current curves for a given irradiance and temperature were studied. D. Pelin et al. [12] described a PV emulator using DC programmable sources and conducted a 6-day emulation. In [13], a real-time emulator for control and diagnosis purposes of solar PV generation was developed in a Matlab/Simulink environment. I-V characteristics of generator model were emulated with a programmable DC/AC power source. A PV emulator with quasi-sliding mode controller to realize PV characteristics under rapidly varying environmental conditions was proposed in [14]. In their design case, a user-friendly graphical user interface (GUI) was developed. In [15], a PV emulator using dSPACE controller with simple control was designed to achieve fast response. Current mode controller and a modified look-up table were combined for the control mechanism.

In most of the papers reviewed above, the theoretical models using V-I curves of the given PV modules and the conventional silicon (Si)-based switching devices were commonly used for designing converter-based PV emulators. In the literature, a conference paper [16] presented a hardware design case of the PV module emulator, in which GaN MOSFETs were used; however, the details of controller design issues were not discussed. Comprehensive analysis and comparison of performance between Si- and GaN-based DC-DC converters can be found in [17]. In this paper, a gallium nitride (GaN)-based and DSP-controlled DC-DC converter is presented to achieve a flexible and highly efficient PV emulator with the capability to closely simulate any V-I characteristic of a PV module using parameters either from manufacture's datasheets or measured data, including normal operating conditions and shading conditions. Flexible emulation features have been implemented with the proposed digital control scheme and verified with a 1 kW hardware prototype. This paper is organized into four sections. In the next section, the PV model is firstly derived, from the cell level to the module level. Then, the modeling of the DC/DC converter and controllers, along with the implementation methodology, are explained. Section 3 presents results from the simulation and implementation of the proposed emulator with an electronic load to demonstrate the performance of the proposed PV emulator. The test of the system efficiencies of the proposed PV emulator using GaN-based synchronous buck converter operated at different switching frequencies is presented in Section 4. Finally, a summary of the paper is given in the last section.

2. PV Module Modeling and Design of Synchronous Buck Converter

2.1. PV Module Modeling

A PV module consists of a number of PV cells in series and parallel. Each PV cell is made of a P-N junction semiconductor diode. As the equivalent circuit shown in Figure 1, I_{ph} represents equivalent current source (A); D_j represents the P-N junction diode; R_{sh} represents equivalent parallel resistance of the material (Ω); R_s represents equivalent series resistance of the material (Ω), and V_p and I_p represent output voltage (V) and current (A), respectively.

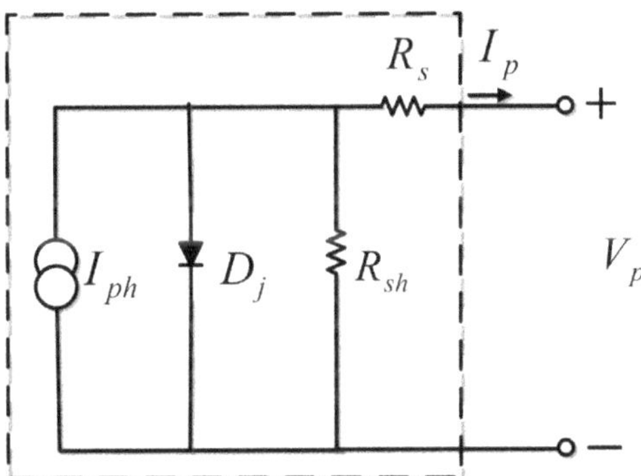

Figure 1. Equivalent circuit of a PV cell.

From Figure 1, the relationship between output voltage and current can be derived as follows:

$$I_p = I_{ph} - I_{rs} \cdot \left\{ \exp\left[\frac{q(V_p + R_s I_p)}{AkT}\right] - 1 \right\} - \frac{V_p + R_s I_p}{R_{sh}}, \tag{1}$$

where I_{rs} represents equivalent reverse saturated current of D_j (A); q represents elementary charge (1.60×10^{-19} C); A represents ideal factor (ranging from 1 to 5); k represents Boltzmann constant (1.38×10^{-23} J/K), and T represents surface temperature (K).

Because the voltage, current, and power generated by a single PV cell are low, many PV cells are usually connected in series and parallel to form a PV module in order to boost the output, as the equivalent circuit shown in Figure 2, where n_p represents the number of paralleled diode paths, and n_s represents the number of series-connected diodes in each path.

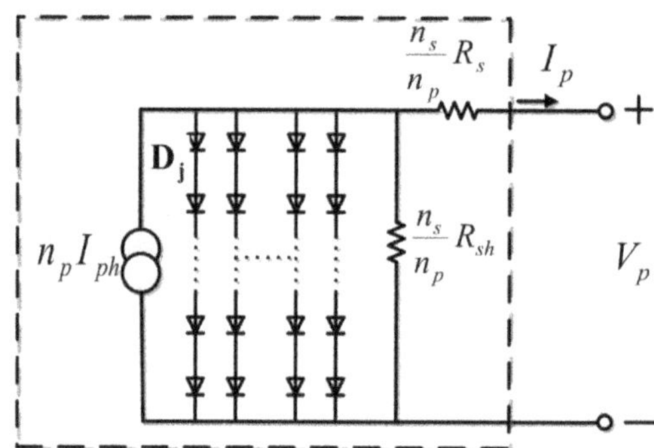

Figure 2. Equivalent circuit of a PV module.

Under normal conditions, R_{sh} is large, and R_s is small. Ignoring these two parameters, we can simplify (1) to obtain (2).

$$I_p = I_{ph} - I_{rs} \cdot [\exp\left(\frac{qV_p}{AkT}\right) - 1], \tag{2}$$

where

$$I_{rs} = I_{rr} \cdot \left(\frac{T}{T_r}\right)^3 \cdot \exp[\frac{qE_g}{kA}(\frac{1}{T_r} - \frac{1}{T})], \tag{3}$$

I_{rr} represents reverse saturated current (A) at reference temperature (T_r), and

$$E_g = 1.16 - (7.02 \times 10^{-4}) \cdot \frac{T^2}{T - 1108}, \tag{4}$$

Also,

$$I_{ph} = [I_{scr} + \frac{\alpha}{1000}(T - T_r)] \cdot S, \tag{5}$$

where I_{scr} represents short-circuit current (A) at reference temperature and irradiance; α represents short-circuit temperature coefficient of a PV cell (mA/°C), and S represents solar irradiance (kW/m^2). The output power of a PV cell can then be expressed as follows:

$$P_p = V_p I_p = I_{ph} V_p - I_{rs} V_p \cdot [\exp(\frac{qV_p}{AkT}) - 1]. \tag{6}$$

In Figure 2, R_{sh} and R_s can be ignored so that the relationship between output voltage and current can be expressed as follows:

$$I_p = n_p I_{ph} - n_p I_{rs} \cdot [\exp(\frac{qV_p}{AkTn_s}) - 1]. \tag{7}$$

As a result, the output power of the PV module can then be expressed as follows:

$$P_p = V_p I_p = n_p I_{ph} V_p - n_s I_{rs} V_p \cdot [\exp(\frac{qV_p}{AkTn_s}) - 1] \tag{8}$$

In this paper, two PV modules are evaluated for the design of the PV emulator. The specifications of a single PV module are as follows: rated power P_{mp} = 300 W, number of PV cells = 60, rated current I_{mp} = 7.87 A, rated voltage V_{mp} = 38.4 V, short-circuit current I_{sc} = 8.7 A, open-circuit voltage V_{oc} = 48 V, and idea factor A = 1.5. The ambience temperature is set at 25 °C. To evaluate the V-I characteristics of multiple PV modules connected in series under normal and partial shading scenarios, two PV modules are used, as shown in Figure 3, where the output terminals of each module are shunted with a bypass diode.

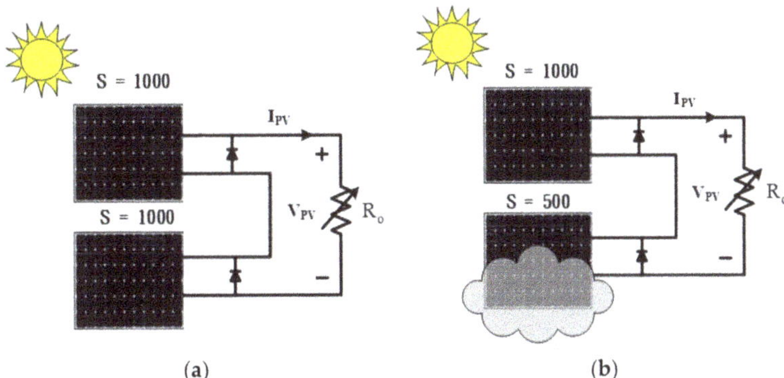

Figure 3. Schematic configuration of two PV modules connected in series, (**a**): $S = 1000/1000$, (**b**): $S = 1000/500$.

By applying the specifications to the derived equations and designating load and irradiance conditions, theoretical P-V characteristics of the PV modules can be calculated. Figure 4 shows theoretical and measured P-V characteristics of a single PV module under 0–150 Ω load conditions and two irradiance conditions (1000 and 500 W/m^2, respectively). Next, Figure 5 shows theoretical and measured P-V characteristics of two PV modules connected in series under the same load conditions and one module receives 500 W/m^2 irradiance, and the other receives 1000 W/m^2 irradiance.

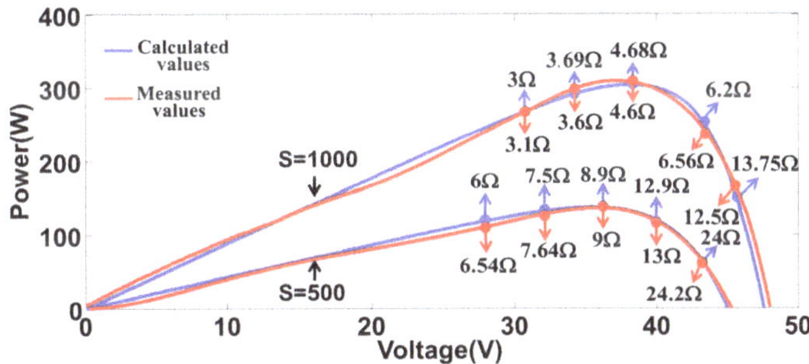

Figure 4. Comparison of calculated and measured P-V characteristics of a single PV module under 0–150 Ω load conditions and with the same irradiance conditions (S = 1000 and 500).

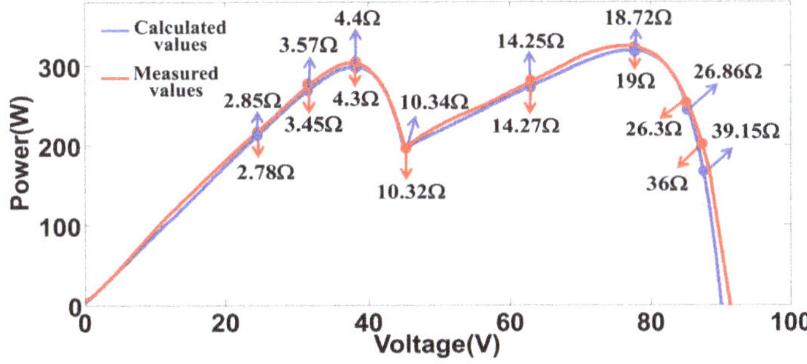

Figure 5. Comparison of calculated and measured P-V characteristics of two PV modules connected in series under 0–150 Ω load conditions and irradiance condition of S = 1000 and 500 for the two modules, respectively.

As can be seen in Figures 4 and 5, the measured value and calculated values are not perfectly meet. This is dure to the inevitable error in curve fitting a limited number of measured points. Some advanced algorithms can be used to improve the quality of fitting but the calculating burden of digital controller will be increased.

2.2. Design of Synchronous Buck Converter

2.2.1. Operating Principle and Mathematical Model of Synchronous Buck Converter

Figure 6 shows the circuit configuration of the synchronous buck converter used in this paper, including two power switches S_1 and S_2, an inductor L, an input capacitor C_{in}, and an output capacitor C_o. The operating principle of this converter is as follows: S_1 acts as the main switch; when it is on, V_{in} charges L, and when S_1 is off, the energy stored in L is transmitted through S_2 to C_o.

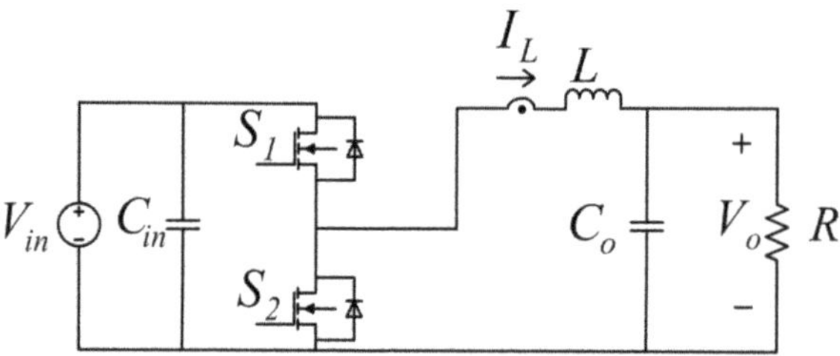

Figure 6. Configuration of synchronous buck converter.

State average method can be used to derive the mathematical model of the synchronous buck converter. When S_1 is on, and S_2 is off, the voltage across L can be expressed as follows:

$$V_L = L\frac{di_L}{dt} = V_{in} - V_o. \tag{9}$$

When S_1 is off, and S_2 is on, the voltage across L can be expressed as follows:

$$V_L = L\frac{di_L}{dt} = -V_o. \tag{10}$$

As a result, average inductor voltage can be expressed as follows:

$$\begin{aligned}
\overline{V_L(t)} &= L\overline{\frac{di_L}{dt}} = \frac{1}{T_s}\int_0^{T_s} V_L(\tau)d\tau \\
&= V_L(\tau1)\frac{t_1}{T_s} + V_L(\tau2)\frac{T_s - t_1}{T_s} \\
&= V_L(\tau1)d + V_L(\tau2)d''
\end{aligned} \tag{11}$$

Substituting (9) and (10) into (11) gives the following:

$$\begin{aligned}
\overline{V_L(t)} &= (V_{in} - V_R)d + (-V_R)d'' \\
&= dV_{in} - V_R
\end{aligned} \tag{12}$$

Next, we consider the following disturbance terms:

$$i_L = I_L + \hat{i}_L(t), |I_L| >> |\hat{i}_L(t)|; \tag{13}$$

$$d = D + \hat{d}(t), |D| >> |\hat{d}(t)|; \tag{14}$$

$$v_{in} = V_{in} + \hat{v}_{in}(t), |V_{in}| >> |\hat{v}_{in}(t)|; \tag{15}$$

$$v_o = V_o + \hat{v}_o(t), |V_o| >> |\hat{v}_o(t)|. \tag{16}$$

As a result, (12) can be expressed as follows:

$$L\frac{d(I_L+\hat{i}_L(t))}{dt} = (D+\hat{d}(t))\cdot(V_{in}+\hat{v}_{in}(t)) - (V_o+\hat{v}_o(t))$$
$$= [DV_{in}-V_o] + [D\hat{v}_{in}(t)+V_{in}\hat{d}(t)-\hat{v}_o(t)] + [\hat{d}(t)\hat{v}_{in}(t)] \tag{17}$$

where $\hat{d}(t)\hat{v}_{in}(t)$ is very small and thus can be neglected, and

$$L\frac{d\hat{i}_L(t)}{dt} = D\hat{v}_{in}(t)+V_{in}\hat{d}(t)-\hat{v}_o(t). \tag{18}$$

Performing Laplace transform on (18) gives the following:

$$sL\hat{i}_L(s) = D\hat{v}_{in}(s)+V_{in}\hat{d}(s)-\hat{v}_o(s). \tag{19}$$

As a result, we get the mathematical model shown in Figure 7.

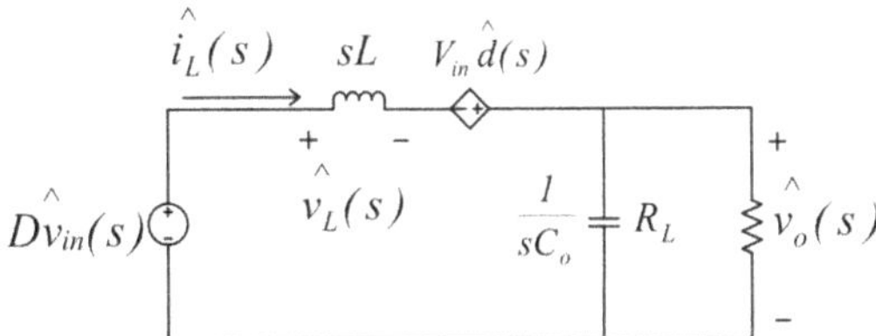

Figure 7. Mathematical model of synchronous buck converter.

2.2.2. Design of Controllers for the Proposed PV Emulator

The synchronous buck converter proposed in this paper adopts single-loop inductor current control. A signal generator is first used to generate a triangular wave, which is used to calculate inductor current command, and the current command is then compared with current feedback to yield the error signal, which is used to generate PWM control voltage. Finally, trigger signal is generated to adjust the duty cycle of the switches. Figure 8 shows the converter architecture, where k_s and k_v represent current and voltage sensing scales, respectively; I_L and $i_L{}^*$ represent current feedback and command, respectively, and v_{con} represents PWM voltage.

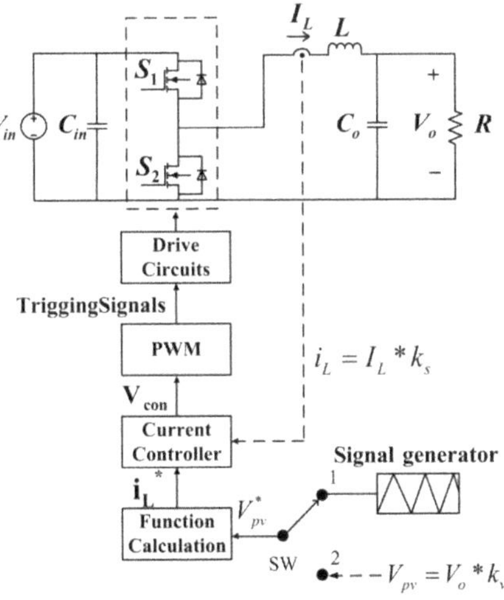

Figure 8. Block diagram of current mode control of synchronous buck converter.

Letting $\hat{v}_{in}(s) = 0$, we get the following:

$$sL\hat{i}_L(s) - V_{in}\hat{d}(s) + \hat{i}_L(s)\left(\frac{1}{sC_O}//R_L\right) = 0. \tag{20}$$

As a result,

$$\frac{\hat{i}_L(s)}{\hat{d}(s)} = \frac{V_{in}}{\left(sL + \frac{R_L}{1+sC_OR_L}\right)} = \frac{V_{in}(1+sC_OR_L)}{sL(1+sC_OR_L) + R_L} = \frac{V_{in}}{L} \cdot \frac{s + \frac{1}{C_OR_L}}{s^2 + \frac{1}{C_OR_L}s + \frac{1}{C_OL}}. \tag{21}$$

Consequently, the block diagram of inductor current control loop can be graphed as shown in Figure 9, where type II proportional-integral (PI) controller and K factor are used to perform further quantification design.

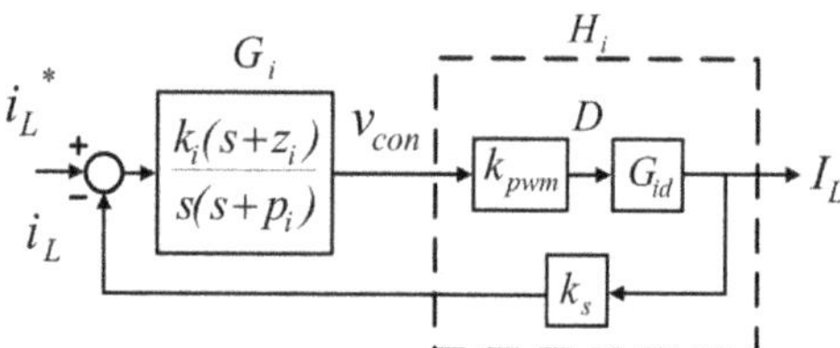

Figure 9. Block diagram of inductor current control loop.

According to Figure 9, we can obtain plant transfer function of the current control loop:

$$\begin{aligned} H_i(s) &= k_{PWM} \cdot G_{id} \cdot k_s = \frac{1}{v_t} \times \frac{\hat{i}_b(s)}{\hat{d}(s)} \times k_s \\ &= \frac{k_s V_{in}}{v_t L} \times \frac{s + \frac{1}{R_L C_O}}{s^2 + \frac{1}{R_L C_O}s + \frac{1}{LC_O}} \end{aligned} \tag{22}$$

where R_L represents load resistance. Substituting the related design specifications into (22) gives the following:

$$H_i(s) = \frac{6400s + 5.77 \times 10^6}{s^2 + 901.6s + 2 \times 10^7}. \tag{23}$$

Crossover frequency is chosen at $1/8$ times the switching frequency:

$$\omega_{ci} = f_S \times \frac{1}{8} \times 2\pi = 50K \times \frac{1}{8} \times 2\pi = 39250(\text{rad/s}). \tag{24}$$

Consequently, we can calculate the gain and phase angle of the plant and the phase boost required at the crossover frequency:

$$H_i(\omega_i) = Gain_{Hi}\angle Angle_{Hi} = 0.16551\angle - 90°. \tag{25}$$

$$\begin{aligned} PhaseBoost &= PhaseMargin - Angle_{Hi} - 90° \\ &= 80° - (-90) - 90° = 80° \end{aligned} \tag{26}$$

K factor can then be calculated as follows:

$$K_{factor} = tan\left(\frac{PhaseBoost}{2} + 45°\right) = tan\left(\frac{80°}{2} + 45°\right) = 11.43. \tag{27}$$

As a result, we get controller zero and pole:

$$z = \frac{\omega_{ci}}{K_{factor}} = \frac{39250}{11.43} \cong 3433.95(\text{rad/S}). \tag{28}$$

$$p = \omega_{ci} \times K_{factor} = 39250 \times 11.43 \cong 448627.5(\text{rad/S}). \tag{29}$$

Required gain compensation at the crossover frequency is as follows:

$$k_i = \frac{p}{Gain_{Hi}} = \frac{448627.5}{0.16551} \cong 2710576.4. \tag{30}$$

Finally we obtain the controller transform function according to (28)–(30):

$$G_i(s) = \frac{k_i(s+z)}{s(s+p)} = \frac{2710576.4(s+3433.95)}{s(s+448627.5)}. \tag{31}$$

Figure 10 shows Bode plot of the current control loop. The phase margin is 80 degrees.

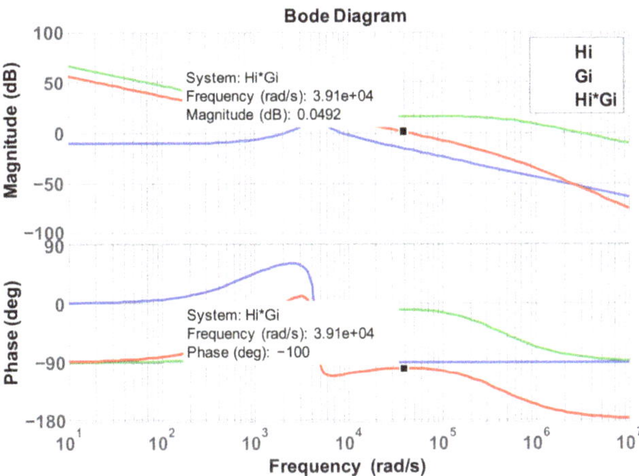

Figure 10. Bode plot of inductor current control loop.

3. Simulation and Implementation

To validate the performance of the designed PV emulator, both simulation studies with computer software and hardware implementation are performed.

Figure 11 shows the complete software simulation model of the proposed PV emulator and the arrangement of control signals. In this paper, two control methods, the direct referencing method (case 1 and case 2) and output voltage feedback method (case 3), are used to implement the control of the proposed PV emulator with a digital control scheme. In Figure 11, block (a) is the power circuit of the proposed synchronous buck converter. Block (b) is an analog to digital (AD) module in TI's DSP. Block (c) is the designed type-II current controller for tracking the current commands sent from PV model blocks (d) or (e). Block (d) is a program block interpreting the theoretical voltage-current relationship of emulated PV module as given in equation (7). Block (e) has two program blocks respectively describing the shedding conditions of two PV modules having different irradiances, e.g., S = 1000 and S = 500. In this study, a simple curve fitting method is used to fit the measured points (V & I pairs) of a given PV module and irradiance condition. In this study case, the number of measured voltage and current data pair is 10. It can be predicted that the measured and theoretically calculated values cannot be perfectly meet. This is due to the inevitable error in curve fitting a limited number of measured points. Some advanced algorithms can be used to improve the quality of fitting but the calculating burden of digital controller will be increased. In block (f), a voltage ramp generator is programmed to send out the emulated terminal voltages of the PV module to the PV model blocks, (e) and (d), for calculating the corresponding currents. As can be seen in Figure 11, the current commands of the PV emulator can be any combination of the outputs from (d) and (e), depending on the desired emulation scenario. It should be noted that, normally, a properly designed user interface is required for the practical application of the PV emulator; however, it is out of the investigation scope of the present paper.

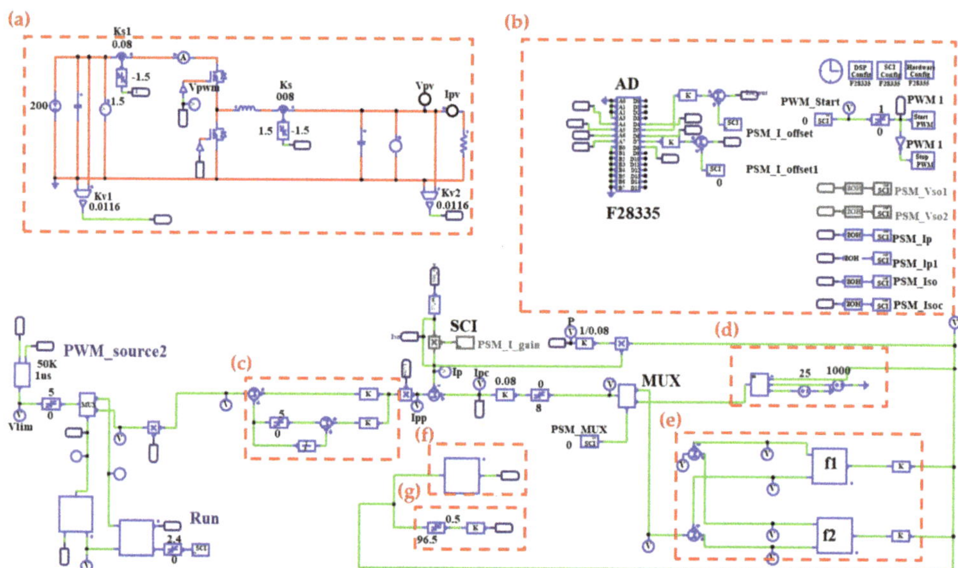

Figure 11. Complete simulation model of the proposed PV emulator and the arrangement of control signals. (**a**): buck-boost converter, (**b**): analog to digital module, (**c**): current controller, (**d**) and (**e**): PV model blocks, (**f**) and (**g**): voltage command blocks.

A photo of the complete hardware implementation of the DC-DC synchronous buck converter-based PV module emulator is shown in Figure 12, including (1) DC power supply, (2) DSP-controlled GaN based synchronous buck converter, (3) auxiliary power, (4) function generator, (5) personal computer, (6) digital oscilloscope, (7) voltage and current probes, and (8) resistive load. The synchronous buck converter is built using GaN power switches (TPH3207WS) with properly designed output inductor and capacitor values. Laboratory-made sensing circuits are used for voltage and current measurements. During operation, sensed current and voltage feedbacks are directly fed into the ADC ports of TI's TMS320F28335 controller. Corresponding PWM pulses are then sent out via the PWM port. Gate pulses are isolated from the buck converter's power circuitry with proper driving IC before being applied to the gate of the GaN switches. The major advantage of using DSP and a digital control scheme is online parameter tuning and better flexibility in implementing possible changes of conditions. To verify the effectiveness and performance of the proposed PV emulator, two sets of dynamic V-I characteristics and conditions, i.e., normal irradiance and partial shading conditions, are tested. It is found that the response time of the proposed controller is very short and that the absolute error between experimental results and calculated values is quite small. Typical results are presented in the following paragraphs for comparison.

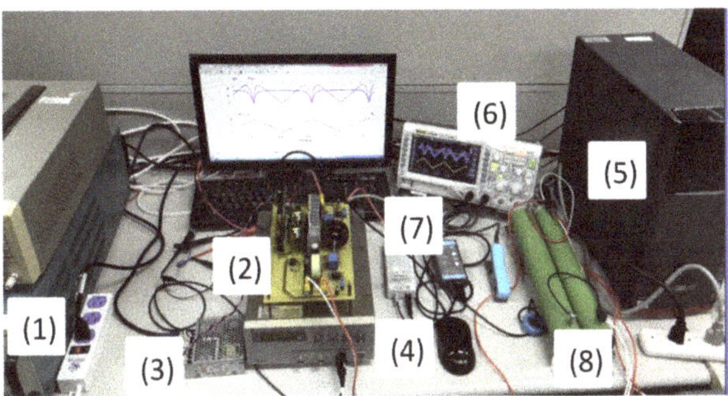

Figure 12. Complete hardware implementation of the proposed PV module emulator.

In case 1, theoretical characteristics of the two PV modules under normal irradiance (S = 1000 & 1000) are used. Figure 13a,b shows a set of simulation and measured results from the PV emulator (I_{PV}: 5 A/div; V_o: 40 V/div; T: 2 s/div; V_{PV}: 48 V/div; P_{PV}: 270 W/div).

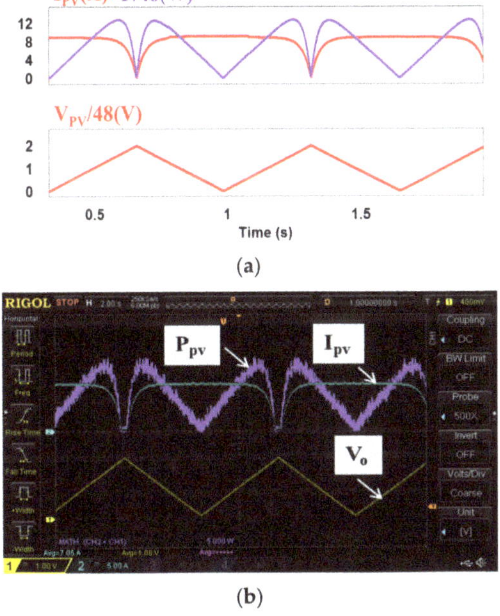

Figure 13. Experimental results of the proposed PV emulator under S = 1000 & 1000: (**a**) simulated power and current outputs and PV voltage; (**b**) implemented power and current outputs and PV voltage.

In case 2, both theoretical and measured characteristics of the two PV modules under partial shading (S = 500 & 1000) are used. Figure 14a,b and Figure 15a,b respectively show the two sets of simulation and measured results from the proposed PV emulator (I_{PV}: 5 A/div; V_o: 40 V/div; T: 2 s/div; V_{PV}: 48 V/div; P_{PV}: 270 W/div).

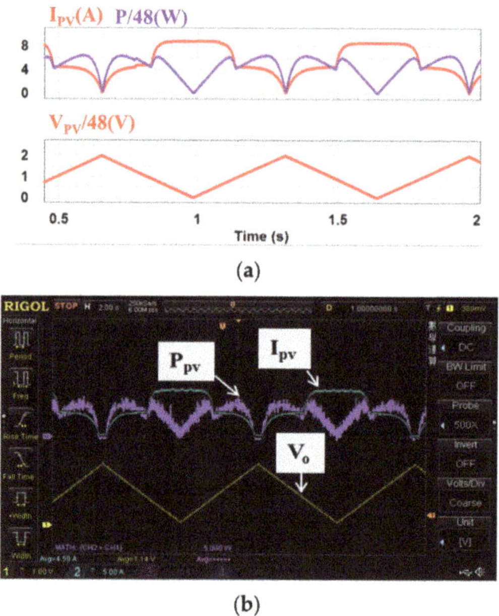

Figure 14. Output results of the proposed PV emulator under S = 500 & 1000 with theoretical PV values: (**a**) simulated power and current outputs and PV voltage; (**b**) implemented power and current outputs and PV voltage.

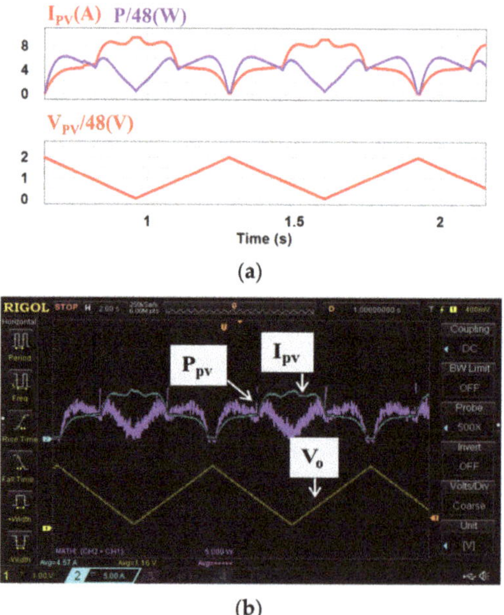

Figure 15. Output results of proposed PV emulator under *S* = 500 & 1000 with practical measured PV data: (**a**) simulated power and current outputs and PV voltage; (**b**) implemented power and current outputs and PV voltage.

In case 3, the control transition between the conditions of cases 1 and 2 is performed. Figure 16a shows simulated transition from normal condition (A) to shading condition (B), and Figure 16b shows the corresponding measured waveforms. Figure 17 shows the results of control transition from shading condition (B) to normal condition (A) (I_{PV}: 5 A/div; V_{pv}: 40 V/div; T: 20 ms/div).

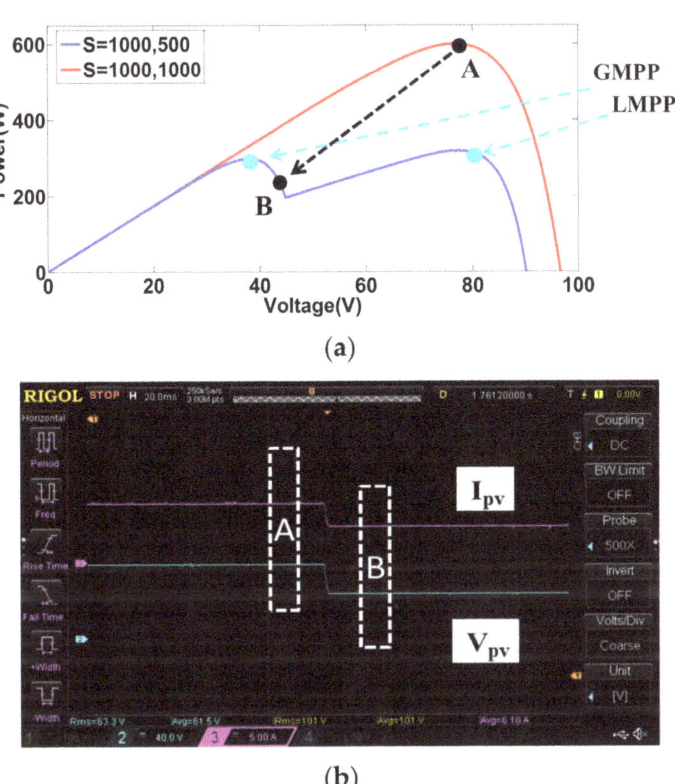

Figure 16. Results of transition from normal condition (A) to shading condition (B): (**a**) simulation; (**b**) implementation.

As can be seen in Figures 13–17, performance in tracking the voltage-current parameters of a given PV module is high. It has been observed that the control error in all test cases is almost negligible. In control practice, the control error can be affected by a number of factors. Typical ones include the type of controller used, the precision of sensors used (the current sensors in this specific design case), the sampling speed, the level of EMI and the quality of the AD device, etc. In this design case, a type II controller, equivalent to a PI controller with a low-pass filter, is utilized to guarantee a theoretical zero-control-error. The accuracy of the Hall current sensor used in our design is about 0.2–1%. To eliminate the 0.2–1% sensing error, a homemade digital compensator is used at the AD configuring module of the TI's DSP controller with a pre-tuning step.

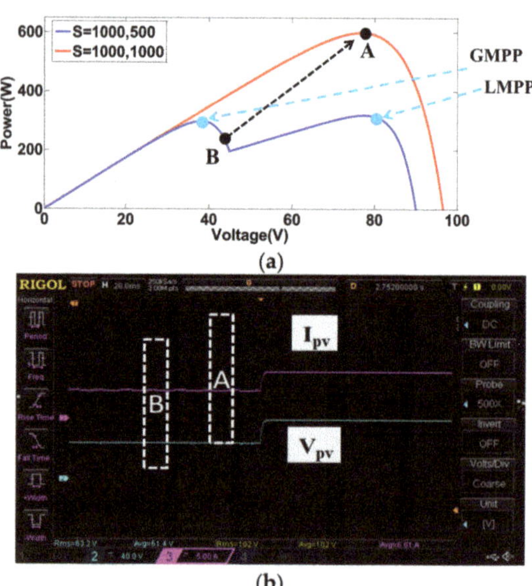

Figure 17. Results of transition from shading condition (B) to normal condition (A): (**a**) simulation; (**b**) implementation.

4. The Analysis of System Efficiency

In this paper, the system efficiency of the proposed PV emulator using a GaN-based synchronous buck converter operated at different switching frequencies (50 kHz, 80 kHz) and different output power levels is practically tested. Figure 18 shows the system block diagram of the efficiency tests carried out in this study. In this test, the DC terminals of the proposed GaN-based synchronous buck converter are connected to a programmable DC power supply having the output voltage set to 200 V, and the DC output terminal voltage of the proposed GaN-based synchronous buck converter is regulated at a fixed 100 V by the proposed voltage controller. For testing the converter efficiency under different output power levels, a programmable electronic load is connected to the DC output terminal of the converter. By setting different P_{out} and measuring the corresponding P_{in} of the converter, the system efficiency at a specific power level and switching frequency can be readily calculated. In this paper, two switching frequencies, i.e., 50 and 80 kHz, are tested at five load levels. The calculated results are graphically shown in Figure 19. As can be seen in Figure 19, a maximum efficiency of 99.05% appears at about 80% of the converter's rated capacity (1 kW), with a switching frequency of 50 kHz, and it is found that, when the switching frequency increases, the efficiency decreases. This is mainly due to the increase in switching losses. In practice, other factors may also affect the performance of efficiency; typical factors include layout construction, the noise level of sensing devices, and driving and control techniques used.

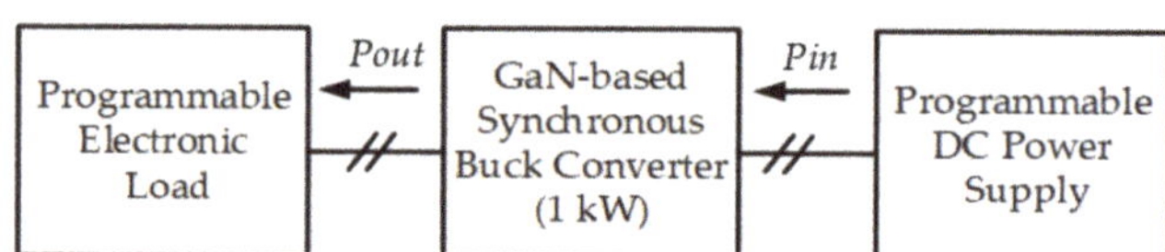

Figure 18. The system block diagram of the efficiency tests.

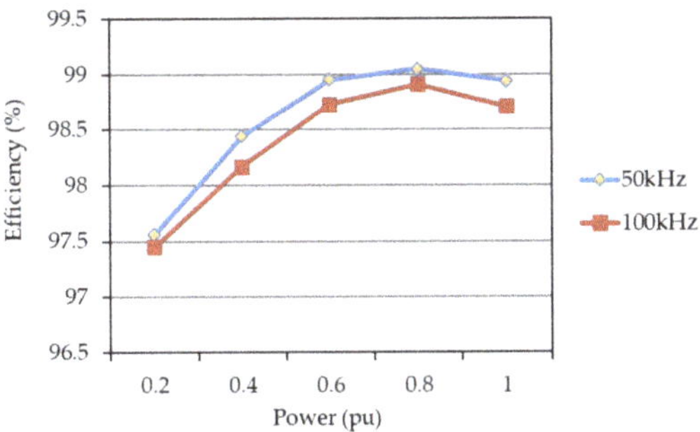

Figure 19. Efficiencies of the proposed GaN-based synchronous buck converter at different switching frequencies.

5. Conclusions

This paper has presented the systematic design procedure of a GaN-based and digitally controlled high-performance PV emulator that can be widely used in a variety of PV converter studies. Detailed characteristics of PV cells and PV modules have been explained and modeled. The issues regarding controller design and various emulation conditions in terms of normal irradiance and partial shading phenomena in practical solar PV generation applications have been investigated. Based on comprehensive simulation studies and experimental implementation, the proposed GaN-based PV emulator with 50 kHz switching speed exhibits a highest efficiency of 99.05% and excellent dynamic response and accurate current/voltage tracking capabilities. Through this design case, it has been demonstrated that wide-bandgap (WBG) switching devices such as the gallium nitride-based high electron mobility transistors (HEMTs) used in this study offer huge potential for outperforming conventional silicon devices, especially in terms of switching speed and conduction losses. Typical results have been presented to verify the feasibility and effectiveness of the proposed PV emulator.

Author Contributions: The corresponding author, C.-T.M., conducted the research work, verified the results of simulation cases and hardware tests, wrote the draft and polished the final manuscript. Z.-Y.T., a postgraduate student in the department of EE, CEECS, National United University, Taiwan, assisted in paper searching and organized the test results. H.-H.K. and C.-L.H. provided technical discussion during the research period. All authors have read and agreed to the published version of the manuscript.

Funding: This research was funded by both MOST, Taiwan, with grant numbers MOST 110-2221-E-239-032 and the Institute of Nuclear Energy Research (INER), Atomic Energy Council, Taiwan, with grant number NL1100393.

Institutional Review Board Statement: Not applicable.

Informed Consent Statement: Not applicable.

Data Availability Statement: No new data were created or analyzed in this study. Data sharing is not applicable to this article.

Acknowledgments: The author would like to thank the Ministry of Science and Technology (MOST) of Taiwan and the Institute of Nuclear Energy Research (INER), Taiwan, for financially supporting the energy-related research regarding key technologies in microgrids and the design of advanced power and energy systems.

Conflicts of Interest: The authors declare no conflict of interest.

References

1. US Energy Information Administration. Available online: https://www.eia.gov/ieo (accessed on 15 September 2017).
2. Adefarati, T.; Bansal, R.C. The Impacts of PV-Wind-Diesel-Electric Storage Hybrid System on the Reliability of a Power System. *Energy Procedia* **2017**, *105*, 616–621. [CrossRef]
3. BP. Available online: https://www.bp.com (accessed on 1 October 2017).
4. Ayop, R.; Tan, C.W. Rapid Prototyping of Photovoltaic Emulator Using Buck Converter Based on Fast Convergence Resistance Feedback Method. *IEEE Trans. Power Electron.* **2019**, *34*, 8715–8723. [CrossRef]
5. Balato, M.; Costanzo, L.; Gallo, D.; Landi, C.; Luiso, M.; Vitelli, M. Design and implementation of a dynamic FPAA based photovoltaic emulator. *Sol. Energy* **2016**, *123*, 102–115. [CrossRef]
6. Erkaya, Y.; Moses, P.; Flory, I.; Marsillac, S. Development of a solar photovoltaic module emulator. In Proceedings of the 2015 IEEE 42nd Photovoltaic Specialist Conference (PVSC), New Orleans, LA, USA, 14–19 June 2015.
7. Azharuddin, S.M.; Vysakh, M.; Thakur, H.V.; Nishant, B.; Babua, T.S.; Muralidhar, K.; Paul, D.; Jacob, B.; Balasubramanian, K.; Rajasekar, N. A near accurate solar PV emulator using dSPACE controller for real-time control. *Energy Procedia* **2014**, *61*, 2640–2648. [CrossRef]
8. Atoche, A.C.; Castillo, J.V.; Ortegón-Aguilar, J.; Carrasco-Alvarez, R.; Gío, J.S.; Colli-Menchi, A. A high-accuracy photovoltaic emulator system using ARM processors. *Sol. Energy* **2015**, *120*, 389–398. [CrossRef]
9. Heredero-Peris, D.; Capó-Lliteras, M.; Miguel-Espinar, C.; Lledó-Ponsati, T.; Montesinos-Miracle, D. Development and implementation of a dynamic PV emulator with HMI interface for high power inverters. In Proceedings of the 2014 16th European Conference on Power Electronics and Applications, Lappeenranta, Finland, 26–28 August 2014.
10. Koran, A.; LaBella, T.; Lai, J. High Efficiency Photovoltaic Source Simulator with Fast Response Time for Solar Power Conditioning Systems Evaluation. *IEEE Trans. Power Electron.* **2014**, *29*, 1285–1297. [CrossRef]
11. Xenophontos, A.; Rarey, J.; Trombetta, A.; Bazzi, A.M. A flexible low-cost photovoltaic solar panel emulation platform. In Proceedings of the 2014 Power and Energy Conference at Illinois (PECI), Champaign, IL, USA, 28 February–1 March 2014.
12. Pelin, D.; Opačak, M.; Pal, M. PV Emulation by using DC Programmable Sources. *Int. J. Electr. Comput. Eng. Syst.* **2017**, *8*, 11–17. [CrossRef]
13. Belaout, A.; Krim, F.; Talbi, B.; Feroura, H.; Laib, A.; Bouyahia, S.; Arabi, A. Development of real time emulator for control and diagnosis purpose of Photovoltaic Generator. In Proceedings of the 2017 6th International Conference on Systems and Control (ICSC), Batna, Algeria, 7–9 May 2017.
14. Shinde, U.K.; Kadwane, S.G.; Gawande, S.P.; Keshri, R. Solar PV emulator for realizing PV characteristics under rapidly varying environmental conditions. In Proceedings of the 2016 IEEE International Conference on Power Electronics, Drives and Energy Systems (PEDES), Trivandrum, India, 14–17 December 2016.
15. Mai, T.D.; De Breucker, S.; Baert, K.; Driesen, J. Reconfigurable emulator for photovoltaic modules under static partial shading conditions. *Sol. Energy* **2017**, *141*, 256–265. [CrossRef]
16. Erkaya, Y.; Marsillac, S. A PV module emulator based on GaN devices with over 99% peak efficiency. In Proceedings of the 2016 IEEE 43rd Photovoltaic Specialists Conference (PVSC), Portland, OR, USA, 5–10 June 2016.
17. Mikołaj, A.K.; Piotr, G.; Bartosz, N.; Kornel, W. Design, Analysis and Comparison of Si- and GaN-Based DC-DC Wide-Input-Voltage-Range Buck-Boost Converters. *Int. J. Electron. Telecommun.* **2021**, *67*, 337–343.

 *micromachines*

Article

Improved Performance of GaN-Based Light-Emitting Diodes Grown on Si (111) Substrates with NH$_3$ Growth Interruption

Sang-Jo Kim [1,†], **Semi Oh** [2,†], **Kwang-Jae Lee** [3], **Sohyeon Kim** [4] and **Kyoung-Kook Kim** [4,*]

[1] School of Materials Science and Engineering, Gwangju Institute of Science and Technology, Gwangju 61005, Korea; prokimsj@gmail.com

[2] Department of Electrical Engineering and Computer Science, University of Michigan, Ann Arbor, MI 48109, USA; ohsemi1230@gmail.com

[3] Department of Electrical Engineering, Stanford University, Stanford, CA 94305, USA; kwangjae@stanford.edu

[4] Department of Advanced Convergence Technology, Research Institute of Advanced Convergence Technology, Korea Polytechnic University, 237 Sangidaehak-ro, Siheung-si 15073, Korea; sohyeon.kim@kpu.ac.kr

* Correspondence: kim.kk@kpu.ac.kr

† These authors equally contributed.

Abstract: We demonstrate the highly efficient, GaN-based, multiple-quantum-well light-emitting diodes (LEDs) grown on Si (111) substrates embedded with the AlN buffer layer using NH$_3$ growth interruption. Analysis of the materials by the X-ray diffraction omega scan and transmission electron microscopy revealed a remarkable improvement in the crystalline quality of the GaN layer with the AlN buffer layer using NH$_3$ growth interruption. This improvement originated from the decreased dislocation densities and coalescence-related defects of the GaN layer that arose from the increased Al migration time. The photoluminescence peak positions and Raman spectra indicate that the internal tensile strain of the GaN layer is effectively relaxed without generating cracks. The LEDs embedded with an AlN buffer layer using NH$_3$ growth interruption at 300 mA exhibited 40.9% higher light output power than that of the reference LED embedded with the AlN buffer layer without NH$_3$ growth interruption. These high performances are attributed to an increased radiative recombination rate owing to the low defect density and strain relaxation in the GaN epilayer.

Keywords: AlN buffer layer; NH$_3$ growth interruption; strain relaxation; GaN-based LED; low defect density

Citation: Kim, S.-J.; Oh, S.; Lee, K.-J.; Kim, S.; Kim, K.-K. Improved Performance of GaN-Based Light-Emitting Diodes Grown on Si (111) Substrates with NH$_3$ Growth Interruption. *Micromachines* **2021**, *12*, 399. https://doi.org/10.3390/mi12040399

Academic Editor: Giovanni Verzellesi

Received: 10 March 2021
Accepted: 2 April 2021
Published: 5 April 2021

Publisher's Note: MDPI stays neutral with regard to jurisdictional claims in published maps and institutional affiliations.

Copyright: © 2021 by the authors. Licensee MDPI, Basel, Switzerland. This article is an open access article distributed under the terms and conditions of the Creative Commons Attribution (CC BY) license (https://creativecommons.org/licenses/by/4.0/).

1. Introduction

Substantial progress in fabricating highly efficient, GaN-based, multiple-quantum-well (MQW) light-emitting diodes (LEDs) has proven these materials useful in a variety of applications, such as micro-displays, automobiles, general lighting, and optoelectronics [1–3]. Furthermore, GaN-based epitaxial-layers for LED devices are conventionally fabricated on sapphire and SiC substrates. However, the sapphire substrate has poor thermal conductivity—as low as 25 Wm^{-1}K^{-1}—that encourages heat dissipation. Therefore, it is a critical issue in high-output LED operation. Additionally, GaN and SiC substrates are expensive and still limitedly used in large-scale LED fabrication than a sapphire substrate [4,5].

The large-scale Si substrate has attracted significant interest in growing GaN-based devices owing to lower cost than the traditional sapphire and SiC substrates [6,7]. Moreover, Si substrate can be easily integrated with electronic and optical devices [8].

However, growing a high-quality GaN epitaxial layer on a Si (111) substrate presents several key challenges. First, when the GaN layer is grown directly on the Si substrate, the Si surface easily reacts with NH$_3$ to form the SiN$_x$, which the GaN layer cannot grow. Subsequently, the Si substrate reacts with Ga to promote melt-back etching of Ga-Si eutectic alloys [9]. Second, the large lattice mismatch (~17%) between GaN and Si (111) causes a high dislocation density in the GaN layer leading to lower LED performance [10]. Third, the difference in thermal expansion coefficients (~56%) between Si and GaN introduces large tensile stress in the GaN layer during the cooling process from the high growth temperature, which causes the wafer bowing and cracks generation [11].

Therefore, many researchers have introduced various methods to reduce threading dislocation, stress mitigation, and remove cracks, such as epitaxial lateral overgrowth, nanoporous GaN layers, graded AlGaN interlayers, and Al(Ga)N/GaN superlattices [12–15]. Particularly, the AlN layer, which acts as a bottom buffer layer on the Si substrate, significantly affects crystalline quality and stress management of the GaN layer.

AlN layers with rough surfaces and poor crystalline qualities lead to GaN layers with poor crystalline quality. Therefore, many studies have revealed that high-quality AlN buffer layers minimize crystal misorientations and dislocation density in the GaN layer. Krost et al. developed a low-temperature (LT)-AlN layer—that is, a novel method to reduce stress [16].

Comparatively, high-temperature (HT)-AlN layers yield reduced dislocation densities and promote the relaxation of compressive strain due to an increased Al migration length. In addition, many studies are also investigated the high-quality GaN layer grown on a Si substrate using the HT- and LT-AlN growth process [17,18]. Moreover, Hirayama et al. used an AlN buffer layer grown with pulsed NH_3 flow on a sapphire substrate [19], which is an effective way to enhance the lateral migration of Al atoms and produce a smooth epitaxial surface.

However, there have been no reports on the effects of ammonia (NH_3) growth interruption for the AlN layer in GaN-based LEDs grown on Si (111) substrates.

We demonstrate that AlN layers prepared with NH_3 growth interruption to be served as a buffer layer improve the crystalline quality and strain relaxation in GaN-based LEDs grown on Si (111) substrates. The Al migration time of the AlN was controlled by NH_3 pulse timing.

The optical output power of LEDs grown using NH_3 growth interruption was 40.9% greater than that of the reference LED embedded with the AlN buffer layer without NH_3 growth interruption (injection current of 300 mA). Such remarkable improvements of optical output power are predominantly attributed to reduced dislocation densities and strain relaxation, which originated from an increased Al migration in the AlN buffer layer.

2. Materials and Methods

InGaN/GaN MQW LEDs were grown on Si (111) substrates using metal-organic chemical vapor deposition. Trimethylaluminum (TMAl) and NH_3 were used as the Al and N sources, respectively, and high-purity hydrogen was employed as the carrier gas. Si (111) substrates were first annealed at 1050 °C for 10 min to remove the native oxide. The substrates were subsequently passivated by TMAl with a pre-dose time of 10 s to prevent Si melt-back etching. A 200 nm-thick AlN layer was subsequently deposited at 1050 °C as a reference. Comparatively, for samples prepared by NH_3 growth interruption, TMAl was constantly introduced into the chamber while NH_3 was injected into the reactor following a specified pulsed regimen. Specifically, the halted (t_1) NH_3 flow time was changed: 3 s for sample A, 5 s for sample B, 7 s for sample C, and 9 s for sample D, as schematically shown in Figure 1.

After that, to prevent crack formation, two pairs of 12 nm-thick low-temperature AlN layers and a 1.2 μm-thick undoped GaN layer were deposited.

Furthermore, a 1.5 μm-thick n-type GaN layer ($n = 6 \times 10^{18}/cm^3$), six periods of 2.3 nm-thick $In_{0.18}Ga_{0.82}N$ layer, and 7.7 nm-thick GaN-based MQWs layers were subsequently grown. The 15 nm-thick $Al_{0.15}Ga_{0.85}N$ electron blocking layer (EBL) and the 200 nm-thick p-type GaN layer (($n = 5 \times 10^{19}/cm^3$) were finally deposited.

To fabricate the n-electrode, the epilayers were partially etched until the *n*-type GaN layer was exposed. The 200 nm-thick ITO layer was deposited using an electron-beam evaporator on the remaining parts of the p-type GaN layer and annealed at 600 °C in O_2 atmosphere for 1 min using the rapid thermal annealing. The Ti/Al (50/200 nm) layers were deposited as an n-electrode. Finally, the Cr/Al (30/200 nm) layers were deposited on the *p*- and *n*-electrodes and annealed at 300 °C for 1 min.

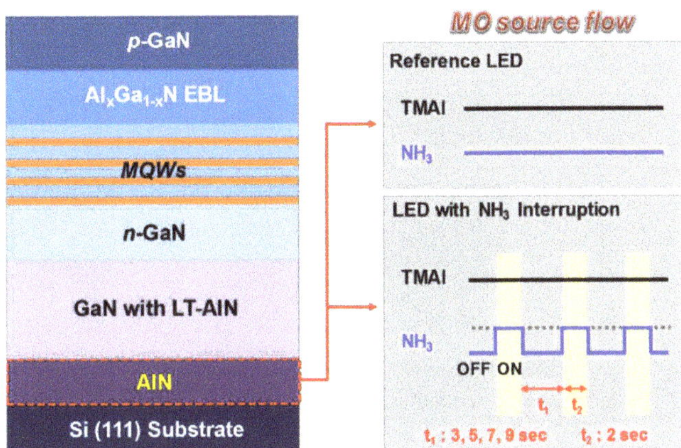

Figure 1. Schematics of light-emitting diodes (LEDs), including the sequence of all growth steps prepared with and without the NH$_3$ growth interruption method.

3. Results and Discussion

3.1. Epitaxial Characteristics

The effect of NH$_3$ growth interruption on the crystal quality of GaN was explored using X-ray diffraction (XRD) (PANalytical X'Pert PRO, Almelo, Netherlands) (reference sample and Sample C), as shown in Figure 2a,b. In addition, all samples of full-width-at-half-maximum (FWHM) values of the GaN (0002), GaN (10–12), and AlN (0002) reflections are plotted in Figure 3a.

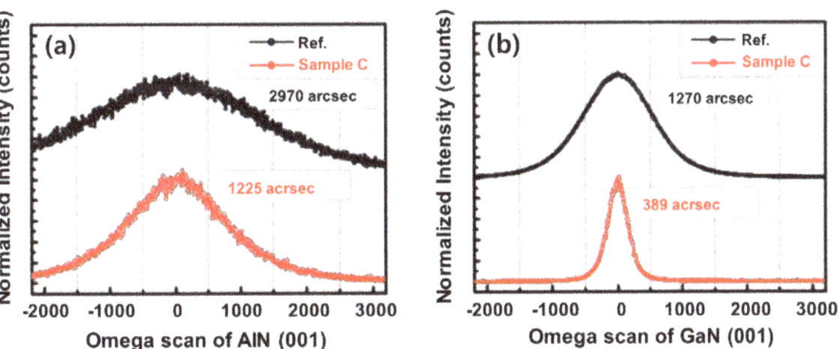

Figure 2. X-ray diffraction (XRD) Omega scan of reference and sample C (**a**) AlN (001) and (**b**) GaN (001).

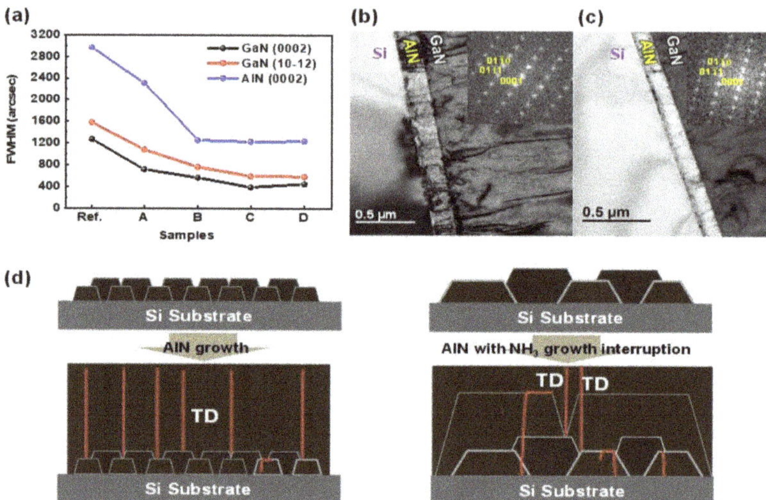

Figure 3. (**a**) The full-width-at-half-maximum (FWHM) values of the GaN (0002), GaN (10–12), and AlN (0002) reflections for all the samples. TEM images of (**b**) a reference sample and (**c**) a sample C. (**d**) Schematic growth mechanism of AlN with and without the NH$_3$ growth interruption.

The full-width-at-half-maximum (FWHM) of the (0002) and (10–12) reflections of GaN and AlN layers typically indicate imperfections in the crystal showing the densities of the screw (D_s) and edge (D_e) dislocations in the epitaxial layer, respectively. The parameters D_s and D_e can be obtained using the following equations [20,21]:

$$D_S = \beta^2_{(0002)}/4.35b_c^2 \tag{1}$$

$$D_e = \beta^2_{(10-12)}/4.35b_a^2 \tag{2}$$

where $\beta^2_{(0002)}$ and $\beta^2_{(10-12)}$ denote the FWHM of the (0002) and (10–12) reflections, respectively; b_c and b_a represent the Burgers vector lengths of the c- and a-axial lattice constant, respectively. The FWHM values of the (0002) and (10–12) reflections of GaN were 714 and 1079 arcsec for sample A, 563.4 and 760.3 arcsec for sample B, and 389 and 589.3 arcsec for sample C, respectively. These FWHM values are much lower than the corresponding FWHM values (1270 and 1580 arcsec) of a reference sample. The corresponding D_s and D_e are tabulated in Table 1. These results contain similar values to the other papers mentioned in introduction of HT-AlN, and LT-AlN growth for high-quality GaN layer grown on a Si substrate [17,18].

Table 1. The full-width-at-half-maximum (FWHM) values of the X-ray diffraction (XRD) rocking curve and calculated dislocation densities for all the samples.

Sample	t_1/t_2	XRD FWHM (arcsec)			Dislocation Density ($\times 10^9$ cm^{-2})		
		GaN (001)	GaN (102)	AlN (001)	D_s (GaN)	D_e (GaN)	D_s (AlN)
Ref.	t_1	1270	1580	2970	13.8	56	81.7
Sample A	3/2	714	1079	2316	4.36	26.3	49.7
Sample B	5/2	563	760	1255	2.71	13	14.6
Sample C	7/2	389	589	1225	1.29	7.85	13.9
Sample D	9/2	448	581	1237	1.71	7.63	14.2

Al adatoms have a high sticking coefficient and short migration time on Si substrates; therefore, they have a significantly low probability of moving from their point of impact. This gives rise to many nucleation sites for growth with a high density of extended defects, such as dislocations and grain boundaries [22]. Moreover, growth interruption by halting NH_3 gas flow provides Al adatoms with sufficient residence time to incorporate the most energetic favorable lattice sites on the Si substrate. It promotes a large number of small nucleation sites, which further minimizes coalescence-related defects. However, significantly long migration time for sample D promotes AlN layers with lower crystalline quality and increased dislocation densities compared to those of sample C due to the deteriorating surface roughness, as shown in Table 1 [23–25].

We analyzed transmission electron microscopy (TEM) to deeply investigate the crystalline quality of AlN layers produced using NH_3 growth interruption. Figure 3b,c show the cross-sectional bright-field TEM images of the reference sample and sample C, respectively, which were acquired near the GaN [0001] zone axis. Dislocations were clearly observed at the interface of the AlN and Si in the reference material; however, they are remarkably reduced in sample C because the longer Al migration time (as shown in Figure 3d) decreases coalescence related defects and dislocation densities.

The impact of the NH_3 growth interruption process on the residual stress in the GaN epilayer was quantified by photoluminescence (PL) and Raman measurements, as shown in Figure 4a,b, respectively. The PL intensities of samples prepared by the NH_3 growth interruption process increased with Al migration time (t_1), indicating higher GaN crystal quality than the other samples. However, the PL intensity decreased with longer Al migration times. Moreover, the PL peak position of a sample C blue-shifted from 366.7 nm to 365.3 nm and subsequently red-shifted to 365.6 nm in sample D, indicating that the residual tensile strain in sample C is smaller than that in the other samples. Figure 4c shows the Raman spectra acquired from the reference sample and sample C, specifically using a 514 nm light with the excitation power of 2.4 mW.

The E_2 (high) vibrational mode is highly sensitive to strain. Therefore, it is widely used to quantify the stress in GaN epilayers [25]. The wavenumber of the E_2 (high) mode of the reference sample and sample C was 566.03 cm^{-1} and 567.18 cm^{-1}, respectively. Both the values were red-shifted from the E_2 (high) wavenumber from a standard free-standing bulk GaN (567.5 cm^{-1}), proving that the GaN epilayers in the reference and sample C were under tensile stress [26]. Shifts of the E_2 (high) phonon peak are related to the relaxation of residual strain; it can be calculated using the following equation [27]:

$$\Delta\omega_\gamma - \omega_0 = K_\gamma \cdot \sigma_{xx} \tag{3}$$

where ω_γ and ω_0 represent the Raman wavenumbers of the E_2 (high) phonon peak of sample C and reference sample, respectively. A proportionality factor K_γ (4.2 cm^{-1} GPa^{-1}) originates from hexagonal GaN [28]. The E_2 (high) phonon peak of sample C was blue-shifted by 1.15 cm^{-1} from that of the reference sample, corresponding to the impressive different relaxation of tensile stress (σ_{xx}) (0.274 GPa). The relaxation of tensile strain in GaN layer for sample C causes the shrinking the lattice constant, which shifts the PL peak position to higher energy [29]. The reduced tensile stress in sample C arises predominantly from the coalescence of large-sized grains due to the NH_3 growth interruption method [30,31].

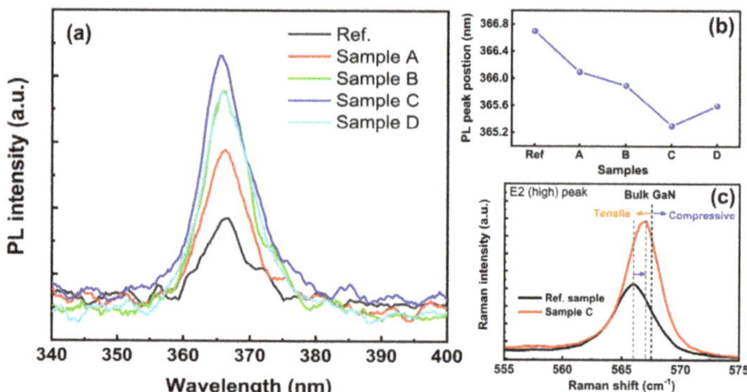

Figure 4. (**a**) Photoluminescence (PL) intensity and (**b**) PL peak position of all the studied samples. (**c**) Raman spectra of the reference sample and sample C.

3.2. Device Characteristics

Figure 5 shows the current-voltage (I-V) characteristics and light output power of the reference sample and sample C. Figure 5a reveals that the forward voltage of sample C at an injection current of 20 mA was 3.76 V, which is lower than that of the reference sample (3.91 V) under the same condition.

Furthermore, the series resistances estimated from the I-V curves of the reference sample and sample C were 37.2 ohm and 31.6 ohm, respectively. These values indicate that decreasing the density of defects, such as threading dislocations by NH_3 growth interruption method, improves the crystal quality of GaN epi-layer; moreover, it results in a decreased series resistance and forward voltage. Figure 5b shows that the light output power of sample C at an injection current of 300 mA is 40.9% greater than that of the reference sample; it is attributed to enhanced radiative recombination due to the reduced dislocation densities and relaxation of internal tensile strain. We could not measure the light output power above 300 mA by the limitation of measurement system. However, normally, as the injection current is increased, the carrier overflow and Auger recombination are also increased. Therefore, at high current, the portion of increasing light output power will be decreased [32,33].

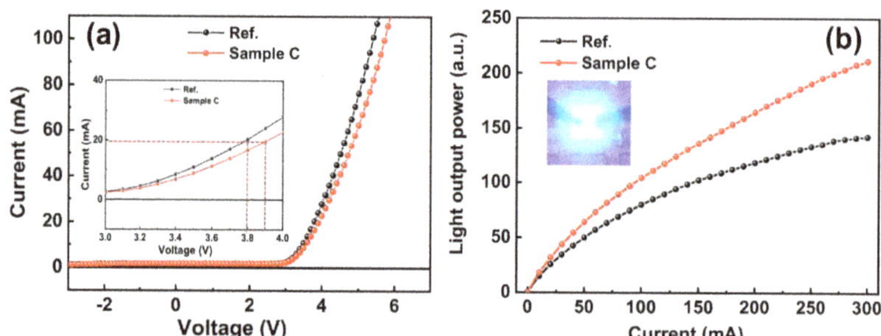

Figure 5. (**a**) Current-voltage (I-V) characteristics and (**b**) optical output power of the reference sample and a sample C.

4. Conclusions

We demonstrated the NH_3 growth-interruption process to prepare the high-performance InGaN/GaN MQW LEDs grown on Si (111) substrates. The XRD results revealed low FWHM values for GaN grown on AlN layers using the NH_3 growth interruption method; it established substantially improved crystal quality compared to the reference LED embedded with the AlN buffer layer without NH_3 growth interruption.

Improved crystalline quality is further corroborated by comparative TEM analyses, which is attributed to the effectively reduced dislocation densities and coalescence by longer Al migration times. A lower forward voltage of 3.76 V was observed at an injection current of 20 mA for the LED fabricated by the NH_3 growth interruption method; however, the reference LED had a forward voltage of 3.91 V. The optical output power of the LED prepared using the NH_3 growth interruption was 40.9% greater than that of the reference LED.

Such enhanced optical output is attributed to increased radiative recombination rates due to the decreased dislocation densities and relaxation of internal tensile strain, which arise from the longer Al migration time on the Si (111) substrate. These results demonstrate that the NH_3 growth interruption method is an important technique for growing high-performance LEDs grown on Si (111) substrate. Therefore, these results ultimately represent a step towards realizing high-efficiency and high-power LEDs.

Author Contributions: Conceptualization, S.-J.K. and S.O.; methodology, S.-J.K. and K.J.L.; formal analysis, S.-J.K., K.J.L. and S.K; writing—original draft preparation, S.-J.K. and S.O.; writing—review and editing, S.O and K.-K.K.; supervision, K.-K.K. All authors have read and agreed to the published version of the manuscript.

Funding: This research was supported by the Ministry of Science and ICT (MIST), Korea, under the Information Technology Research Center (ITRC) support program (IITP-2020-2018-0-01426) supervised by the Institute for Information and Communications Technology Planning and Evaluation (IITP) by the Korea Institute for Advancement of Technology (KIAT) grant funded by the Korea Government (MOTIE) (P0008458, The Competency Development Program for Industry), and by the Technology Innovation Program (or Industrial Strategic Technology Development Program) (20002694, Gas sensor) funded by the Ministry of Trade, Industry, and Energy (MOTIE, Korea).

Conflicts of Interest: The authors declare no conflict of interest.

References

1. Schubert, E.F.; Kim, J.K. Solid-State Light Sources Getting Smart. *Science* **2005**, *308*, 1274–1278. [CrossRef] [PubMed]
2. Ponce, F.A.; Bour, D.P. Nitride-based semiconductors for blue and green light-emitting devices. *Nature* **1997**, *386*, 351–359. [CrossRef]
3. Rogers, J.A.; Someya, T.; Huang, Y. Materials and Mechanics for Stretchable Electronics. *Science* **2010**, *327*, 1603–1607. [CrossRef] [PubMed]
4. Liu, X.; Wang, H.Y.; Chiu, H.C.; Chen, Y.; Li, D.; Huang, C.R.; Kao, H.L.; Kuo, H.C.; Chen, S.W.H. Analysis of the back-barrier effect in AlGaN/GaN high electron mobility transistor on free-standing GaN substrates. *J. Alloys Compd.* **2020**, *814*, 152293. [CrossRef]
5. Li, C.; Li, Z.; Peng, D.; Yang, Q.; Zhang, D.; Luo, W.; Pan, C. Growth of thin AlN nucleation layer and its impact on GaN-on-SiC heteroepitaxy. *J. Alloys Compd.* **2020**, *838*, 155557. [CrossRef]
6. Zhang, X.; Li, P.; Zou, X.; Jiang, J.; Yuen, S.H.; Tang, C.W.; Lau, K.M. Active Matrix Monolithic LED Micro-Display Using GaN-on-Si Epilayers. *IEEE Photonics Technol. Lett.* **2019**, *31*, 865–868. [CrossRef]
7. Sun, Y.; Zhou, K.; Feng, M.; Li, Z.; Zhou, Y.; Sun, Q.; Liu, J.; Zhang, L.; Li, D.; Sun, X.; et al. Room-temperature continuous-wave electrically pumped InGaN/GaN quantum well blue laser diode directly grown on Si. *Light Sci. Appl.* **2018**, *7*, 13. [CrossRef]
8. Lee, K.H.; Bao, S.; Zhang, L.; Kohen, D.; Fitzgerald, E.; Tan, C.S. Integration of GaAs, GaN, and Si-CMOS on a common 200 mm Si substrate through multilayer transfer process. *Appl. Phys. Express* **2016**, *9*, 086501. [CrossRef]
9. Khoury, M.; Tottereau, Q.; Feuillet, G.; Vennegues, P.; Zuniga-Perez, J. Evolution and prevention of meltback etching: Case study of semipolar GaN growth on patterned silicon substrates. *J. Appl. Phys.* **2017**, *122*, 105108. [CrossRef]
10. Tran, C.A.; Osinski, A.; Karlicek, R.F.J. Growth of InGaN/GaN multiple-quantum-well blue light-emitting diodes on silicon by metalorganic vapor phase epitaxy. *Appl. Phys. Lett.* **1999**, *75*, 1494–1496. [CrossRef]
11. Dadgar, A.; Blasing, J.; Diez, A.; Alam, A.; Heuken, M.; Krost, A. Metalorganic Chemical Vapor Phase Epitaxy of Crack-Free GaN on Si (111) Exceeding 1 µm in Thickness. *Jpn. J. Appl. Phys.* **2000**, *39*, L1183–L1185. [CrossRef]
12. Kim, B.; Lee, K.; Jang, S.; Jhin, J.; Lee, S.; Baek, J.; Yu, Y.; Lee, J.; Byun, D. Epitaxial Lateral Overgrowth of GaN on Si (111) Substrates Using High-Dose, N+ Ion Implantation. *Chem. Vap. Depos.* **2010**, *16*, 80–84. [CrossRef]
13. Lee, K.J.; Chun, J.; Kim, S.J.; Ha, C.S.; Park, J.W.; Lee, S.J.; Song, J.C.; Baek, J.H.; Park, S.J. Enhanced optical output power of InGaN/GaN light-emitting diodes grown on a silicon (111) substrate with a nanoporous GaN layer. *Opt. Express* **2016**, *24*, 4391–4398. [CrossRef]
14. Able, A.; Wegscheider, W.; Engl, K.; Zweck, J. Growth of crack-free GaN on Si (1 1 1) with graded AlGaN buffer layers. *J. Cryst. Growth* **2005**, *276*, 415–418. [CrossRef]

15. Ni, Y.; He, Z.; Yang, F.; Zhou, D.; Yao, Y.; Zhou, G.; Shen, Z.; Zhong, J.; Zhen, Y.; We, Z.; et al. Effect of AlN/GaN superlattice buffer on the strain state in GaN-on-Si (111) system. *Jpn. J. Appl. Phys.* **2015**, *54*, 015505. [CrossRef]
16. Reiher, A.; Blasing, J.; Dadgar, A.; Diez, A.; Krost, A. Efficient stress relief in GaN heteroepitaxy on Si (111) using low-temperature AlN interlayers. *J. Cryst. Growth.* **2003**, *248*, 563–567. [CrossRef]
17. Lin, P.; Tien, C.; Wang, T.; Chen, C.; Ou, S.; Chung, B.; Wuu, D. On the role of AlN insertion layer in stress control of GaN on 150-mm Si (111) substrate. *Crystals* **2017**, *7*, 134. [CrossRef]
18. Lin, Y.; Yang, M.; Wang, W.; Lin, Z.; Li, G. A low-temperature AlN interlayer to improve the quality of GaN epitaxial films grown on Si substrates. *CrystEngComm* **2016**, *18*, 8926. [CrossRef]
19. Hirayama, H.; Yatabe, T.; Noguchi, N.; Ohashi, T.; Kamata, N. 231–261 nm AlGaN deep-ultraviolet light-emitting diodes fabricated on AlN multilayer buffers grown by ammonia pulse-flow method on sapphire. *Appl. Phys. Lett.* **2007**, *91*, 071901. [CrossRef]
20. Kaganer, V.M.; Brandt, O.; Trampert, A.; Ploog, K.H. X-ray diffraction peak profiles from threading dislocations in GaN epitaxial films. *Phys. Rev. B.* **2005**, *72*, 045423. [CrossRef]
21. Vickers, M.E.; Kappers, M.J.; Datta, R.; McAleese, C.; Smeeton, T.M.; Rayment, F.D.; Humphreys, C.J. In-plane imperfections in GaN studied by X-ray diffraction. *J. Phys. D Appl. Phys.* **2005**, *38*, A99–A104. [CrossRef]
22. Jain, R.; Sun, W.; Yang, J.; Shatalov, M.; Hu, X.; Sattu, A.; Lunev, A.; Deng, J.; Shturm, I.; Bilenko, Y.; et al. Migration enhanced lateral epitaxial overgrowth of AlN and AlGaN for high reliability deep ultraviolet light emitting diodes. *Appl. Phys. Lett.* **2008**, *93*, 051113. [CrossRef]
23. Rahman, M.N.A.; Talik, N.A.; Abdul, M.I.M.; Sulaiman, A.F.; Allif, K.; Zahir, N.M.; Shuhaimi, A. Ammonia flux tailoring on the quality of AlN epilayers grown by pulsed atomic-layer epitaxy techniques on (0001)-oriented sapphire substrates via MOCVD. *CrystEngComm* **2019**, *21*, 2009–2017. [CrossRef]
24. Kum, D.; Byun, D. The Effect of Substrate Surface Roughness on GaN Growth Using MOCVD Process. *J. Electron. Mater.* **1997**, *26*, 1098–1102. [CrossRef]
25. Guillaume, G.; Gael, G.; Marc, P.; Eric, F.; Daniel, A.; Yvon, C.; Favrice, S. A detailed study of AlN and GaN grown on Silicon-on-porous Silicon substrate. *Phys. Status Solidi A* **2017**, *214*, 1600450.
26. Wang, K.; Yu, T.; Wei, Y.; Li, M.; Zhang, G.; Fan, S. Coordinated stress management and dislocation control in GaN growth on Si (111) substrates by using a carbon nanotube mask. *Nanoscale* **2019**, *11*, 4489–4495. [CrossRef]
27. Chen, Y.; Chen, Z.; Li, J.; Chen, Y.; Li, C.; Zhan, J.; Yu, T.; Kang, X.; Jiao, F.; Li, S.; et al. A study of GaN nucleation and coalescence in the initial growth stages on nanoscale patterned sapphire substrates via MOCVD. *CrystEngComm* **2018**, *20*, 6811–6820. [CrossRef]
28. Kisielowski, C.; Kruger, J.; Ruvimov, S.; Suski, T.; Ager, J.W.; Jones, E.; Liliental-Weber, Z.; Rubin, M.; Weber, E.R.; Bremser, M.D.; et al. Strain-related phenomena in GaN thin films. *Phys. Rev. B* **1996**, *54*, 17745. [CrossRef]
29. Chowdhury, S.; Biswas, D. Impact of varying buffer thickness generated strain and threading dislocations on the formation of plasma assisted MBE grown ultra-thin AlGaN/GaN heterostructure on silicon. *AIP Adv.* **2015**, *5*, 057149. [CrossRef]
30. Tanoto, H.; Yoon, S.F.; Loke, W.K.; Chen, K.P.; Fitzgerald, E.A.; Dohrman, C.; Narayanan, B. Heteroepitaxial growth of GaAs on (100) Ge/ Si using migration enhanced epitaxy. *J. Appl. Phys.* **2008**, *103*, 104901. [CrossRef]
31. Tan, B.; Hu, J.; Zhang, J.; Zhang, Y.; Long, H.; Chen, J.; Du, S.; Dai, J.; Chen, C.; Xu, J.; et al. AlN gradient interlayer design for the growth of high-quality AlN epitaxial film on sputtered AlN/sapphire substrate. *CrystEngComm* **2018**, *20*, 6557–6564. [CrossRef]
32. Akyol, F.; Nath, D.; Krishnamoorthy, S.; Park, P.; Rajan, S. Suppression of electron overglow and efficiency droop in N-polar GaN green light emitting diodes. *Appl. Phys. Lett.* **2012**, *100*, 111118. [CrossRef]
33. Piprek, J.; Romer, F.; Witzigmann, B. On the uncertainly of the Auger recombination coefficient extracted from InGaN/GaN light-emitting diode efficiency droop measurements. *Appl. Phys. Lett.* **2015**, *106*, 101101. [CrossRef]

 *micromachines*

Article

A Novel Ultrasonic TOF Ranging System Using AlN Based PMUTs

Yihsiang Chiu [1], Chen Wang [1], Dan Gong [2], Nan Li [1], Shenglin Ma [2,*] and Yufeng Jin [1,*]

[1] School of Electronic and Computer Engineering Bldg. A, Peking University Shenzhen Graduate School, University Town, Xili, Nanshan, Shenzhen 518055, China; 1601111224@pku.edu.cn (Y.C.); wangchen2019@pku.edu.cn (C.W.); 1701213526@sz.pku.edu.cn (N.L.)

[2] Department of Mechanical & Electrical Engineering, School of Aeronautics and Astronautics, Xiamen University, Xiamen 361005, China; 19920171150940@stu.xmu.edu.cn

* Correspondence: mashenglin@xmu.edu.cn (S.M.); yfjin@pku.edu.cn (Y.J.)

Abstract: This paper presents a high-accuracy complementary metal oxide semiconductor (CMOS) driven ultrasonic ranging system based on air coupled aluminum nitride (AlN) based piezoelectric micromachined ultrasonic transducers (PMUTs) using time of flight (TOF). The mode shape and the time-frequency characteristics of PMUTs are simulated and analyzed. Two pieces of PMUTs with a frequency of 97 kHz and 96 kHz are applied. One is used to transmit and the other is used to receive ultrasonic waves. The time to digital converter circuit (TDC), correlating the clock frequency with sound velocity, is utilized for range finding via TOF calculated from the system clock cycle. An application specific integrated circuit (ASIC) chip is designed and fabricated on a 0.18 μm CMOS process to acquire data from the PMUT. Compared to state of the art, the developed ranging system features a wide range and high accuracy, which allows to measure the range of 50 cm with an average error of 0.63 mm. AlN based PMUT is a promising candidate for an integrated portable ranging system.

Keywords: aluminum nitride; piezoelectric micromachined ultrasonic transducers; ranging; time of flight (TOF); time to digital converter circuit (TDC)

Citation: Chiu, Y.; Wang, C.; Gong, D.; Li, N.; Ma, S.; Jin, Y. A Novel Ultrasonic TOF Ranging System Using AlN Based PMUTs. *Micromachines* **2021**, *12*, 284. https://doi.org/10.3390/mi12030284

Academic Editor: Giovanni Verzellesi

Received: 12 February 2021
Accepted: 4 March 2021
Published: 8 March 2021

Publisher's Note: MDPI stays neutral with regard to jurisdictional claims in published maps and institutional affiliations.

Copyright: © 2021 by the authors. Licensee MDPI, Basel, Switzerland. This article is an open access article distributed under the terms and conditions of the Creative Commons Attribution (CC BY) license (https://creativecommons.org/licenses/by/4.0/).

1. Introduction

Various piezoelectric materials, such as polyvinylidene fluoride (PVDF) [1], piezoelectric lead zirconate titanate (PZT) [2], zinc oxide (ZnO) [3] and aluminum nitride (AlN) [4], have been widely applied in piezoelectric micromachined ultrasonic transducers (PMUTs) and bulk ultrasonic transducers. Although PVDF is flexible with good processibility and fast dynamic response, mechanical deformation is required to covert the non-polarized phase to polarized phase. Therefore, it is hard to be integrated in the conventional micro processing. The disadvantages of PZT lie in its high deposition temperature (e.g., 600 °C) and high voltage required to be polarized. Although processing temperature is low, ZnO is poorly compatible with CMOS processing due to its proneness to chemical reaction with acid and base and the pollution problems during CMOS processing owing to fast diffusibility of zinc ions. In comparison, AlN can be deposited at low temperatures (<400 °C) and is compatible with complementary metal oxide semiconductor (CMOS) process [5,6]. Although AlN has a lower piezoelectric constant than that of PZT, AlN based PMUT is considered to be able to achieve comparable performance because of its lower dielectric constant than that of PZT. In the last decade, AlN PMUT has aroused an increasing global research interests and breakthrough has been achieved in its commercialization for ranging sensing. For instance, application-oriented R&D interest is increasing in recent years. In 2009, Shelton et al. [7], presented an AlN based PMUT with a film radius of 175, 200 and 225 μm operating in the 200 kHz range for short-range air coupling ultrasonic applications. When 1V input voltage was used to excite at a resonant frequency of 220 kHz, a film deflection of 210 nm was achieved. In 2014, Lu et al. [8], presented a high-frequency (10–55 MHz), fine-pitch (45–70 μm) PMUT array based on a customized silicon-on-insulator (SOI) wafer with buried cavities for medical imaging., The fabricated 9×9 array samples made of 40 μm diameter PZT PMUTs had a 3.4 MHz bandwidth at the center frequency of 10.4 MHz and measured pressure sensitivity of 2 kPa/V at a distance of 1.25 mm. In 2015, Horsley et al. [9], proposed an AlN PMUT for human-machine interfaces. In 2017, Goh et al. [10], proposed an AlN PMUT based ultrasonic measurement method for stainless steel tubes.

With the continuous development and wide application of computer, automation and industrial robots, the ranging technology, such as radar ranging [11], infrared ranging [12], laser ranging [13] and ultrasonic ranging [14], is becoming critical to ensuring accurate positioning. Compared with other ranging methods, ultrasonic ranging is not affected by the transparency and color of the target object, is insensitive to environmental noise, and can be used in direct sunlight. Besides, because of the relatively low speed of sound, ultrasonic transducer based ranging technology mitigates the high-speed electronics requirements confronted by RF (radio frequency) and optical ranging. Hence, ultrasonic ranging has the characteristics of low power consumption, which makes it an attractive alternative at short-range ranging (<10 m).

The traditional ultrasonic rangefinders use bulk piezoelectric ceramic with high output power. However, the mismatch of the acoustic impedance with air results in poor conversion efficiency between the electric field and the sound field [15]. Besides, their bulky size limits their use in portable devices. AlN has a quartzite crystal structure with a good piezoelectric effect along its c-axis orientation. It does not need to be polarized like lead PZT [4], and the AlN membrane has a lower deposition temperature than PZT (the former is less than 400 °C, and the latter is 600 °C), which results in its better compatibility with standard CMOS processes. Although the piezoelectric constant of AlN is not as high as PZT, its lower dielectric constant makes it possible to achieve performance comparable to PZT. Therefore, AlN based PMUT is a promising candidate for an integrated portable ranging system.

Regarding to ultrasonic ranging, either frequency modulated continuous wave mode (FMCW) or pulse echo mode (PE) is used. FMCW mode includes binary frequency shift keying (BFSK) [16], two frequencies continuous wave (TFCW) [17] and multi-frequencies continuous wave (MFCW) [18]. PE mode includes time of flight (TOF) [19]. Theoretically, frequency modulated continuous wave (FMCW) does not have dead zone while it is not the case for the PE mode [20]. However, measuring range of FMCW mode is short, and it is difficult to do Doppler coupling or isolate the transmitter and receiver. In comparison, the traditional flight (TOF) methods suffer from high levels of systematic errors due to degradation of the amplitude of received signal. To address this issue, the ultrasonic signal received by the PMUT is amplified by the high pass filter (HPF) and then being read by analog to digital converter (ADC) [14].

In this paper, two pieces of PMUTs are applied. One is used to transmit and the other is used to receive ultrasonic waves. The time to digital converter circuit (TDC), correlating the clock frequency with sound velocity, is utilized for range finding via TOF calculated from the system clock cycle. A prototype is fabricated by integrating two AlN PMUTs with an ASIC chip that is designed and fabricated on a 0.18 μm CMOS process. Its performance is studied experimentally. Compared to state of the art, the developed time to digital converter circuit (TDC) ranging system features a wide range and high accuracy.

2. Design of the Ranging System

The block diagram of the proposed ranging system is shown in Figure 1. The system consists of two PMUTs and a transceiver circuit. The ultrasonic transducer consists of a square AlN piezoelectric film sandwiched between two Mo electrodes suspended with buried vacuum cavity in the substrate. The voltage applied to the two electrodes forms an electric field in the thickness direction causing a transverse stress, which generates an acoustic wave from out-of-plane bending of the membrane. Similarly, the incident pressure wave causes the membrane to bend and generates a charge on the electrodes for receivers.

When the PMUT is stimulated at a resonant frequency, the electromechanical conversion efficiency will be maximized.

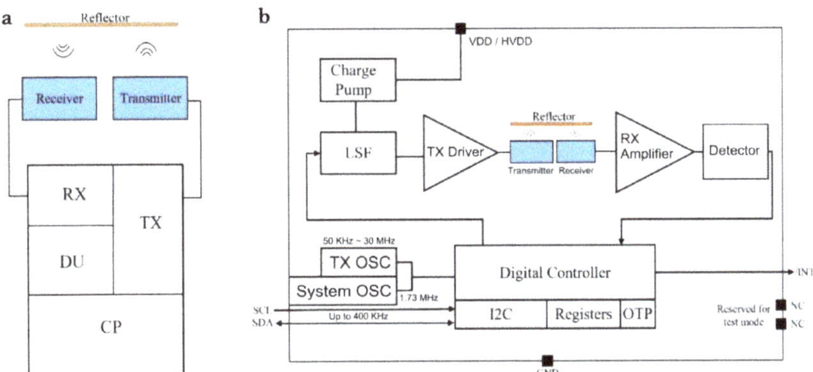

Figure 1. The block diagram of ultrasonic time of flight (TOF) ranging system using aluminum nitride (AlN) piezoelectric micromachined ultrasonic transducers (PMUTs). (**a**) One PMUT is a transmitter and the other is a receiver. (**b**) The function block diagram of the system.

The resonant frequency of the PMUT has a great influence on the ranging range and resolution. The resonance frequency of 100 kHz is chosen for the PMUT for modeling. If the PMUT is modeled as a rectangular membrane of uniform material with fully clamped boundaries (dimension: L_x and L_y), the modal frequencies of the membrane are defined by Equation (1) [21]:

$$f_{m,n} = \frac{1}{2} \times \sqrt{\frac{T}{\sigma}} \times \sqrt{\frac{m^2}{L_x^2} + \frac{n^2}{L_y^2}} \quad m, n = 1, 2, 3, \dots \tag{1}$$

where T is the surface tension, and σ is the area density. If $L_x = L_y = L$, the resonant frequency in the first mode (m = 1, n = 1) is as follows:

$$f_0 = \frac{1}{2L} \sqrt{\frac{2T}{\sigma}} \tag{2}$$

It can be seen from Equation (2) that as the length of the side increases, the resonant frequency of PMUT decreases.

As shown in Figure 1a, the circuit used to transmit and receive signal from PMUTs fabricated by TSMC (Taiwan Semiconductor Manufacturing Company, Ltd., Taiwan, China) 0.18 µm CMOS process consists of four main portions: a charge pump circuit (CP), a digital unit (DU), a transmitter circuit (TX) and a receiver circuit (RX). As shown in Figure 1b, the CP will rise the voltage of 1.8 V to 32 V needed to drive the level shifter (LSF). The oscillator (OSC) will generate a system clock (system OSC) of 1.73 MHz and a drive clock (TX OSC) of 97 kHz. The voltage of TX OSC will rise to 32 V by LSF, which is then amplified by the TX driver to stimulate the PMUT. Next, the PMUT transforms the driving signal to acoustic wave and receives the reflected signal after the time of flight (TOF). The goal of the receiver circuit is to amplify and detect the received echo signal, and send it to digital part to derive the distance info from TOF. The digital unit will handle the control of the ranging process and support the setting adjustment and status reported through the I2C interface. The calibration functions are designed in this chip as well, and the results are stored in the OTP (one time programmable). The transceiver circuit only requires a single 1.8 V supply source and can be controlled via I2C interface. HV charge pump and clock generator are all integrated without adding external components. An interrupt output pin is arranged to minimize the polling bandwidth/burden of baseband chip. Once the proximity detection value is beyond the lower or upper thresholds, the interrupt is asserted to alert the host about the situation.

The distance (L) between the transducer and the object is figured out by TOF using ultrasound's echo and receive roundtrip time Δt using Equation (3).

$$L_{min} = \frac{C_{us}}{2f_{sys}} \tag{3}$$

where c_{us} is the velocity of ultrasound in air and f_{sys} is the frequency of system. It is worth noting that sound velocity in air is not constant, which is affected by environment condition such as humidity and temperature. High humidity environment can quickly dampen the ultrasound, resulting in a very short transmission distance. The relationship between sound velocity and temperature can be described in Equation (4):

$$c_{us} = (331.3 + 0.606 \times T) \text{ m/s} \tag{4}$$

where T is absolute temperature in unit of Celsius. The transmission speed of ultrasound in air at 25 °C is 346 m/s, which gives the ultrasound one–way travel time of 2.89 μs for 1 mm transmission distance. As shown in Figure 1a, the total travel distance of ultrasound from transmitter to receivers is twice that of the distance between the PMUT and the object (1 mm here). Hence, the total travel time of ultrasound is 5.89 μs, which corresponds to the system frequency of 1.73 MHz and minimum distance resolution of 0.1 mm using Equation (3). A 3D finite elements model of the PMUT was created using COMSOL Multiphysics software (Version 5.0, COMSOL Co. Ltd., Stockholm, Sweden) to study the mode shape and the time-frequency characteristics. The first vibration mode of the PMUT is shown in Figure 2a. PMUT operating in the first mode has maximum amplitude, and thus is more suitable for ranging applications. Figure 2b,c shows the simulation result of frequency characteristics of the PMUT. It can be seen that the resonant frequency is 100.17 kHz when the side length is 3500 μm, and the maximum displacement is about 75 μm at resonant frequency.

The propagation process of ultrasound is shown in Figure 2d. Starting from t = 0 s, the film begins vibrating and radiating ultrasound outward with a fan angle of 180° (Figure 2(di–diii)). At t = 120 μs, a part of the ultrasound bypasses the baffle and propagates downward, then reaches the receiver before the reflected waves (Figure 2(div,dv)). Finally, ultrasounds reach the reflector and reflects downward until they are detected by the receiver (Figure 2(dvi,dvii)).

Figure 2e,f show the simulated results of near-field and far-field sound pressure versus time. As seen from the Figure 2e, the envelopes of the ultrasonic waves at different points are very similar, their frequency is also the same, but the amplitude of the waves isometrically decrease, and the phase also moves equidistantly to the right. It is worth mentioning that the time difference of arrival (TDOA) of the three waves Δt is approximately 6×10^{-7} s, and the distance difference $\Delta s = v \cdot \Delta t = 0.2076$ mm, where v = 346 m/s is sound speed at room temperature, is exactly equal to the distance between two points. This shows that the ultrasonic wave is basically not affected by the external environment and is radiating out without distortion. For far-field sound pressure in Figure 2f, except for the 10 cm point, the envelopes of the other two waves have changed and the amplitudes will no longer attenuate after reaching the maximum, which may be caused by the superposition of the original wave and the reflected wave.

The received voltage of the receiver when the sensing distance is 7 cm and 8 cm is shown in Figure 2g,h, respectively. The ultrasounds shown in Figure 2(div,dv) that bypass the baffle and reach the receiver are the waves on the left of the red dashed line, and the right is the reflected waves for sensing. Therefore, the received voltage is the superposition of two waves, and the minimum amplitude between the two waves is the arrival time of the reflected waves. We can calculate the measured distance $s_{(g)} = v \cdot t_{(g)}/2 = 7.0065$ cm and $s_{(h)} = v \cdot t_{(h)}/2 = 8.0099$ cm from the arrival time. Since the results were very close to the actual distance, it shows that the method has extremely high accuracy.

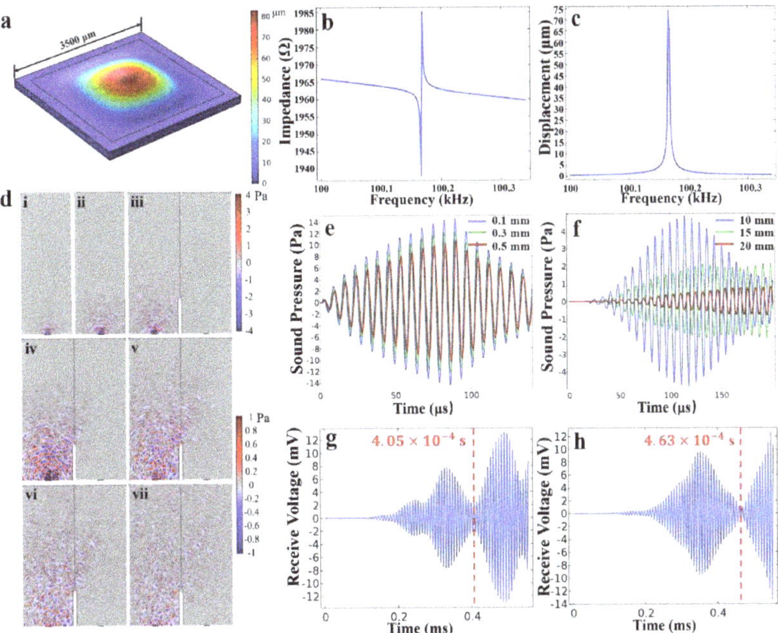

Figure 2. The finite element simulation results of PMUTs. (**a**) The first vibration mode diagram of PMUT. (**b**) Impedance and (**c**) displacement versus the frequency characteristics of the PMUT. (**d**) Propagation state of ultrasound when the time is (**i**) 30 μs, (**ii**) 60 μs, (**iii**) 90 μs, (**iv**) 120 μs, (**v**) 150 μs, (**vi**) 180 μs and (**vii**) 210 μs. The grayness in the picture is the air domain. The upper part of the air is the reflector. PMUTs are located on both sides of the bottom. The left is the transmitter and the right is the receiver. In the middle of the air is a baffle to prevent interference between the transmitter and receiver. (**e**) Near−field and (**f**) far−field sound pressure versus time, where the points of near−field and far−field sound pressure are 0.1, 0.3, 0.5 mm and 10, 15, 20 mm from the center of the film, respectively. The receiving voltage of the receiver when the sensing distance is (**g**) 7 cm and (**h**) 8 cm. The red dashed line is the arrival time of the reflected wave, and the specific time is marked on the left.

3. System Implementation

3.1. Fabrication and Characterization of PMUTs

The steps to fabricate a PMUT device are shown in Figure 3a. The process starts with a bare silicon wafer with a thickness of 200 μm, followed by a layer of 1 μm SiO_2 deposition using chemical vapor deposition (CVD). Afterwards, 0.1 μm-thickAlN seeds layer is deposited by atomic layer deposition (ALD) to ensure the formation of the (002) crystal orientation of AlN during the subsequent deposition Then, 0.1 μm-thick bottom Mo layer, 1 μm-thick AlN layer and 0.1 μm-thick Mo layer are deposited by physical vapor deposition (PVD) and patterned using SiO_2 as hard masks, as shown in Figure 3(aIV–aVII). Subsequently, PE-SiO_2 is deposited (Figure 3(aVIII)) and patterned (Figure 3(aIV)). Deposition of the top Al/Cu connection line and pads is shown in Figure 3(aX). Finally, a trench is etched from the back of the silicon wafer by reactive-ion etching (DRIE) to release the membrane structure, and a protective material is used to protect the front side membrane from damage as shown in Figure 3(aXI). In order to increase the sound pressure of the PMUT in the air, the bottom silicon is bonded to the glass using an anodic bonding process, and a vacuum cavity is formed by evacuation as shown in Figure 3(aXII). The top-view of PMUT is shown in Figure 3(bi) by optical microscopy (OM). The fabricated device is a 3500 μm-long square membrane with a cavity having dimensions of 2800 μm × 2800 μm × 167 μm. The top and bottom electrodes of the PMUT are connected to the signal and ground terminals via aluminum wires, respectively. Figure 3(bii) shows the secondary electron microscopy (SEM) imaging of the membrane with a close-up detail of the thickness of each layer.

Two PMUTs with similar frequencies are selected as the transmitter and receiver for better ranging performance. As shown in Figure 3c,d, the impedance characteristics of two PMUTs are measured using an impedance analyzer (Agilent Technologies 4294A, Santa Clara, CA, USA). It can be seen that the resonant frequencies of the transmitter and receiver are 97 kHz and 96 kHz, which are very close to the simulation results. The electromechanical coupling coefficient k_{eff}^2 can be derived by resonant frequency f_r and anti-resonant frequency f_a through Equation (5) [22]:

$$k_{eff}^2 = \frac{f_a^2 - f_r^2}{f_r^2} \tag{5}$$

The calculated electromechanical coupling coefficients of the transmitter and receiver are 2.0% and 2.6%, respectively. Due to errors in the manufacturing process, it is difficult to ensure that the resonant frequencies and electromechanical coupling coefficients of the two PMUTs are identical.

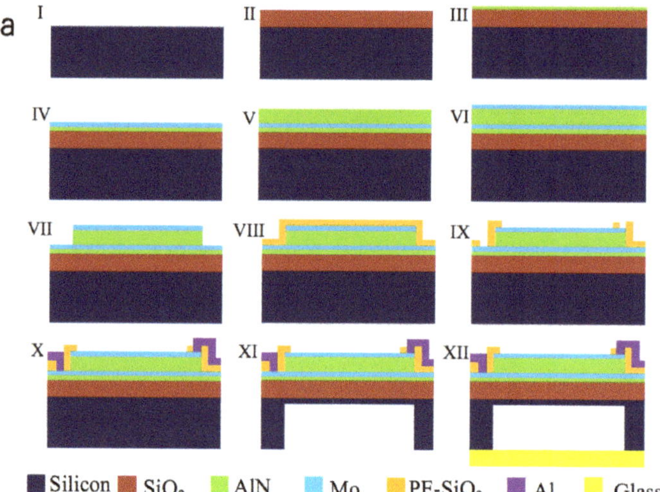

Figure 3. *Cont.*

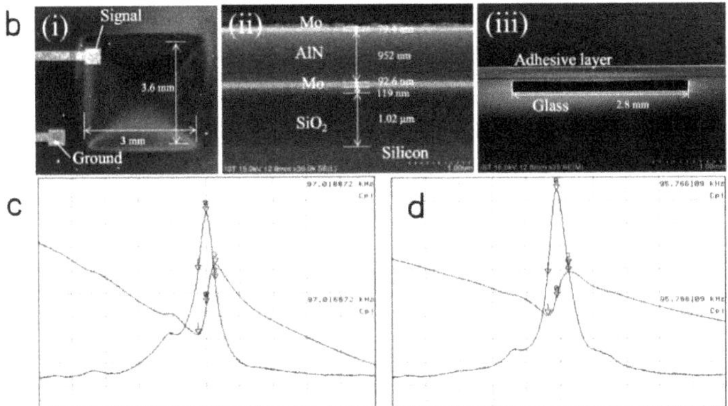

Figure 3. The fabrication and characterization of PMUTs. (**a**) The process flow of the air-coupled PMUT based on AlN. (**b**) Optical microscopy (OM) and SEM imaging of the PMUT (the adhesive layer is added for SEM imaging). (**c**) Analysis of impedance characteristics of transmitter (97 kHz), and (**d**) receiver (96 kHz).

3.2. System Clock Design and Implementation

The oscillator is used to create a main/system clock which is applied to ranging controller and TOF calculation. In order to easily represent or derive the distance result from TOF, the intuitive way is to correlate the clock frequency with sound velocity. The equation of sound speed is defined in Equation (4). At 25 °C, it can be derived that the TOF is 5.78 μs/mm. Adopting 1.73 MHz as the clock frequency of TOF calculator, we can get 0.1 mm resolution in theory, which is reasonable in the usage of ranging controller.

Since relaxation oscillator has the advantages of supply insensitive clocking, low power consuming and compatible with frequency trimming, it is adopted to generate the clock of the ranging system, as shown in Figure 4a. The oscillation frequency is derived using Equation (6):

$$f_{osc} = \frac{I_B}{2C_{RAMP} \times V_{REF}}$$ (6)

In our design, $I_B = 4$ μA, $C_{RAMP} \sim 2$ pF, $V_{REF} \sim 0.5$ V were selected to obtain the desired clock frequency. Through simulation, the effects of different V_{DD} on the clock frequency variation were studied.

At a temperature of 25 °C, when the power supply V_{DD} rises from 1.62 V to 1.98 V, the clock frequency gradually decreases with a variation of 1.330% for post-sim, and the simulation results are shown in Figure 4b. The green dashed line indicates the pre-simulation results, while the red solid line indicates the post-simulation results. The former is higher than the latter because the post-sim considers the influence of parasitic capacitance.

Four chips were selected as samples for testing. The measurement results are summarized in Figure 4c. It can be seen from the experimental results that as V_{DD} increases, the clock frequency gradually decreases, and the tendency is the same as the simulation results. After frequency adjustment, the calibrated frequency can reach about 1.73 MHz, and the frequency error after calibration is well within 0.5% of the design target.

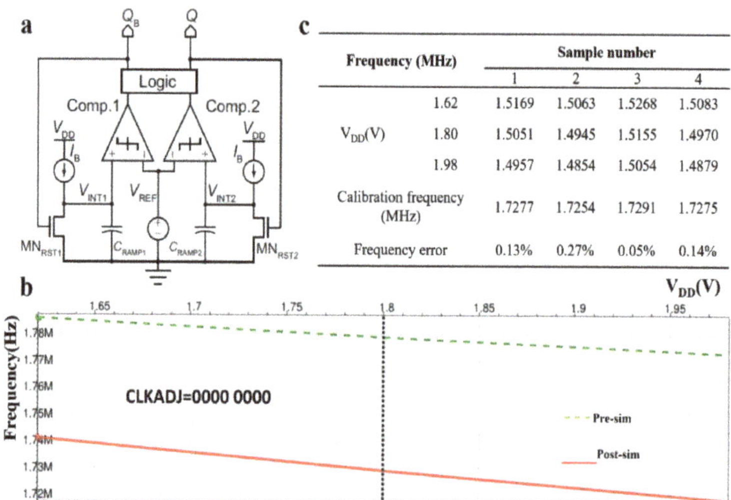

c

Frequency (MHz)		Sample number			
		1	2	3	4
V_DD(V)	1.62	1.5169	1.5063	1.5268	1.5083
	1.80	1.5051	1.4945	1.5155	1.4970
	1.98	1.4957	1.4854	1.5054	1.4879
Calibration frequency (MHz)		1.7277	1.7254	1.7291	1.7275
Frequency error		0.13%	0.27%	0.05%	0.14%

Figure 4. Realizing 1.73 MHz frequency of system (OSC) for TOF calculation. (**a**) The block diagram of the relaxation oscillator. (**b**) System frequency vs. power supply voltage by simulation, and (**c**) measurement results of the frequency vs. power supply.

3.3. Charge Pump (CP)

The charge pump circuit is utilized to rise the voltage of 1.8 V to 32 V to drive the level shifter (LSF). It is a DC-DC converter that uses a capacitor as an energy storage element to generate an output voltage greater than the input voltage. When the drive voltage is 32 V, the PMUT can be fully activated. The output voltage of the CP is affected by the frequency and load. Through experimental measurement, the relationship between the output voltage and frequencies/load is determined, as shown in Figure 5a. The black line indicates the change of output voltage with frequency when there is no load, while the red line indicates the case when the load is 5.6 MΩ. It can be seen that the output voltage with load is lower than that without load, but the output voltage is positively related to frequency, from which can be estimated that the output voltage is about 29 V when the frequency is 97 kHz, as shown by the dotted line in the Figure 5a. Low output voltage is due to the parasitic cap, which will be improved in the future.

3.4. Transmitter (TX)

The two terminals of the circuit in this design were connected to the transmitter and receiver, respectively. As shown in Figure 5b, the TX circuit includes three blocks of TXOSC, Charge-Pump (CP), and HV Output Driver. Firstly, TXOSC with VREF circuit provides the internal voltage reference and generated clock supports charge-pump, CP5_CLK, CP32_CLK, and TXDATA of the HV output driver, respectively. It can provide the frequency range from 40 kHz to 30 MHz, and the frequency of TXOSC used in this work was 97 kHz.

Then, Charge-Pump block pumped high voltage from 1.8 V to 32 V with two blocks, one was 1.8 V to 5 V and another was 5 V to 32 V, for output driver to stimulate the transmitter. The operation frequency was 2 MHz for CP5V and 60 kHz for CP32V, respectively. Relative bypass capacitor for CP5V was an internal capacitor of 450 pF or co-package with external capacitor of 1 nF and CP32V was used with an external bypass capacitor of 1 nF or 10 nF.

Finally, HV Output Driver drives PMUT sensor with TXDATA with operation frequency and voltage magnitude. Figure 5b describes in detail the function block diagram of TX, and Figure 5c shows the signal simulation of the driver circuit.

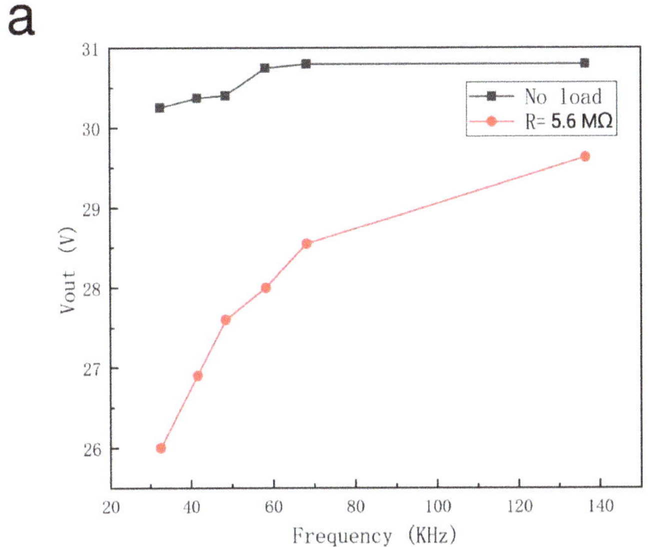

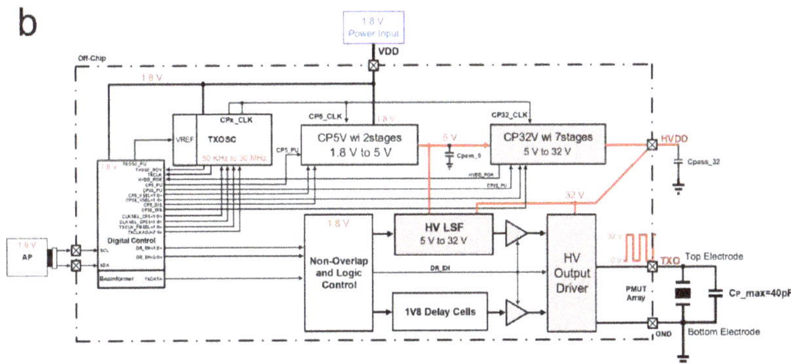

Figure 5. *Cont.*

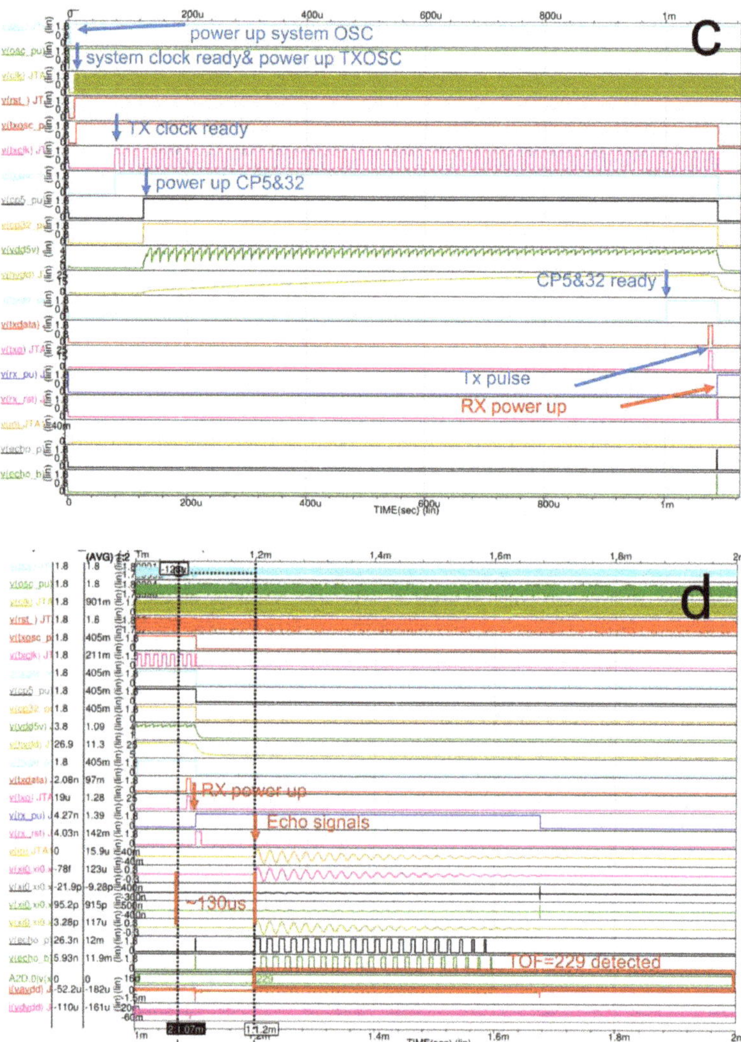

Figure 5. Charge pump, transmitter and receiver circuits. (**a**) The relationship between the output voltage of the charge pump circuit and the driving frequency. (**b**) The function block diagram of TX. Signal simulation of (**c**) the drive circuit and (**d**) the receiver circuit.

3.5. Receiver (RX)

There are multiple receiver implementation solutions. In this design, pre-amplifier and echo comparators were adopted in consideration of the flexibility and supportability. In order to obtain the required gain, three stage capacitive feedback amplifiers were used to form the pre-amplifier. After the amplifier, there were two threshold programmable comparators which were used to detect the echo signal. The whole receiver was realized by using fully differential structure in order to reduce the noise effect.

As shown in Figure 5d, when it passes 229 system clock cycles from the transmission to the reception of the ultrasonic signal, the distance between the PMUT and the target is 22.9 mm.

4. System Evaluation

4.1. Power Consumption of the Chip

When the power supply voltage is 1.8 V, the chip can run in different modes by controlling the value of the flag, including CHIPON, TXOSCON, CP5_ON, CP32_ON and RXON. The chip has different power consumption in different working modes. By measuring the current consumption in different modes, the chip consumption was obtained, as shown in Figure 6b. Because there was a current leakage path, the idle and standby power consumption was high, which will be improved in the future.

4.2. Ranging Experiments and Results

Two PMUTs were connected to a fixed printed circuit board (PCB) through wire bonding as the transmitter and receiver, respectively. The experimental setup is shown in the Figure 6a. A reflector, which is a piece of smooth aluminum plate, was fixed on the measuring claw of a Vernier caliper with a resolution of 0.02 mm. A 1.8 V power supply was applied to power the transceiver chip. The oscilloscope was connected to the TX and RX terminals of the chip to monitor the transmitted electrical pulse and the received ultrasonic echo signal.

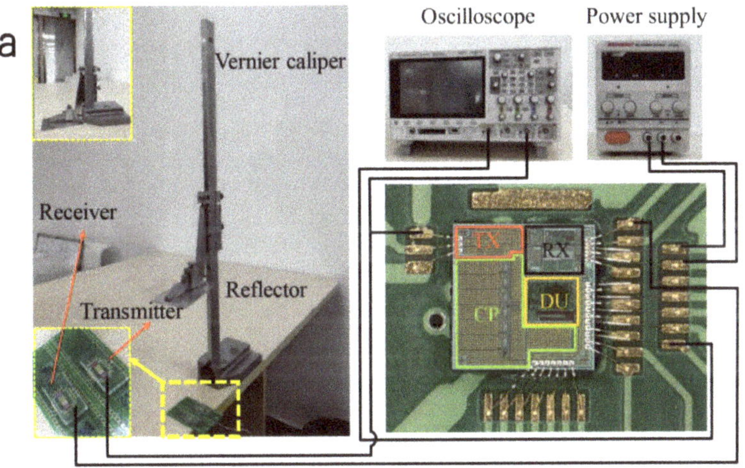

b	Power consumption(µW) (V_{DD}=1.8V)	Sample number				Flag
		1	2	3	4	
	Idle mode	166.14	165.24	165.06	166.32	No setting required
	Standby mode	242.64	241.2	241.2	242.64	CHIPON = 1
Operation mode	TX mode	532.26	536.94	529.92	541.8	CHIPON = 1; TXOSC_ON = 1; CP5_ON = 1; CP32_ON = 1
	RX mode	437.04	434.7	434.34	435.96	CHIPON = 1; TXOSC_ON = 1; RXON = 1

Figure 6. *Cont.*

C (i)

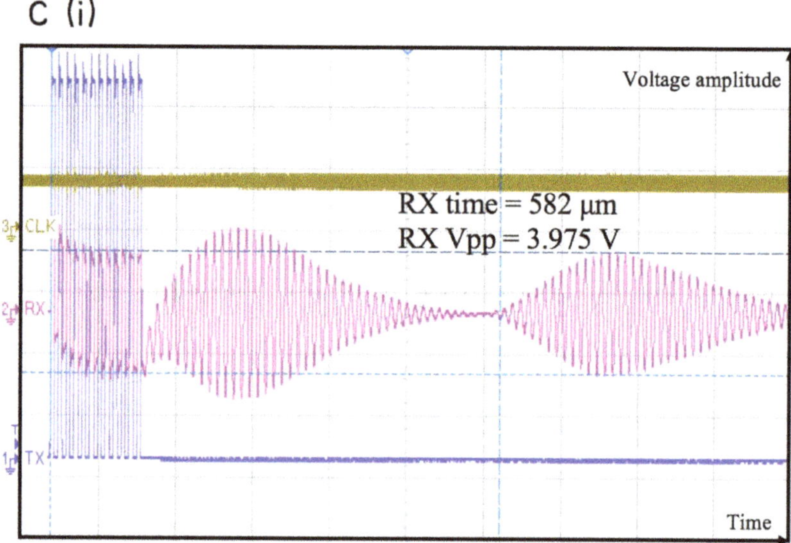

C (ii)

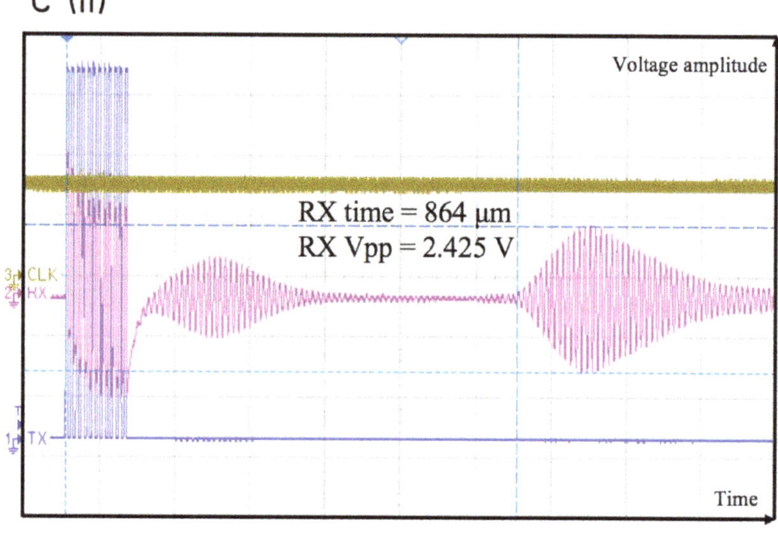

Figure 6. *Cont.*

C (iii)

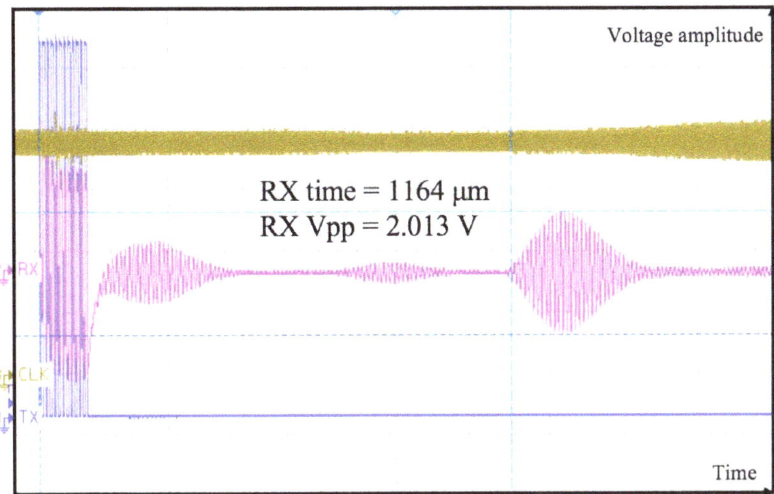

C (iv)

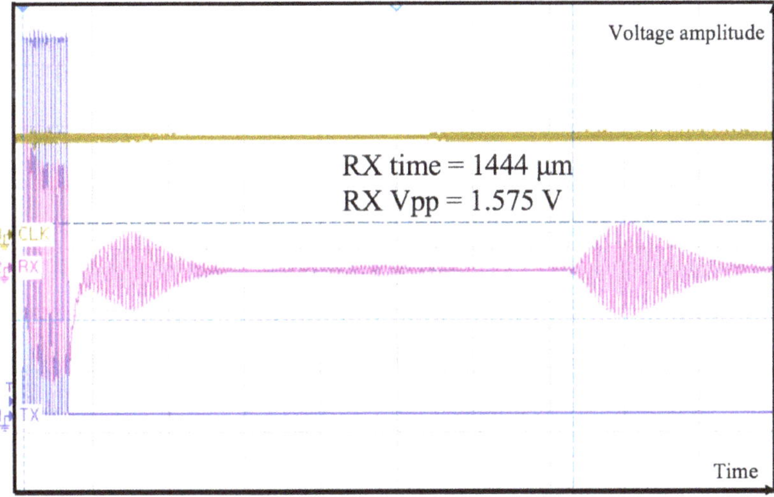

Figure 6. *Cont.*

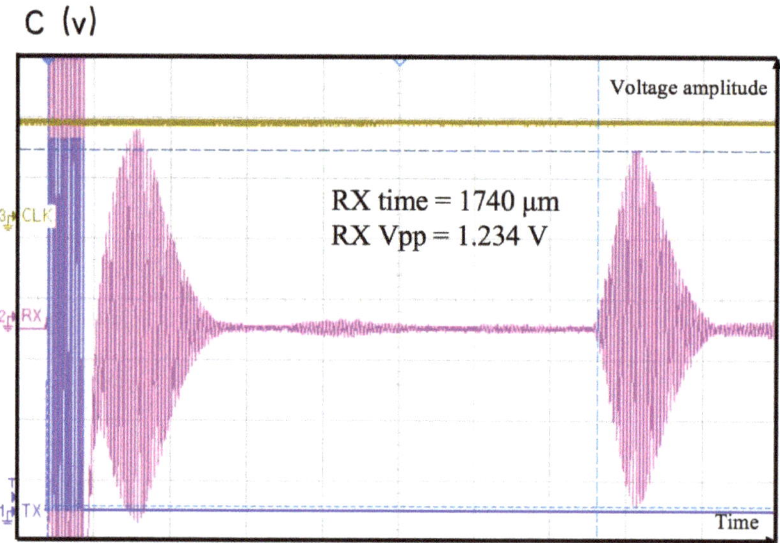

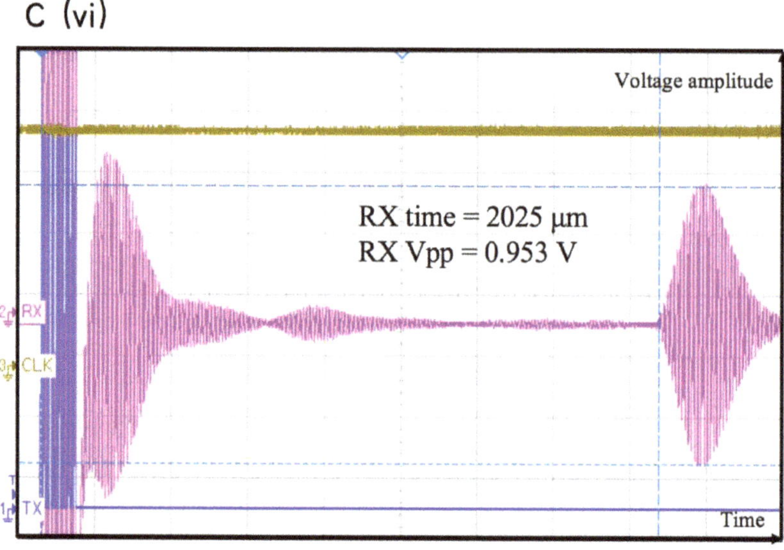

Figure 6. *Cont.*

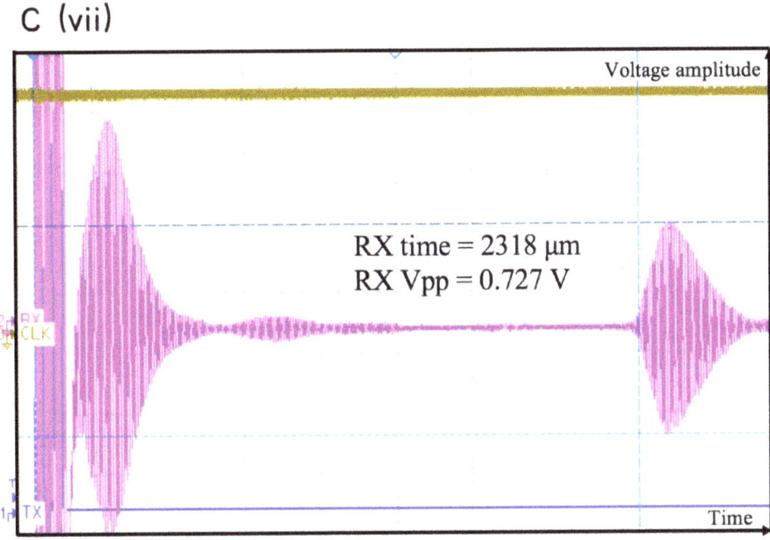

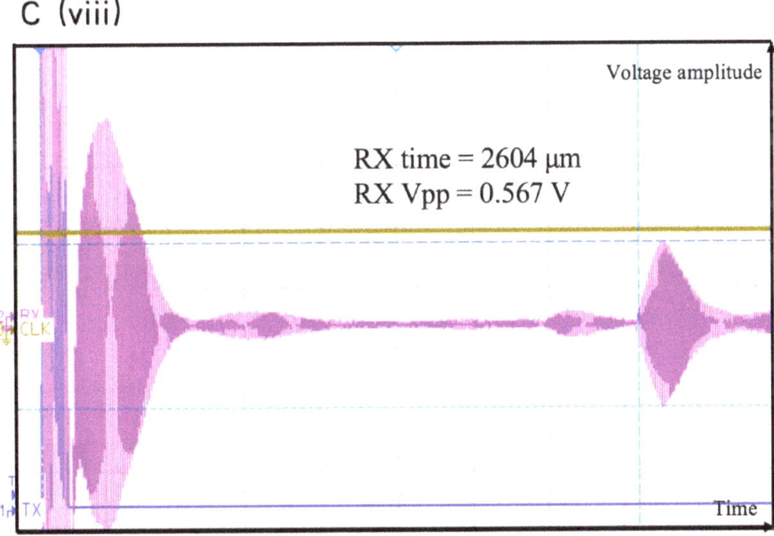

Figure 6. *Cont.*

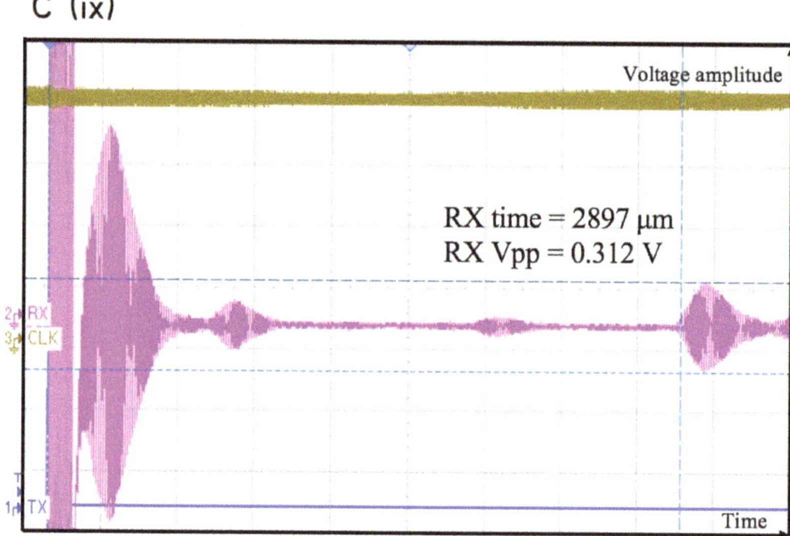

Figure 6. Measuring setup and results of ranging system. (**a**) The setup for range measurement (100–500 mm). (**b**) Power consumption of the chip in different working modes. (**c**) The experimental results of measuring waveforms. The distance from (**i**) to (**ix**) is from 10 cm to 50 cm in the increment of 5 cm.

It was found that when the number of the TX pulses was 12, the PMUT can be fully oscillated. If the number continues to increase, the amplitude can no longer be increased significantly. Therefore, 12 TX pulses were used to stimulate the transmitter. The measurement results are shown in the Figure 6c. The blue line indicates the twelve electrical pulse excitation signals, and the brown line indicates the waveform of system clock with a frequency of 1.73 MHz. The purple line indicates the waveform of the receiver. Due to the crosstalk between the transmitter and the receiver, the receiver has an electrical signal fluctuation before the ultrasonic echo arrives, which can be filtered out by hardware isolation or filtering algorithms.

The experimental results of average of five measurements are summarized in Table 1. n_{clk} refers to the number of cycles of received voltage, L is the sensing distance. RX Vpp is the voltage amplified by the pre-amplifier, which is 500 times the actual received voltage. This data will determine the maximum test distance, because the longer the distance, the smaller the voltage of the reflected wave. Since the receiver can only sense certain limit of voltage, the sensing range is determined by the received voltage.

Table 1. The results of the ranging (100–500 mm).

Results	The Readings of Vernier Calipers (mm)	n_{clk}	L (mm)	Range Errors (mm)	RX Vpp (mV)
1	100.24	1004	100.42	0.18	4012.8
2	150.12	1499	149.86	−0.26	2403
3	200.14	2008	200.8	0.66	2010.4
4	250.02	2505	250.48	0.46	1566.4
5	300.22	2997	299.68	−0.54	1262.2
6	350.78	3496	349.56	−1.22	952.4
7	400.08	3998	399.76	−0.32	752.4
8	450.22	4495	449.52	−0.7	584
9	500.32	4994	499.4	−0.92	343.2
Standard deviation σ (mm)				0.63	

The performance comparisons with previous works are listed in Table 2. For the same TOF method in [14], the range and accuracy of this work both have been improved due to the usage of more advanced fabrication process and customized circuits. Compared with the phase-shift method [23], the TOF method has a larger sensing range, but the accuracy is slightly lower, especially at short distances (e.g., <100 mm), while the phase shift method has extremely high accuracy.

Table 2. Performance comparison with previous works.

Reference	Method	Transducers	Max Range	Range Error
[20]	Phase-shift sound	Si Thermal ultrasound	0.11 m	4 mm
[14]	TOF sound	AlN PMUT	0.45 m	3.9 mm (3σ)
[23]	Phase-shift sound	AlN PMUT	0.1 m/0.3 m	71.1 μm/1.82 mm (3σ)
This work	TOF sound	AlN PMUT	0.5 m	0.63 mm (3σ)

5. Conclusions

In this paper, we present a high-accuracy CMOS driven ultrasonic ranging system based on air coupled piezoelectric micromachined ultrasonic transducers (PMUTs) using time of flight (TOF). The mode shape and the time-frequency characteristics of PMUTs were firstly simulated and analyzed. PMUTs with frequencies of 97 kHz and 96 kHz were chosen as the transmitter and receiver, achieving both high range accuracy and ranging field. Based on the 0.18 μm CMOS process, the time to digital converter circuit simplifies the ranging method by system clock counting. The experimental results show that the accuracy of the ranging between 10 cm and 50 cm is 0.63 mm (one standard deviation), which is better than that of the previously reported rangefinder. The TOF based ranging system using PMUTs and CMOS circuit shows great application potentials in short-range communication, such as ultrasonic positioning, gesture recognition, medical assistance, etc. Future work will focus on the integration of CMOS and PMUTs, as well as applications based on ultrasonic rangefinders.

Author Contributions: Conceptualization, Y.C. and S.M.; methodology, Y.C.; software, Y.C.; validation, Y.C., C.W. and D.G.; formal analysis, Y.C.; investigation, N.L.; resources, Y.C., S.M. and Y.J.; data curation, Y.C.; writing—original draft preparation, N.L.; writing—review and editing, Y.C.; visualization, C.W., and D.G.; supervision, Y.C., S.M. and Y.J.; project administration, S.M. and Y.J.; funding acquisition, Y.J. All authors have read and agreed to the published version of the manuscript.

Funding: This work was funded by the National Natural Science Foundation of China with grant number No. U1613215, and the TSV 3D Integrated Micro/Nanosystem Laboratory (ZDSYS201802061 805105).

Institutional Review Board Statement: Not applicable.

Informed Consent Statement: Not applicable.

Data Availability Statement: All data, models, and code generated or used during the study appear in the submitted article.

Conflicts of Interest: The authors declare that they have no conflict of interest.

References

1. Sabry, R.S.; Hussein, A.D. PVDF: ZnO/BaTiO$_3$ as high out-put piezoelectric nanogenerator. *Polym. Test.* **2019**, *79*, 106001. [CrossRef]
2. Baborowski, J.; Ledermann, N.; Maralt, P.; Schmitt, D. Simulation and characterization of piezoelectric micromachined ultrasonic transducers (PMUTs) based on PZT/SOI membranes. *Integr. Ferroelectr.* **2010**, *54*, 557–564. [CrossRef]
3. Li, J.; Ren, W.; Fan, G.; Wang, C. Design and fabrication of piezoelectric micromachined ultrasound transducer (pMUT) with partially-etched ZnO film. *Sensors* **2017**, *17*, 1381.
4. Lu, Y.; Horsley, D.A. Modeling, fabrication, and characterization of piezoelectric micromachined ultrasonic transducer arrays based on cavity SOI wafers. *J. Microelectromech. Syst.* **2015**, *24*, 1142–1149. [CrossRef]

5. Lu, Y.; Heidari, A.; Shelton, S.; Guedes, A.; Horsley, D.A. High frequency piezoelectric micromachined ultrasonic transducer array for intravascular ultrasound imaging. In Proceedings of the 2014 IEEE 27th International Conference on Micro Electro Mechanical Systems (MEMS), San Francisco, CA, USA, 26–30 January 2014.

6. Ruby, R.C.; Bradley, P.; Oshmyansky, Y.; Chien, A.; Larson, J.D. Thin film bulk wave acoustic resonators (FBAR) for wireless applications. In Proceedings of the 2001 IEEE Ultrasonics Symposium. Proceedings. An International Symposium (Cat. No.01CH37263), Atlanta, GA, USA, 7–10 October 2001.

7. Shelton, S.; Chan, M.; Park, H.; Horsley, D.; Boser, B.; Izyumin, I.; Przybyla, R.; Frey, T.; Judy, M.; Nunan, K.; et al. CMOS-compatible AlN piezoelectric micromachined ultrasonic transducers. In Proceedings of the 2009 IEEE International Ultrasonics Symposium, Rome, Italy, 20–23 September 2009.

8. Lu, Y.; Heidari, A.; Horsley, D.A. A high fill-factor annular array of high frequency piezoelectric micromachined ultrasonic transducers. *J. Microelectromech. Syst.* **2014**, *24*, 904–913. [CrossRef]

9. Horsley, D.A.; Rozen, O.; Lu, Y.; Shelton, S.; Guedes, A.; Przybyla, R.; Tang, H.; Boser, B.E. Piezoelectric micromachined ultrasonic transducers for human-machine interfaces and biometric sensing. In Proceedings of the 2015 IEEE Sensors, Busan, Korea, 1–4 November 2015.

10. Goh, C.L.; Rahim, R.A.; Rahiman, H.F.; Cong, T.Z.; Wahad, Y.A. Simulation and experimental study of the sensor emitting frequency for ultrasonic tomography system in a conducting pipe. *Flow Meas. Instrum.* **2017**, *54*, 158–171. [CrossRef]

11. Yuan, Q.; Fan, L.; Tong, Z.; Yang, X.; Cao, Y. A study on time-voltage conversion method for pulse radar ranging. *Mod. Radar* **2012**, *34*, 69–73.

12. Wu, J.; Yu, X.; Shi, S.; Zheng, L.; Sun, W. Infrared ranging technology by using single photon APD array readout integrated circuit. *Infrared Laser Eng.* **2017**, *46*, 69–74.

13. Shi, G.; Wang, W.; Zhang, F. Precision improvement of frequency-modulated continuous-wave laser ranging system with two auxiliary interferometers. *Opt. Commun.* **2018**, *411*, 152–157. [CrossRef]

14. Przybyla, R. In-air rangefinding with an ALN piezoelectric micromachined ultrasound transducer. *IEEE Sens. J.* **2011**, *11*, 2690–2697. [CrossRef]

15. Wang, T.; Lee, C. Zero-bending piezoelectric micromachined ultrasonic transducer (pMUT) with enhanced transmitting performance. *J. Microelectromechanical Syst.* **2015**, *24*, 2083–2091. [CrossRef]

16. Li, C.; Hutchins, D.A.; Green, R.J. Short-range ultrasonic digital communications in air. *IEEE Trans. Ultrason. Ferroelectr. Freq. Control* **2008**, *55*, 908–918. [PubMed]

17. Huang, C.F.; Young, M.S. Multiple-frequency continuous wave ultrasonic system for accurate distance measurement. *Rev. Sci. Instrum.* **1999**, *70*, 1452. [CrossRef]

18. Testi, N.; Xu, Y.; Jin, Y. Successive-MFCW modulation for ultra-fast narrowband radar. In Proceedings of the 2013 IEEE International Conference on Acoustics, Speech and Signal Processing, Vancouver, BC, Canada, 26–31 May 2013.

19. Khyam, M.O.; Ge, S.S.; Li, X.; Pickering, M.R. Highly accurate time-of-flight measurement technique based on phase-correlation for ultrasonic ranging. *IEEE Sens. J.* **2017**, *17*, 434–443. [CrossRef]

20. Kuratli, C.; Huang, Q. A CMOS ultrasound range finder microsystem. In Proceedings of the 2000 IEEE International Solid-State Circuits Conference, San Francisco, CA, USA, 7–9 February 2000; pp. 180–181.

21. Wang, T.; Kobayashi, T.; Lee, C. Micromachined piezoelectric ultrasonic transducer with ultra-wide frequency bandwidth. *Appl. Phys. Lett.* **2015**, *106*, 013501. [CrossRef]

22. Jung, J.; Kim, S.; Lee, W.; Choi, H. Fabrication of a two-dimensional piezoelectric micromachined ultrasonic transducer array using a top-crossover-to-bottom structure and metal bridge connections. *J. Micromech. Microeng.* **2013**, *23*, 125037. [CrossRef]

23. Chen, X.; Xu, J.; Chen, H.; Ding, H.; Xie, J. High-accuracy ultrasonic rangefinders via pMUTs arrays using multi-frequency continuous waves. *J. Microelectromech. Syst.* **2019**, *28*, 634–642. [CrossRef]

 *micromachines*

Article

Performance Optimization of Nitrogen Dioxide Gas Sensor Based on Pd-AlGaN/GaN HEMTs by Gate Bias Modulation

Van Cuong Nguyen [1], Kwangeun Kim [2] and Hyungtak Kim [1,*]

1. School of Electronic and Electrical Engineering, Hongik University, Seoul 04066, Korea; nvcuong@mail.hongik.ac.kr
2. School of Electronics and Information Engineering, Korea Aerospace University, Gyeonggi 10540, Korea; kke@kau.ac.kr
* Correspondence: hkim@hongik.ac.kr; Tel.: +82-2-320-3013

Abstract: We investigated the sensing characteristics of NO_2 gas sensors based on Pd-AlGaN/GaN high electron mobility transistors (HEMTs) at high temperatures. In this paper, we demonstrated the optimization of the sensing performance by the gate bias, which exhibited the advantage of the FET-type sensors compared to the diode-type ones. When the sensor was biased near the threshold voltage, the electron density in the channel showed a relatively larger change with a response to the gas exposure and demonstrated a significant improvement in the sensitivity. At 300 °C under 100 ppm concentration, the sensor's sensitivities were 26.7% and 91.6%, while the response times were 32 and 9 s at $V_G = 0$ V and $V_G = -1$ V, respectively. The sensor demonstrated the stable repeatability regardless of the gate voltage at a high temperature.

Keywords: gate bias modulation; palladium catalyst; gallium nitride; nitrogen dioxide gas sensor; high electron mobility transistor

Citation: Nguyen, V.C.; Kim, K.; Kim, H. Performance Optimization of Nitrogen Dioxide Gas Sensor Based on Pd-AlGaN/GaN HEMTs by Gate Bias Modulation. *Micromachines* **2021**, *12*, 400. https://doi.org/10.3390/mi12040400

Academic Editor: Giovanni Verzellesi

Received: 2 March 2021
Accepted: 2 April 2021
Published: 5 April 2021

Publisher's Note: MDPI stays neutral with regard to jurisdictional claims in published maps and institutional affiliations.

Copyright: © 2021 by the authors. Licensee MDPI, Basel, Switzerland. This article is an open access article distributed under the terms and conditions of the Creative Commons Attribution (CC BY) license (https://creativecommons.org/licenses/by/4.0/).

1. Introduction

Nanotechnology is being employed in the research of sensor devices to a great extent. Owing to the unique properties of the nanomaterials such as great adsorptive capacity due to the large surface-to-volume ratio, the ability of tuning electrical properties by controlling the composition and the size of the nanomaterial, the easy configuration and integration in low-power microelectronic systems, the gas sensors based on nanotechnology are excellent candidates for sensitive detection of chemical and biological species [1–4].

Nitrogen dioxide (NO_2) is one of the most harmful gases released into the atmosphere from both natural sources and human activities. Prolonged exposure with a high concentration of NO_2 can cause lung tissue inflammation, aggravate respiratory diseases, particularly asthma, leading to respiratory symptoms. The LC50 (the lethal concentration for 50% of those exposed) for a 1-h exposure of NO_2 for humans has been estimated to be 174 ppm [5]. In industry, nitrogen dioxide gas is emitted from the burning of fuel, the exhaust of furnaces and power plants, etc. Therefore, the development of ppm-level NO_2 gas sensors that can operate at very high temperatures is necessary.

Gas sensors based on metal oxide have been intensively investigated for several decades [6–9]. Despite high sensitivity and easy fabrication, they are not considered a promising candidate for extreme environment electronics due to a long response time, poor selectivity towards any specific gas, and unstable operation at harsh environmental conditions [10]. To develop a gas sensor that operates at high temperatures, gallium nitride-based gas sensors [11–17] recently have attracted great attention thanks to the wide bandgap of 3.4 eV, high thermal, and chemical stability.

Currently, NO_2 gas sensors based on AlGaN/GaN high electron mobility transistors (HEMTs) have been the subject of intense research [18–24]. The large variation of the sensitivity of these sensors was explained by the trade-off between the sensitivity (S) and the base drain current (I_0). For example, a Pt-AlGaN/GaN sensor showed a significant current change of 2.8 mA measured at 400 °C under 450 ppm NO_2, but the high base current of 16.5 mA led to 17% of sensitivity [21]. Another Pt-AlGaN/GaN sensor showed 5.5% of sensitivity while it had a high current change of 1.801 mA and a high base current of 33 mA [22]. On the other hand, a NH_3 gas sensor archived an ultra-high sensitivity of 18,300% at 150 °C, but the base current was limited in the pA range [25].

There are some approaches to optimize the sensitivity of sensors based on HEMTs. Firstly, choosing an appropriate catalyst is very important. Since the sensors work based on the chemical mechanism, the catalyst directly affects the sensor characteristics such as working temperature, sensitivity, response, and recovery times. Our previous work proved that sensors based on Pd-AlGaN/GaN HEMT showed better performance than the Pt one [24]. Second, the surface treatment, like hydrogen peroxide treatment [26] or plasma treatment [27], can improve the gas sensor's sensitivity. Third, since HEMTs are 3-terminal devices, they provide the possibility of adjusting the gate voltage to control the drain current, thereby optimizing the sensor's performance, which is a superior advantage when compared to diode-type sensors.

In this paper, we comprehensively investigated the sensitivity enhancement of the NO_2 gas sensor based on Pd-AlGaN/GaN HEMT by gate bias modulation. When the sensor was biased close to the threshold voltage, the sensitivity and the response time were much improved. The Technology Computer-Aided Design (TCAD) simulation revealed that the improvement of sensitivity was attributed to the larger change in the channel electron concentration near the threshold.

2. Materials and Methods

The fabrication process of the sensor is shown in Figure 1. AlGaN/GaN HEMT-type sensors were fabricated at the Inter-University Semiconductor Research Center (ISRC), Seoul, Korea. The AlGaN/GaN-on-Si substrate consisted of a 10 nm in situ SiN_x layer, a 13 nm $Al_{0.3}Ga_{0.7}N$ barrier layer, a 4.2 µm i-GaN layer, and AlGaN/AlN buffer layers. AlGaN layer parameters such as Al-content and the thickness were optimized for the maximum transconductance by increasing Al-content and thinning the thickness. The source and drain contacts with Ti/Al/Ni/Au (20/120/25/50 nm) were formed by e-beam evaporation with a lift-off process and followed by rapid thermal annealing (RTA) at 830 °C for 30 s in N_2 ambient. Then, 300 nm-depth mesa isolation was formed by inductively coupled plasma (ICP) etching with BCl_3/Cl_2 to define the active region. The 30 nm Pd layer as a gate electrode was then formed by e-beam evaporation and a lift-off process. Afterward, the interconnect bi-layer probing pads of Ti/Au with thickness 20/300 nm were formed by e-beam evaporation and lift-off. A passivation layer of 200 nm SiN_x was deposited using plasma-enhanced chemical vapor deposition (PECVD) at 190 °C in order to protect the sensor's surface. Finally, the SiN_x layer was patterned and etched to open the Pd-gate to the ambient and the contact pads for measurement. The dimensions of the Pd-gate electrode were 24 µm × 120 µm, the source-gate and gate-drain spacings were 2 µm. The fabricated AlGaN/GaN HEMT exhibited sheet resistance, 2DEG mobility, and sheet carrier density of 493 $\Omega/\square$, 1420 $cm^2/(V·s)$, and 8.9×10^{12} cm^{-2}, respectively. The microscope image of the fabricated Pd-AlGaN/GaN HEMT device is shown in Figure 2a.

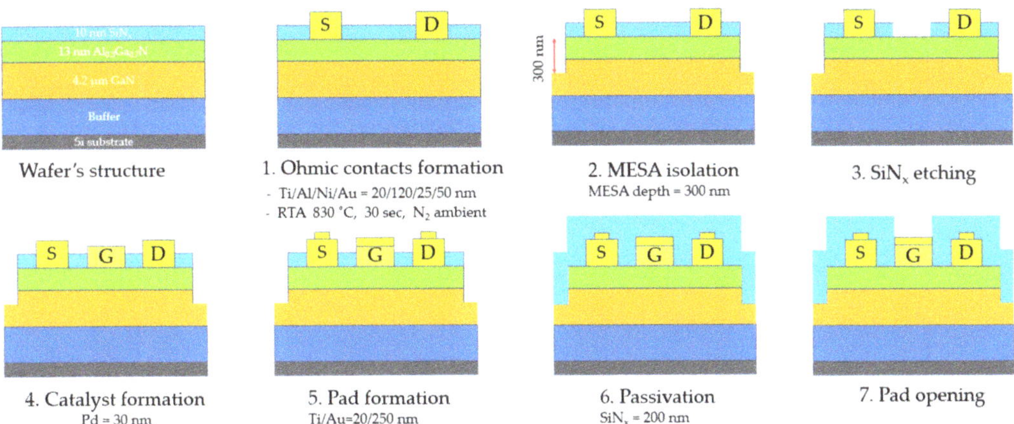

Figure 1. Fabrication process of Pd-AlGaN/GaN high electron mobility transistors (HEMT) sensor.

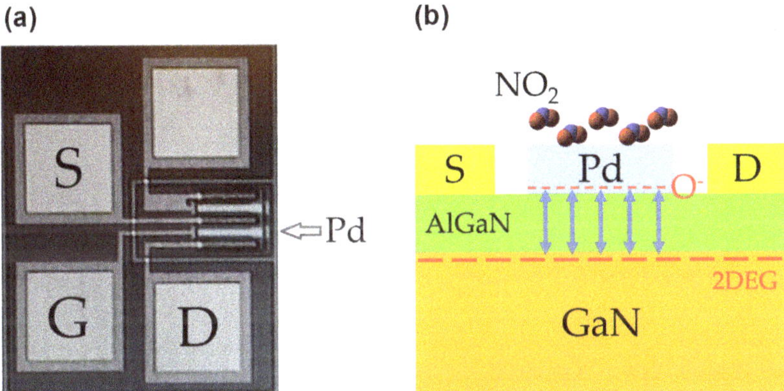

Figure 2. Microscope image of fabricated Pd-AlGaN/GaN HEMT device (**a**), and the sensing mechanism of the sensor (**b**).

Gas sources consist of synthesized dry air as the background gas, and 100 ppm NO_2 as target gas. The background and the target gases were mixed by mass flow controllers (MFCs) to archive the different concentrations of NO_2. The combined total gas flow was set to 200 sccm in all measurements. The sensors were loaded in a chamber containing a hot chuck to control the operating temperature. The DC and transient characteristics of the sensor were measured using the HP 4155A semiconductor parameter analyzer.

The sensitivity is defined as the ratio between the change of drain current and initial current:

$$Sensitivity(\%) = \frac{I_0 - I_{NO_2}}{I_0} = \frac{\Delta I}{I_0}, \tag{1}$$

where I_0 and I_{NO_2} are drain currents under the flow of dry air and NO_2 gas, respectively. The response and recovery times were calculated from transient characteristic, where they showed 90% of the total change (ΔI) in drain current [28].

The sensing mechanism of nitrogen dioxide sensors based on Pd-AlGaN/GaN was investigated and reported in our previous work [24]. When nitrogen dioxide gas is adsorbed on the Pd catalyst layer, the nitrogen dioxide molecules are dissociated in nitrogen monoxide (NO) going to the gas phase and oxygen ions on the Pd surface [29,30]. Then, the negatively charged oxygen ions diffuse through the gate and reach the surface of AlGaN layers. Here, they affect the number of mobile carriers in two-dimensional electron gas (2DEG) of the HEMT structure, leading to a reduction of drain current (Figure 2b).

3. Results

The transfer characteristics of the fabricated HEMT at different temperatures were shown in Figure 3. The device exhibited stable operations up to 300 °C due to GaN's thermal stability.

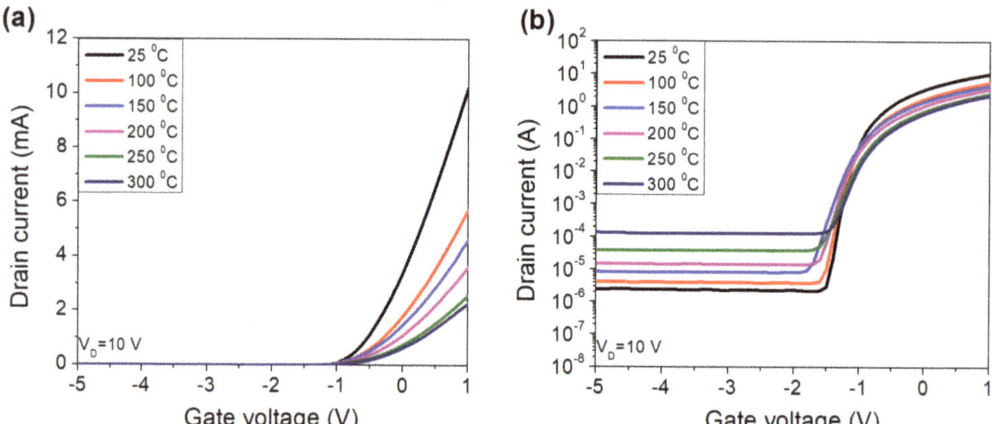

Figure 3. Transfer characteristics of fabricated device in linear (**a**) and log (**b**) scales.

Figure 4 showed the response of the sensor at different gate voltages at different biases and temperatures. While the adsorption of oxygen ions on the Pd/AlGaN interface occurs even at room temperature, the desorption totally takes place at high temperatures beyond 200 °C, which directly affects the recovery of the sensor. This indicates that this sensor is more suitable for high temperature operation. At a higher temperature, the adsorption and desorption of negatively charged oxygen ions took place more efficiently, resulting in faster response and recovery times. Starting from 300 °C, the sensor fully recovered to the base current, which agreed with other studies [21–23]. Additionally, the sensitivity was improved at higher temperatures (Figure 4d). When the device is biased closed to the threshold voltage, the reduction of the current level leads to a higher sensitivity, which is similar to the gate recess approach [31]. Since the gate bias modulation is a damage-free technique, it exhibited the advantage over the gate recess method to improve the sensor's performance.

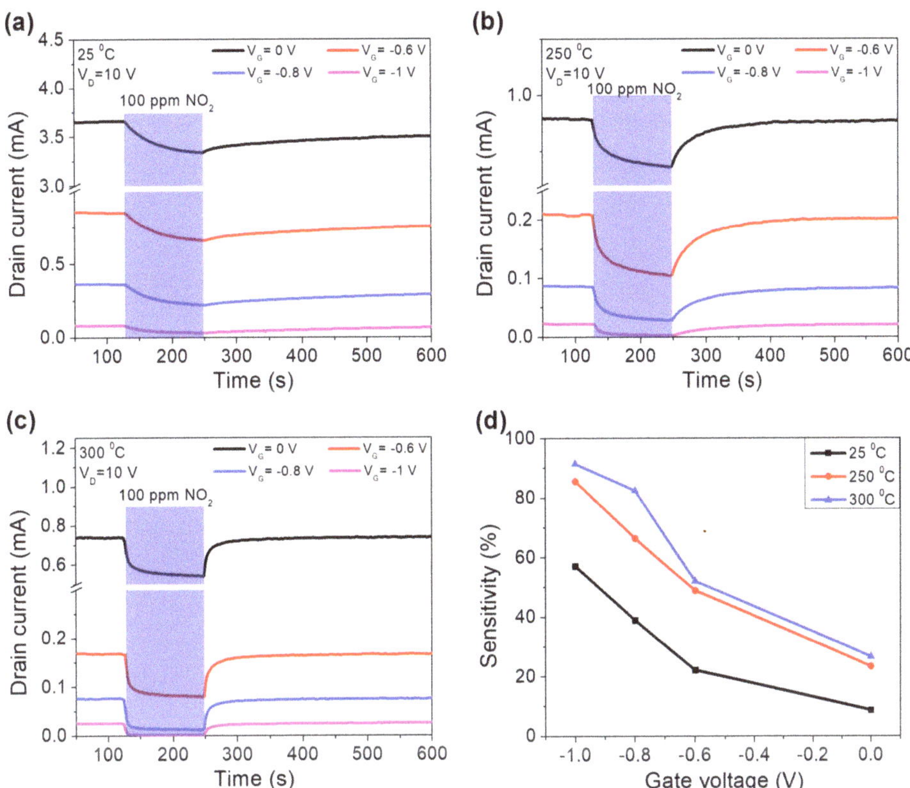

Figure 4. Sensor's response at 25 °C (**a**), 250 °C (**b**), 300 °C (**c**), at different gate biases and sensitivity as a function of gate voltage (**d**).

To discover the physical mechanism of the sensitivity improvement, the simulation on Silvaco TCAD was performed with our wafer parameters. The gas injection was simulated by the variation of the gate work function because NO_2 incorporation through Pd-catalyst eventually introduces oxygen ions onto AlGaN barrier surface. The oxygen ions provided a negative charge, leading to an increase of the Pd work function. The simulation results showed that when the device was biased at -1 V, the difference of the conduction band and electron quasi-Fermi level was much smaller than that at 0 V (Figure 5a,b), which resulted in a remarkable reduction of channel electron density (Figure 5c). Relative change of electron concentration responding to the gas exposure was increased from 14.5% ($V_G = 0$ V) to 30% ($V_G = -1$ V).

When HEMT sensor is biased at a high negative gate voltage, the drain current became too low, which may affect the stability of sensing characteristics. Our sensor showed good repeatability regardless of the gate voltage. In 10 continuous cycles at 300 °C, the sensor totally recovered to the initial current, even at a high negative gate voltage of -1 V, when the drain current was limited in the μA range (Figure 6a). There was an improvement in response time, from 32 s ($V_G = 0$ V) to 9 s ($V_G = -1$ V), while the recovery times were 36 and 48 s, respectively. The repeatable response of the sensor indicated that the gate bias modulation did not interact with the sensing mechanism. The sensor also exhibited good stability in time. No significant change in transfer characteristics was observed when the measurement was done as-fabricated sensor and 6 months later (Figure 6b).

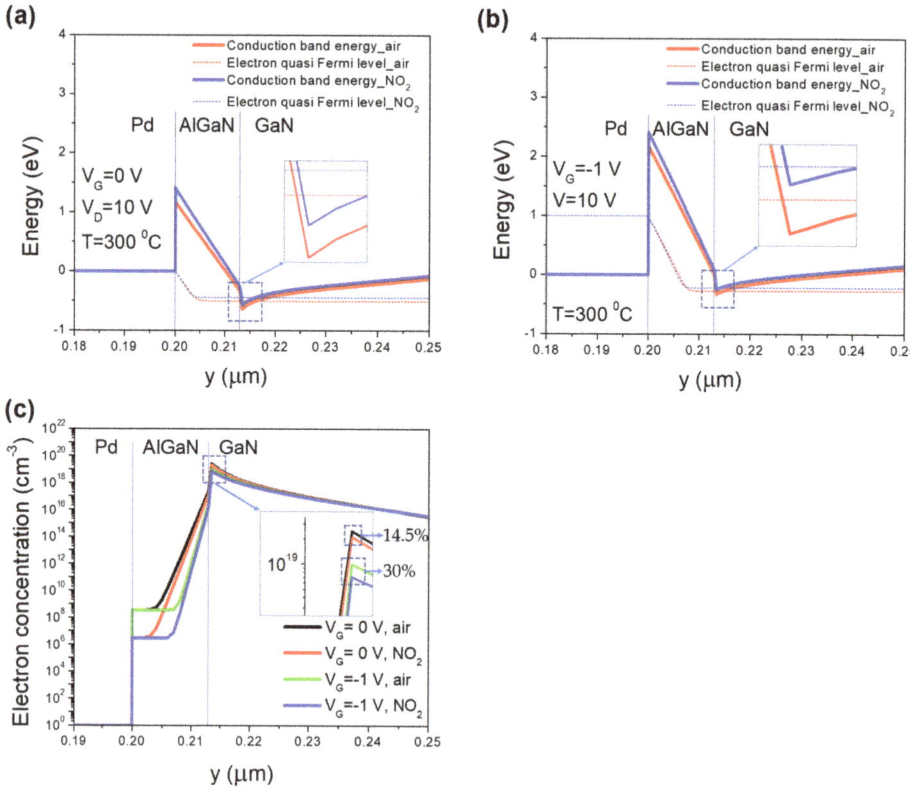

Figure 5. Silvaco TCAD simulation: (**a**) conduction band energy at $V_G = 0$ V and (**b**) $V_G = -1$ V, (**c**) electron concentration.

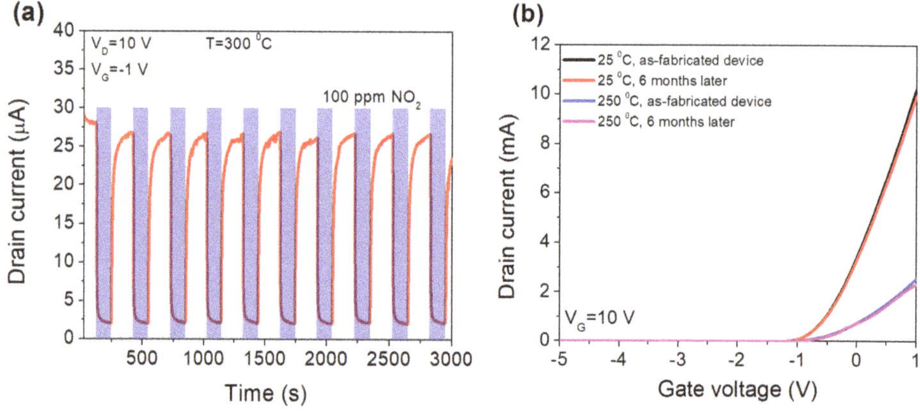

Figure 6. Repeatability of the sensor's response at gate voltages -1 V (**a**) at 300 °C and the stability of the sensor (**b**).

Figure 7a–d showed the response of the sensor under different concentrations of NO_2 from 10 to 100 ppm at different gate voltages from 0 to -1 V at 300 °C. The sensor exhibited a huge improvement in sensitivity for all concentrations of NO_2 (Figure 7e). The sensitivity under 10 ppm of NO_2 was 6% at $V_G = 0$ V and 45.4% at $V_G = -1$ V, which is much improved when compared to other sensors based on AlGaN/GaN HEMTs (Table 1). The improvement of response time at lower gate voltages was also recognized (Figure 7f).

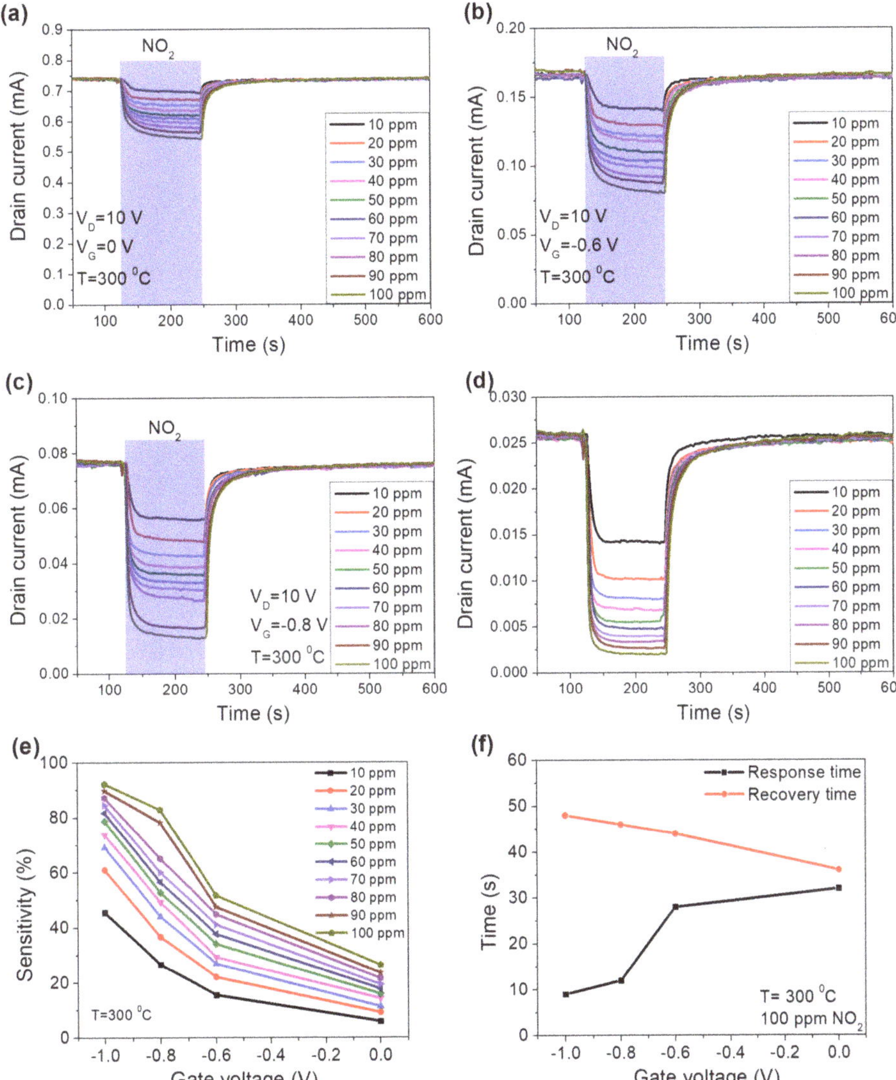

Figure 7. Response of the sensor under the different concentrations of NO_2 at different gate voltages of 0 V (**a**), −0.6 V (**b**), −0.8 V (**c**), −1 V (**d**), sensitivity at different concentrations (**e**), the response and recovery times as a function of gate voltage (**f**), at 300 °C.

Table 1. The comparison of key parameters of NO_2 gas sensor.

Sensing Materials	Structure	NO_2 (ppm)	T (°C)	I_0 (mA)	Sensitivity (S)/ Response (R)	Response Time (s)	Recovery Time (s)	References
Pt	HEMT ($V_G = 0$ V)	10	300	35	S = 1%	~4 min	~4 min	[21]
Pt	HEMT (floating gate)	10	300	33	S = 5.5%	~2 min	~5 min	[22]
Pt	HEMT ($V_G = 0$ V)	100	300	43.8	S = 7%	~3 min	~2 min	[23]
SnS_2/RGO	3D	8	RT	-	R = 49.8%	153	76	[32]
F-SWCNTs	Thin film	50	RT	-	S = 37%	4 min	~8 min	[33]
MoS_2 p-n junction	Thin film	20	RT	-	R ≈ 90	150	30	[34]
Pt	HEMT (floating gate)	10	275	33.9	S = 5.1%	56	285	[35]
Pd	HEMT ($V_G = 0$ V)	10	300	0.74	S = 6%	32	36	This work
Pd	HEMT ($V_G = -1$ V)	10	300	26 μA	S = 45.4%	9	48	This work

4. Conclusions

We demonstrated the optimization of NO_2 gas sensor's performance by adjusting the gate bias of NO_2 gas sensor based on Pd-AlGaN/GaN HEMTs, which is a superior advantage compared to Schottky diode-type sensors. The sensor exhibited a notable improvement of the sensitivity and response time when biased at -1 V. The physical mechanism of this phenomenon was explained by the reduction of channel electron density in TCAD simulation.

Author Contributions: Experiment, analysis, simulation, measurement, V.C.N.; methodology, validation, K.K.; review, methodology, funding acquisition and supervision, H.K. All authors have read and agreed to the published version of the manuscript.

Funding: This work was supported by Korea Institute for Advancement of Technology (KIAT) grant funded by the Korea Government (MOTIE, P07820002101) and the grant funded by Korea Electric Power Corporation (R18XA06-05).

Conflicts of Interest: The authors declare no conflict of interest.

References

1. Pandey, S. Highly sensitive and selective chemiresistor gas/vapor sensors based on polyaniline nanocomposite: A comprehensive review. *J. Sci. Adv. Mater. Dev.* **2016**, *1*, 431–453. [CrossRef]
2. Pandey, S.; Nanda, K.K. Au Nanocomposite Based Chemiresistive Ammonia Sensor for Health Monitoring. *ACS Sens.* **2016**, *1*, 55–62. [CrossRef]
3. Barako, M.T.; Gambin, V.; Tice, J. Integrated nanomaterials for extreme thermal management: A perspective for aerospace applications. *Nanotechnology* **2018**, *29*, 154003. [CrossRef]
4. Hu, L. Nanowire-Assembled Hierarchical $ZnCo_2O_4$ Microstructure Integrated with a Low-Power Microheater for Highly Sensitive Formaldehyde Detection. *ACS Appl. Mater. Interfaces* **2016**, *8*, 31764–31771. [CrossRef]
5. Book, S.A. Scaling toxicity from laboratory animals to people: An example with nitrogen dioxide. *J. Toxicol. Environ. Health* **1982**, *9*, 719–725. [CrossRef] [PubMed]
6. Hotovy, I.; Rehacek, V.; Siciliano, P.; Capone, S.; Spiess, L. Sensing characteristics of NiO thin films as NO_2 gas sensor. *Thin Solid Films* **2002**, *418*, 9–15. [CrossRef]
7. Fine, G.F.; Cavanagh, L.M.; Afonja, A.; Binions, R. Metal Oxide Semi-Conductor Gas Sensors in Environmental Monitoring. *Sensors* **2010**, *10*, 5469–5502. [CrossRef] [PubMed]
8. Rai, P.; Majhi, S.M.; Yu, Y.T.; Lee, J.H. Noble metal@metal oxide semiconductor core@shell nano-architectures as a new platform for gas sensor applications. *RSC Adv.* **2015**, *5*, 76229–76248. [CrossRef]
9. Watson, J. The tin oxide gas sensor and its applications. *Sens. Actuators* **1984**, *5*, 29–42. [CrossRef]
10. Korotcenkov, G. Gas response control through structural and chemical modification of metal oxide films: State of the art and approaches. *Sens. Actuators B Chem.* **2005**, *107*, 209–232. [CrossRef]
11. Luther, B.P.; Wolter, S.D.; Mohney, S.E. High temperature Pt Schottky diode gas sensors on n-type GaN. *Sens. Actuators B Chem.* **1999**, *56*, 164–168. [CrossRef]
12. Yun, F.; Chevtchenko, S.; Moon, Y.-T.; Morkoç, H.; Fawcett, T.J.; Wolan, J.T. GaN resistive hydrogen gas sensors. *Appl. Phys. Lett.* **2005**, *87*. [CrossRef]

13. Anderson, T.; Ren, F.; Pearton, S.; Kang, B.S.; Wang, H.-T.; Chang, C.-Y.; Lin, J. Advances in Hydrogen, Carbon Dioxide, and Hydrocarbon Gas Sensor Technology Using GaN and ZnO-Based Devices. *Sensors* **2009**, *9*, 4669–4694. [CrossRef] [PubMed]
14. Lee, D.-S.; Lee, J.-H.; Lee, Y.-H.; Lee, D.-D. GaN thin films as gas sensors. *Sens. Actuators B Chem.* **2003**, *89*, 305–310. [CrossRef]
15. Kang, B.S.; Ren, F.; Gila, B.P.; Abernathy, C.R.; Pearton, S.J. AlGaN/GaN-based metal-oxide-semiconductor diode-based hydrogen sensor. *Appl. Phys. Lett.* **2004**, *84*, 1123–1125. [CrossRef]
16. Lundstrom, I.; Sundgren, H.; Winquist, F.; Eriksson, M.; Krantz-Rulcker, C.; Lloyd-Spetz, A. Twenty-five years of field effect gas sensor research in Linköping. *Sens. Actuators B Chem.* **2007**, *121*, 247–262. [CrossRef]
17. Choi, J.H.; Park, T.H.; Hur, J.H.; Cha, H.Y. AlGaN/GaN heterojunction hydrogen sensor using ZnO-nanoparticles/Pd dual catalyst layer. *Sens. Actuators B Chem.* **2020**, *325*, 128946. [CrossRef]
18. Vitushinsky, R.; Crego-Calama, M.; Brongersma, S.H.; Offermans, P. Enhanced detection of NO$_2$ with recessed AlGaN/GaN open gate structures. *Appl. Phys. Lett.* **2013**, *102*, 172101. [CrossRef]
19. Schalwig, J.; Muller, G.; Eickhoff, M.; Ambacher, O.; Stutzmann, M. Group III-nitride-based gas sensors for combustion monitoring. *Mater. Sci. Eng. B* **2002**, *93*, 207–214. [CrossRef]
20. Sun, J.; Sokolovskij, R.; Iervolino, E.; Liu, Z.; Sarro, P.M.; Zhang, G. Suspended AlGaN/GaN HEMT NO$_2$ Gas Sensor Integrated with Microheater. *J. Microelectromech. Syst.* **2019**, *28*, 997–1004. [CrossRef]
21. Bishop, C.M.; Halfaya, Y.; Soltani, A.; Sundaram, S.; Li, X.; Streque, J.; Gmili, Y.E.; Voss, P.L.; Salvestrini, J.P.; Ougazzaden, A. Experimental Study and Device Design of NO, NO$_2$, and NH$_3$ Gas Detection for a Wide Dynamic and Large Temperature Range Using Pt/AlGaN/GaN HEMT. *Sensors* **2016**, *16*, 6828–6838. [CrossRef]
22. Ranjan, A.; Agrawal, M.; Radhakrishnan, K.; Dharmarasu, N. AlGaN/GaN HEMT-based high-sensitive NO$_2$ gas sensors. *Jpn. J. Appl. Phys.* **2019**, *58*, SCCD23. [CrossRef]
23. Halfaya, Y.; Bishop, C.; Soltani, A.; Sundaram, S.; Aubry, V.; Voss, P.L.; Salvestrini, J.P.; Ougazzaden, A. Investigation of the Performance of HEMT-Based NO, NO$_2$ and NH$_3$ Exhaust Gas Sensors for Automotive Antipollution Systems. *Sensors* **2016**, *16*, 273. [CrossRef]
24. Nguyen, C.V.; Kim, H. High Performance NO$_2$ Gas Sensor Based on Pd-AlGaN/GaN High Electron Mobility Transistors with Thin AlGaN Barrier. *J. Semicond. Technol. Sci.* **2020**, *20*. [CrossRef]
25. Chen, T.Y.; Chen, H.I.; Liu, Y.J.; Huang, C.C.; Hsu, C.S.; Chang, C.F.; Liu, W.C. Ammonia Sensing Properties of a Pt/AlGaN/GaN Schottky Diode. *IEEE Trans. Electron. Devices* **2011**, *58*, 1541–1547. [CrossRef]
26. Chen, C.C.; Chen, H.I.; Liu, I.P.; Liu, H.Y.; Chou, P.C.; Liou, J.K.; Liu, W.C. Enhancement of hydrogen sensing performance of a GaN-based Schottky diode with a hydrogen peroxide surface treatment. *Sens. Actuators B Chem.* **2015**, *211*, 303–309. [CrossRef]
27. Wen, X.; Schuette, M.L.; Gupta, S.K.; Nicholson, T.R.; Lee, S.C.; Lu, W. Improved Sensitivity of AlGaN/GaN Field Effect Transistor Biosensors by Optimized Surface Functionalization. *IEEE Sens. J.* **2011**, *11*, 1726–1735. [CrossRef]
28. Arafat, M.M.; Dinan, B.; Akbar, S.A.; Haseeb, A.S.M.A. Gas Sensors Based on One Dimensional Nanostructured Metal-Oxides: A Review. *Sensors* **2012**, *12*, 7207–7258. [CrossRef]
29. Wickham, D.T.; Banse, B.A.; Koel, B.E. Adsorption of nitrogen dioxide and nitric oxide on Pd(111). *Surf. Sci.* **1991**, *243*, 83–95. [CrossRef]
30. Zheng, G.; Altman, E.I. The oxidation of Pd(111). *Surf. Sci.* **2000**, *462*, 151–168. [CrossRef]
31. Sokolovskij, R.; Zhang, J.; Zheng, H.; Li, W.; Jiang, Y.; Yang, G.; Yu, H.; Sarro, P.M.; Zhang, G. The Impact of Gate Recess on the H$_2$ Detection Properties of Pt-AlGaN/GaN HEMT Sensors. *IEEE Sens. J.* **2020**, *20*, 8947–8955. [CrossRef]
32. Wu, J.; Wu, Z.; Ding, H.; Wei, Y.; Huang, W.; Yang, X.; Li, Z.; Qiu, L.; Wang, X. Flexible, 3D SnS$_2$/Reduced graphene oxide heterostructured NO$_2$ sensor. *Sens. Actuators B Chem.* **2020**, *305*, 127445. [CrossRef]
33. Kumar, S.; Pavelyev, V.; Mishra, P.; Tripathi, N. Thin film chemiresistive gas sensor on single-walled carbon nanotubes-functionalized with polyethylenimine (PEI) for NO2 gas sensing. *Bull. Mater. Sci.* **2020**, *43*, 61. [CrossRef]
34. Zheng, W.; Xu, Y.S.; Zheng, L.L.; Yang, C.; Pinna, N.; Liu, X.H.; Zhang, J. MoS$_2$ Van der Waals p–n Junctions Enabling Highly Selective Room-Temperature NO$_2$ Sensor. *Adv. Funct. Mater.* **2020**, *30*, 2000435. [CrossRef]
35. Radhakrishnan, K.; Ranjan, A.; Lingaparthi, R.; Dharmarasu, N. Enhanced NO$_2$ Gas Sensing Performance of Multigate Pt/AlGaN/GaN High Electron Mobility Transistors. *J. Electrochem. Soc.* **2021**. accepted manuscript. [CrossRef]

micromachines

MDPI

Review

GaN Heterostructures as Innovative X-ray Imaging Sensors—Change of Paradigm

Stefan Thalhammer [1,2,*], Andreas Hörner [1], Matthias Küß [1], Stephan Eberle [1], Florian Pantle [3], Achim Wixforth [1] and Wolfgang Nagel [2]

1   Experimental Physics I, University of Augsburg, Universitätsstrasse 1, D-86159 Augsburg, Germany; andreas.hoerner@physik.uni-augsburg.de (A.H.); matthias.kuess@physik.uni-augsburg.de (M.K.); stephan.eberle@student.uni-augsburg.de (S.E.); achim.wixforth@physik.uni-augsburg.de (A.W.)
2   Senray Technologies GbR, Ebenböckstrasse 23, D-81241 Munich, Germany; wwnagel@senray.de
3   Walter Schottky Institut, Technical University Munich, Am Coulombwall 4, D-85748 Garching, Germany; florian.pantle@wsi.tum.de
*   Correspondence: stefan.thalhammer@physik.uni-augsburg.de

Abstract: Direct conversion of X-ray irradiation using a semiconductor material is an emerging technology in medical and material sciences. Existing technologies face problems, such as sensitivity or resilience. Here, we describe a novel class of X-ray sensors based on GaN thin film and GaN/AlGaN high-electron-mobility transistors (HEMTs), a promising enabling technology in the modern world of GaN devices for high power, high temperature, high frequency, optoelectronic, and military/space applications. The GaN/AlGaN HEMT-based X-ray sensors offer superior performance, as evidenced by higher sensitivity due to intensification of electrons in the two-dimensional electron gas (2DEG), by ionizing radiation. This increase in detector sensitivity, by a factor of 10^4 compared to GaN thin film, now offers the opportunity to reduce health risks associated with the steady increase in CT scans in today's medicine, and the associated increase in exposure to harmful ionizing radiation, by introducing GaN/AlGaN sensors into X-ray imaging devices, for the benefit of the patient.

Keywords: GaN-HEMT mesa structures; 2DEG; X-ray sensor; X-ray imaging

Citation: Thalhammer, S.; Hörner, A.; Küß, M.; Eberle, S.; Pantle, F.; Wixforth, A.; Nagel, W. GaN Heterostructures as Innovative X-ray Imaging Sensors—Change of Paradigm. *Micromachines* **2022**, *13*, 147. https://doi.org/10.3390/mi13020147

Academic Editor: Giovanni Verzellesi

Received: 7 December 2021
Accepted: 13 January 2022
Published: 19 January 2022

Publisher's Note: MDPI stays neutral with regard to jurisdictional claims in published maps and institutional affiliations.

Copyright: © 2022 by the authors. Licensee MDPI, Basel, Switzerland. This article is an open access article distributed under the terms and conditions of the Creative Commons Attribution (CC BY) license (https://creativecommons.org/licenses/by/4.0/).

1. Introduction

Until recently, the commercialization of gallium nitride (GaN) devices was hindered by economic difficulties in the scale fabrication of GaN crystals. However, with the improvement of GaN fabrication, from 2010 to 2016, to a high-quality material, these difficulties have been overcome by various advances in metal-organic chemical vapour deposition (MOCVD) and molecular beam epitaxy (MBE) processing techniques; GaN-based devices are now even able to significantly outperform their Si-based counterparts. With the improvement of GaN material in form of GaN-wafers, the detection of X-rays by GaN sensors can be further explored, and is now on the threshold of technology transfer to society. Due to the inherent features of two-dimensional electron gas (2DEG) in GaN high-electron mobility transistors (HEMT), the intrinsic enhancement by a factor of 10^4 for the number of electrons that can be collected from single ionization events by direct conversion photon counting reveals the tremendous potential of GaN-sensors compared to current technologies, which use indirect conversion systems. These breakthrough findings paved the way for a paradigm shift in X-ray detector technology, while enabling a breakthrough in X-ray and computer tomography (CT) examinations by reducing the effective dose, used today in routine X-ray and CT examinations, to 1/20.

2. GaN Manufacture and *GaNification*

Polycrystalline GaN material was synthesized for the first time around 1930 by flowing ammonia (NH_3) over liquid gallium (Ga), at around 1000 °C [1]. The first single crystalline

GaN was epitaxially grown in 1969 by Maruska and Tietjen [2] by hydride vapor phase epitaxy (HVPE) on a sapphire substrate [3]. In this process, gaseous hydrogen chloride initially reacts with liquid Ga at a temperature of approximately 880 °C to form gallium chloride (GaCl$_3$). In a reaction zone, the GaCl$_3$ is brought close to a GaN crystal nucleus at temperatures between 1000 and 1100 °C. Here, the GaCl$_3$ reacts with the inflowing ammonia, releasing HCl to form crystalline GaN, which crystallizes preferably in the (hexagonal) wurtzite structure [4].

GaN crystals can be grown on a variety of substrates, including sapphire, silicon carbide (SiC) and silicon (Si). By growing a GaN epi layer on top of silicon, the existing silicon manufacturing infrastructure can be used, eliminating the need for costly specialized production sites, and leveraging readily available large-diameter silicon wafers at low cost.

In 1993, using metal organic chemical vapor epitaxy (MOVPE), Asif Khan et al. [5] grew the first AlGaN/GaN heterojunction. Despite a moderate crystallographic quality, the mobility of the 2DEG, at the AlGaN/GaN interface was around 600 cm^2/Vs. This achievement can be regarded as the start of a quest for nitride high-electron mobility transistor technology [6].

In 1999, Ambacher and colleagues [7] proposed a model to analytically describe the 2DEG properties of AlGaN/GaN heterostructures. Today, this model has been widely adopted by the GaN community. In the same year, Sheppard et al. [8] demonstrated high-power microwave HEMTs based on AlGaN/GaN heterostructures grown on a silicon carbide (SiC) substrate.

In 2000, the nature of the 2DEG was further clarified by Ibbetson et al. [9], attributing a key role to the surface states present in nitride materials as source of electrons. A listed historical summary can be found by Roccaforte and Leszczynski [10]. Binary AlN/GaN HEMTs are preferably grown by plasma-assisted molecular beam epitaxy (PA-MBE), as well as metal-organic chemical vapour deposition (MOCVD), which is also known as metal-organic chemical vapour phase epitaxy (MOCVPE) [11,12]. The latter method, which is likely to dominate by more than one magnitude, has gained increased attraction, since MOCVD offers simple fabrication of large quantities, as requested in commercial use for the production of complex semiconductor structures [13] (Figure 1).

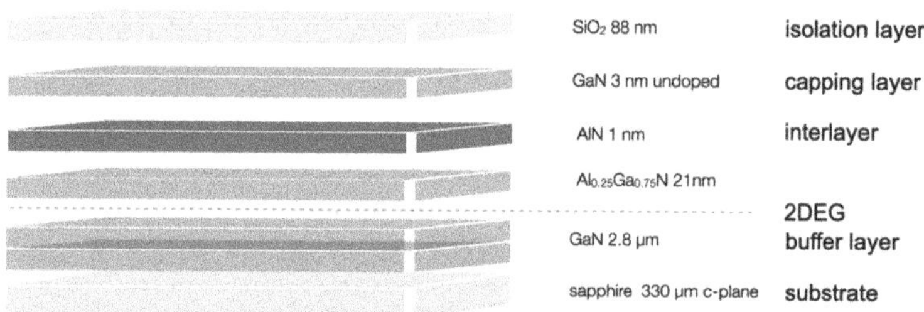

Figure 1. Schematic drawing of the GaN HEMT heterostructure. Dotted line shows the 2DEG area.

In contrast to the well-known silicon, the structure of these HEMT comprise not only from a single element, but from at least two, or more, including gallium arsenide (GaAs), gallium nitride (GaN), indium phosphide (InP) and aluminium nitride (AlN). Since they are arranged according to the periodic table, from rows III. and V. of the main group, they are called III-V semiconductors. A 2DEG is inherently present in the AlGaN/GaN heterostructure, the presence of which is the basis for the operating principle of HEMT devices [10]. HEMT devices were structured from 3µm thick Ga-face GaN films that were grown by MOCVD (TopGaN Ltd., Warsaw, Poland) on a 330 µm thick c-plane sapphire substrate, followed by a 2.8 µm GaN layer, a 25 nm Al$_{0.25}$Ga$_{0.75}$N layer, a 3nm nominally undoped GaN capping layer, and an 88 nm isolation layer. The 2DEG formed just below the heterojunction interface had a sheet carrier concentration of n$_{2DEG}$ $\approx$ 8 Å~1012 cm^{-2} and a mobility of µ $\approx$ 1100 cm^2/Vs at 300 K. In-depth fabrication methods and physical properties of the HEMT devices are described elsewhere [14,15]. Photolithography and reactive ion etching steps were performed according to standard procedure described in detail in Howgate et al. [16]. The ohmic contacts with 200/800/100/900 Å stacks of Ti/Al/Ti/Au defined by lithography were fabricated by evaporation steps [15] (Figure 1). This process produces contact electrodes without affecting the 2DEG conduction.

One of the key features of these heterostructures is the formation of the 2DEG and the resulting properties. The simulation of the Al content in the Al-rich heterostructure (software "Nextnano") showed dependence on the Al concentration—the lower the Al content in the channel, the higher the electron density—and reached significant values well-above $2 \times 10^{13}/cm^3$ [17]. The band gap of the AlGaN layer can be varied by changing the ration of Al to Ga present within the AlGaN layer. The bandgap varies approximately in accordance with Vengard's Law [18]. A striking feature of the lateral GaN HEMTs is the zero-charge feedback. Due to the absence of p-n junctions and current flow in a polarization induced 2DEG, reverse operation starts when the drain voltage falls below the sum of the gate potential and the threshold voltage, thus creating a reverse channel. Furthermore, due to the expansion of the space charge layer along GaN/AlGaN interface within a basically undoped substrate, the output capacitance shows a very linear characteristic [19].

Since the 1990s, GaN has been used in opto-electronics, commonly in light emitting diodes (LED). GaN emits a blue light, used for disc-reading in Blu-ray. Additionally, GaN is used in semiconductor power devices, RF components, lasers, and photonics with an overall reduction of electric power consumption. However, superior features of GaN-based HEMTs compared to Si-based devices, such as faster switching speed, higher thermal conductivity, and lower on-resistance, outcompete Si-based solutions in many areas. Hence, the overall nitride device market forecasts for the next years are much brighter compared to those of the other compound semiconductors. According to the market analysis report of Grand View Research, the global GaN semiconductor devices market size was valued at USD 1.65 billion in 2020 and is expected to expand at a compound annual growth rate (CAGR) of 21.5% from 2021 to 2028 [20]. Their wide bandgap and high stability make them a perfect material for high-power and high frequency transistors present in the market. For those reasons, the term *"GaNification"* was created for the revolution expected in modern electronics and opto-electronics [10].

In our view, and by analogy, a change of paradigm can be expected from GaN-based X-ray detector technology for the development of GaN-based devices described above. In contrast to the notion that GaN layers are poorly suited for X-ray dosimetry, held back in 2008, it can be demonstrated that a GaN thin film arranged on a carrier substrate with a thickness of <50 μm, provided with ohmic contacts, allows a sensitive, reproducible direct detection of X-rays with an energy of 1-300 keV (up to values in the μGy range) [21,22]. The functional principle of the GaN thin film radiation detector fundamentally differs from the conventionally available direct semiconductor detectors for X-ray radiation, as it requires a simple resistance or conductivity measurement, such as a photoconductor [23].

With regard to the application of X-rays in today's medicine, CT scanning is the method of choice. CT scanners use computer-controlled X-rays to create images of the body. At the same time, the effects of radiation and increasing health concerns hinder the market growth of CT scanner devices and equipment. Radiation coming from manmade sources, such as CT scans, nuclear medicine scans, and PET scans, carry major health hazards and risks. Accumulated low doses can cause cancer in the long run. The medical justification for performing CT imaging is that the anticipated benefit exceeds the anticipated radiation exposure risk. Thus, an optimal dose of radiation to accomplish the diagnostic task should be used. However, this optimal dose is not always known to the practitioner, resulting in a wide spectrum of radiation doses being used, even for common diagnostic tasks [24,25].

The increasing awareness of risks associated with radiation exposure triggered a variety of dose-reduction techniques, such as tube current or tube voltage modulation, automatic exposure control, and low-kilovolt scanning [26], where a dose reduction between 10% and 30% is possible, depending on examination type. Compared to the real-time analytical reconstruction method, filtered back projection (FBP) [27], which has reached its limitation and does not allow for further dose reductions, vendor-specific iterative reconstruction algorithms were introduced, which allow further dose reductions [28,29].

Altogether, there remains a strong medical need for means to reduce the effective dose in CT-scanners. We address this need with a proprietary detector technology based on GaN thin film. Compared to current CT detectors that use indirect converter systems, direct converter photon counting results in a breakthrough for X-ray and CT-scans in reducing the photon energy, from up to 130 keV [30] to 5 keV [16]. The effective dose can be minimized in the state-of the art dose applied in the routine X-ray and CT-scans of today. In the following section we describe the working principle and imaging potential of GaN-HEMT mesa structures.

3. X-ray Detection Using Semiconductor Material

In 1946, R.S. Ohl discovered the p-n junction in silicon, which can be used for particle detection [31]. Electron–hole pairs are the basic information carriers in semiconductor detectors. Electromagnetic radiation (X-ray and gamma ray) and particle radiation (alpha and beta radiation) generate free charge carriers, electrons, and holes in the semiconductor along the pathway taken through the detector. By capturing these electron–hole pairs, the detection signal is formed (Figure 2). Whether, and how many, electron–hole pairs are generated by the incident ionizing radiation depends largely on the size of the band gap energy. This band gap, or forbidden zone, describes the energetic distance between the valence and conduction band. When photonic energy is supplied to the band gap area, valence electrons are excited into the conduction band. These electrons in the conduction band can absorb energy from an electric field and, together with the resulting defect electrons, i.e., holes from the valence band, make the material conductive.

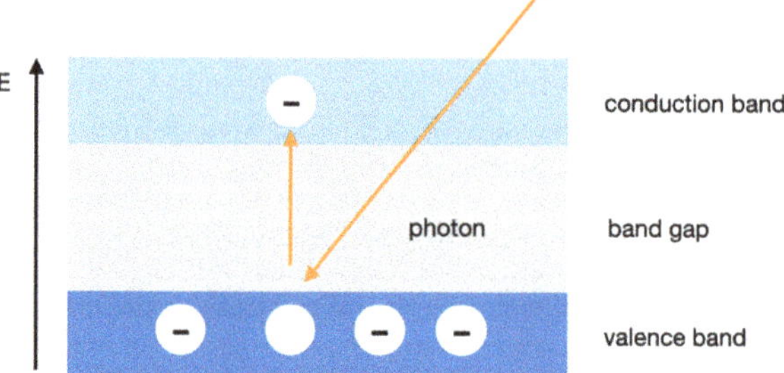

Figure 2. Principle of the formation of electron–hole pairs with ionizing radiation. Bands determine the density of available energy states. E = energy of electrons.

In X-ray detection, the photon emits the entire charge corresponding to its energy at one point in the matter due to a photoelectric effect, which leads to the formation of secondary electrons [32]. A primary electron is lifted from the valence band into the conduction band. Due to its high kinetic energy, numerous secondary electrons and phonons are formed in the process. At higher photon energies (50 keV to 1 MeV), the Compton effect also occurs, in which only part of the energy is transferred to the electron and deposited in the detector.

These generated electron–hole pairs are electrically amplified and can be read out. In the manufacture and structuring of semiconductor detectors, the combination of different conductive areas, doping, is used. Electric fields are applied across these, and the charge carriers generated by the radiation can be transported or amplified within the semiconductor. The principle of operation is that of a diode (Figure 3).

There is a much higher density of conduction electrons in the n-type region of the semiconductor and vice versa in the p-type region. The junction represents a discontinuity in electron density, and a net diffusion from the high-density side to low density occurs for both electrons and holes. The effect of the diffusion from each side of the junction is to build up a net negative space charge on the p-type side and a net positive space charge on the n-type side of the junction, compared to the rest of the p- and n-type. The accumulated space charge creates an electric field that reduces the tendency for further diffusion. At equilibrium, the field is just adequate to prevent additional diffusion across the junction and a steady state charge distribution is established. The region over which the imbalance occurs is called the depletion region [32].

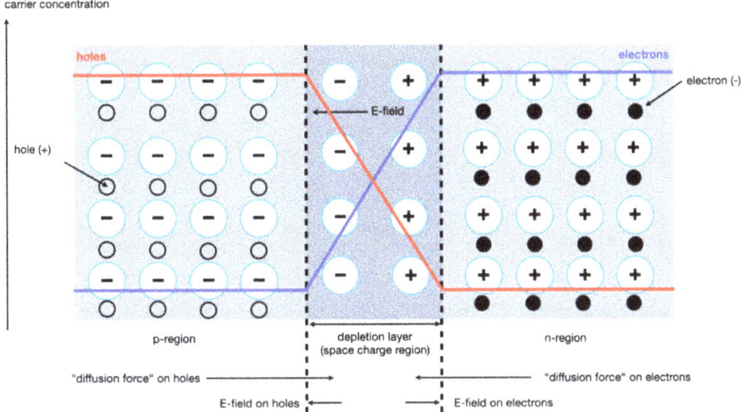

Figure 3. Schematic sketch of a p-n junction.

This depletion region acts as a radiation detector. The electric field sweeps electrons created in or near the junction back toward the n-type material, and any holes are swept back toward the p-type material; the motion of these electrons and holes creates the electric signal. However, the contact potential is too low to generate electric fields that will move charge carriers quickly. This can lead to incomplete charge collection, as charges have time to be trapped or recombine. This means that the noise characteristics of an unbiased junction is poor. In addition, the thickness of the depletion region is quite small, so only a small volume of the crystal acts as a radiation detector, which requires an extension of the depletion region [33].

As a general rule, semiconductors with larger atomic masses absorb more X-ray photons, which enhances the response of the detector, but tend to have a smaller band gap energy, which degrades performance in terms of dark current. In addition, low band gap materials are generally more mechanically fragile and prone to damage by irradiation than wide band gap materials [34]. The collection of electrons is better than that of holes, due to their larger mobility and the residual n-type doping in GaN. Some holes become trapped, resulting in a positive charge in the layer. When the bias voltage is large enough to produce a dark current comparable to the photocurrent, the contacts inject additional electrons to balance the positive trapped charge, until electrons recombine with the trapped holes. The longer the electron trapping time, the greater the additional current, the greater the photoconductive gain and the longer the response time. This behaviour is general and common. It must be distinguished from what was observed in [35], where strong nonlinearities and quenching by visible light were observed, and explained by defects activated by X-ray illumination. The GaN was undoped, and the dark currents were low and completely dominated by tunnel currents, which are highly sensitive to defects. This is not the case when the GaN is doped, and the currents are quasi-ideally described by thermionic emission [36]. A low energy is required to generate an electron–hole pair in semiconductor materials (−3 eV for germanium) compared to the energy required to generate an electron-ion pair in gases (−30 eV for typical gas detectors) or to generate an electron–hole pair in scintillators (−100 eV) [32].

As a result, a large number of electron–hole pairs are produced and reach the electrodes, increasing the number of pairs per pulse and reducing both statistical fluctuation and the signal/noise ratio in the intensifier. This is a major advantage over other detectors, and the output pulse provides much better energy resolution. Moreover, the small sensitive area used to detect radiation (a few millimetres) and the high velocity of charge carriers provide an excellent charge collection time (-10^{-7} s).

At this moment, when ionizing radiation interacts with the semiconductor in this depleted region, electrons are lifted into the conduction band, leaving holes in the valence band and creating a large number of electron–hole pairs. When a voltage is applied to the semiconductor, these carriers are readily attracted to the electrodes, and a current flows into the circuit, resulting in a pulse. The size of the pulse is directly proportional to the number of carriers collected, which in turn is proportional to the energy deposited in the material by the incident radiation. In semiconductors, electrons can be also thermally excited from the valence band to the conduction band as the temperature increases. Therefore, some semiconductor detectors must be cooled to reduce the number of electron–hole pairs in the crystal in the absence of radiation. Table 1 summarizes and compares relevant features characterizing a radiation sensor based on the diode principle, the GaN-HEMT and the 2DEG-intensifier.

Table 1. Comparison of the functional principle of the diode, GaN-HEMT, and 2DEG intensifier.

Feature	Diode	GaN-HEMT	2DEG Intensifier
Contacts	Schottky ohmic, with p-n, pin-diodes	ohmic	ohmic
Detection principle	diode	FET	MOSFET
Depletion layer	yes	no	no
Current flow	no	yes	yes
Aging	yes	no	no
Quality feature	dark current	signal/noise ratio	signal/noise ratio
Increasing sensitivity	enlarge depletion layer	adaption of electrical conductivity	alteration of gate voltage

Another important parameter for qualifying a semiconductor material for X-ray imaging is the signal to noise ratio (S/N)—the higher the S/N ratio, the better. The band gap is a critical factor. On the one hand a large signal is produced by many electron–hole pairs. This can be achieved by low ionizing energies in small band gap materials. On the other hand, only a few intrinsic charge carriers are generated in materials with wide band gap, resulting in low noise.

Before using the GaN-HEMT devices for X-ray imaging, these sensors showed their potential in particle and electromagnetic radiation sensing. A summary review of AlGaN sensing applications can be found in Upadhyay et al. [37]. A comprehensive review was published by Wang et al. [22]. α-particle detection was first realized using a GaN double Schottky contact device. Detection was performed using a 5.48 MeV α-particles emitted from an ^{241}Am source [38]. In addition, 500 µm thick bulk GaN-sensors in a sandwich structure showed α-particle detection [39].

When operating the open-gate GaN-HEMT sensor in a mesa structure formation, where an ohmic contact is realized on the cap layer, the device can detect β^--radiation and functions as a dosimeter [40]. During exposure of the device to β^--particles, energy is transferred in either ionizing or non-ionizing interactions [41]. In ionizing events, (e^+/e^-) pairs can be produced within the sensor. The pairs generated in the vicinity of the heterojunction are separated by the internal electrical field, and the electrons subsequently accumulate in the 2DEG, contributing to its charge carrier density. Positive charges, on the other hand, drift into the bulk or towards the sensor surface, in turn contributing to the bulk conductivity. The resulting charge separation causes a photovoltage, that affects the position of the 2DEG relative to the Fermi level, thereby capable of considerably altering the charge carrier density considerably and thus giving rise to high gains [16].

Detection of X-rays with GaN thin film is difficult due to the low absorption coefficient, and both theoretical calculations and experimental measurements show that the absorption coefficient is only large for photon energies between 10 and 20 keV [21]. Our GaN thin film high-mobility heterostructures were tested with X-ray sources from 40 to 300 keV Bremsstrahlung. We found that the photoconductive device response exhibits a large gain, is nearly independent of the angle of radiation, and is constant within 2% of the signal throughout this medical diagnostic X-ray range, indicating that these sensors do not require recalibration for geometry or energy [42].

4. GaN-HEMT Based X-ray Imaging

Our GaN-HEMT sensors can operate in two different modes, depending on the X-ray tube voltage. For detection of higher acceleration voltages, tested from 40 to 150 kV, the GaN thin film works as the detector, while for low voltages, tested from 5 to 7 kV, the HEMT structure is employed. A detailed description and characterization of the GaN-HEMT device can be found in Hofstetter et al. [42] and Howgate et al. [16].

The imaging potential of the GaN-HEMT mesa structure was demonstrated using human phantom segments. For this purpose, individual sensors were attached to a two-dimensional translation stage (SH4018L1704 stepping motors, Telco, Houston, TX, USA) and scanned under human phantom segments, while simultaneously recording the device response. An acceleration voltage of 150 kV was used. Positioning of the translation stage, source meter, and data acquisition were controlled by a LabView program 8.6 code (National Instruments, Austin, TX, USA). The source meter provides the possibility of simultaneously biasing the detector (U_{SD}) and measuring the current (I_{SD}). The integration time for a single data reading is characterized by the number of power line cycles, NPLC. Here, one PLC at 50 Hz results in 20 ms, and integration time was set to NPLC one. Measurements were performed with a data acquisition frequency in the 10 Hz regime. Unlike other semiconductor detectors, images were recorded at room temperature without an additional cooling of the sensor material. Although such devices have the potential to be easily miniaturized to form high-density pixel arrays for imaging purposes, the chips we fabricated were 5×3 mm^2, with an internal active area of 2.8×1 mm^2, and only contained one single device, thus limiting the pixel dimensions to this size. Several hundred pictures were acquired at different offsets from the centre position of each pixel, with an acquisition time of 200 ms, and the recorded data were recombined into a single image by interpolation and averaging. The two-dimensional data were plotted on an 8-bit X-ray scale contour map (Figure 4) [23,43]. A detailed description of the used mesa HEMT sensors and experimental results on X-ray dosimetry and imaging are presented in Hofstetter et al. [42].

Unlike most established semiconductor X-ray detectors, in which photo-excited electron–hole pairs are separated and collected via an internal space charge region, our devices are operated without the intentional formation of a depletion layer. The photocurrents of GaN-sensors are monitored with the application of a small DC voltage between two ohmic contacts. Although space charge regions are present, due to surface band bending, interface band alignment, and extended defects, the measured current primarily flows in parallel to such depletion regions, rather than through them. Illumination leads to non-equilibrium concentrations of free carriers, which reduce the total volume of the depletion regions, and, in turn, increase the effective volume of the material through which carriers can be transported, i.e., both conductivity and conductance are increased. As a result, the photocurrent response is not directly limited by the direct conversion of absorbed energy into free charge carriers, and the devices can function as X-ray sensitive sensors with significant internal intensification. Due to the light and temperature sensitivities of the devices, all measurements were performed with careful exclusion of stray light and at room temperature [42].

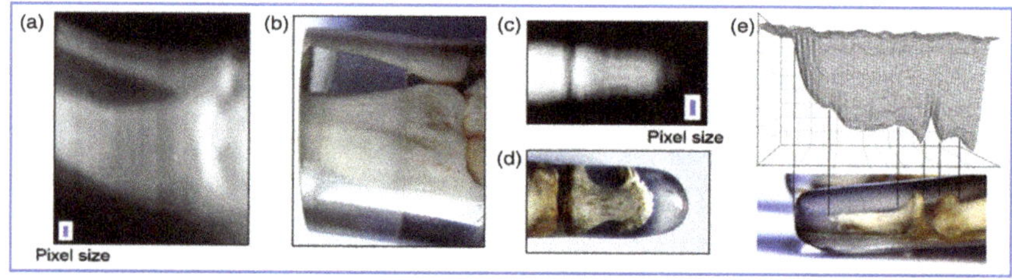

Figure 4. The 8-bit X-ray grey-scale contour maps recorded using GaN-HEMTs (the total device size is indicated (white) and the active area (blue)) of a human (**a**) wrist and (**c**) index finger phantom. Optical pictures of the scanned (**b**) wrist and (**d**) finger phantom. (**e**) Side view of a 3D plot of the data in (**c**) compared to an optical image in the same orientation ([42]; copyright IOP Publishing, 2011).

Compared to other semiconductor detectors, which are based on a measurable current due to the separation of electron–hole pairs, our GaN sensors show effects that can be described with this simple model of charge carrier generation, as in photovoltaics.

The GaN-HEMT works as a direct X-ray converter, and the mode of operation is based on the photo effect (photoconductive or photoelectric effect). The photo effect summarizes three closely related but different processes of photon interaction with matter. In all three cases, an electron is released from a bond, i.e., in an atom, in the valence band, or in the conduction band of a solid material, dissolved by absorbing a photon. The energy of the photon must be at least as high as the binding energy of the electron (photo- or Compton effect).

A distinction is made between the external and internal photoelectric effect (Auger effect). The latter is crucial for the GaN-HEMT. In the case of the internal photo effect, a bound shell electron is released in a solid through photon absorption, but without leaving the body. This occurs in semiconductor materials and insulators. The photoelectrons and the holes created at the same time increase the electrical conductivity.

Photoconductivity is the increase in the electrical conductivity of semiconductor materials due to the formation of unbound electron–hole pairs during irradiation. The electrons are lifted from the valence band into the energetically higher conduction band by means of the energy of the photons, for which the energy of the individual photon must at least correspond at least to the band gap of the irradiated semiconductor. Since the size of the band gap depends on the nature of the material, the maximum wavelength of the light up to which photoconduction occurs differs depending on the semiconductor.

5. GaN/AlGaN HEMT—2DEG Intrinsic Intensifier

The GaN-HEMT sensor follows the basic principle of a metal-oxide-semiconductor field effect transistor (MOS-FET), a voltage-controlled circuit element whose conductance is controlled via the gate-source voltage. The intrinsic intensifier, such as a HEMT consisting of a heterostructure of GaN ($E_g = 3.4$ eV) and $Al_{0.26}Ga_{0.74}N$ ($E_g = 4$ eV), operates accordingly. Since the band gap of AlGaN is larger than that of GaN, a 2DEG is formed at the interface of these two materials on the GaN side, which can serve as a conductive channel. One of the most interesting features of $Al_xGa_{1-x}N$ is the possibility of tailoring the energy gap, and thus the inherent 2DEG, by varying the Al concentration, with Al working as a dopant. When using the energy band model, the piezoelectric polarization and the resulting internal electric fields occurring in the AlGaN/GaN interface cause the conduction band to bend below the Fermi level, resulting in the accumulation of free electrons in a potential well (Figure 5). Typically, the 2DEG is characterized by sheet carrier density values in the order of 10^{13} cm^{-2} and mobility in the range of 1000-2000 cm^2/Vs [10].

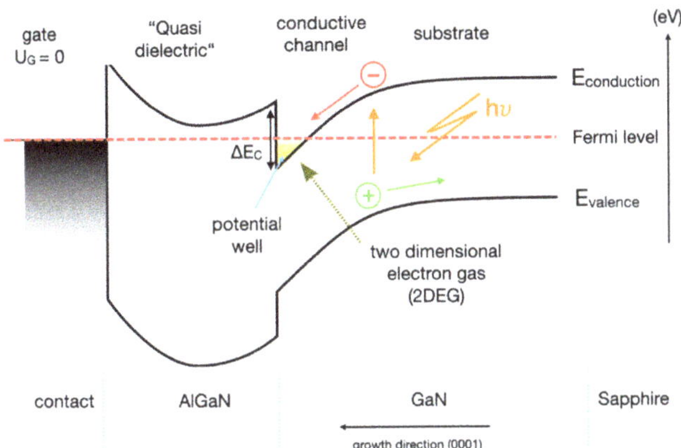

Figure 5. Energy band diagram of a HEMT with 2D electron gas. Schematic representation of the band-bending of the conductive band at the GaN/AlGaN material interface. The conduction band is bent below the Fermi level, which leads to an accumulation of free electrons in the potential well.

With regard to the operating principle of AlGaN/GaN heterostructure, X-rays change the conductivity of the 2DEG. When an electron–hole pair is generated, this results in charge separation due to the built-in electric field, perpendicular to the heterojunction plane. As a result, electrons drift into the 2DEG channel, and the holes drift into the volume and surface regions of the sensor. Since the GaN-HEMT sensors are more or less transparent, the absorbed energy is roughly constant as a function of depth, comparable to the linear energy transfer, through the sensor structure, and most of the electron–hole pairs are generated inside the much thicker GaN-buffer layer.

This charge separation, and the associated accumulation of electrons in the 2DEG channel, leads to an internal photovoltaic effect that causes a shift in the threshold potential. As a result, the current response is proportional to the transconductance of the detector. At a constant applied voltage, the current through the 2DEG forms a signal that can be assigned to a dose rate. The layer structure is only a few nm thick, and the 2DEG has no three-dimensional extension. Charge carriers generated by photons during X-ray imaging in the GaN-buffer layer can now additionally fall into this potential well and form an inversion layer with high carrier density and mobility. This has a direct impact on the photo-generated current.

Thus, an area-dependent, or volume-independent, measurement takes place in the HEMTs, which is a decisive criterion, especially for high-resolution dosimetry (e.g., the formation of artifacts, or distortion of the measurement signal). The measurement signal in the 2DEG is generated in $<10^{-3}$ s. Due to the specific band structure and the piezoelectric nature of both semiconductor materials, a triangular potential well is generated at the interface. An important boundary condition is the constant Fermi level over the entire structure (dotted red line, Figure 5).

Since the 2DEG is naturally present in the AlGaN/GaN heterostructure, and the Fermi level at the interface lies above the conduction band minimum (Figure 5), a current flows between source and drain, even when the gate bias is zero. The output current of a HEMT can be modulated by applying a negative bias voltage to the gate, until a "threshold voltage" is reached, at which the Fermi level is pulled below the conduction band edge of the AlGaN layer and the 2DEG channel is depleted (Figure 5).

In summary, the triangular potential well is our tuneable switch for sensitivity; the intrinsic intensifier (Figure 6). This means that the collecting of electrons and the constant Fermi level can be changed by gate voltage [44].

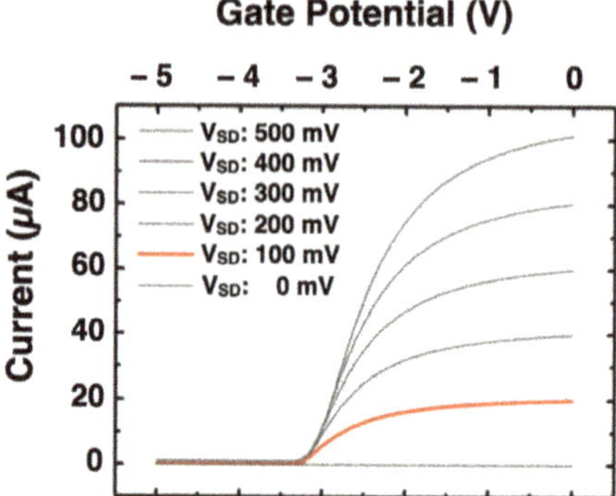

Figure 6. A plot of the source–drain current, as a function of gate–drain potential (reproduced with permission from [16]; published by Wiley Online Library, 2012).

6. GaN/AlGaN HEMT Based X-ray Imaging

To demonstrate the potential of the intrinsic intensification principle of the GaN-HEMT sensor, a single GaN/AlGaN HEMT sensor was connected to a computer-controlled two-dimensional translation stage, to image the light bulb of a small flashlight. Scans were performed using a Pt-gated device with a 500 μm HEMT channel. Source–drain bias was set to 120 mV, and the gate was biased to $V_G = 3.5$ V for intrinsic intensification. Control and data acquisition were the same as described in Section 4. As indicated in Figure 7, the spiral filament of the light bulb is clearly visible, and in the surrounding area the increase of current shows an absence of material, indicating that the centre is hollowed out to form a sealed cavity. In this proof-of-concept, we were able to achieve a detection range down to 50 μm, with a photon rate in the 10^6 counts per second regime. The acquisition time for a single pixel was about 1 s due to motor movement and data acquisition time.

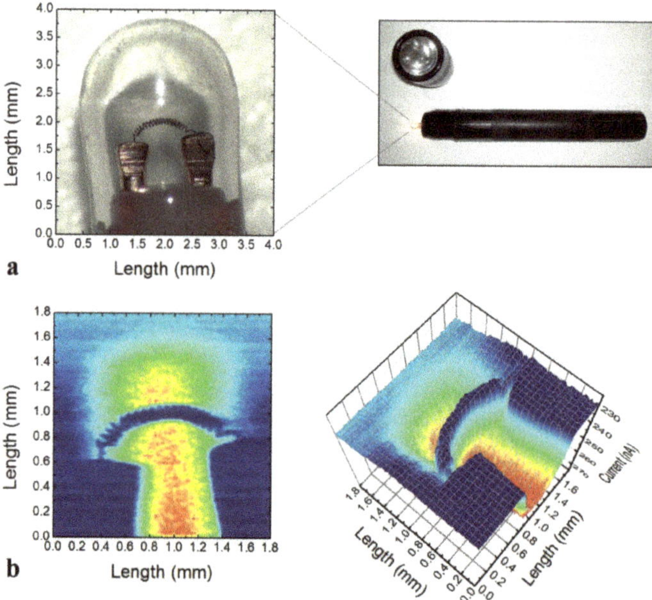

Figure 7. (**a**) A microscope image of the bulb shows the approximately 3 × 3 mm object with a 50 μm thick spiral filament. (**b**) High resolution HEMT X-ray image of a flashlight bulb. The 2D and 3D images

show the plotted X-ray data. Scans were performed with a Pt-gated device with a 500 μm HEMT channel. The source–drain bias was set to 120 mV and the gate was biased to $V_G = -3.5$ V. All measurements were performed with 5 keV X-rays at room temperature. Reproduced with permission from [43].

7. Summary and Conclusions

The fabrication of GaN heterostructures with improved quality, at economic scale, is now mature. GaN-based devices are capable of significantly outperforming their silicon-based counterparts. To date, little attention has been paid to GaN-based X-ray detection. However, we believe that the introduction of a 2DEG channel, by adding AlGaN/GaN heterointerfaces into the GaN thin film, will open another chapter in *GaNification* [10], as the tremendous sensitivity of GaN HEMT sensors will reach a new level of hardware innovation, reducing the effective dose used today in routine X-ray and CT scans. Device dimensions, i.e., pixel size, can be reduced by orders of magnitude using modern standard microfabrication techniques. Thus, there should be no practical constraints on the production of large-area X-ray sensors based on pixel arrays. High-resolution imaging using the shown integrated GaN HEMT pixel sensors should be possible based on our proof-of-principle measurements [44,45]. The material properties of GaN heterostructures such as robustness, light weight, and features such as room temperature operation without cooling, excellent signal-to-noise ratio, and the 2DEG-based tuneable intrinsic sensitivity, altogether form the basis for enabling a paradigm shift in X-ray detector technology.

Author Contributions: Conceptualization, S.T., A.W. and W.N.; methodology, A.H., M.K., S.E. and F.P.; writing—original draft preparation, S.T. and W.N.; writing—review and editing, A.W. and W.N.; All authors have read and agreed to the published version of the manuscript.

Funding: This research received no external funding.

Acknowledgments: S.T. and W.N. would like to acknowledge the excellent work of M. Hofstetter and J. Howgate during their Ph.D. theses.

Conflicts of Interest: The authors declare no conflict of interest.

References

1. Juza, R.; Hahn, H. Über die Kristallstrukturen von Cu3N, GaN und InN Metallamide. *Z. Anorg. Allg. Chem.* **1938**, *239*, 282–287. [CrossRef]
2. Maruska, H.P.; Tietjen, J.J. Preparation and properties of vapor deposited single crystalline GaN. *Appl. Phys. Lett.* **1969**, *15*, 327–329. [CrossRef]
3. Voronenkov, V.; Bochkareva, N.; Zubrilov, A.; Lelikov, Y.; Gorbunov, R.; Latyshev, P.; Shreter, Y. Hydride Vapor-Phase Epitaxy Reactor for Bulk GaN Growth. *Phys. Status Solidi* **2019**, *217*, 1900629–1900631. [CrossRef]
4. Bockowski, M.; Iwinska, M.; Amilusik, M.; Fijalkowski, M.; Lucznik, B.; Sochacki, T. Challenges and future perspectives in HVPE-GaN growth on ammonothermal GaN seeds. *Semicond. Sci. Technol.* **2016**, *31*, 093002. [CrossRef]
5. Asif Khan, M.; Van Hove, J.M.; Kuznia, J.N.; Olson, D.T. High electron mobility GaN/AlxGa1−xN heterostructures grown by low-pressure metalorganic chemical vapor deposition 1991. *Appl. Phys. Lett.* **1991**, *58*, 2408–2410. [CrossRef]
6. Asif Khan, M.; Bhattarai, A.R.; Kuznia, J.N.; Olson, D.T. High electron mobility transistor based on a GaN/AlxGa1−xN heterojunction. *Appl. Phys. Lett.* **1993**, *63*, 1214–1215. [CrossRef]
7. Ambacher, O.; Foutz, B.; Smart, J.; Shealy, J.R.; Weimann, N.G.; Chu, K.; Murphy, M.; Sierakowski, A.J.; Schaff, W.J.; Eastman, L.F.; et al. Two-dimensional electron gases induced by spontaneous and piezoelectric polarization in undoped and doped AlGaN/GaN Heterostructures. *J. Appl. Phys.* **2000**, *87*, 3222–3234. [CrossRef]
8. Sheppard, S.T.; Doverspike, K.; Pribble WLAllen, S.; Palmour, J.; Kehias, L.; Jenkins, T. High-power microwave GaN/AlGaN HEMT's on semi-insulating silicon carbide sub-strates. *IEEE Electron. Device Lett.* **1999**, *20*, 161–163. [CrossRef]
9. Ibbetson, J.P.; Fini, P.T.; Ness, K.D.; Ness, K.D.; DenBaars, S.P.; Speck, J.S.; Mishra, U.K. Polarization effects, surface states, and the source of electrons in AlGaN/GaN heterostructure field effect transistors. *Appl. Phys. Lett.* **2000**, *77*, 250–252. [CrossRef]
10. Roccaforte, F.; Leszczynski, M. Introduction to Gallium Nitride properties and applications. In *Nitride Semiconductor Technology: Power Electronics and Optoelectronic Devices*, 1st ed.; Roccaforte, F., Leszczynski, M., Eds.; Wiley-VCH Verlag GmbH&Co. KGaA: Hoboken, NJ, USA, 2008; pp. 1–38.
11. Hu, H.; Zhang, B.; Liu, L.; Xu, D.; Shao, Y.; Wu, Y.; Hao, X. Growth of freestanding Gallium Nitride (GaN) through polyporous interlayer formed directly during successive hydride vapor phase epitaxy (HVPE) process. *Crystals* **2020**, *10*, 141. [CrossRef]

12. Liu, A.C.; Tu, P.T.; Langpoklakpam, C.; Huang, Y.W.; Chang, Y.T.; Tzou, A.J.; Hsu, L.H.; Lin, C.H.; Kuo, H.C.; Chang, E.Y. The evolution of manufacturing technology for GaN electronic devices. *Micromachines* **2021**, *12*, 737. [CrossRef] [PubMed]

13. Godejohann, B.J.; Ture, E.; Müller, S.; Prescher, M.; Kirste, L.; Aidam, R.; Polyakov, V.; Bruckner, P.; Breuer, S.; Köhler, K.; et al. AlN/GaN HEMTs grown by MBE and MOCVD: Impact of Al distribution. *Phys. Status Solidi* **2017**, *254*, 1600715. [CrossRef]

14. Murphy, M.J.; Chu, K.; Wu, H.; Yeo, W.; Schaff, W.J.; Ambacher, O.; Eastman, L.F.; Eustis, T.J.; Silcox, J.; Dimitrov, R.; et al. High-frequency AlGaN/GaN polarization-induced high electron mobility transistors grown by plasma-assisted molecular-beam epitaxy. *Appl. Phys. Lett.* **1999**, *75*, 3653–3655. [CrossRef]

15. Dimitrov, R.; Murphy, M.; Smart, J.; Schaff, W.; Shealy, J.R.; Eastman, L.F.; Ambacher, O.; Stutzmann, M. Two-Dimensional Electron Gas in Ga-face and N-face Al/GaN/GaN heterostructures grown by plasma-induced molecular beam epitaxy and metalorganic chemical vapor deposition on sapphire. *J. Appl. Phys.* **2000**, *87*, 3375–3380. [CrossRef]

16. Howgate, J.D.; Hofstetter, M.; Schoell, S.J.; Schmid, M.; Schäfer, S.; Zizak, I.; Hable, V.; Greubel, C.; Dollinger, G.; Thalhammer, S.; et al. Ultrahigh gain AlGaN/GaN high energy radiation detectors. *Phys. Status Solidi* **2012**, *209*, 1562–1567. [CrossRef]

17. Abid, I.; Metha, J.; Cordier, Y.; Derluyn, J.; Degroote, S.; Miyake, H.; Medjdoub, F. AlGaN Channel High Electron Mobility Transistors with Regrown Ohmic Contacts. *Electronics* **2021**, *10*, 635. [CrossRef]

18. Dridi, Z.; Bouhafs, B.; Ruterana, P. First-principle investigation of lattice constants and bowing parameters in wurtzite AlxGa1-xN, InxGa1-xN and InxAl1-xN alloys. *Semicond. Sci. Technol.* **2003**, *18*, 850–856. [CrossRef]

19. Deboy, G.; Treu, M.; Haeberlen, O.; Neumayer, D. Si, SiC and GaN power devices: An unbiased view on key performance indicators. In Proceedings of the 62nd IEEE International Electron Devices Meeting (IEDM 2016), San Francisco, CA, USA, 3–7 December 2016; pp. 20.2.1.–20.2.4.

20. Grand View Research. Available online: https://www.grandviewresearch.com (accessed on 5 March 2020).

21. Duboz, J.Y.; Laügt, M.; Schenk, D.; Beaumont, B.; Reverchon, J.L.; Wieck, A.D.; Zimmerling, T. GaN for X-ray detection. *Appl. Phys. Lett.* **2008**, *92*, 263501–263503. [CrossRef]

22. Wang, J.; Mulligan, P.; Brillson, L.; Cao, L.R. Review of using gallium nitride for ionizing radiation detection. *Appl. Phys. Lett.* **2015**, *2*, 031102–0311012. [CrossRef]

23. Thalhammer, S.; Hofstetter, M.; Howgate, J.D.; Stutzmann, M. Radiation Detector and Measurement Device for Detecting X-ray Radiation. U.S. Patent No. 9,402,548, 2 August 2016.

24. Smith-Bindman, R.; Lipson, J.; Marcus, R.; Kim, K.P.; Mahesh, M.; Gould, R.; Berrington de González, A.; Miglioretti, D.L. Radiation dose associated with common computed tomography examinations and the associated lifetime attributable risk of cancer. *Arch Intern. Med.* **2009**, *169*, 2078–2086. [CrossRef]

25. Fletcher, J.G.; Yu, L.; Fidler, J.L.; Levin, D.L.; DeLone, D.R.; Hough, D.M.; Takahashi, N.; Venkatesh, S.K.; Sykes, A.G.; White, D.; et al. Estimation of Observer Performance for Reduced Radiation Dose Levels in CT: Eliminating Reduced Dose Levels That Are Too Low Is the First Step. *Acad. Radiol.* **2017**, *24*, 876–890. [CrossRef] [PubMed]

26. Lell, M.M.; Wildberger, J.E.; Alkadhi, H.; Damilakis, J.; Kachelriess, M. Evolution in computed tomography: The battle for speed and dose. *Investig. Radiol.* **2015**, *50*, 629–644. [CrossRef] [PubMed]

27. Geyer, L.L.; Schoepf, U.J.; Meinel, F.G.; Nance, J.W., Jr.; Bastarrika, G.; Leipsic, J.A.; Paul, N.S.; Rengo, M.; Laghi, A.; De Cecco, C.N. State of the Art: Iterative CT Reconstruction Techniques. *Radiology* **2015**, *276*, 339–357. [CrossRef] [PubMed]

28. Lell, M.; Wucherer, M.; Kachelrieß, M. Dose and dose reduction in computed tomography. *Radiol. Up2date* **2017**, *17*, 163–178.

29. Kataria, B.; Nilsson Althén, J.; Smedby, Ö.; Persson, A.; Sökjer, H.; Sandborg, M. Image quality and potential dose reduction using advanced modeled iterative reconstruction (ADMIRE) in abdominal CT—A review. *Radiat. Prot. Dosim.* **2021**, *195*, 177–187. [CrossRef] [PubMed]

30. Willenmink, M.J.; Persson, M.; Pourmorteza, A.; Pelc, N.P.; Fleischmann, D. Photon-counting CT: Technical Principles and Clinical Prospects. *Radiology* **2018**, *289*, 293–312. [CrossRef] [PubMed]

31. Ohl, R.S. Light-Sensitive Electric Device. U.S. Patent 2,402,662, 25 June 1946.

32. Knoll, G.F. *Radiation Detection and Measurement*, 4th ed.; John Wiley & Sons: Hoboken, NJ, USA, 2010; Chapter 11.

33. MED PHYS 4RA3,4RB3,6R03: Radiation & Radioisotope Methodology I, II. Available online: https://www.science.mcmaster.ca/radgrad/programs/courses.html?id=224 (accessed on 12 January 2022).

34. Bale, D.S.; Szeles, C. Nature of polarization in wide-band gap semiconductor detectors under high-flux irradiation: Application to semi-insulating Cd1−xZnxTe. *Phys. Rev. B* **2008**, *77*, 035205–035216. [CrossRef]

35. Duboz, J.Y.; Beaumont, B.; Reverchon, J.L.; Wieck, A.D. Anomalous photoresponse of GaN X-ray Schottky detectors. *Appl. Phys.* **2009**, *105*, 114512–114517. [CrossRef]

36. Duboz, J.Y.; Frayssinet, E.; Chenot, S.; Reverchon, J.L.; Idir, M. X-ray detectors based on GaN Schottky diodes. *Appl. Phys. Lett.* **2010**, *97*, 163504–163507. [CrossRef]

37. Upadhyay, K.T.; Chattopadhyay, M.K. Sensor applications based on AlGaN/GaN heterostructures. *Mater. Sci. Eng. B* **2021**, *263*, 114849. [CrossRef]

38. Vaitkus, J.; Cunningham, W.; Gaubas, E.; Rahman, M.; Sakai, S.; Smith, K.M.; Wang, T. Semi-insulating GaN and its evaluation for α-particle detection. *Nucl. Instrum. Methods Phys. A* **2002**, *509*, 60–64. [CrossRef]

39. Lee, I.H.; Polyakov, A.Y.; Smirnov, N.B.; Govorkov, A.V.; Kozhukhova, E.A.; Zaletin, V.M.; Gazizov, I.M.; Kolin, N.G.; Pearton, S.J. Electrical properties and radiation detector performance of free-standing bulk n-GaN. *J. Vac. Sci. Technol. B* **2012**, *30*, 021205-7. [CrossRef]
40. Schmid, M.; Howgate, J.D.; Rühm, W.; Thalhammer, S. High mobility AlGaN/GaN devices for β⁻ dosimetry. *Nucl. Instrum. Methods Phys. Res. A* **2016**, *819*, 14–19. [CrossRef]
41. Look, D.C.; Farlow, G.C.; Drevinsky, P.J.; Bliss, D.F.; Sizelove, J.R. On the nitrogen vacancy in GaN. *Appl. Phys. Lett.* **2003**, *83*, 3525–3527. [CrossRef]
42. Hofstetter, M.; Howgate, J.; Sharp, I.D.; Stutzmann, M.; Thalhammer, S. Development and evaluation of gallium nitirde-based thin films for X-ray dosimetry. *Phys. Med. Biol.* **2011**, *56*, 3215–3231. [CrossRef] [PubMed]
43. Howgate, J. GaN Heterostrucutres for Biosensing and Radiation Detection. PhD Thesis, TU Munich, Munich, Germany, 2012.
44. Thalhammer, S.; Hofstetter, M.; Howgate, J.D.; Stutzmann, M. X-ray Camera for the High-Resolution Detection of X-rays. U.S. Patent No. 9,354,329, 31 May 2016.
45. Thalhammer, S.; Hofstetter, M.; Howgate, J.D.; Stutzmann, M. Method for Detecting Radiation and Examination Device for the Radiation-Based Examination of a Sample. U.S. Patent No. 9,354,330, 31 May 2016.

 *micromachines*

Article

Laser Processing of Transparent Wafers with a AlGaN/GaN Heterostructures and High-Electron Mobility Devices on a Backside

Simonas Indrišiūnas [1,*], Evaldas Svirplys [1], Justinas Jorudas [2] and Irmantas Kašalynas [2]

[1] Laser Microfabrication Laboratory, Center for Physical Sciences and Technology (FTMC), Savanoriu Ave. 231, LT-02300 Vilnius, Lithuania; evaldas.svirplys@ftmc.lt

[2] Terahertz Photonics Laboratory, Center for Physical Sciences and Technology (FTMC), Saulėtekio 3, LT-10257 Vilnius, Lithuania; justinas.jorudas@ftmc.lt (J.J.); irmantas.kasalynas@ftmc.lt (I.K.)

* Correspondence: simonas.indrisiunas@ftmc.lt

Abstract: Sapphire and silicon carbide substrates are used for growth of the III-N group heterostructures to obtain the electronic devices for high power and high frequency applications. Laser micromachining of deep channels in the frontside of the transparent wafers followed by mechanical cleavage along the ablated trench is a useful method for partitioning of such substrates after the development of the electronics on a backside. However, in some cases damage to the component performance occurs. Therefore, the influence of various parameters of the laser processing, such as fluence in the spot size, substrate thickness, orientation, and the polarization of focused laser beam, to the formation of damage zones at both sides of the transparent substrate with thin coatings when ablating the trenches from one side was investigated. The vicinity effect of the ablated trenches on the performance of the electronics was also evaluated, confirming the laser micromachining suitability for the dicing of transparent wafers with high accuracy and flexibility.

Keywords: laser micromachining; sapphire; silicon carbide; AlGaN/GaN heterostructures; high-electron mobility devices

Citation: Indrišiūnas, S.; Svirplys, E.; Jorudas, J.; Kašalynas, I. Laser Processing of Transparent Wafers with a AlGaN/GaN Heterostructures and High-Electron Mobility Devices on a Backside. *Micromachines* **2021**, *12*, 407. https://doi.org/10.3390/mi12040407

Academic Editor: Giovanni Verzellesi

Received: 28 February 2021
Accepted: 2 April 2021
Published: 6 April 2021

Publisher's Note: MDPI stays neutral with regard to jurisdictional claims in published maps and institutional affiliations.

Copyright: © 2021 by the authors. Licensee MDPI, Basel, Switzerland. This article is an open access article distributed under the terms and conditions of the Creative Commons Attribution (CC BY) license (https://creativecommons.org/licenses/by/4.0/).

1. Introduction

In high power, high temperature resistance, high frequency electronics applications wide bandgap III-N group semiconductors (nitrides) have an advantage over conventional silicon electronics [1]. The III-N group heterostructure layers are usually grown on a wide-bandgap (transparent for visible light) materials, such as sapphire, silicon carbide (SiC), or gallium nitride (GaN) which provide small or even no lattice mismatch for the epitaxial layers in order to obtain superior high power electronic devices based on two-dimensional electron gas (2DEG) with high-electron mobility [2,3]. After the growth of required heterostructure layers on the substrate and manufacturing of the electronics, the wafer has to be partitioned into smaller pieces which contain separate electronic components and circuits. For that, mechanical scribing and cleavage or dicing with a diamond saw are widely used in industry providing relatively fast and cheap solutions, however, applicability of these tools for accurate and flexible shape partitioning of closely situated (distance < 100 μm) electrical components is challenging or even impossible. Laser micromachining of transparent materials can be invoked for accurate processing since the positioning accuracy and the size of ablation zone as small as several tens of micrometers to several micrometers can be achieved [4].

Having in mind that the substrate side with the electrical components should be hindered as little as possible, the ablation of deep channels in the backside of the substrate by direct laser ablation (DLA) followed by mechanical cleavage along the ablated trench line has been proposed for partitioning the samples while avoiding the damage or contamination to the front side [5]. However, in this work we observed that in some cases this method may result in the formation of the damage zones on both sides of the wafer near the ablated trench line, in the material layers on the "good" side of the substrate. Since backside damage formation is undesirable effect of DLA, it may act as a limiting factor for a wider adoption of DLA methods hindering the application of laser scribing for the separation of electronics components on the transparent wafers. Thus, the investigation of various operation regimes under destructive light mater interaction is crucial for processing of transparent materials exploiting all advantages of the laser micromachining.

Damage phenomena were reported in the literature, regarding laser cutting of thin sheets of transparent materials (glasses, fused silica) but without the electronic devices on a backside. In particular, it was demonstrated that some damage to the rear side and volume of the glass slab was made during the laser scribing process [6]. The damage on the rear side of the Shott glass substrate was attributed to the laser radiation, escaping from the ablation channel by refraction from the crater walls. In this case damage to the rear side of the substrate can be avoided or significantly reduced by selecting S polarization (polarization vector parallel to the laser scribing direction) which at oblique propagation angle has a higher reflectance from the air-glass interface, compared to P polarization, resulting in the peak fluence of the refracted radiation too low to reach glass damage threshold.

In [7], band-like damage was observed on the backside of laser cut 100 μm-thick aluminum-borosilicate-glass. It was reported that the damage was observed for both S and P polarizations, but was more pronounced in the case of S polarization (polarization vector parallel to the cut line).

Some authors [8] attribute backside damage to the collision of laser-induced plasma-generated stress waves: collision of longitudinal (compression) wave and Rayleigh surface wave, excited on the back surface by the transverse (distortion) stress wave. In this case, it was reported that when ablating the channel in 156 μm-thick borosilicate glass two scans (producing 20–30 μm depth channel) was enough to produce the damage lines on the back surface. With increasing scan number, the depth of the channel increased and the damage lines were appearing closer to the channel plane. The 117 μm distance from the channel to the damage zone was reported. In this case, 150 fs pulses at 800 nm wavelength, 50 μJ pulse energy, and 1 kHz repetition rate were used.

In [9], pump-probe experiments of laser irradiated glass, supplemented by simulations using linear beam propagation method, demonstrated that when ablated crater reaches a certain depth, part of the irradiation energy is deposited in the relatively narrow regions extending from crater sidewalls to the glass volume, at an angle (approximately 24°) to the glass surface. In this case, 80 fs, 47 μJ, 1 Hz, 800 nm laser irradiation was used.

In [10], the distance from the channel to the damage zone in a 90 μm-thick fused silica sheet was reported as 40.6 μm. The rear damage mechanism was explained as local intensity enhancements, caused by the interference of the laser radiation transmitted and reflected in the glass/air interface on the backside of the sheet. Laser irradiation reaches the back surface by refraction from the ablated crater wall and propagation to the rear interface of the sheet. The irregularity of the damage zone was explained by local modifications of the refractive index near the ablated crater. It was reported that P polarized laser beam produces a weaker damage zone due to the low reflectivity of the P polarization at the fused silica/air interface (compared to S polarization). It was proposed to put the fused silica sheet in the distilled water during the laser processing thus further reducing the reflectance at the glass/air interface.

In [11], 10 ps laser pulses were used to ablate Corning Eagle XG and Gorilla glasses. In this case, several types of damage to the rear side of the glass plate were observed. At low scan numbers, circular damage areas with a diameter similar to the focused laser beam diameter, which can be attributed to the diffraction pattern from the laser, formed opaque zone on the front side of the glass plate. At higher (several hundred) scan numbers line-like damage zones on the rear side of the sample at both sides of the cut were manifesting. The distance from the cut line (for 0.7 mm-thick sample) varied with the scan number from 170 μm to 500 μm. At large scan numbers, several distinct damage lines could be observed.

In [12], ablation of channels in fused silica were compared with theoretical model and provided quite good agreement. It was suggested that laser radiation experiences interference in the material surrounding the ablation channel due to interference of radiation refracted from the channel walls, reflected and refracted from the channel walls, and entering the material from the region surrounding the crater, where fluence is too low for nonlinear absorption. Spike-like damage regions emanating at an angle to the crater walls were observed experimentally and also replicated in the model as regions of high free electron density.

The aim of this paper was to investigate the formation of damage zones in the coating, deposited on the backside of the transparent substrate, during the laser scribing, using various laser processing parameters. Formation of damage zones in backside coatings in a broad range of various thickness substrates was performed. A method to reduce the area of the damage zones by oblique laser beam scanning was presented and validated by laser-machining of SiC wafers with real AlGaN/GaN heterostructures and high-electron mobility devices on a backside.

The effect of laser beam damage to the surface morphology and the performance of real electronic devices made on GaN/AlGaN heterostructures, grown on the back side of the sapphire or SiC substrates [13], was investigated. Additional research was also performed on the transparent glass (soda-lime) substrates with a thin gold film deposited on the back side, to avoid unnecessarily high costs of material.

2. Experimental Setup

The samples were the 350–500 µm-thick SiC and sapphire wafers with developed semiconductor layers and metal contacts and the 0.15–4.4 mm thick soda-lime glasses with a gold layer of 30 nm thickness, deposited on one of the surfaces by a DC sputter coater Q150T ES (Quorum Technologies). The SiC and sapphire samples under processing were attached to the glass plate with an optical cleaning tissue in between, trying to avoid scratching of the semiconductor layers and metal contacts. Soda-lime samples were processed using a special holder so that the processed zones had no physical contact with any substance on both sides (sample was hanging in the air).

The experiments were conducted using picosecond laser Atlantic (Ekspla): fundamental harmonics wavelength 1064 nm, pulse energy up to 150 µJ, pulse repetition rate up to 1 MHz, pulse duration 10 ps, second (532 nm) and third (355 nm) harmonics available. A laser beam was focused using one of the focusing 50 mm or 100 mm focal length lenses for 355 nm, and a 50 mm focal length focusing lens for 1064 and 532 nm. Spot size radii using 1064 nm and 532 nm wavelengths were 19.5 µm and 12.5 µm, respectively. A laser beam was scanned by displacing the sample with respect to the laser beam using XYZ translation stages (ALS10020, Aerotech). Scanning speed of 10 mm/s and 100 kHz pulse repetition rate was used for all scribes in the gold coated soda-lime samples. Scanning speed of 100 mm/s and 400 kHz pulse repetition rate using the third harmonics was employed for trench ablation in the SiC and sapphire wafers. Polarization was controlled using an appropriate half wave or quarter wave plate (see Figure 1).

The cross-section of the ablation channel was investigated by performing a perpendicular scribe from the backside and breaking the sample along this scribe. After breaking, the samples were cleaned with a deionized water in an ultrasonic bath.

Channel shape in soda-lime glass and damage to the layers deposited on the backside of various transparent substrates were investigated using an optical microscope.

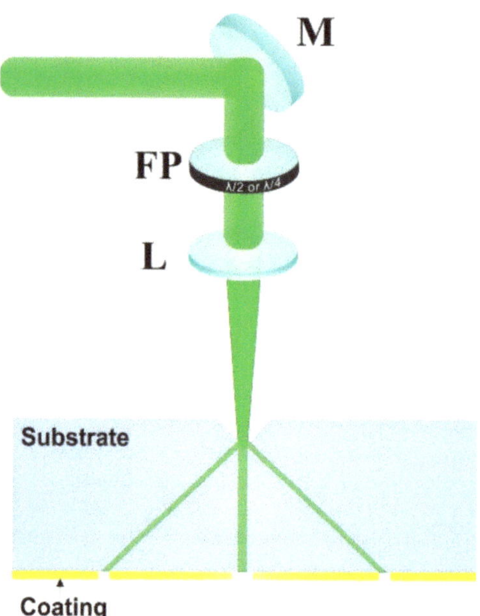

Figure 1. Experimental setup. M—mirror, FP—phase plate, L—focusing lens.

3. Results and Discussion

3.1. Laser Scribing of SiC and Sapphire Wafers with Electronic Devices on a Backside

Figure 2 shows optical microscope images of laser ablated trenches in SiC and sapphire substrates, which have AlGaN/GaN heterostructures with electronic devices on a backside. Ablation of trenches was performed always from the substrate side. Damaged areas on the opposite side of the substrate near the cut line were found in the sample shown in Figure 2b only.

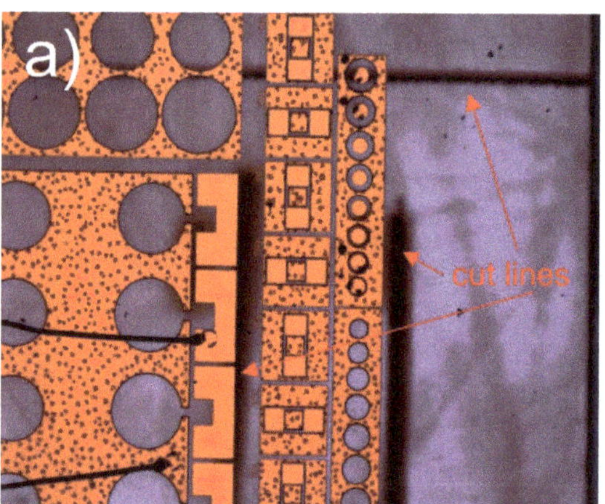

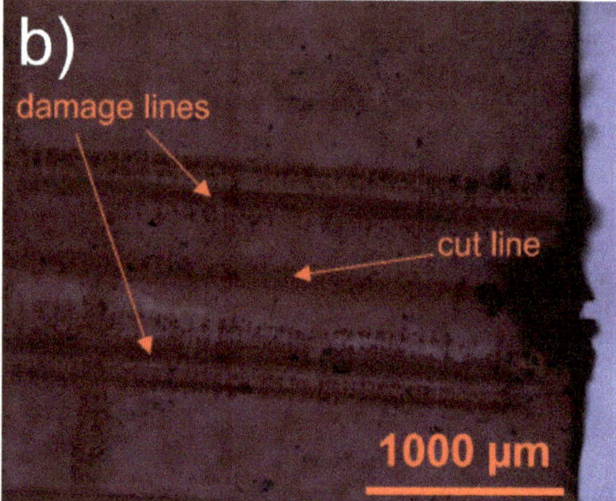

Figure 2. Laser scribe trenches in (**a**) SiC and (**b**) sapphire substrates with semiconductor and metal layers on the backside. Laser cutting was performed from the substrate side. In (**a**), the lenses with the focal length of 50 mm and 100 mm were used to process vertical and horizontal trench lines, respectively. In (**b**), 100 mm focusing lens was used. Formation of the damage lines was observed on both sides of the wafer along the trench but on a sapphire substrate only.

3.2. Point Damages

The properties of the damage generation were investigated in detail, employing the 1 mm thick glass samples with a 30 nm thick gold films deposited on the backside. Samples were irradiated by a focused laser beam for some duration without moving the beam or the sample. Figure 3 shows typical damage shapes in a gold coating for several polarization states. The damage areas formed at 800–1050 μm distance from the irradiation spot (black dot in the image center). It was seen that the damage zones formed in those sections of the circular areas, surrounding the irradiation spots, which lie roughly parallel to the polarization direction. The polarization is indicated by the double arrows above the microscope images shown in Figure 3. The relatively large distance from the irradiated spot to the damage zone, the ratio of distance/sample thickness ≈ 1, was in agreement with the results reported in [6,10], where refraction from the crater walls was proposed as a damage formation mechanism. The ratio damage distance/sample thickness ≈ 0.5 was considerably larger than those reported considering other known damage formation mechanism by laser beam; for example, the beam diffraction from the ablated crater. The diffraction of radiation from the opaque zone in the irradiated spot (diffraction from the inverse aperture) in the case of 1 mm thick substrate would produce high intensity ring much closer to the irradiated spot (<50 μm) with the ratio damage distance/sample thickness < 0.05 [11].

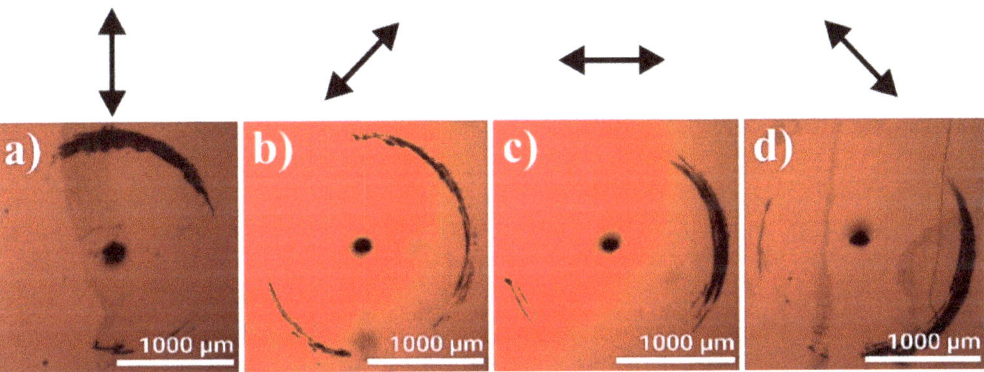

Figure 3. Damage in the 30 nm gold coating on the backside of the 1 mm thick glass plate for various polarization angles, indicated as black arrows above the images (**a–d**). The approximate polarization direction is illustrated by the arrows above each image. Laser wavelength 1064 nm, fluence 16 J/cm², 1000 pulses per spot. Measurement labels are in micrometers.

3.3. Line Damages

The influence of various parameters, such as fluence in the spot size, substrate thickness, orientation, and polarization of the laser beam, to the formation of damage zones when ablating the trench in the back-side of the glass with thin metal coating was investigated. Figure 4a shows a dependence of the distance d from the ablation channel to the damaged area on the substrate thickness. It must be noted that in samples containing thin glass substrates (less than 1.6 mm thick), higher-order damage zones also appear and the number of damage zones increases with decreasing substrate thickness. The higher-order damages can be explained by the internal reflection of the radiation, escaping from the ablation channel, on the air/coating/glass and air/glass/interfaces (Figure 4b). The absence of the higher-order damages in thick substrates can be explained by the relatively large optical path length, resulting in the absorption of the escaped radiation. For example, for soda-lime glass, if radiation escapes from the channel at 40° angle to the substrate, in 1 mm thick substrate and is reflected from the glass/coating/air interface back into the substrate it will propagate 2.6 mm, before hitting the glass/coating/air interface again. By using the well-known relation, $I = I_0 \cdot \exp(-4\pi k/\lambda \cdot h)$ for the intensity attenuation inside material thickness h, losses inside the substrates of various thickness can be evaluated. I_0 is initial intensity, I—intensity after propagating material thickness h, k is the extinction coefficient, λ is the wavelength. The extinction coefficient for soda-lime at 1064 nm wavelength is $k = 4.9 \times 10^{-6}$ [14]. After propagation of 2.6 mm distance, the beam intensity is reduced by 14%, compared to the beam reaching the interface the first time. Note that losses due to the transmission through the interface were not accounted. Having the same propagation angle but a 3 mm thick substrate, it will result in a 7.8 mm propagation length and 37% intensity attenuation.

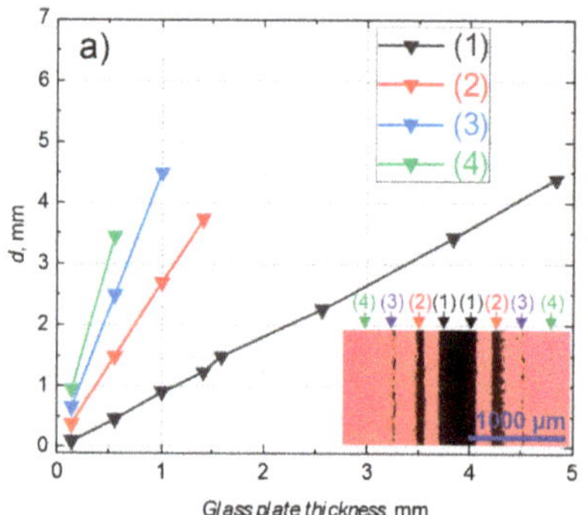
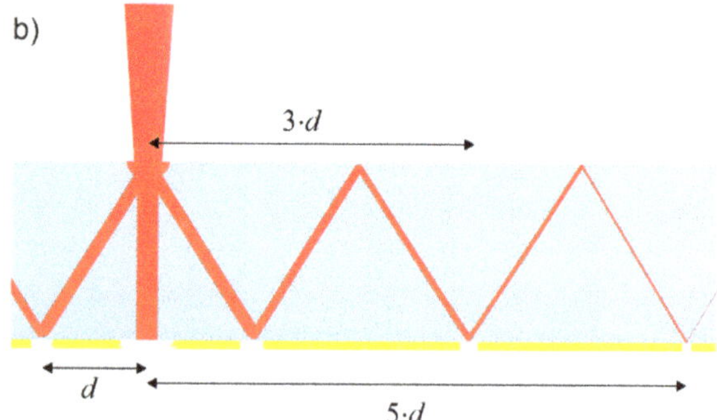

Figure 4. (**a**) distance d from the laser scan line to the damage area in the gold coating on various thickness glass plates, (1)–(4) denotes the first, second, etc. damage zones, insert shows damage zones in the coating on 0.15 mm-thick substrate; (**b**) illustration of the formation of high order damage zones by multiple reflections.

It is worth noting that no higher-order damage zones were observed in the sapphire sample, as shown in Figure 2b. This can also be the result of the relatively low ablation threshold of gold coating directly deposited on soda-lime substrate, due to low adhesion: 0.14 J/cm^2 and 0.04 J/cm^2 for 1064 nm and 532 nm wavelengths, respectively.

In Figure 5 experimental results on the variation of polarization state for 532 nm laser wavelength are provided. Figure 5a shows the dependence of the distance from the plane, containing ablated trench, to the damage zone on the laser spot size fluence for S, P, and circular polarization states. The S and P polarizations, in this case, are defined as follows: for S polarization the polarization vector is parallel to the ablated trench and for the P polarization the polarization vector is perpendicular to the ablated trench (vector lies in the plane containing trench cross-section and laser beam).

Orienting the polarization vector parallel to the trench (S polarization), we found a significant reduction of the damage to the backside coating as shown in Figure 5b–e. It can be seen that at low fluence the damage to the coating using S polarization is significantly smaller in comparison to those made using the P polarization. Explanation is a higher transmittance of P polarized radiation through the air/glass interface at ablated trench walls. Our results were in agreement with previous results reported in [6], demonstrating the similar laser induced damage in the back surface of the Schott glass plate, However, in [10] the opposite result (lower damage using S polarization) was reported for fused silica samples. The result was explained by lower reflectance of P polarization at the backside glass/air interface, resulting in the reduction of interference between incident and reflected radiation at this interface, however, our experimental results unambiguously showed lower damage generation using S polarization.

Some alterations in the backside coating could be seen even for S polarization at low fluencies (shown in Figure 5a as hollow symbols). Also, it can be noted that the distance from the trench to the damaged zone depends on the fluence and reaches a peak at about 5.5 J/cm^2 independent of polarization. This can be related to the change in crater shape with increasing fluence. Figure 5f shows dependence of the distance to the damage zone on the ratio of ablated trench width and depth when using S polarized laser beam. It is evident that the distance to the damage zone changes with the final shape of the ablation crater.

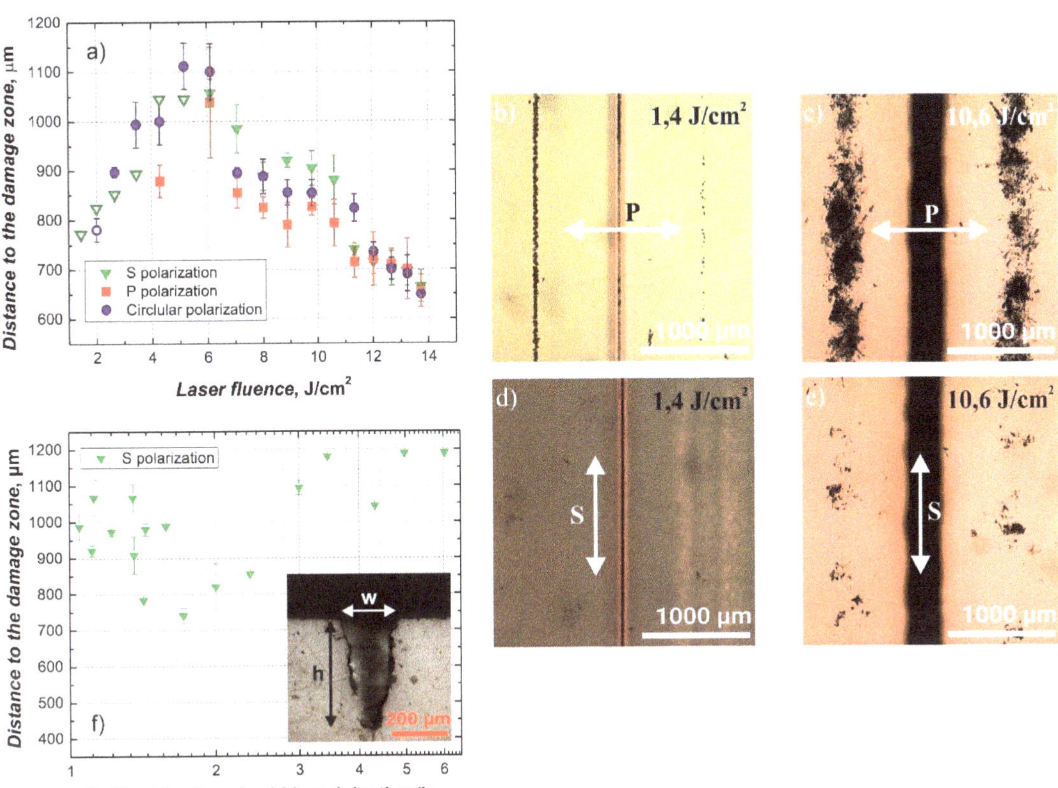

Figure 5. (**a**) Dependence of the distance from the cut line to the damage zone on the laser fluence for S, P and circular polarizations. The distance was measured from the location, corresponding to the center of the ablation channel, to the center of one of the side damage zones. (**b–e**) Optical microscope images, showing the coating on the backside of the soda-lime substrate after ablation of the trench, visible in the middle of the image, from the substrate side. (**f**) Dependence of the distance to the damage zone on the ratio between ablated trench width w and depth h. Laser beam polarization directions are indicated by the arrows. All samples were prepared using 532 nm wavelength, 100 scans, 10,000 pulses per mm.

In case the polarization selection is not sufficient to prevent damage in the coating, the "dead zone" in the backside coating along both sides of the scan line can be reduced by orienting a substrate at an angle to the plane, which contains the scanned laser beam. In Figure 6 this is illustrated by comparing two cases: when the substrate is perpendicular to the laser beam ($\Psi = 0°$; Figure 6a) and when the substrate is rotated so that the angle between the substrate and the beam scanning plane is 40° ($\Psi = 40°$; Figure 6b). The crater is shown as a triangle shape due to its simplicity. The rays, showing the laser radiation, propagates to the crater wall and are refracted, according to the Snell's law. In the case of $\Psi = 0°$ zones containing the refracted radiation and reaching the backside coating are formed on both sides of the crater. In the case of $\Psi = 40°$ at one side refracted radiation is contained very near the ablation channel, and on the other side propagates into the substrate at a shallow angle, resulting in both losses due to the absorption in the substrate material and in a widening of the zone at which radiation reaches the backside coating, so reducing the probability to damage it.

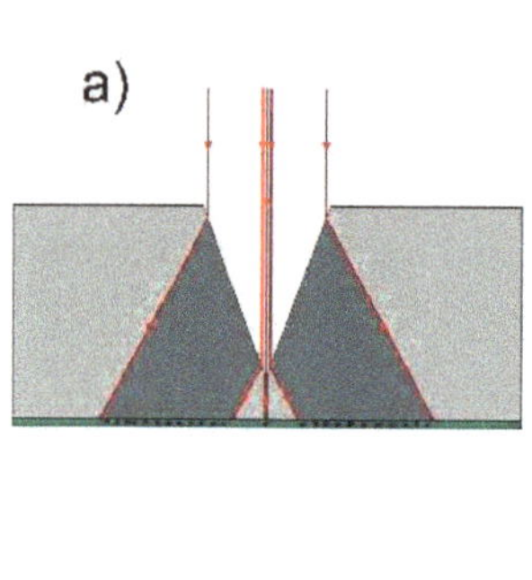 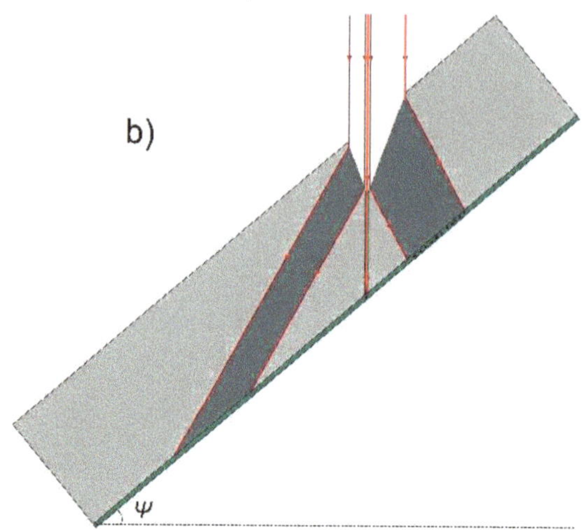

Figure 6. Illustration of damage to the coating, deposited on the backside of laser scribed transparent substrate. (**a**) Schematically shows damage area when a trench is ablated in substrate oriented perpendicularly to the plane in which laser beam is scanned; (**b**) shows damage area when the substrate is oriented at 40° angle to the plane in which laser beam is scanned.

In Figure 7a, experimentally obtained dependence of the distance from the ablation trench to the damage zones at both sides of the cut line on the angle between the substrate and the beam scanning plane Ψ is shown. When $\Psi = 0°$, two damage zones on both sides of the ablation plane are formed at approximately 1000 µm distance from the beam scanning plane. While Ψ increases to about 20°, both of these damage zones remain present, although the damage zone on left-hand side becomes weaker, reduced to the isolated islands of damaged or removed coating. Also, all damage zones shift to the left from the cut line resulting in a moderate increase of the sum damage zone width (Figure 7b). At the same time, an increase of the Ψ results in the formation of damage area, caused by irradiation transmitted straight through the substrate, not directly beneath the cut line, but at some distance from it due to the substrate orientation. When Ψ is further increased the damage zone at the left-hand side becomes even weaker (Figure 7c) and completely disappears at angles equal or larger than 28°, resulting in a sharp drop of the sum damage zone width from 3000 µm to less than 750 µm. When $28° \leq \Psi \leq 44°$, a slight decrease of the sum damage zone width can be observed (Figure 7b), caused by the right damage zone moving beneath the channel, and impeded by the increasing width and distance from the cut line of the damage zone caused by irradiation transmitted straight through the substrate (Figure 7a,c).

It can be concluded that when the selection of the beam polarization is not possible or does not provide sufficiently good results, rotation of a substrate at an angle to the plane which contains the scanned laser beam can be used for reduction of the "dead zone" width 2–3 times. However, it must also be considered that fabrication time required to ablate the channel, deep enough for breaking the substrate along it, in such case would increase. Figure 8 shows the dependence of trench depth on the sample rotation angle Ψ for 532 nm wavelength. The trench depth is reduced by 35%, when the substrate rotation angle is changed from 0° to 42°, keeping other parameters constant.

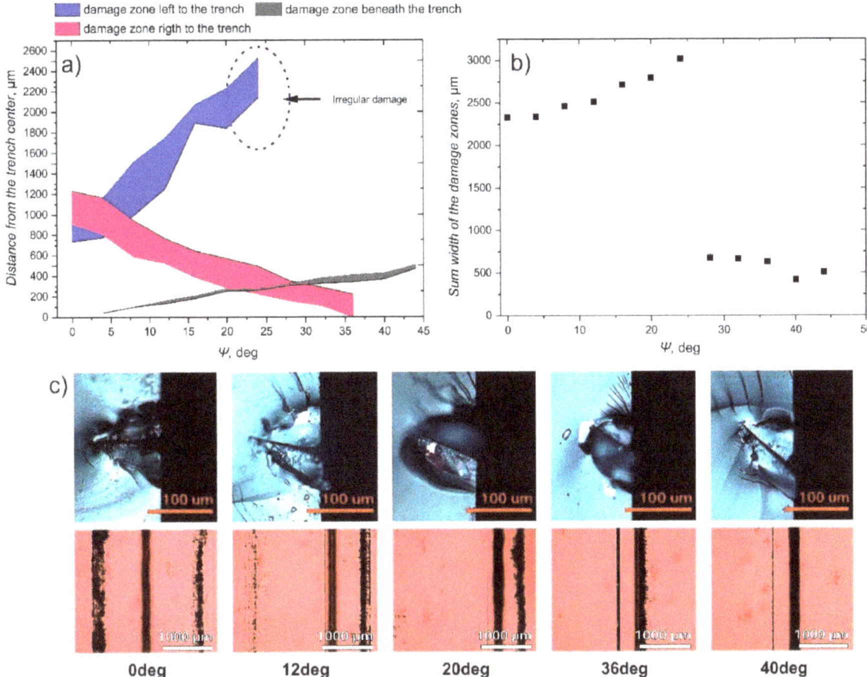

Figure 7. Evaluation of damage zones in the backside coating when changing the angle between the substrate and the beam scanning plane Ψ. (**a**) shows the dependence of the distance from the ablation channel to the near and far limits of damage zones formed in both sides of the ablation channel and beneath it (colored bands correspond to the particular damage zone); (**b**) shows the dependence of the sum width of the coating containing damage areas in the vicinity of the ablation channel; (**c**) shows cross-sections of the ablation channels and backside coating beneath the ablation channel for various angles Ψ. Laser wavelength 532 nm, polarization direction perpendicular to the scan line (P), 100 scans, 10,000 pulses per mm, 6.1 J/cm^2.

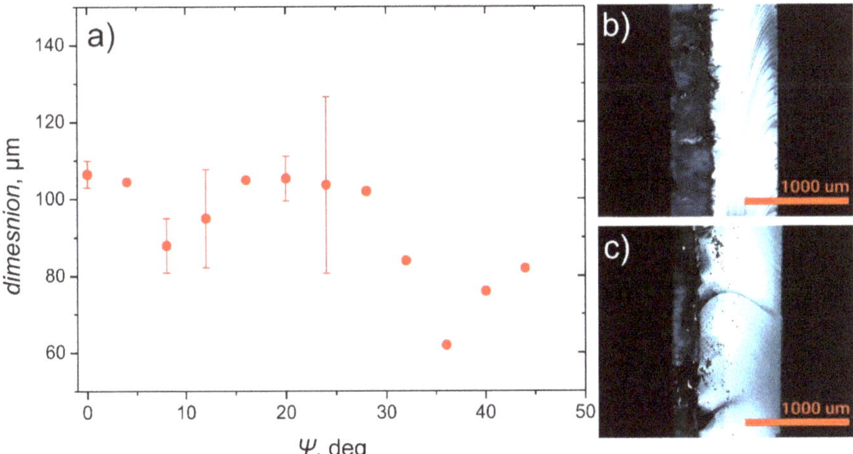

Figure 8. (**a**) Dependency of the ablated crater depth on the sample rotation angle Ψ. Laser wavelength 532 nm, polarization direction perpendicular to the scan line (P polarization), 100 scans, 10,000 pulses per mm, 6.1 J/cm^2; (**b,c**) Channels, ablated using Ψ = 0° and Ψ = 42°, respectively. Other parameters were kept constant: wavelength 532 nm, P polarization, 100 scans, 10,000 pulses per mm, 13.8 J/cm^2.

4. Performance of Electronic Devices

Finally, the laser ablation process was optimized and used to dice transparent wafers with AlGaN/GaN heterostructures with various electronic devices. Before laser processing, some of Schottky barrier diodes (SBDs) and High-electron-mobility transistors (HEMTs) were selected for detailed investigation by measuring the I–V characteristics in the EPS150 probe station (Cascade Microtech, Beaverton, OR, USA) equipped with the source-measure-unit SMU Keithley 2400 (Tektronix, Beaverton, OR, USA). Selected devices are indicated by red color rectangle in Figure 9 and their respective electrical characteristics before the processing are shown in Figure 10 by solid lines. Visible trenches appeared after the laser ablation on the back side and are seen in Figure 9 due to the transparency of the substrate. Using the optimized processing parameters, no visible damage to the AlGaN/GaN heterostructures was observed along the ablation lines (see Figure 9). The Schottky barrier diodes (SBDs) used for investigation are indicated by red color rectangle. The depth of the trench was found to be about 270 µm, as it is shown in Figure 11 (right), and the width at the top was found to be about 30 µm with obvious reduction down to a few microns at the maximum depths.

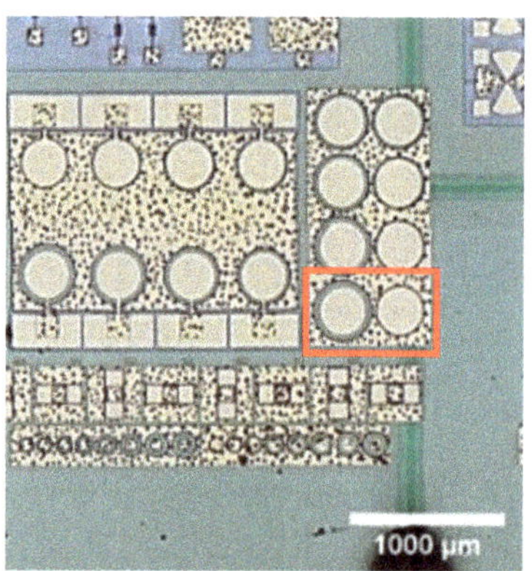
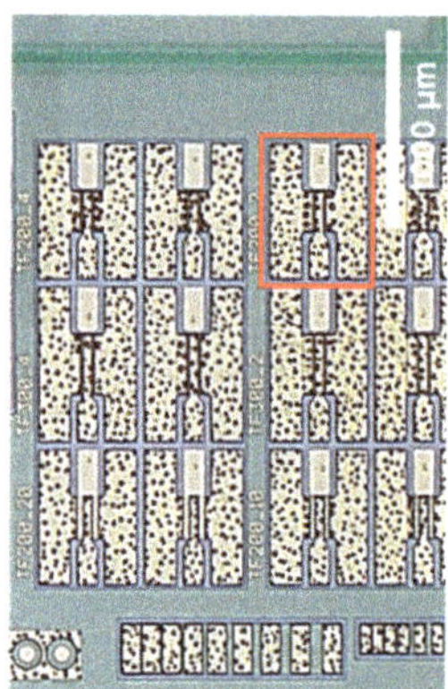

Figure 9. Microscope images of the transparent SiC wafer with AlGaN/GaN heterostructures and electronic devices being in a focus and laser ablated trenches being on the back side. Red color rectangles indicate selected devices for investigations: Schottky barrier diodes (**left**) and High-electron mobility transistor (**right**). Note that the trench line was made directly under one of the diodes. Scale bar is 1000 µm.

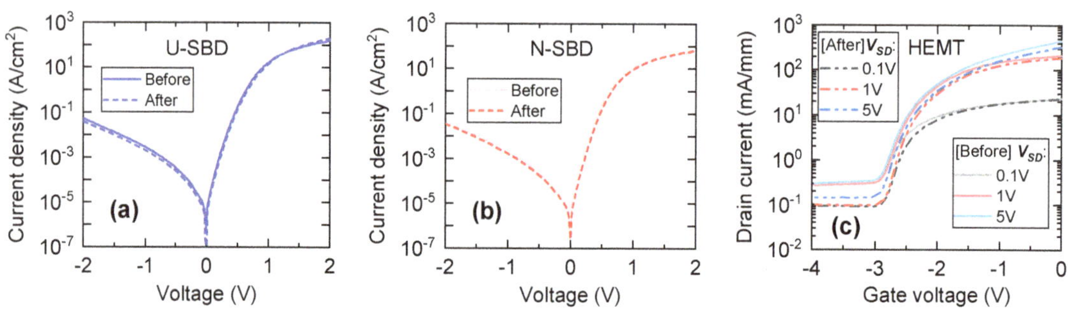

Figure 10. Current-voltage (I-V) characteristics of Schottky barrier diodes (SBDs) (**a**,**b**) and high-electron-mobility transistors (HEMT) (**c**) located under (**a**) and near (approximately 250 µm away) the trenches (**b**,**c**) formed in the back side of the substrate by laser microfabrication. The samples were characterized before and after the trenches were processed on the back side of SiC wafer in a depth of 270 µm.

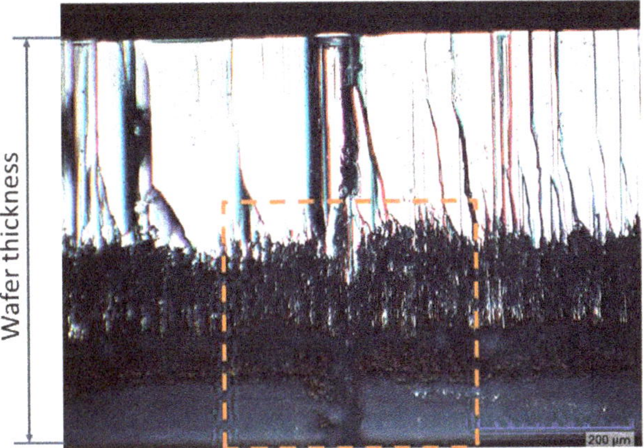
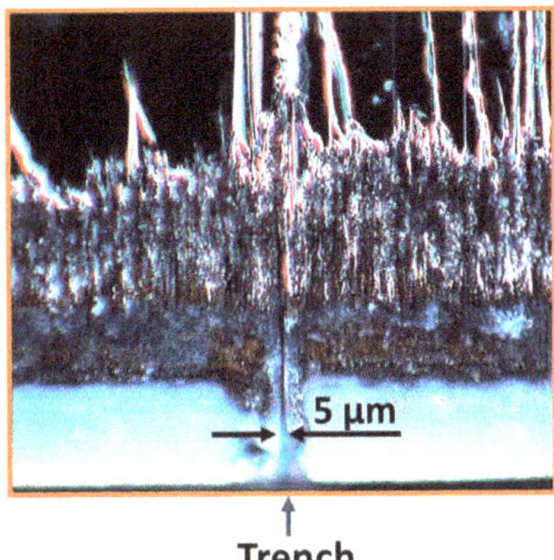

Figure 11. Microscope image of edge of SiC wafer after cleavage along the laser processed trench line (**left**) and zoomed in area (**right**). The depth of the trench formed by laser microfabrication in a perpendicular direction to image plane is of about 270 μm while the width varies from 30 μm down to 5 μm or even smallest values going deeper inside the trench. Scale bar is 200 μm.

Current-voltage (I-V) measurements were performed on Schottky barrier diodes (SBDs) before and after laser cutting in order to investigate possible laser-induced damage to the active layers of the AlGaN/GaN high electron mobility (HEMT) structure. As seen in Figure 9, two SBDs were chosen in close proximity to the laser cutting lines. One SBD was directly under the cutting line (labelled as U-SBD), while the other selected for the investigation SBD (N-SBD) was approximately 250 μm away from it. The I-V characteristics of the first and second diode after laser dicing are shown in Figure 10a,b by dashed lines, respectively. The transfer characteristics of HEMT were measured in a similar way and results are shown in Figure 10c. All results clearly demonstrate, that the laser micromachining did not affect the I–V characteristics of different electronic devices. Moreover, by fitting the low voltage region of forward current voltage characteristics and using thermionic emission model, the ideality factor, n, and Schottky barrier height, φ_b, were extracted. The values of both parameters were found to be about n = 1.4 and 0.45 eV, respectively, without a noticeable change due to the laser ablation of trenches on the SiC substrate. A small modification of I–V characteristics of all devices was attributed to different ambient conditions during the measurements, estimating its deviation to be within a range of 2%.

5. Conclusions

The influence of various parameters, such as fluence in the spot size, substrate thickness, orientation, and polarization of the laser beam, to the formation of damage zones when ablating the trench in the back-side of the transparent substrate with electronic devices was investigated. The experimental results regarding minimization of the damage to the backside coating were in agreement with the radiation refraction from the ablated crater model described in literature for the back surface damage in a transparent material. We found that the selection of laser beam polarization is not always sufficient to prevent the damage in the coating. The "dead zone" in the backside coating along both sides of the scan line can be reduced up to 2–3 times by using an optimal orientation of substrate, found to be at a 28° angle to the incident plane. However, in this case, the trench depth should be additionally optimized. Proper settings of laser ablation for the selected SiC wafer with back side heterostructures and electronic devices is validated demonstrating laser-based microfabrication and substrate dicing without modification on the electrical characteristics.

Author Contributions: Conceptualization, S.I. and I.K.; methodology, S.I. and I.K.; validation and investigation, S.I., E.S., J.J., and I.K.; writing—original draft preparation, S.I. and I.K.; writing—review and editing, S.I. and I.K.; visualization, S.I., E.S. and J.J.; supervision, S.I. and I.K.; project administration and funding acquisition I.K. All authors have read and agreed to the published version of the manuscript.

Funding: This research received funding from the Research Council of Lithuania (Lietuvos mokslo taryba) through the project "T-HP" (Grant No. DOTSUT-184) under the European Regional Development Fund measure No. 01.2.2-LMT-K-718-03-0096.

Data Availability Statement: The data that support the findings of this study are available on request from the corresponding author.

Acknowledgments: This work was supported from the Research Council of Lithuania (Lietuvos mokslo taryba) through the "T-HP" Project (Grant No. DOTSUT-184) funded by the European Regional Development Fund according to the supported activity "Research Projects Implemented by World-class Researcher Groups" under the Measure No. 01.2.2-LMT-K-718-03-0096.

Conflicts of Interest: The authors declare no conflict of interest.

References

1. Ma, C.-T.; Gu, Z.-H. Review on Driving Circuits for Wide-Bandgap Semiconductor Switching Devices for Mid- to High-Power Applications. *Micromachines* **2021**, *12*, 65. [CrossRef] [PubMed]
2. Flack, T.J.; Pushpakaran, B.N.; Bayne, S.B. GaN Technology for Power Electronic Applications: A Review. *J. Electron. Mater.* **2016**, *45*, 2673–2682. [CrossRef]
3. Sai, P.; Jorudas, J.; Dub, M.; Sakowicz, M.; Jakštas, V.; But, D.B.; Prystawko, P.; Cywinski, G.; Kašalynas, I.; Knap, W.; et al. Low frequency noise and trap density in GaN/AlGaN field effect transistors. *Appl. Phys. Lett.* **2019**, *115*, 183501. [CrossRef]
4. Varel, H.; Ashkenasi, D.; Rosenfeld, A.; Wähmer, M.; Campbell, E.E.B. Micromachining of quartz with ultrashort laser pulses. *Appl. Phys. A* **1997**, *65*, 367–373. [CrossRef]
5. Gu, E.; Jeon, C.W.; Choi, H.W.; Rice, G.; Dawson, M.D.; Illy, E.K.; Knowles, M.R.H. Micromachining and dicing of sapphire, gallium nitride and micro LED devices with UV copper vapour laser. *Thin Solid Film.* **2004**, *453–454*, 462–466. [CrossRef]
6. Collins, A.; Rostohar, D.; Prieto, C.; Chan, Y.K.; O'Connor, G.M. Laser scribing of thin dielectrics with polarised ultrashort pulses. *Opt. Lasers Eng.* **2014**, *60*, 18–24. [CrossRef]
7. Shin, H.; Kim, D. Cutting thin glass by femtosecond laser ablation. *Opt. Laser Technol.* **2018**, *102*, 1–11. [CrossRef]
8. Vanagas, E.; Kawai, J.; Tuzhilin, D.; Kudryashov, I.; Mizuyama, A.; Nakamura, K.; Kondo, K.-I.; Koshihara, S.-Y.; Takesada, M.; Matsuda, K.; et al. Glass cutting by femtosecond pulsed irradiation. *J. Micro/Nanolithographymemsand Moems* **2004**, *3*, 358–363. [CrossRef]
9. Kalupka, C.; Großmann, D.; Reininghaus, M. Evolution of energy deposition during glass cutting with pulsed femtosecond laser radiation. *Appl. Phys. A* **2017**, *123*, 376. [CrossRef]
10. Sun, X.; Zheng, J.; Liang, C.; Hu, Y.; Zhong, H.; Duan, J.a. Improvement of rear damage of thin fused silica by liquid-assisted femtosecond laser cutting. *Appl. Phys. A* **2019**, *125*, 461. [CrossRef]
11. Russ, S.; Siebert, C.; Eppelt, U.; Hartmann, C.; Faißt, B.; Schulz, W. Picosecond laser ablation of transparent materials. In Proceedings of the SPIE LASE, San Jose, CA, USA, 8–13 June 2013. [CrossRef]
12. Aldana, J.R.V.D.; Mendez, C.; Roso, L. Saturation of ablation channels micro-machined in fused silica with many femtosecond laser pulses. *Opt. Express* **2006**, *14*, 1329–1338. [CrossRef] [PubMed]
13. Jorudas, J.; Šimukovič, A.; Dub, M.; Sakowicz, M.; Prystawko, P.; Indrišiūnas, S.; Kovalevskij, V.; Rumyantsev, S.; Knap, W.; Kašalynas, I. AlGaN/GaN on SiC Devices without a GaN Buffer Layer: Electrical and Noise Characteristics. *Micromachines* **2020**, *11*, 1131. [CrossRef] [PubMed]
14. Rubin, M. Optical properties of soda lime silica glasses. *Sol. Energy Mater.* **1985**, *12*, 275–288. [CrossRef]

 *micromachines*

Article

Dual Laser Beam Asynchronous Dicing of 4H-SiC Wafer

Zhe Zhang [1,2], Zhidong Wen [1,2], Haiyan Shi [1], Qi Song [3], Ziye Xu [3], Man Li [1], Yu Hou [1,*] and Zichen Zhang [1,*]

[1] Microelectronics Instruments and Equipment R&D Center, Institute of Microelectronics, Chinese Academy of Sciences, Beijing 100029, China; zhangzhe1@ime.ac.cn (Z.Z.); wenzhidong@ime.ac.cn (Z.W.); shihaiyan@ime.ac.cn (H.S.); liman@ime.ac.cn (M.L.)

[2] School of Microelectronics, University of Chinese Academy of Sciences, No. 19(A) Yuquan Road, Beijing 100049, China

[3] International Research Centre for Nano Handling and Manufacturing of China, Changchun University of Science and Technology, Changchun 130022, China; songqi@ime.ac.cn (Q.S.); xuziye@ime.ac.cn (Z.X.)

* Correspondence: houyu@ime.ac.cn (Y.H.); zz241@ime.ac.cn (Z.Z.)

Abstract: SiC wafers, due to their hardness and brittleness, suffer from a low feed rate and a high failure rate during the dicing process. In this study, a novel dual laser beam asynchronous dicing method (DBAD) is proposed to improve the cutting quality of SiC wafers, where a pulsed laser is firstly used to introduce several layers of micro-cracks inside the wafer, along the designed dicing line, then a continuous wave (CW) laser is used to generate thermal stress around cracks, and, finally, the wafer is separated. A finite-element (FE) model was applied to analyze the behavior of CW laser heating and the evolution of the thermal stress field. Through experiments, SiC samples, with a thickness of 200 μm, were cut and analyzed, and the effect of the changing of continuous laser power on the DBAD system was also studied. According to the simulation and experiment results, the effectiveness of the DBAD method is certified. There is no more edge breakage because of the absence of the mechanical breaking process compared with traditional stealth dicing. The novel method can be adapted to the cutting of hard-brittle materials. Specifically for materials thinner than 200 μm, the breaking process in the traditional SiC dicing process can be omitted. It is indicated that the dual laser beam asynchronous dicing method has a great engineering potential for future SiC wafer dicing applications.

Keywords: silicon carbide; wafer dicing; stealth dicing; laser thermal separation; dry processing; laser processing

Citation: Zhang, Z.; Wen, Z.; Shi, H.; Song, Q.; Xu, Z.; Li, M.; Hou, Y.; Zhang, Z. Dual Laser Beam Asynchronous Dicing of 4H-SiC Wafer. *Micromachines* **2021**, *12*, 1331. https://doi.org/10.3390/mi12111331

Academic Editor: Aiqun Liu

Received: 30 September 2021
Accepted: 28 October 2021
Published: 29 October 2021

Publisher's Note: MDPI stays neutral with regard to jurisdictional claims in published maps and institutional affiliations.

Copyright: © 2021 by the authors. Licensee MDPI, Basel, Switzerland. This article is an open access article distributed under the terms and conditions of the Creative Commons Attribution (CC BY) license (https://creativecommons.org/licenses/by/4.0/).

1. Introduction

SiC power devices have continuously increased their share in the high-power semiconductor market in the last decade and are used in a series of applications such as electric vehicles and urban rail transit. However, due to their hardness and brittleness characteristics, there is one bottleneck in the SiC device manufacturing field, which is the wafer dicing process. Currently, SiC wafers are mainly mechanically diced by diamond-coated blades with low feed rates, in the range of 5–10 mm/s, and a high risk of side chipping at the edges of the diced chips. Furthermore, the diameter of the 4H-SiC wafer has increased from 25 to 100 mm, and the 150 mm transition is upcoming. With the recent smaller and thinner trend in semiconductor manufacturing, mechanical sawing has reached its limit in SiC wafer dicing.

To improve the dicing quality of SiC wafers, many novel dicing technologies have been developed to fulfill the requirements of throughput, edge quality, and costs, such as laser ablation cutting, plasma cutting, high-pressure water cutting, electrical discharge wire cutting, and water-jet guided laser cutting. The thermal separation method is a critical technology that is suitable for the dicing of hard-brittle materials [1]. Typically, the generation and extension of cracks in materials are critical issues in the field of materials science and engineering [2]. However, the thermal separation method developed the theory of utilizing crack generation and extension along a predetermined trajectory. The development of the theory in this field will further enrich the fracture behavior of micro-nano manufacturing. Based on this method, green and efficient cutting technologies for a wide range of materials have been developed [3].

Currently, there are two major processes based on the thermal separation method: non-premade trajectory cutting (NPTC) [4] and premade trajectory cutting (PTC) [5]. The process of NPTC is: firstly, prefabricate a micro notch at one point on the edge of the sheet; then, the thermal stress generated by the heat source scanning drives the force for crack growth along the scanning track until the whole sheet is fractured. The process of PTC is: firstly, a depth of cutting trajectory is performed on the upper or lower surface of the sheet; then, the thermal stress generated by the heat source scanning drives the cutting track to extend to the depth of the plate.

NPTC is mainly used for the rough machining of thicker glass and ceramic plates and other thick materials, with no chips and microcracks in the middle of the cut trajectory. The objectives of the researchers are to increase the cutting speed, reduce the trajectory deviation, and improve the surface cutting quality. A practical method based on the principle of the NPTC to increase the cutting speed is to change the shape of the heat source energy distribution and application of cooling. Yamamoto et al. [6] used the elliptical distribution CO_2 laser + water-cooled method for the thermal fracture cutting of glass. It was shown that increasing laser heat source power or applying cooling measures could increase the stress intensity factor at the initial crack to reach the threshold value quickly, to increase the cutting speed. Abramov et al. [7] of Corning used laser-induced thermal cracking to cut chemically strengthened glass. Another method to improve the cutting speed is to use a body heating source. The researchers of the LEMI company from Japan used a surface heating source—a tubular infrared lamp with an output power of 1 kW (for body heating) was placed 200 mm above the material and superimposed on the LD laser scan. However, the cutting speed without the IR lamp was only 23 mm/s [8].

The major problem in NTPC is trajectory deviation. Salman et al. [9] used a diode laser to cut 5-mm-thick soda-lime glass at a speed of 33 mm/s. It was found that there were severe trajectory shifts at the entrance and exit of the material. They simulated the stress field of the workpiece during the cutting process. It was found that the reason for the shifted trajectory was that the tensile stress at the entrance and exit of the cut was huge. Salman et al. [10] also studied the stress distribution at the entrance and exit of the cut using simulation. The artificial neural network model and finite element model were applied, respectively, for different thicknesses and laser scanning speeds. The results showed that the prediction results of the artificial neural network model were better than those of the finite element model.

The surface quality obtained by the thermal cracking method effectively improves bending strength. Kondratenko et al. [11] investigated laser-induced thermal cracking of cutting glass with thicknesses of 4–19 mm. They compared the strength of 6-mm-thick glass cut by mechanical, grinding, and laser. Finally, the quality of glass cut by the laser-induced thermal cracking method was better, and the edge strength was 5.5 times higher than that of conventional mechanical cutting.

Premade trajectory cutting is mainly used for the processing of liquid crystal display (LCD), plasma display (PDP), and flat panel display (FPD). PTC is characterized by faster cutting speeds and higher accuracy of the cutting trajectory. Many researchers have expanded the applicability of the PTC method for material dicing. Kang et al. [12] performed high-speed cutting of laminated glass. For PDP cutting, the authors used a two-step cutting method of scribing and thermal cracking, which is more efficient than the one-step cutting method. A series of processing devices were developed based on this principle. Huang et al. [13] created an implicit crack inside the glass using a 10 W 355 nm Nd:YAG laser. Huang et al. [14] also introduced ultrasonic vibration into the Nd:YAG UV laser and continuous CO_2 laser in the cutting system of LCD glass substrates. The results showed that the introduction of ultrasonic vibration could improve the cutting speed greatly (three times the original speed).

Cross-sectional quality is significantly improved by using PTC. KIM et al. [15] used a femtosecond laser for etching and a CO_2 thermal laser for cracking to separate LCD glass. The experimental results showed that at low numbers of femtosecond laser pulses, the glass damage was small and the groove depth was not deep enough. When the number of femtosecond laser pulses was increased to six, the following CO_2 laser was more effective in separating the glass by thermal stress. Wang et al. [16] conducted a study on laser thermal cleavage-cutting crystal glass substrates and proposed a new grooving and thermal cracking cutting method. Firstly, micro-cracks were created on the surface of liquid crystal glass using the instantaneous high energy of a YAG laser. Then, the glass was heated by CO_2 laser and cooled by Ar gas.

Stealth dicing [17–19] is another state-of-the-art dicing method where a pulsed laser, at a wavelength capable of penetrating the material, is focused inside the substrate. Focused laser spots cause an extremely high power density, both temporally and spatially, at localized points. By moving the laser along the desired path at different depths, several passes of the laser ablation points are formed. When external tensile stress is subsequently applied, the dies are separated. The process is fast, clean, and has zero kerfs. However, stealth dicing is typically combined with a mechanical breaking process. When it comes to hard-brittle material such as SiC [20], this mechanical breaking process can cause serious edge breakage and even cause wrong crack propagation and, finally, reduce the dicing yield.

In this work, we investigate a dual laser beam asynchronous dicing method by combining stealth dicing and premade trajectory cutting. The laser-based cutting method proposed in this work is clean, fast, and efficient and does not involve any chemical agent or liquid. The dependence of the laser process parameters on the cutting quality was theoretical and experimentally investigated. The quality of the cut edge was thoroughly analyzed by optical microscopy.

2. Materials and Methods

Figure 1 shows the whole set-up used in this study, which includes a 5 W-532 nm femtosecond pulsed laser manufactured by NKT working under the condition of 750 fs pulse duration, and a 10 W-1040 nm CW laser manufactured by IPG with customized wavelength. The pulsed laser is focused on the interior of the SiC wafer by an objective microscope lens to create bottom-up stealth dicing layers, while the continuous laser is sent to a Galvano scanner system and used as a heat source to create thermal stress.

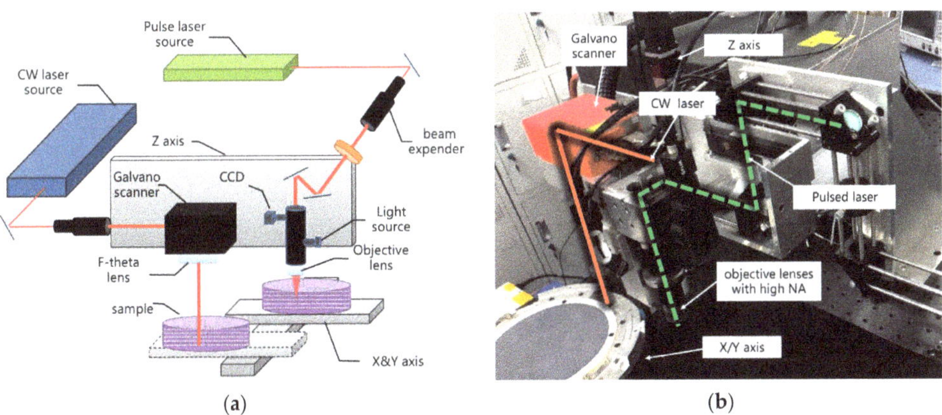

Figure 1. Configuration of a dual laser beam asynchronous dicing system: (**a**) schematic diagram of experimental set-up; (**b**) experimental platform with dual laser path.

2.1. Dicing Method

The operation of the DBAD method in our experiments is as follows: firstly, stealth dicing is operated on SiC. The pulse duration of the laser used in this process is 750 fs, and the scanning speed is 3000 µm/s. Through the fine control of focus depth and single pulse energy, the surface has no visible cracks after the process. Then, a CW laser sweeps through the trace of SD, and the internal cracks from the stealth dicing are extended vertically due to the thermal stress. The parameters of the two lasers are listed in Table 1.

Table 1. Main parameters of the lasers.

Laser Parameters	Pulsed Laser	CW Laser
Wavelength	532 nm	1040 nm
Max power	5 W	10 W
Repetition rate	20~200 kHz	N/A
Focal length	4 mm	140 mm
Beam diameter	8 µm	20 µm
Quality factor	M2 < 1.1	M2 < 1.2
Beam mode	TEM$_{00}$ Gaussian	

2.2. Samples

Briefly, 4-inch diameter silicon carbide (4H-SiC) wafers of 200 µm thickness were selected in this study; the physical properties of the material are shown in Table 2. The key principle of this dicing method is to minimize cracks and chippings and realize the cutting track as straight as possible. As a result of the wafers' brittle and hard characteristics, different laser parameters lead to very different results.

Table 2. Physical properties of 4H-SiC.

Material Properties	Value
Density	3210 kg/m^3
Thermal conductivity	490 W/(m·K)
Constant pressure heat capacity	690 J/(kg·K)
Coefficient of thermal expansion	4.3×10^{-6} 1/K
Poisson's ratio	0.185
Young's modulus	7×10^{11} Pa
Absorption coefficient	30 cm^{-1} [21]

2.3. Numerical Molding

To better understand the heat accumulation process and stress concentration process caused by the continuous laser, a two-dimensional finite element model (FEM) was established. The thermal stress produced by the moving CW laser around the interior hole produced by the previous pulsed laser is shown schematically in Figure 2. The height of the voids produced by the pulse laser set in this model is 20 µm, while the width is 5 µm. The size is approximately the same as the size of the hole created in the experiment.

High-intensity lasers, incident upon a material that is partially transparent, will deposit power into the material itself. The absorption of the incident light can be described by the Beer–Lambert law as it can be written in differential form for the light intensity I as:

$$\partial I / \partial z = \alpha(T)I, \tag{1}$$

where z is the coordinate along the beam direction, and $\alpha(T)$ is the temperature-dependent absorption coefficient of the material. As the heating and subsequent cooling process can vary in space and time, the evolution of temperature distribution is predicted by solving the time-dependent partial differential equation:

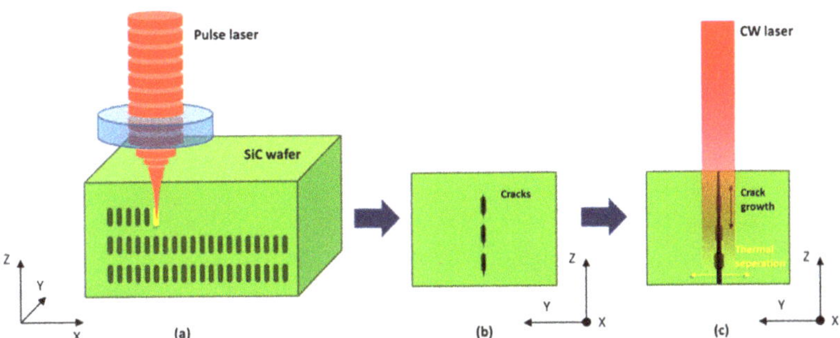

Figure 2. The schematic of the FEM model. (**a**) The stealth dicing process with the pulse laser, (**b**) the internal void produced from (**a**,**c**). The crack growth process caused by the CW laser.

$$\rho C_p \partial T/\partial t - \nabla \cdot (k \nabla T) = Q = \alpha(T)I, \tag{2}$$

where Q is the heat source, which is equal to the absorbed light; ρ and C_p are the density and constant pressure heat capacity of the material, respectively. The formula of thermal stress is given by:

$$F = Y(\varepsilon \Delta T)/L_0, \tag{3}$$

where Y is Young's modulus of the given material, ε is the coefficient of linear thermal expansion of the given material, and L_0 is the original length of the material before the expansion. These three equations are coupled with each other and are resolved using COMSOL Multiphysics.

The component structure with mesh is shown in Figure 3. A cuboid model is established with three microvoids inside, which are shaped like an ellipsoid to simulate the micro-cracks ablated during the stealth dicing process. For balancing the demand for simulating precision and computational efficiency, the model is simplified to a mirror-symmetrical model in the x–z plane, and infinity element layers are used for thermal diffusion simulation of large wafers. The energy deposition is assumed to be a moving Gaussian profile and is modeled by a boundary heat flux in the x–y plane (z = 0); energy depositions generate heat in the material, which can cause local stress concentrations resulting from thermal expansion.

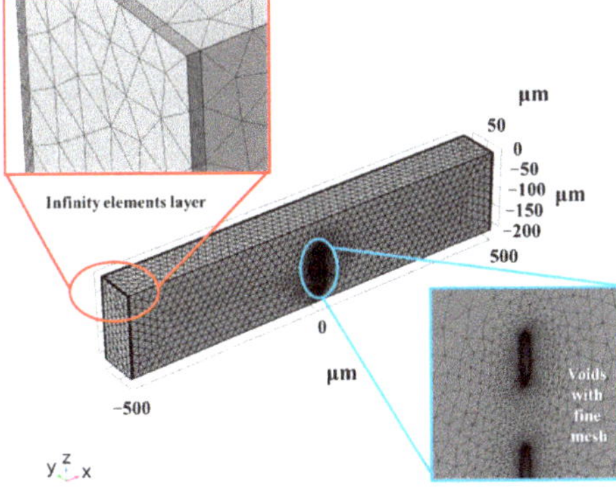

Figure 3. The diagram of the component structure with mesh.

3. Results

3.1. Numerical Molding Results

Figure 4 shows the dynamic temperature change of the 4H-SiC material during CW laser scanning when P = 10 W, v = 1000 mm/s. The maximum temperature inside the sample is only 312 K, which is far below the melting point of 4H-SiC. It can be concluded that the CW laser scanning process does not induce thermal damage to the substrate.

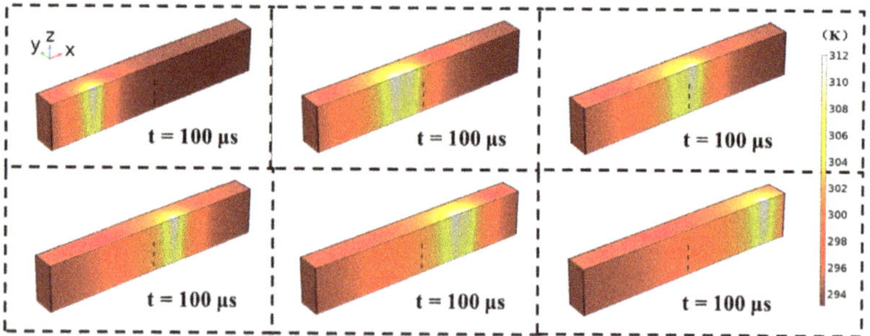

Figure 4. Moving laser heating material (10 W; 1000 mm/s).

Figure 5 shows the thermal stress at the endpoint of the long axis of the void-changing process during laser scanning. The maximum thermal stress, 48 MPa, presents at t = 400 μs, which indicates that 10 W of CW laser power is capable of generating enough thermal stress for the separation process.

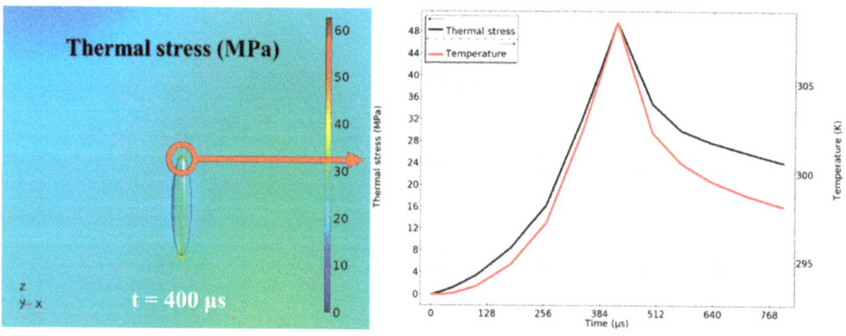

Figure 5. Thermal stress at the endpoint of the long axis of voids.

Figure 6 shows the thermal stress distribution change along the z-direction through the center of three voids, as we can see a large stress gradient at both the endpoints of the long axis of the ellipsoid; this is the main cause for the cracks to spread along the z-direction.

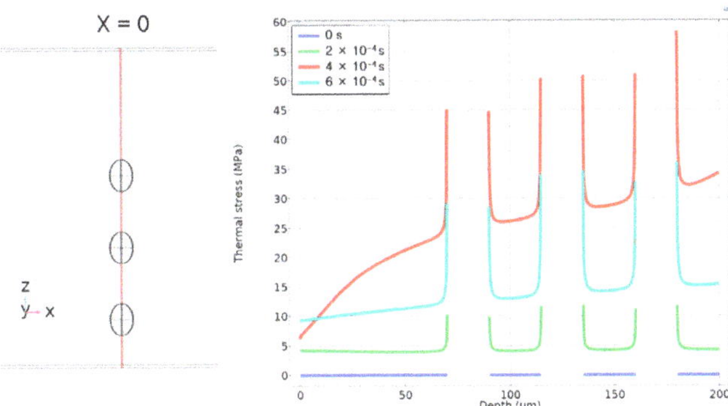

Figure 6. Thermal stress distribution along the z-direction.

3.2. Experimental Results

In this study, a series of experiments were conducted to cut SiC using a dual laser beam asynchronous dicing method based on the finite element simulation. Based on the principle of DBAD, single-pass stealth dicing is processed on the material, and then the thermal stress is generated using the 8 W-1040 nm continuous laser to extend the crack. Finally, the material is cut completely through the simple wafer expanding process. If the stealth cutting operation and thermal cracking are performed simultaneously, it will lead to misalignment and defects inside hard-brittle materials such as SiC, resulting in large errors in the positioning accuracy of the subsequent process.

Figure 7 shows the experimental results of cutting the SiC wafer with a thickness of 200 µm using DBAD compared with the simulation results. The surface of the material after stealth dicing is illustrated in Figure 7a. There were processing traces with no remarkable cracks. Then, a clear crack was performed through scanning using the continuous laser due to the thermal separation, as shown in Figure 7b. Figure 7c shows the cutting profile of SiC with a thickness of 200 µm. There was a line of three craters because of the process of SD. The height of the craters was about 25 µm, and the width was almost 9 µm. Additionally, it can be seen that the cracks from the thermal press could almost separate the material. Finally, in Figure 7d, the wafer of SiC is completely separated after the simple wafer-expanding process.

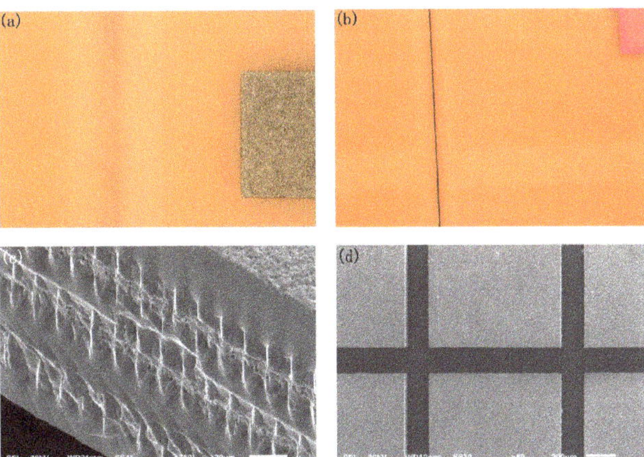

Figure 7. The results of cutting 200 µm SiC. (**a**) The surface after the stealth dicing; (**b**) the thermal crack from the continuous laser; (**c**) the cutting profile; and (**d**) the final separation after the expanding process.

The experimental results are in agreement with the above FEM simulation results, proving the correctness of the FEM simulation model and the effectiveness of the DBAD method. The quality of processing is perfectly expected. Compared with traditional stealth dicing, there was no serious edge breakage and no wrong crack propagation occurred from the mechanical breaking process. During the stealth dicing operation, there was no remarkable crack trace on the surface, so that the processing window was expanded with an improved production rate. In addition, the following wafer-expanding process maybe not be necessary, especially for the processing of thin hard-brittle materials. It is also suitable for the processing of hard-brittle materials with a thickness less than 200 μm.

4. Discussion

In our experiments, the effect of the changing continuous laser power on the DBAD system was also studied. The thermal press is generated using a 1040 nm continuous laser. However, the crack trajectory may be bent with the power of the laser rising to 10 W compared with the crack using an 8 W continuous laser, as seen in Figure 8. The reason for this phenomenon is still unknown. The effect of continuous laser parameters on the DBAD process will be investigated deeply in subsequent studies. It is beneficial to find better laser parameters during the processing.

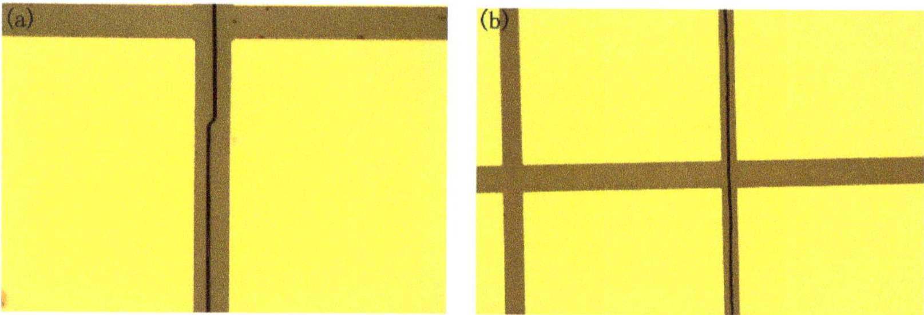

Figure 8. The crack from cutting the 200 μm SiC using (**a**) a 10 W/1040 nm continuous laser and (**b**) an 8 W/1040 nm continuous laser.

Miniaturization and performance enhancement drives the development of wafer separation technologies. Especially for hard-brittle materials, chippings occur with a poor cutting quality using mechanical dicing, while traditional stealth dicing with a mechanical breaking process leads to edge breakage. It is shown that the validity of the DBAD method is certified because of the above finite element simulation and experiments. The processing flow for cutting hard-brittle materials is simplified by improving processing quality compared with traditional processing. An effective method is provided, using dual laser beam asynchronous dicing to cut hard-brittle materials such as SiC.

5. Conclusions

In this paper, a dual laser beam asynchronous separation method for SiC wafer dicing has been put forward. In this method, a series of micro-cracks is firstly formed inside the wafer through a stealth dicing process by controlling the focal depth; the SD process will not induce visible cracks on the surface. Then, a CW laser is loaded on the dicing street, and the thermal stress leads to the wafer separation process.

The absorption of the CW laser and the resulting thermal stress was calculated using a finite element model. The simulation indicated that the tensile stress produced by the CW laser heating in the upper and lower ends of the voids is the main mechanism of vertical crack propagation. To get better separation quality, the moving speed and the power of CW laser should be properly adjusted. Insufficient laser power will weaken thermal accumulation, and the SiC wafer will not reach its fractural strength. In contrast, excess laser power can cause unnecessary heat accumulation and potentially harm the circuit. Based on the simulation analysis, an experimental machine was built, and several 4H-SiC wafers with a thickness of 200 μm were cut; a neat cutting side wall without chipping was obtained. Through the experimental study, it can be determined that the proper depth of the last SD layer can be 50 μm below the surface, and the optimal moving speed of the CW laser is 1000 mm/s; acting with 8 W of laser power, the fracture will propagate upward stably, and the SiC wafer can be separated along the expected SD path.

This novel DBAD method provides an effective solution for wafer cutting, specifically for hard-brittle materials with a thickness less than 200 μm, compared with other traditional processing methods.

Author Contributions: Conceptualization, Z.Z. (Zhe Zhang), Z.W., Q.S., Z.X., M.L., Y.H. and Z.Z. (Zichen Zhang); Formal analysis, Z.Z. (Zhe Zhang); Funding acquisition, H.S., M.L., Y.H. and Z.Z. (Zichen Zhang); Investigation, Z.Z. (Zhe Zhang), Z.W., H.S., Z.X., M.L., H.Y and Z.Z. (Zichen Zhang); Methodology, Z.Z. (Zhe Zhang), Z.W., H.S., Q.S., Z.X., M.L., Y.H. and Z.Z. (Zichen Zhang); Project administration, H.S., M.L., Y.H. and Z.Z. (Zichen Zhang); Resources, H.S., M.L., and Z.Z. (Zichen Zhang); Software, Z.Z. (Zhe Zhang) and Y.H.; Supervision, Y.H. and Z.Z. (Zichen Zhang); Validation, Z.Z. (Zhe Zhang), Y.H. and Z.Z. (Zichen Zhang); Visualization, Z.Z. (Zhe Zhang); Writing—original draft, Z.Z. (Zhe Zhang), Z.W.; Writing—review & editing, Z.Z. (Zhe Zhang), Z.W. All authors have read and agreed to the published version of the manuscript.

Funding: This research was funded by the Beijing Municipal Science & Technology Commission, grant number Z211100004821002.

Conflicts of Interest: The authors declare no conflict of interest.

References

1. Sharifi, H.; Larouche, D. An automatic granular structure generation and finite element analysis of heterogeneous semi-solid materials. *Model. Simul. Mater. Sci. Eng.* **2015**, *23*, 065013. [CrossRef]
2. Mhamdia, R.; Serier, B.; Albedah, A.; Bouiadjra, B.B.; Kaddouri, K. Numerical analysis of the influences of thermal stresses on the efficiency of bonded composite repair of cracked metallic panels. *J. Compos. Mater.* **2017**, *51*, 3701–3709. [CrossRef]
3. Karlsson, M.N.A.; Deppert, K.; Wacaser, B.A.; Karlsson, L.S.; Malm, J.O. Size-controlled nanoparticles by thermal cracking of iron pentacarbonyl. *Appl. Phys. A-Mater. Sci. Process.* **2005**, *80*, 1579–1583. [CrossRef]
4. Ueda, T.; Yamada, K.; Oiso, K.; Hosokawa, A. Thermal stress cleaving of brittle materials by laser beam. *Cirp Ann.-Manuf. Technol.* **2002**, *51*, 149–152. [CrossRef]
5. Tsai, C.-H.; Huang, B.-W. Diamond scribing and laser breaking for LCD glass substrates. *J. Mater. Process. Technol.* **2008**, *198*, 350–358. [CrossRef]
6. Yamamoto, K.; Hasaka, N.; Morita, H.; Ohmura, E. Thermal stress analysis on laser scribing of glass. *J. Laser Appl.* **2008**, *20*, 193–200. [CrossRef]
7. Terakado, N.; Uchida, S.; Sasaki, R.; Takahashi, Y.; Fujiwara, T. CO_2 laser-induced weakening in chemically strengthened glass: Improvement in processability and its visualization by Raman mapping. *Ceram. Int.* **2018**, *44*, 2843–2846. [CrossRef]
8. Karube, K.; Karube, N. Laser-induced cleavage of LCD glass as full-body cutting. In Proceedings of the Conference on Laser-Based Micro- and Nano-Packaging and Assembly II, San Jose, CA, USA, 22–24 January 2008.
9. Nisar, S.; Sheikh, M.A.; Li, L.; Safdar, S. Effect of thermal stresses on chip-free diode laser cutting of glass. *Opt. Laser Technol.* **2009**, *41*, 318–327. [CrossRef]
10. Nisar, S.; Li, L.; Sheikh, M.A.; Pinkerton, A.J. The effect of continuous and pulsed beam modes on cut path deviation in diode laser cutting of glass. *Int. J. Adv. Manuf. Technol.* **2010**, *49*, 167–175. [CrossRef]
11. Buchanov, V.V.; Kaptakov, M.O.; Murav'ev, E.N.; Revenko, V.I.; Solinov, E.F. Theoretical Research of Laser Glass Cutting. *Glass Phys. Chem.* **2018**, *44*, 254–259. [CrossRef]
12. Kang, H.S.; Hong, S.K.; Oh, S.C.; Choi, J.Y.; Song, M.G. A study of cutting glass by laser. In Proceedings of the 2nd International Symposium on Laser Precision Microfabrication, Singapore, 16–18 May 2002.
13. Huang, K.-C.; Wu, W.-H.; Tseng, S.-F.; Hwang, C.-H. The Mixed Processing Models Development of Thermal Fracture and Laser Ablation on Glass Substrate. In Proceedings of the International Conference on Advances in Materials and Processing Technologies. Ctr Arts & Metiers ParisTech, Paris, France, 24–27 October 2010.
14. Huang, K.-C.; Hsiao, W.-T.; Hwang, C.-H.; Lin, R.-L.; Yeh, J.-L.A. The laser ablation model development of glass substrate cutting assisted with the thermal fracture and ultrasonic mechanisms. *Opt. Lasers Eng.* **2015**, *67*, 31–35. [CrossRef]
15. Shin, J. Investigation of the surface morphology in glass scribing with a UV picosecond laser. *Opt. Laser Technol.* **2019**, *111*, 307–314. [CrossRef]

16. Xuhuang, W.; Jianhua, Y.A.O.; Guobin, Z.; Chenghua, L.O.U.; Yuansi, Y. Research of the technology of laser cutting LCD glass substrates based on thermal cracking method. *Laser Technol.* **2011**, *35*, 472–476.
17. Zhong, Z.-W. Recent advances and applications of abrasive processes for microelectronics fabrications. *Microelectron. Int.* **2019**, *36*, 150–159. [CrossRef]
18. Yang, Y.-L.; Ito, H.; Kim, Y.S.; Ohba, T.; Chen, K.-N. Evaluation of metal/polymer adhesion and highly reliable four-point bending test using stealth dicing method in 3-d integration. *IEEE Trans. Compon. Packag. Manuf. Technol.* **2020**, *10*, 956–962. [CrossRef]
19. Ohmura, E.; Fukuyo, F.; Fukumitsu, K.; Morita, H. Internal modified-layer formation mechanism into silicon with nanosecond laser. *J. Achiev. Mater. Manuf. Eng.* **2006**, *17*, 381–384.
20. Okada, T.; Tomita, T.; Katayama, H.; Fuchikami, Y.; Ueki, T.; Hisazawa, H.; Tanaka, Y. Local melting of Au/Ni thin films irradiated by femtosecond laser through GaN. *Appl. Phys. A-Mater. Sci. Process.* **2019**, *125*, 1–6. [CrossRef]
21. Biedermann, E. The optical absorption bands and their anisotropy in the various modifications of sic. *Solid State Commun.* **1965**, *3*, 343–346. [CrossRef]

micromachines

MDPI

Article

Temperature-Controlled Crystal Size of Wide Band Gap Nickel Oxide and Its Application in Electrochromism

Muyang Shi [1], Tian Qiu [2], Biao Tang [3], Guanguang Zhang [1], Rihui Yao [1,*], Wei Xu [1], Junlong Chen [1], Xiao Fu [1], Honglong Ning [1,*] and Junbiao Peng [1]

[1] State Key Laboratory of Luminescent Materials and Devices, Institute of Polymer Optoelectronic Materials and Devices, South China University of Technology, Guangzhou 510640, China; 201430320229@mail.scut.edu.cn (M.S.); msgg-zhang@mail.scut.edu.cn (G.Z.); xuwei@scut.edu.cn (W.X.); msjlchen@gmail.com (J.C.); 201630343721@mail.scut.edu.cn (X.F.); psjbpeng@scut.edu.cn (J.P.)
[2] Department of Intelligent Manufacturing, Wuyi University, Jiangmen 529020, China; timeqiu@hotmail.com
[3] Guangdong Provincial Key Laboratory of Optical Information Materials and Technology & Institute of Electronic Paper Displays, South China Academy of Advanced Optoelectronics, South China Normal University, Guangzhou 510006, China; biao.tang@guohua-oet.com
* Correspondence: yaorihui@scut.edu.cn (R.Y.); ninghl@scut.edu.cn (H.N.); Tel.: +86-20-8711-4525 (H.N.)

Abstract: Nickel oxide (NiO) is a wide band gap semiconductor material that is used as an electrochromic layer or an ion storage layer in electrochromic devices. In this work, the effect of annealing temperature on sol-gel NiO films was investigated. Fourier transform infrared spectroscopy (FTIR) showed that the formation of NiO via decomposition of the precursor nickel acetate occurred at about 300 °C. Meanwhile, an increase in roughness was observed by Atomic force microscope (AFM), and precipitation of a large number of crystallites was observed at 500 °C. X-ray Diffraction (XRD) showed that the NiO film obtained at such a temperature showed a degree of crystallinity. The film crystallinity and crystallite size also increased with increasing annealing temperature. An ultraviolet spectrophotometer was used to investigate the optical band gap of the colored NiO films, and it was found that the band gap increased from 3.65 eV to 3.74 eV with the increase in annealing temperature. An electrochromic test further showed that optical modulation density and coloring efficiency decreased with the increase in crystallite size. The electrochromic reaction of the nickel oxide film is more likely to occur at the crystal interface and is closely related to the change of the optical band gap. An NiO film with smaller crystallite size is more conducive to ion implantation and the films treated at 300 °C exhibit optimum electrochromic behavior.

Keywords: nickel oxide; annealing temperature; crystallite size; optical band gap; electrochromic device

Citation: Shi, M.; Qiu, T.; Tang, B.; Zhang, G.; Yao, R.; Xu, W.; Chen, J.; Fu, X.; Ning, H.; Peng, J. Temperature-Controlled Crystal Size of Wide Band Gap Nickel Oxide and Its Application in Electrochromism. *Micromachines* **2021**, *12*, 80. https://doi.org/10.3390/mi12010080

Received: 16 December 2020
Accepted: 11 January 2021
Published: 14 January 2021

Publisher's Note: MDPI stays neutral with regard to jurisdictional claims in published maps and institutional affiliations.

Copyright: © 2021 by the authors. Licensee MDPI, Basel, Switzerland. This article is an open access article distributed under the terms and conditions of the Creative Commons Attribution (CC BY) license (https://creativecommons.org/licenses/by/4.0/).

1. Introduction

Nickel oxide (NiO) is an important wide band gap semiconductor material with a band gap width of approximately 3.6 to 4.0 eV. As one of the rare p-type semiconductors in transition metal oxides, nickel oxide has a stable band gap and excellent electrochromic properties [1]. As a new functional material, NiO has applications in many fields, such as a hole-transporting layer in solar cells [2,3], as a p-type transparent semiconductor [4,5], and as an electrochromic layer or an ion storage layer in electrochromic devices [6–8]. An electrochromic device (ECD) can adjust the modulating optical transmittance in the visible light region through ion implantation and extraction in response to the switching of an externally applied potential [9,10]. Figure 1 shows a generic ECD structure called a five-layer "battery-type" structure [11]. In the middle of two transparent substrates (generally a glass substrate), there are five superimposed thin layers, which are respectively an electrochromic layer, an ion storage layer, an electrolyte layer, and two transparent conductive layers. Energy-saving features include low power consumption, large optical modulation, high coloring efficiency, and good optical storage, making ECD useful for smart Windows, skylights, etc. [12,13]. Since the discovery of the electrochromic properties of NiO, extensive studies have been conducted on NiO-based electrochromic devices [14]. As one of the best anode electrochromic materials, NiO is often used to form complementary electrochromic devices with cathode electrochromic materials like WO_3, and this type of electrochromic device has better optics modulation capability [15,16].

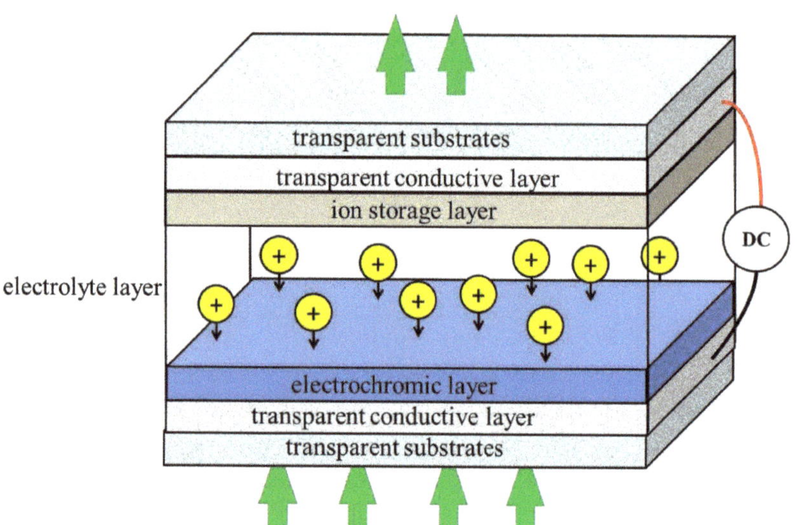

Figure 1. A typical ECD design. Arrows indicate the movement of ions and electrons in an applied electric field.

So far, various methods, include sputtering [17], chemical vapor deposition [18], electron-beam evaporation [19], and sol-gel [20] have been used for preparing NiO films. Currently, magnetron sputtering is used as a commercial technology because it can produce films with high uniformity and reliability. However, such a method requires expensive equipment. In contrast, the sol-gel method is much more desirable as far as the cost is concerned. More importantly, sol-gel technology is promising for future commercial device manufacturing via the low-cost printing route. However, at present, the sol gel prepared films still have problems like non-uniformity and poor repeatability. With the development of new sol-gel technology, such as inkjet printing [13], sol-gel technology is promising for commercial applications in the future.

In sol-gel technology, a key parameter is annealing temperature [21], which is critical in controlling the microstructure and electrochromic properties of the films. In this paper, the microstructure, electrical properties, optical modulation capabilities, and coloring efficiency of sol gel prepared nickel oxide films obtained under different annealing temperatures were analyzed, and the results may improve the current understanding of the relationship between the microstructure and performance of such oxide films.

2. Materials and Methods

Nickel acetate ($C_4H_6NiO_4$, Macklin Biochemical Co. Ltd., Shanghai, China) was added to 2-Methoxyethanol ($C_3H_8O_2$, Macklin Biochemical Co. Ltd., Shanghai, China) and mixed, then ethanolamine was added as a stabilizer, and the mixture was heated in an oil bath at 60 °C for 2 h. Finally, it was left to stand at room temperature for 24 h and aged to obtain sol-gel. Spin coating technology was used to prepare NiO films on commercial indium tin oxide (ITO) glass. The thickness of NiO is optimized and controlled by the solution concentration and spin coating parameters, which has an important impact on the electrochromic transmittance modulation ability. In this work, our study focuses on changing annealing temperature while keeping other variables consistent, such as sol concentration, spin coating parameters, substrate, and electrolyte. These deposited films were annealed at 100 °C, 200 °C, 300 °C, 400 °C, and 500 °C for 60 min.

The crystals of the film were analyzed by X-ray diffraction (XRD, PANalytical Empyrean DY1577, PANalytical, Almelo, The Netherlands). Surface morphology was measured with an atomic force microscope (AFM, Nano Instruments BY3000, Nano Instruments, Beijing, China). In the simple device constructed, a 0.5 mol/L KOH solution was used as an electrolyte, and ITO was used as an electrode to measure its electrochromic performance. The transmission spectrum was measured by an ultraviolet spectrophotometer (SHIMADZU UV2600, SHIMADZU, Tokyo, Japan) with ITO glass as a blank. The current of the electrochromic test was recorded by an electrochemical workstation (CH Instruments CHI600E, CH Instruments, Shanghai, China).

3. Results

The FTIR spectra of the films are shown in Figure 2. With increasing temperature, the absorption caused by acetate groups at 1465 cm^{-1} [22] and that due to the δ(H–OH) bonding at 1630 cm^{-1} [23] of the precursor gradually gets weak and even disappears at 300 °C, which indicates that the nickel acetate precursor has largely transformed to NiO at such a temperature. The FTIR spectra of NiO films annealed above 200 °C show that both ν (Ni^{2+}–O) vibration at about 470 cm^{-1} and ν (Ni^{3+}–O) vibration at about 470 cm^{-1} exist simultaneously, which indicates excess oxygen, and leads to a deviation of the stoichiometric ratio. This is a common phenomenon in the preparation of nickel oxide, which leads to lattice distortion and lattice constant changes. This is consistent with the results of lattice constants calculated by XRD below [24]. In addition, the FTIR spectra of the samples after annealing at 300 °C, 400 °C, and 500 °C show similar shapes, indicating that the chemical composition of the films was basically stable when the temperature was higher than 300 °C.

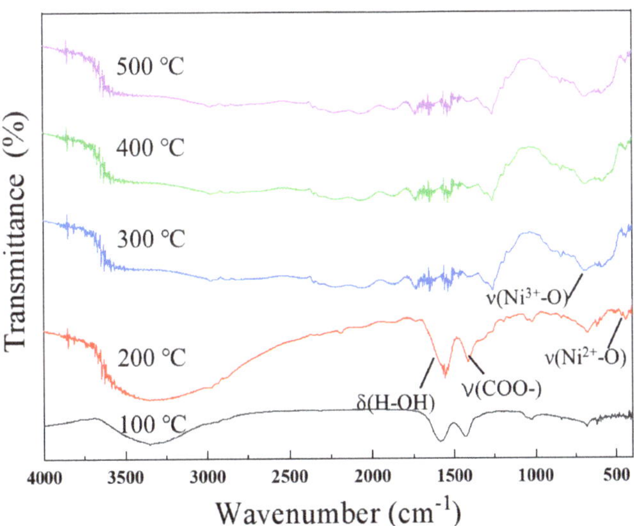

Figure 2. FTIR spectra of NiO films annealed at different temperatures.

The surface morphology of the films measured by AFM are shown in Figure 3a–e. Figure 3f shows a comparison of the roughness of the films. The surface of the NiO film annealed at 100 °C is the smoothest. This is because the solvent ethylene glycol methyl ether, with a boiling point of about 150 °C, still exists in the films. Probably due to the evaporation of the solvent at 200 °C, surface defects such as pores are formed and the sample annealed at 200 °C has a higher roughness than the films annealed at 100 °C. However, heat treatment at a higher temperature of 300 °C largely removes the defects and hence gives films with a reduced roughness, as shown in Figure 3c. As the heat treatment temperature is further increased, the roughness is increased again. This is mainly due to the increase in crystallinity and precipitation of relatively larger crystallites. This is partly consistent with the increased diffraction intensity, as shown in XRD observation (Figure 4).

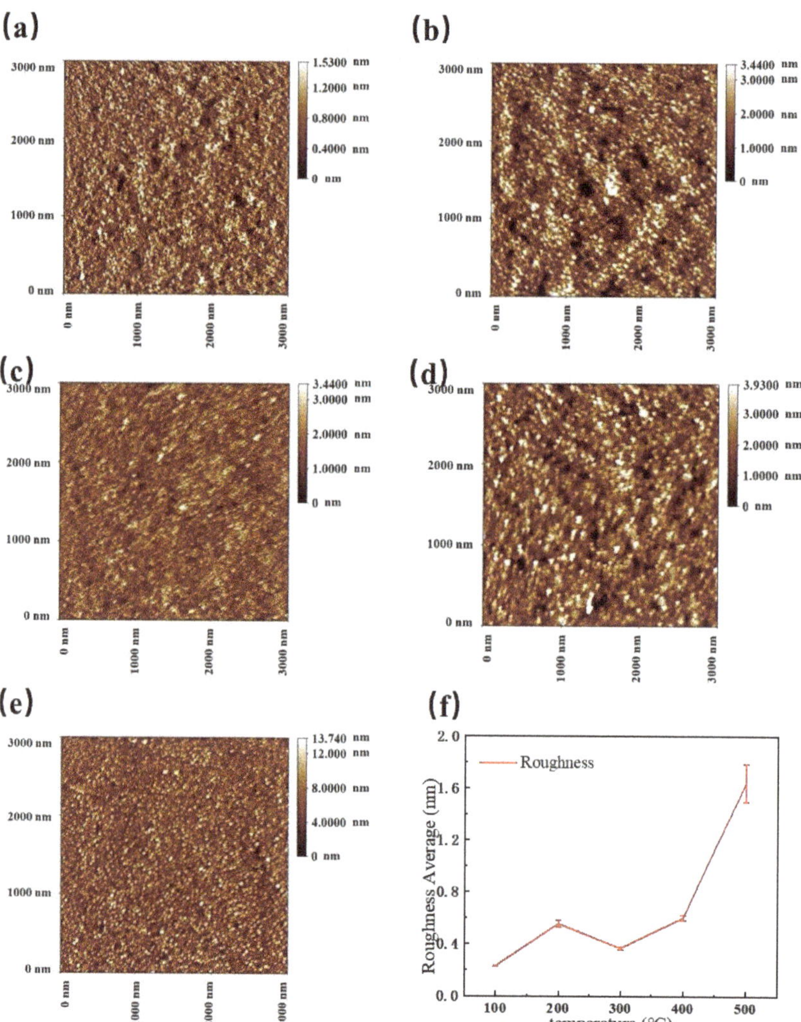

Figure 3. The atomic force microscope (AFM) images 3000 nm × 3000 nm and the roughness of NiO films. (**a**) 100 °C; (**b**) 200 °C; (**c**) 300 °C; (**d**) 400 °C; (**e**) 500 °C; (**f**) the roughness of NiO films, which are read by the support software of AFM.

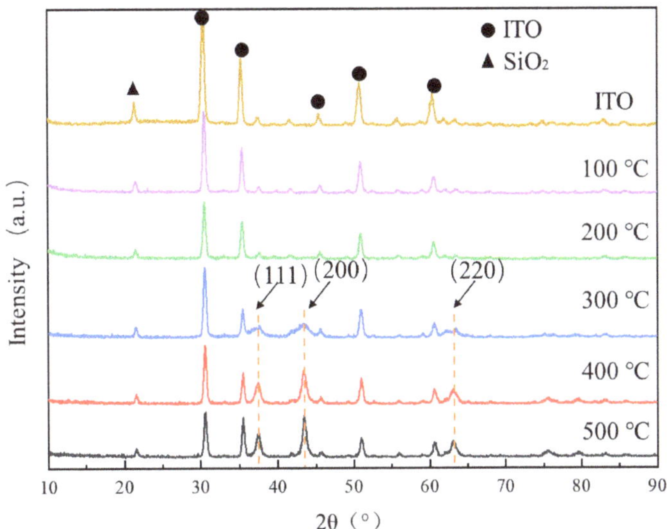

Figure 4. XRD patterns of NiO films annealed at different temperature.

Figure 4 shows the XRD patterns of films annealed at different temperatures and ITO substrates.

The crystal structure of these films was further analyzed by HighScore Plus. The XRD patterns of the samples after annealing at 100 °C and 200 °C show similar shapes with the XRD pattern of ITO glass. Based on the information obtained from the FTIR spectrum, it can be judged that nickel oxide has not yet formed at this stage. Furthermore, there are three diffraction peaks (111) (200) (220), which match with standard card of nickel oxide (Reference code: 01-071-1179 by HighScore Plus 3.0e) of NiO at the patterns of the NiO films annealed at 300 °C, 400 °C, and 500 °C, which demonstrate that these films began to crystalline when the annealing temperature is higher than 300 °C [25]. All three diffraction peaks show a broadening phenomenon, which is caused by insufficient crystallinity or too small crystallite size. As the annealing temperature increased, the three diffraction peaks at the patterns of the sample became sharper, which represent an increase in crystallinity and crystallite size.

In order to measure the degree of crystallinity, we calculated the crystallite size and lattice parameter in different crystal plane directions of NiO films annealed at different temperatures. The lattice parameter could be calculated by using the following equation:

$$a = \frac{\lambda}{2\sin\theta\sqrt{h^2 + k^2 + l^2}} \tag{1}$$

Crystallite size could be calculated by measuring the half-maximum value of the X-ray diffraction peak (FMHM) using the Debye-Scherrer formula [26]:

$$D = \frac{K\lambda}{\beta\cos\theta} \tag{2}$$

Among them, D is the average thickness (nm) of the crystal crystallites in the direction perpendicular to the crystal plane; K is the Scherrer constant, which is 0.89 in spherical particles, 0.94 in cubic particles, and nickel oxide crystals are cubic structures, so K here is 0.94; h, k and l are the Miller index of the corresponding crystal orientation; λ is the X-ray wavelength, 0.154056 nm; β is the half-width of the X-ray diffraction peak (FMHM), θ is the diffraction angle, and the values of β and θ are obtained by HighScore Plus 3.0e.

The lattice parameters of the nickel oxide films at different annealing temperatures in each crystal orientation were shown in Table 1.

Table 1. Lattice parameters of NiO films in different directions at different annealing temperatures.

Annealing Temperature (°C)	$a_{(111)}$ (nm)	$a_{(200)}$ (nm)	$a_{(220)}$ (nm)
300	0.4170	0.4166	0.4178
400	0.4164	0.4168	0.4171
500	0.4162	0.4167	0.4169

As the annealing temperature increases, the lattice constant in the (111) and (220) directions decrease (approximately from 0.417 nm to 0.416 nm), while the lattice constant in the (200) direction hardly changes (about 0.416 nm). The lattice constant of the nickel oxide crystal is about 0.416 nm [27,28], and the lattice constant of the (111) and (220) crystal directions of the nickel oxide annealed at 300 °C is slightly higher than this value, which may be caused by lattice distortion. As the annealing temperature increases, the defects in the crystal decrease, so the lattice parameters of each crystal orientation gradually approach 0.416 nm.

In Table 2, it is observed that the crystallite size increased with increasing annealing temperature. The crystallite boundaries between the crystallites can be regarded as a type of defect, and the annealing process improves the lattice energy and can reduce the defects, thereby promoting the growth of the crystal crystallites. Therefore, as the temperature increases, more energy is provided during the annealing process. The crystallinity of the crystallites increased, and the crystallites grew larger, which is consistent with the above analysis of the lattice parameters.

Table 2. Crystallite sizes of NiO films in different directions at different annealing temperatures.

Annealing Temperature (°C)	$D_{(111)}$ (nm)	$D_{(200)}$ (nm)	$D_{(220)}$ (nm)
300	5.40	6.30	7.72
400	10.12	11.34	13.01
500	13.11	16.21	14.54

The band gap of the NiO film can be measured and analyzed by an ultraviolet spectrophotometer. The optical band gap is distinguished from the band gap measured by other methods. According to Equation (3), the optical band gap can be calculated [29] as:

$$\alpha h v = A(h v - Eg)^{n} \tag{3}$$

where α is the absorption coefficient, which can be measured by the ultraviolet spectrophotometer; h is the Planck constant; n is the light frequency; A is a proportionality constant; Eg is the optical band gap; and n is a number which is 1/2 for the direct band gap semiconductor and 2 for the indirect band gap semiconductor. In this work, n is 1/2 because NiO was a direct semiconductor.

To further investigate the electrochromic effects on the optical band gap of the NiO film, the optical band gap of the NiO film in a bleached state and colored state were analyzed. The electrochromic test was performed on NiO with ITO as the cathode and anode in a 0.5 mol/L KOH solution, and the applied potential was ± 2.5 V. During the electrochromic coloration and bleaching processes, the following electrochemical reactions take place in NiO layers [24,29]:

$$(Bleached)\,NiO + OH^{-} \leftrightarrow (Colored)\,NiOOH + e^{-} \tag{4}$$

$$(Bleached)\,Ni(OH)_{2} \leftrightarrow (Colored)\,NiOOH + H^{+} + e^{-} \tag{5}$$

Figure 5a–c illustrate the relationship between $(\alpha h\nu)^2$ and photon energy $h\nu$. These curves are calculated using the transmission spectra of the NiO film in the colored state and the bleached state. For example, it can be extracted by the start of the optical transition of the NiO film near the band edge, which is equal to the value of the intercept of the fitted line. Figure 5d shows the comparison of the optical band gap values of NiO films annealed at different temperatures and electrochromic states (colored and bleached), indicating that as the annealing temperature increases, the Eg of the colored (initial state) NiO film changes; 3.65 eV increased to 3.74 eV, while the band gap of the bleached NiO film hardly changed.

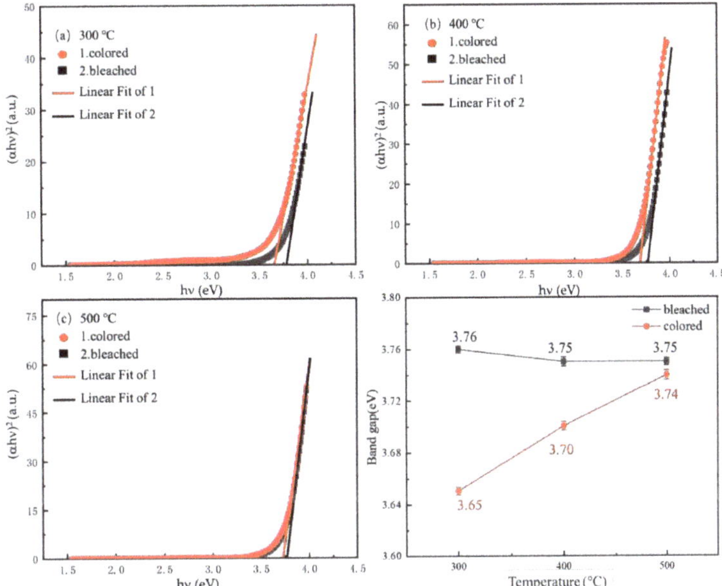

Figure 5. Optical band gap energy of NiO films in a colored state and bleached state. (**a**) 300 °C; (**b**) 400 °C; (**c**) 500 °C; and (**d**) a comparison of optical band gap energy of NiO films annealed at different temperatures and electrochromic states (colored and bleached).

Coloring efficiency (CE (η)) determines the amount of change in optical density (ΔOD) based on the injected/injected charge (Qi) at a specific wavelength, that is, the amount of charge required to maintain optical density. It is given by [30]:

$$\eta = \left(\frac{\Delta OD}{Qi}\right) = \left(\frac{\lg(T_b/T_c)}{Qi}\right) \tag{6}$$

The ΔOD of the thin film was calculated from the transmission spectrum and only films annealed at temperatures above 300 °C had electrochromic response. This can be explained from the test results of FTIR and XRD. Only after being sufficiently annealed can nickel acetate in the precursor be fully converted into nickel oxide to have an electrochromic response. Figure 6 shows the ΔOD—Wavelength variation curves of nickel oxide films with different annealing temperatures. As the annealing temperature increases, the ΔOD of the films decreases significantly. The films annealed at 300 °C exhibit better optical modulation properties than those annealed at 400 °C and 500 °C. The change trend of ΔOD is also consistent with the change trend of the optical band gap, indicating that the change of the band gap is related to the optical change of the NiO film.

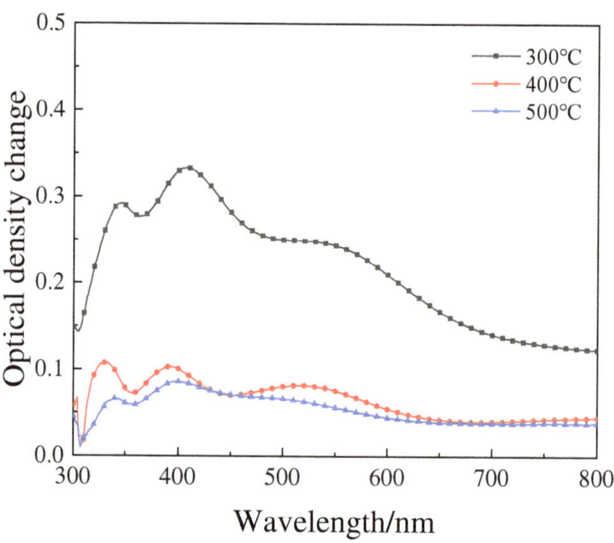

Figure 6. ΔOD of NiO thin films with different annealing temperatures as a function of wavelength.

The electrochemical performance of the film was also measured by an electrochemical workstation. Figure 7a shows the current density-time curves of NiO films with different annealing temperatures in three cycles. By integrating the curves, we can obtain the charge density-time curve (Figure 7b). With the increase in annealing temperature, the charge density in each electrochromic cycle of the film decreases significantly, which means that less electrochromic reactions occur, which can reasonably explain the aforementioned decrease in the light modulation ability of the film.

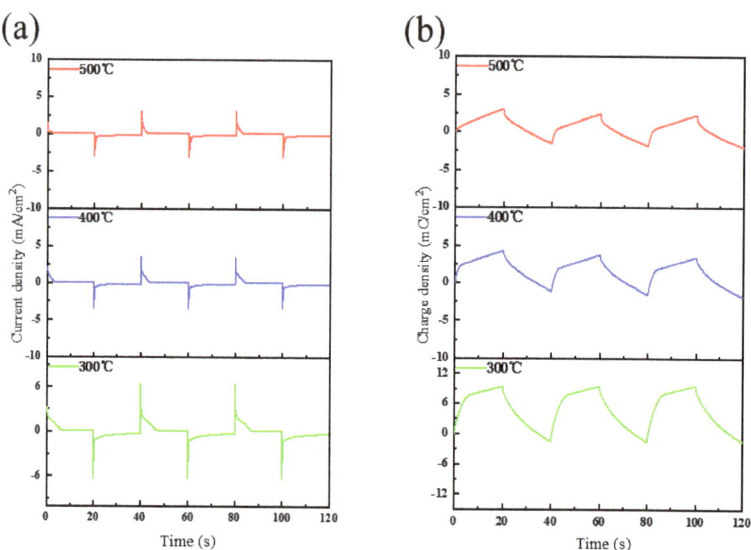

Figure 7. Current density curve (**a**) and charge density curve (**b**) of NiO films with different annealing temperatures.

The coloring efficiency of the film at 550 nm was further calculated by Equation (4), so as to discuss the relationship between the crystallite size and coloring efficiency. Figure 8 shows trends in crystallite size, charge density, and coloring efficiency of NiO films with different annealing temperatures. According to the current research, some researchers believe that the electrochromic reaction of nickel oxide crystals is thought to occur inside the crystal crystallites [31], while other researchers believe that it occurs at the interface of the crystal crystallites [32]. If the electrochromic reaction of nickel oxide occurs at the crystallite interface, within the same unit volume, smaller crystallites have a larger surface area and can react with more ions, so they have a larger charge density. This means that more electrochromic reactions occur, that is, better optical modulation capabilities [33]. A larger surface area also means more efficient ion implantation, that is, higher electrochromic efficiency. By analyzing the change trend between the crystallite size, charge density, and electrochromic efficiency of NiO thin films, as the crystallite size increases, both the charge density and electrochromic efficiency decrease. The discoloration reaction is more likely to occur at the interface of the crystallites.

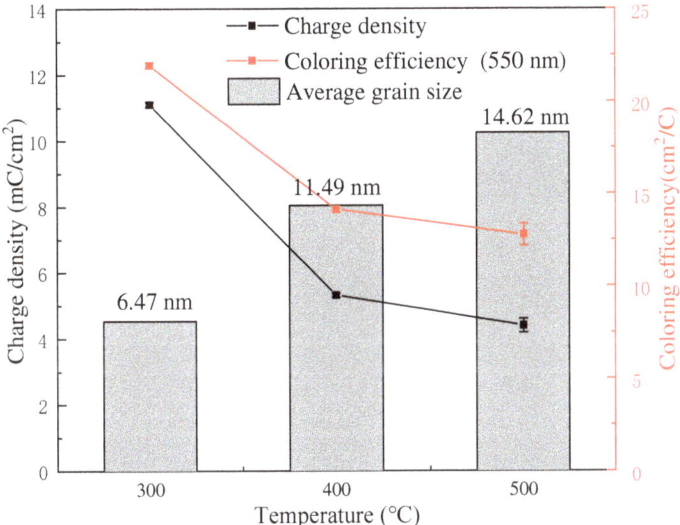

Figure 8. Trends in crystallite size, charge density, and coloring efficiency of NiO films with different annealing temperatures.

On the other hand, from the perspective of optical band gap analysis, the peak current density of these films in the coloring process decreased when the annealing temperature increased (a decrease from 6.2 mA/cm^2 at 300 °C to 2.9 mA/cm^2 at 500 °C). These were due to the decrease of Eg and the decrease of free electrons. Nickel oxide film with a smaller crystallite size has more electrons and ions implanted. When electrons and ions are injected into the nickel oxide film together, the same concentration of electrons and ions will be generated in the film, thereby reducing the optical band gap of the nickel oxide film. The reduction of the optical band gap increases the concentration of free carriers in the conduction band of nickel oxide, and the absorption and reflection of photons by free carriers causes changes in the optical properties of the film.

4. Conclusions

In this work, the effect of annealing temperature on the microstructure and performance of sol gel prepared NiO films were studied. When the annealing temperature is higher than 300 °C, the precursor nickel acetate is largely converted into nickel oxide with a considerable crystallinity and obvious electrochromic response. With the further increase of annealing temperature, the crystallinity film and crystallite size of NiO obviously increases. Analysis of the crystallite size, charge density, (ΔOD), optical band gap, and coloring efficiency of films prepared under different annealing temperatures indicates that the electrochromic reaction of nickel oxide is more likely to occur at the crystallite interface, and smaller crystallite size is more conducive to ion implantation. Therefore, the nickel oxide explosion with a smaller crystal crystallite size can inject more ions, so that the optical band gap of the nickel oxide film is reduced, and more carriers are generated, which results in a stronger optical performance change. Under the premise of forming nickel oxide film, the reduction of annealing temperature is conducive to the formation of smaller crystallite size. The nickel oxide film annealed at 300° C has a small crystal crystallite size (6.47 nm), which is conducive to ion implantation to change the optical band gap and has good electrochromic properties.

Author Contributions: Conceptualization, M.S.; Data curation, M.S., B.T., and G.Z.; Formal analysis, M.S., T.Q., W.X., X.F. and J.C.; Funding acquisition, H.N. and R.Y.; Investigation, X.F. and J.C.; Methodology, W.X., B.T. and G.Z.; Project administration, H.N., R.Y. and J.P.; Supervision, H.N.; Writing—original draft, M.S.; Writing—review & editing, H.N., T.Q. and J.P. All authors have read and agreed to the published version of the manuscript.

Funding: This work was supported by the Key-Area Research and Development Program of Guangdong Province (No.2020B010183002), National Natural Science Foundation of China (Grant No.51771074, 62074059 and 22090024), Guangdong Major Project of Basic and Applied Basic Research (No.2019B030302007), Guangdong Natural Science Foundation (No.2018A0303130211), Fundamental Research Funds for the Central Universities (No. 2020ZYGXZR060 and 2019MS012), Ji Hua Laboratory scientific research project (X190221TF191), the Guangdong Provincial Key Laboratory of Optical Information Materials and Technology (2017B030301007), Sail Plan Special Innovative Entrepreneurial Teams in Guangdong Province (2015YT02C093), South China University of Technology 100 Step Ladder Climbing Plan Research Project (No.j2tw202004035, j2tw202004034 and j2tw202004095), National College Students' Innovation and Entrepreneurship Training Program (No.202010561001, 202010561004 and 202010561009) and 2020 Guangdong University Student Science and Technology Innovation Special Fund ("Climbing Plan" Special Fund).

Conflicts of Interest: The authors declare no conflict of interest.

References

1. Malik, R.; Tomer, V.K.; Mishra, Y.K.; Lin, L. Functional Gas Sensing Nanomaterials: A Panoramic View. *Appl. Phys. Rev.* **2020**, *7*, 021301. [CrossRef]
2. Singh, A.; Ranjan, R.; Ranjan, S.; Singh, A.; Garg, A.; Gupta, R.K. Effect of NIO Precursor Solution Ageing on the per-Ovskite Film Formation and Their Integration as Hole Transport Material for Perovskite Solar Cells. *J. Nanosci. Nanotechnol.* **2020**, *20*, 3710–3717. [CrossRef] [PubMed]
3. Saki, Z.; Sveinbjörnsson, K.; Boschloo, G.; Taghavinia, N. The Effect of Lithium Doping in Solution-Processed Nickel Oxide Films for Perovskite Solar Cells. *ChemPhysChem* **2019**, *20*, 3322–3327. [CrossRef]
4. Sato, H.; Minami, T.; Takata, S.; Yamada, T. Transparent Conducting P-Type NiO Thin Films Prepared by Magnetron Sputtering. *Thin Solid Films* **1993**, *236*, 27–31. [CrossRef]
5. Nishiyama, H.; Saito, N.; Chou, H.; Sato, K.; Inoue, Y. Effects of Surface Acoustic Waves on Adsorptive Properties of ZnO and NiO Thin Films Deposited on Ferroelectric Substrates. *Surf. Sci.* **1999**, *433*, 525–528. [CrossRef]
6. Amirzhanova, A.; Karakaya, I.; Uzundal, C.B.; Karaoğlu, G.; Karadas, F.; Ulgut, B.; Dag, Ö. Synthesis and Water Oxidation Electrocatalytic and Electrochromic Behaviours of Mesoporous Nickel Oxide Thin Film Electrodes. *J. Mater. Chem. A* **2019**, *7*, 22012–22020. [CrossRef]
7. Atak, G.; Coşkun, Ö.D. Effects of Anodic Layer Thickness on Overall Performance of All-Solid-State Electrochromic Device. *Solid State Ionics* **2019**, *341*, 115045. [CrossRef]
8. Zhu, Y.; Xie, L.; Chang, T.; Bell, J.; Huang, A.; Jin, P.; Bao, S. High Performance All-Solid-State Electrochromic Device Based on LixNiOy Layer with Gradient LI Distribution. *Electrochim. Acta* **2019**, *317*, 10–16. [CrossRef]
9. Granqvist, C.G. Electrochromic Devices. *J. Eur. Ceram. Soc.* **2005**, *25*, 2907–2912. [CrossRef]
10. Granqvist, C.G. *Handbook of Inorganic Electrochromic Materials*; Elsevier: Amsterdam, The Netherlands, 1995.
11. Danine, A.; Manceriu, L.M.; Faure, C.; Labrugère, C.; Penin, N.; Delattre, A.; Eymin-Petot-Tourtollet, G.; Rougier, A. Toward Simplified Electrochromic Devices Using Silver as Counter Electrode Material. *ACS Appl. Mater. Interfaces* **2019**, *11*, 34030–34038. [CrossRef]
12. Oukassi, S.; Giroud-Garampon, C.; Dubarry, C.; Ducros, C.; Salot, R. All Inorganic Thin Film Electrochromic Device Using LiPON as the Ion Conductor. *Sol. Energy Mater. Sol. Cells* **2016**, *145*, 2–7. [CrossRef]

13. Cai, G.; Darmawan, P.; Cui, M.; Chen, J.; Wang, X.; Eh, A.L.-S.; Magdassi, S.; Lee, P.S. Inkjet-Printed All Solid-State Electrochromic Devices Based on NiO/WO3 Nanoparticle Complementary Electrodes. *Nanoscale* **2016**, *8*, 348–357. [CrossRef] [PubMed]
14. Ganesh, V.; Haritha, L.; Anis, M.; Shkir, M.; Yahia, I.; Singh, A.; AlFaify, S. Structural, Morphological, Optical and Third Order Nonlinear Optical Response of Spin-Coated NiO Thin Films: An Effect of N Doping. *Solid State Sci.* **2018**, *86*, 98–106. [CrossRef]
15. Xie, Z.; Liu, Q.; Zhang, Q.; Lu, B.; Zhai, J.; Diao, X. Fast-Switching Quasi-Solid State Electrochromic Full Device Based on Mesoporous WO3 and NiO Thin Films. *Sol. Energy Mater. Sol. Cells* **2019**, *200*, 110017. [CrossRef]
16. Huang, Q.; Dong, G.; Xiao, Y.; Diao, X. Electrochemical Studies of Silicon Nitride Electron Blocking Layer for All-Solid-State Inorganic Electrochromic Device. *Electrochim. Acta* **2017**, *252*, 331–337. [CrossRef]
17. Huang, Q.J.; Zhang, Q.Q.; Xiao, Y.; He, Y.C.; Diao, X.G. Improved Electrochromic Performance of NiO-Based Thin Films by Lithium and Tantalum Co-doping. *J. Alloy. Compd.* **2018**, *747*, 416–422. [CrossRef]
18. Choi, D.S.; Han, S.H.; Kim, H.; Kang, S.H.; Kim, Y.; Yang, C.M.; Kim, T.Y.; Yoon, D.H.; Yang, W.S. Flexible Electro-Chromic Films Based on Cvd-Graphene Electrodes. *Nanotechnology* **2014**, *25*, 7.
19. Jaing, C.C.; Tang, C.J.; Chan, C.C.; Lee, K.H.; Kuo, C.C.; Chen, H.C.; Lee, C.C. Optical Constants of Electrochromic Films and Contrast Ratio of Reflective Electrochromic Devices. *Appl. Opt.* **2014**, *53*, A154–A158. [CrossRef]
20. Azevedo, C.F.; Balboni, R.D.; Cholant, C.M.; Moura, E.; Lemos, R.M.; Pawlicka, A.; Gündel, A.; Flores, W.H.; Pereira, M.; Avellaneda, C.O. New Thin Films of NiO Doped with V_2O_5 for Electrochromic Applications. *J. Phys. Chem. Solids* **2017**, *110*, 30–35. [CrossRef]
21. Zhang, G.; Lu, K.; Zhang, X.; Yuan, W.; Shi, M.; Ning, H.; Tao, R.; Liu, X.; Yao, R.; Peng, J. Effects of Annealing Temperature on Optical Band Gap of Sol-gel Tungsten Trioxide Films. *Micromachines* **2018**, *9*, 377. [CrossRef]
22. Korošec, R.C.; Bukovec, P. The Role of Thermal Analysis in Optimization of the Electrochromic Effect of Nickel Oxide Thin Films, Prepared by the Sol–Gel Method: Part II. *Thermochim. Acta* **2004**, *410*, 65–71. [CrossRef]
23. Patra, A.; Auddy, K.; Ganguli, D.; Livage, J.; Biswas, P.K. Sol–Gel Electrochromic WO3 Coatings on Glass. *Mater. Lett.* **2004**, *58*, 1059–1063. [CrossRef]
24. Ren, Y.; Chim, W.K.; Guo, L.; Tanoto, H.; Pan, J.; Chiam, S.Y. The Coloration and Degradation Mechanisms of Electrochromic Nickel Oxide. *Sol. Energy Mater. Sol. Cells* **2013**, *116*, 83–88. [CrossRef]
25. Noh, S.; Lee, E.; Seo, J.; Mehregany, M. Electrical Properties of Nickel Oxide Thin Films for Flow Sensor Application. *Sens. Actuators A Phys.* **2006**, *125*, 363–366. [CrossRef]
26. Atak, G.; Coşkun, Ö.D. Annealing Effects of NiO Thin Films for All-Solid-State Electrochromic Devices. *Solid State Ionics* **2017**, *305*, 43–51. [CrossRef]
27. Chem, A. Joint Committee on Powder Diffraction Standards. *Anal. Chem.* **1973**, *45*, 944A.
28. Zou, Y.; Zhang, Y.; Lou, D.; Wang, H.; Gu, L.; Dong, Y.; Dou, K.; Song, X.; Zeng, H. Structural and Optical Properties of WO3 Films Deposited by Pulsed Laser Deposition. *J. Alloy. Compd.* **2014**, *583*, 465–470. [CrossRef]
29. Carpenter, M.K.; Conell, R.S.; Corrigan, D.A. The Electrochromic Properties of Hydrous Nickel Oxide. *Sol. Energy Mater.* **1987**, *16*, 333–346. [CrossRef]
30. Dalavi, D.S.; Devan, R.S.; Patil, R.S.; Ma, Y.-R.; Patil, P.S. Electrochromic Performance of Sol–Gel Deposited NiO thin Film. *Mater. Lett.* **2013**, *90*, 60–63. [CrossRef]
31. Agrawal, A.; Habibi, H.R.; Agrawal, R.K.; Cronin, J.P.; Roberts, D.M.; Caron-Popowich, R.; Lampert, C.M. Effect of Deposition Pressure on the Microstructure and Electrochromic Properties of Electron-Beam-Evaporated Nickel Oxide Films. *Thin Solid Film.* **1992**, *221*, 239–253. [CrossRef]
32. Yoshimura, K.; Miki, T.; Tanemura, S. Nickel Oxide Electrochromic Thin Films Prepared by Reactive DC Magnetron Sputtering. *Jpn. J. Appl. Phys.* **1995**, *34*, 2440–2446. [CrossRef]
33. Granqvist, C. Window Coatings for the Future. *Thin Solid Film.* **1990**, *193*, 730–741. [CrossRef]

 micromachines

MDPI

Article

Bias Stress Stability of Solution-Processed Nano Indium Oxide Thin Film Transistor

Rihui Yao [1], Xiao Fu [1], Wanwan Li [1], Shangxiong Zhou [1], Honglong Ning [1,*], Biao Tang [2], Jinglin Wei [1], Xiuhua Cao [3,*], Wei Xu [1] and Junbiao Peng [1]

1. State Key Laboratory of Luminescent Materials and Devices, Institute of Polymer Optoelectronic Materials and Devices, South China University of Technology, Guangzhou 510640, China; yaorihui@scut.edu.cn (R.Y.); 201630343721@mail.scut.edu.cn (X.F.); mswanwanli@mail.scut.edu.cn (W.L.); 201820117973@mail.scut.edu.cn (S.Z.); magicwei@foxmail.com (J.W.); xuwei@scut.edu.cn (W.X.); psjbpeng@scut.edu.cn (J.P.)
2. Guangdong Provincial Key Laboratory of Optical Information Materials and Technology & Institute of Electronic Paper Displays, South China Academy of Advanced Optoelectronics, South China Normal University, Guangzhou 510006, China; biao.tang@guohua-oet.com
3. State Key Laboratory of Advanced Materials and Electronic Components, Fenghua Electronic Industrial Park, No. 18 Fenghua Road, Zhaoqing 526020, China
* Correspondence: ninghl@scut.edu.cn (H.N.); xiuhuacao@126.com (X.C.); Tel.: +86-20-8711-4525 (H.N.)

Abstract: In this paper, the effects of annealing temperature and other process parameters on spin-coated indium oxide thin film transistors (In_2O_3-TFTs) were studied. The research shows that plasma pretreatment of glass substrate can improve the hydrophilicity of glass substrate and stability of the spin-coating process. With Fourier transform infrared (FT-IR) and X-ray diffraction (XRD) analysis, it is found that In_2O_3 thin films prepared by the spin coating method are amorphous, and have little organic residue when the annealing temperature ranges from 200 to 300 °C. After optimizing process conditions with the spin-coated rotating speed of 4000 rpm and the annealing temperature of 275 °C, the performance of In_2O_3-TFTs is best (average mobility of 1.288 $cm^2 \cdot V^{-1} \cdot s^{-1}$, I_{on}/I_{off} of 5.93 × 10^6, and SS of 0.84 $V \cdot dec^{-1}$). Finally, the stability of In_2O_3-TFTs prepared at different annealing temperatures was analyzed by energy band theory, and we identified that the elimination of residual hydroxyl groups was the key influencing factor. Our results provide a useful reference for high-performance metal oxide semiconductor TFTs prepared by the solution method.

Keywords: indium oxide thin film; solution method; plasma surface treatment; annealing temperature; bias stability

Citation: Yao, R.; Fu, X.; Li, W.; Zhou, S.; Ning, H.; Tang, B.; Wei, J.; Cao, X.; Xu, W.; Peng, J. Bias Stress Stability of Solution-Processed Nano Indium Oxide Thin Film Transistor. *Micromachines* **2021**, *12*, 111. https://doi.org/10.3390/mi12020111

Academic Editor: Giovanni Verzellesi

Received: 28 December 2020
Accepted: 19 January 2021
Published: 22 January 2021

Publisher's Note: MDPI stays neutral with regard to jurisdictional claims in published maps and institutional affiliations.

Copyright: © 2021 by the authors. Licensee MDPI, Basel, Switzerland. This article is an open access article distributed under the terms and conditions of the Creative Commons Attribution (CC BY) license (https://creativecommons.org/licenses/by/4.0/).

1. Introduction

With the active matrix liquid crystal display (AMLCD) and active matrix organic light-emitting diodes (AMOLED) gradually occupying the mainstream position in the display field [1–4], metal oxide thin film transistors (MOS-TFTs) have been widely studied due to their high mobility, high light transmittance, low processing temperature and low processing cost [3–9]. At present, metal oxide semiconductors are mainly fabricated by vacuum deposition methods, which have strict environmental requirements and relatively high manufacturing cost [10,11]. In contrast, solution-processed deposition offers the advantages of a simple process, high-throughput, high material utilization rate, and easy control of chemical components, which provides the possibility for large-area preparation of metal oxide semiconductor [12–17]. The preparation of metal oxides by the solution method usually requires annealing, which promotes the formation of (M–O–M) structure and the densification of film [18]. Solution methods mainly include the spin-coating method, solvothermal method, microwave assisted growth method, sonochemical method, hydrothermal method, electrodeposition method and so on [19–25]. Among them, the spin-coating method has the advantages of low cost, low pollution, energy saving and low film thickness, but it also has the disadvantages of uneven film thickness and waste of solution.

Nowadays, metal oxide semiconductor materials with $(n-1)d^{10}ns^0$ $(n \geq 4)$ electronic configurations have attracted much attention due to their good electrical properties in the amorphous phase, and indium oxide (In_2O_3) is one of them [26,27]. This meets the requirements of low temperature preparation in the field of flexible display. In_2O_3 has been widely studied in the preparation of TFT active layer due to its wide band gap, high mobility, high carrier concentration and good transparency [1,28–31], but there is still little research on the influence mechanism of amorphous In_2O_3 electrical stability [32].

In this paper, In_2O_3 thin films were prepared by spin coating. The spreading property of precursor solution was improved by plasma surface treatment [33]. The phase composition of In_2O_3 thin films was investigated by different annealing methods. On this basis, In_2O_3-TFTs were fabricated on Al electrodes and Si_3N_4 substrates and the electrical characteristics of devices under different process conditions were analyzed. Furthermore, the possible methods to further improve the performance of TFTs prepared by spin coating were explored.

2. Materials and Methods

A semiconductor precursor solution was prepared using Indium nitrate hydrate ($In(NO_3)_3 \cdot 5H_2O$, CAS No.: 13465-14-0), in a solvent of ethylene glycol monomethyl ether (2-MOE, CAS No.: 109-86-4). The precursor solution was stirred in a magnetic mixer for 30 min and filtered, and then ultrasonic treatment was performed for 10 min. After plasma surface treatment, 35 µL of In_2O_3 precursor solution was added dropwise to the single-crystal silicon (CAS No.: 7440-21-3), and then spin coated with a homogenizer (model: KW-4A) at the speed of 4000 rms for 30 s. In the multilayer spin-coating process, every single-layer was pre-annealed at 100 °C for 10 min to evaporate the solvent. Finally, the In_2O_3 films were annealed at 200 °C, 225 °C, 250 °C, 275 °C and 300 °C for 45 min respectively. For the fabrication of In_2O_3-TFTs, bottom-gate devices were fabricated on Al (CAS No.: 7429-90-5) gate electrode and Si_3N_4 (CAS No.: 27198-71-6) dielectric layer. After the deposition of In_2O_3 thin films, an Al source and drain electrodes were formed by magnetron sputtering through a shadow mask. The schematic structure of the In_2O_3-TFTs is shown in Figure 1.

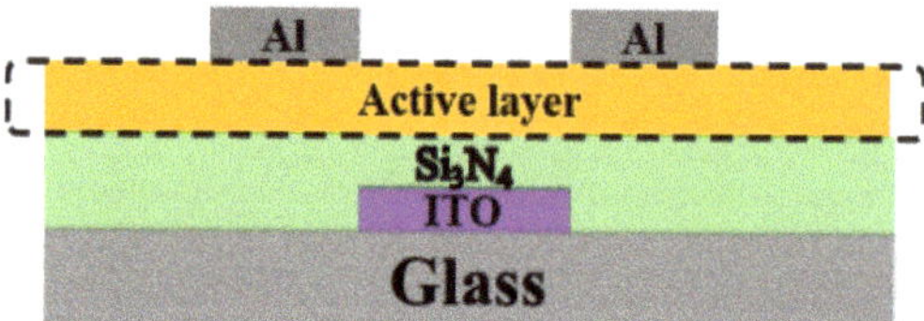

Figure 1. Schematic of indium oxide thin film transistors (In_2O_3-TFTs).

The surface tension of In_2O_3 solutions were measured by an Attension Theta Lite (TL200, Biolin Scientific, Gothenburg, Sweden). The internal chemical composition of In_2O_3 thin films was observed by Fourier transform infrared spectroscopy (FT-IR) (IRprestige21, Shimadzu, Kyoto, Janpan). X-ray reflectivity (XRR) (EMPYREAN, PANalytical, Almelo, The Netherlands) was used to analyze the thickness of the films. X-ray diffraction (XRD) (EMPYREAN, PANalytical, Almelo, The Netherlands) was used to analyze the phase of the films with Cu-kα as the X-ray source, and the scanning speed is $0.1° \cdot s^{-1}$ from 20° to 70°. The surface morphology of the films was measured by atomic force microscopy (AFM) (Being Nano-Instruments BY3000, Being Nano-Instruments, Beijing, China). Semiconductor parameter analyzers (Agilent 4155c, Agilent, Santa Clara, CA, USA) was used under an ambient atmosphere to evaluate the electrical characteristics of TFTs.

3. Results and Discussion

3.1. Effect of Plasma Surface Treatment on Solution Spreading

Figure 2a,b show the spread situation of In_2O_3 precursor solutions without/with plasma surface treatment. It can be found that it is difficult for the In_2O_3 precursor solution to spread uniformly on the substrate without plasma treatment. They agglomerate into many small droplets on the substrate, so the uniformity of the film is poor. However, on the substrate treated by plasma, the precursor solution of In_2O_3 can be spread uniformly without agglomeration, which indicates that plasma surface treatment can significantly improve the film-forming ability of spin coating. The results characterized by Attension Theta Lite show that the surface tension of $In(NO_3)_3$ solution is 39 mN·m^{-1}, while that of 2-MOE is 27.6 mN·m^{-1}, this difference easily leads to the spontaneous agglomeration of micro droplets after spin coating, so the film-forming ability is poor; however, plasma treatment can effectively reduce the difference of surface tension by introducing polarization groups into the surface of the substrate, thus improving the hydrophilicity of the substrate surface, which can improve the uniformity of spin-coating films [34–36]. Based on this result, the subsequent films and TFTs are prepared on substrates with plasma surface treatment.

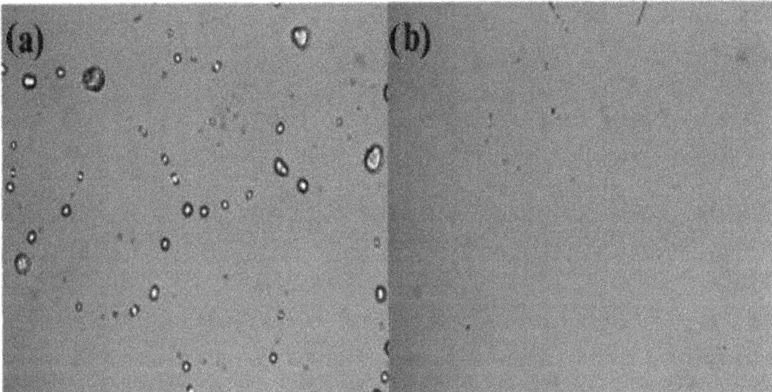

Figure 2. Spreading situation of precursor solution: (**a**) without plasma surface treatment; (**b**) with plasma surface treatment.

3.2. Effect of Annealing Temperature on In_2O_3 Thin Films and TFTs

Figure 3 shows the FT-IR test curves of In_2O_3 thin films prepared at different annealing temperatures. The absorption peak located at 2750 cm^{-1} to 3750 cm^{-1} is the stretching vibration of O–H bond [37], The absorption peak at 1250 cm^{-1} to 1750 cm^{-1} were caused by the bending vibration of the carbon–hydrogen bond and carbon–oxygen bond [38]. The absorption peak at 500–700 cm^{-1} can be attributed to the stretching vibration of the In–O bond. The O-H bond mainly comes from $[In\,(OH)]_{(3-x)}{}^+$, and the hydrolysis reaction of precursor is as follows [18]:

$$In^{3+} + xH_2O \rightarrow [In(OH)]^{(3-x)+} + xH^+ \tag{1}$$

where x is the stoichiometric number.

Excessive concentration of such functional groups easily causes adverse effects on the electrical properties of TFTs, such as high leakage current. It can be seen from Figure 3 that with the increase of annealing temperature, the vibration peak intensity of the O–H bond decreases significantly, which is almost zero at 300 °C. Based on the discussion above, an annealing temperature higher than 250 °C is necessary for promoting metal-oxide bond formation. Combined with the process requirements of device preparation, such as preventing the defects from the hillock of the Al electrode, the appropriate annealing temperature range of In_2O_3 thin films is 250 °C to 300 °C.

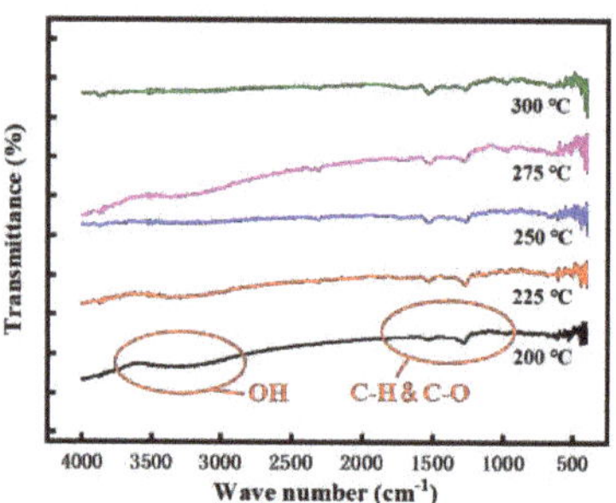

Figure 3. Fourier transform infrared (FT-IR) spectroscopy curves of In_2O_3 thin films annealed at different temperatures.

To investigate the influence of the annealing temperature on the surface morphology and thickness of In_2O_3 thin films, AFM and XRR was used. The AFM scanning area of the images was 4.0 μm × 4.0 μm. Figure 4a–e shows AFM images of In_2O_3 thin films annealed at different temperature, and it can be seen that all samples show a relatively smooth morphology without cracks with low roughness (below 0.4 nm). At the same time, the thickness of In_2O_3 films at different annealing temperatures are in the range of 3–7 nm, which indicates that they are all nano-scale films.

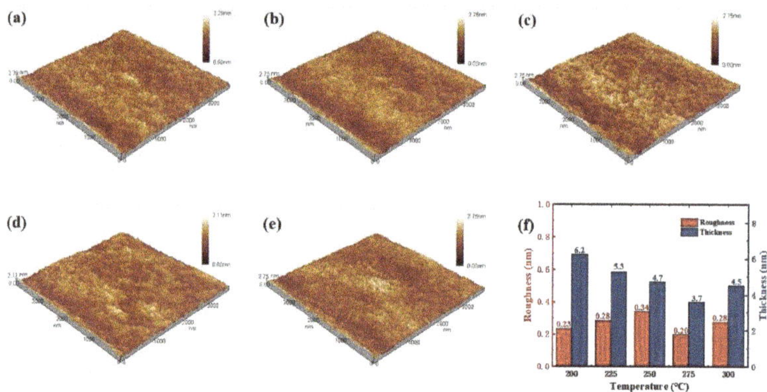

Figure 4. Atomic force microscopy (AFM) 3D images (4.0 × 4.0 μm²) of In_2O_3 thin films annealed at different temperatures: (**a**) 200 °C, (**b**) 225 °C, (**c**) 250 °C, (**d**) 275 °C, (**e**) 300 °C, respectively. (**f**) The surface roughness and thickness of these films.

In our previous studies, it was found that In_2O_3 thin films crystallized only at temperatures above 400 °C [17]. Therefore, we only studied the crystallization of In_2O_3 thin films at 200–300 °C in this work. It can be seen from Figure 5 that as the annealing temperature increases, a diffraction peak related to cubic In_2O_3 gradually appears in the film around 31°, which indicates an increased transformation from amorphous phase into crystalline phase. According to the test results, when the temperature is 275 °C and below, the diffraction peak has a large full width at half maxima (FWHM), weak intensity, and extremely low crystallinity, which indicates that the In_2O_3 thin films prepared by the spin coating can maintain an amorphous structure at this annealing temperature. The amorphous In_2O_3 thin films can achieve better flatness and uniformity than crystalline films, and is conducive to the control of carrier concentration [38].

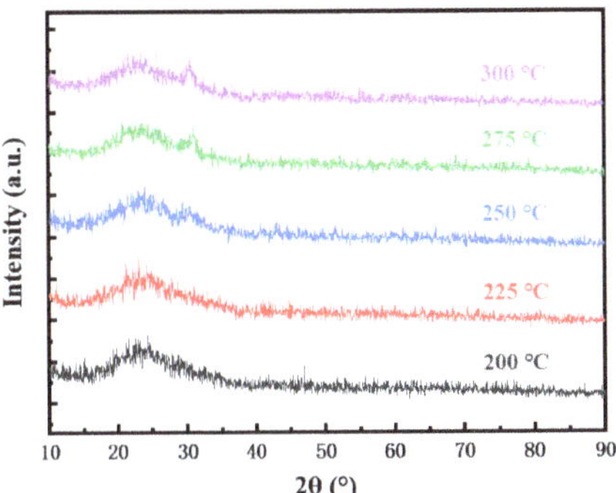

Figure 5. X-ray diffraction (XRD) curves of In_2O_3 annealed at different temperatures.

In$_2$O$_3$-TFTs were prepared under the optimized conditions. Figure 6 shows the transfer I–V characteristics of In$_2$O$_3$-TFTs at different annealing temperatures, and their corresponding electrical characteristics are listed in Table 1. It can be seen from Figure 6 that the devices had a certain negative shift and a large hysteresis, which indicated that there were many defects in the active layer. Combined with the test results in Figure 3, it may be the residual OH$^-$ in the In$_2$O$_3$ films. Another possible factor for the results is that H+ in the Si$_3$N$_4$ gate insulating layer. With the increase of annealing temperature, the hydroxyl group gradually decomposes, and the hysteresis phenomenon gradually weakens [39]. When the annealing temperature is 275 °C, the threshold voltage (V$_{th}$) difference between forward scanning and reverse scanning is the smallest, and the In$_2$O$_3$-TFTs showed a V$_{th}$ of 0.84 V, an I$_{on}$/I$_{off}$ ratio of 5.93×10^6, which was ideal.

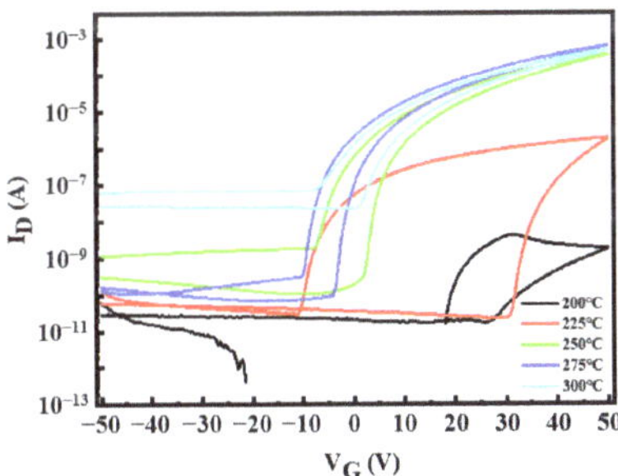

Figure 6. Transfer current-voltage (I-V) characteristics of In$_2$O$_3$-TFTs at different annealing temperature.

Table 1. Properties of In$_2$O$_3$-TFTs at different annealing temperature.

Temperature (°C)	V$_{th}$ (V)	I$_{on}$/I$_{off}$	μ_{sat} (cm^2·V^{-1}·s^{-1})	SS (V·dec^{-1})
200	-	2.83×10^2	-	-
225	39.67	6.70×10^4	-	-
250	6.46	3.00×10^5	0.837	1.77
275	0.84	5.93×10^6	1.288	1.03
300	−2.93	7.27×10^3	1.099	2.69

It can be seen from Table 1 that the V$_{th}$ of In$_2$O$_3$-TFTs gradually negative shift with the increase of annealing temperature, which is the minimum (0.84 V) at 275 °C. Also, the I$_{on}$/I$_{off}$ first increased to a peak value (5.93×10^6) at 275 °C, and then decreased. Furthermore, the saturation mobility (μ_{sat}) and subthreshold swing (SS) reached the maximum

($1.288 \text{ cm}^2 \cdot \text{v}^{-1} \cdot \text{s}^{-1}$) and the minimum ($1.030 \text{ V} \cdot \text{dec}^{-1}$) at 275 °C respectively. The performance is similar to that of In_2O_3-TFTs prepared by Choi at 280 °C (μ_{sat} of $2.4 \text{ cm}^2 \cdot \text{v}^{-1} \cdot \text{s}^{-1}$, I_{on}/I_{off} of 10^6) [40]. According to the FT-IR results, the low carrier concentration inside the In_2O_3 films at low annealing temperature may be due to the existence of undecomposed metal hydroxides in the active layer in the form of various defects, leading to the low μ_{sat}, low I_{on}/I_{off}, large V_{th} and large SS. When the annealing temperature increased from 200 °C to 275 °C, the devices performance gradually improved. However, when the annealing temperature reached 300 °C, the carrier concentration in the In_2O_3 thin films was too high, which made the device unable to turn off normally, thus, the I_{on}/I_{off} was only 7.27×10^3. In addition, the continuous increase of temperature may lead to the degradation of the interface quality of the device, leading to the decrease of other performance.

3.3. Bias Stability of Indium Oxide Thin Film Transistors (In$_2$O$_3$-TFTs)

Figure 7 shows the transfer curves of In_2O_3-TFTs annealed at 275 °C under positive gate bias stress (PBS) and negative gate bias stress (NBS) with a drain-bias stress of $V_{DS} = 20$ V. During the test, a bias stress ($V_{GS} = -50$ V for NBS and $V_{GS} = 50$ V for PBS) was applied to the gate electrode for 5400 s. Figure 8 shows the energy band change of In_2O_3-TFTs under bias voltage. It can be seen from Figure 7a that the V_{th} under PBS drifts forward nearly 20 V. The possible reasons are as follows: (1) the carriers were trapped by the interface defects of In_2O_3/Si_3N_4; (2) the In_2O_3 back channel adsorbed water and oxygen in the environment, and oxygen atoms captured electrons [41]. In both cases, as shown in Figure 8a, the actual carrier concentration will be reduced, so a higher V_{th} is required to form the conductive channel.

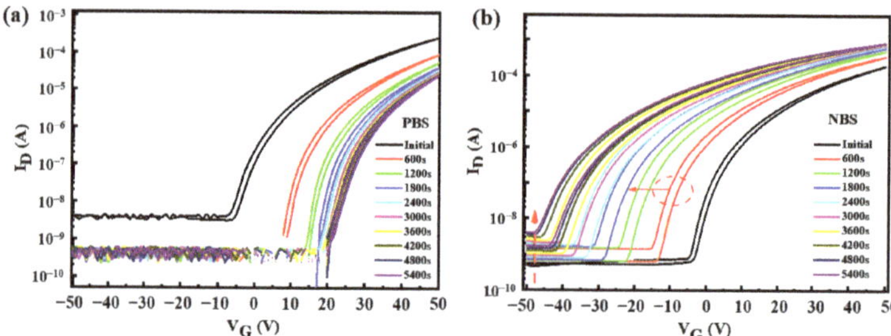

Figure 7. Transfer curves of In_2O_3-TFT annealed at 275 °C under (**a**) positive gate bias stress (PBS) and (**b**) negative gate bias stress (NBS). Measurement conditions: $V_{DS} = 20$ V at room temperature.

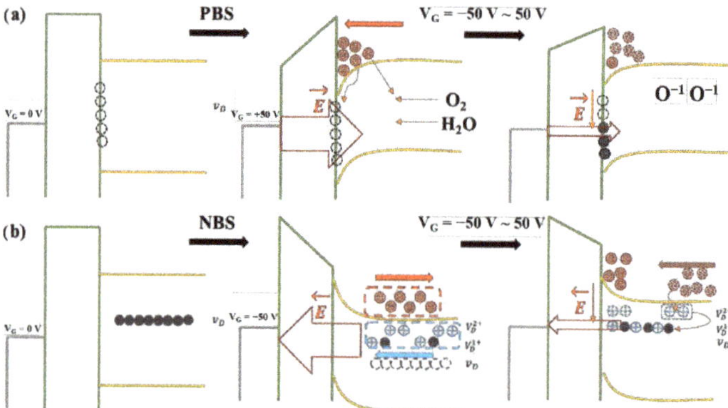

Figure 8. Plot of how about the energy band and carriers of In_2O_3-TFTs changing with the electric field under the (**a**) PBS and (**b**) NBS.

Similarly, V_{th} had a drift of -45 V under NBS, as shown in Figure 7b. In addition, with the increase of NBS time, the I_{off} increased and the hysteresis phenomenon became more obvious, which indicated that there were more defects in the In_2O_3 active layer, or the adsorption of water and oxygen was more serious. This was due to the activation of a large number of donors under NBS, such as impurities in the In_2O_3 films, and these electrons were removed from the active layer by the electric field. At the same time, the donors were positively charged after losing electrons, which were absorbed near the In_2O_3/Si_3N_4 interface. When the bias voltage changed from -50 V to 0 V, the electric field cannot remove all the electrons generated out of the channel. Also, most of the electrons were not easy to compound with donor-like vacancies, but directly formed channel current under voltage, resulting in a negative V_{th}, as shown in Figure 8b.

4. Conclusions

In this study, we fabricated solution-processed In_2O_3 thin films and TFTs, and investigated the factors affecting the stability of the devices. The results show that plasma treatment can significantly improve the spreading of In_2O_3 precursor solution on the substrate surface. The In_2O_3 films without annealing contain more organic residues, and they can be significantly reduced after annealing at 275 °C maintaining the amorphous film structure.

Through the study of the annealing characteristics of In_2O_3-TFTs, it was found that the devices prepared by the solution method have the characteristics of low active layer carrier concentration, high V_{th} and low I_{on}/I_{off} at low temperature. With the increase of annealing temperature, the electrical properties of the devices gradually improve. The optimal performance can be obtained after annealing at 275 °C, which exhibited a high μ_{sat} of 1.288 $cm^2 \cdot V^{-1} \cdot s^{-1}$, high I_{on}/I_{off} of 5.93×10^6, and low SS of 0.84 $V \cdot dec^{-1}$. Also, the bias stability is the best, which may be due to the reduction of organic residues and defects in the films. When further increasing the annealing temperature, the performance deteriorates, which may be due to the film interface degradation. This study provides a useful reference for the improvement and optimization of the performance of electronic devices prepared by the solution method.

Author Contributions: Conceptualization, R.Y. and W.X.; methodology, H.N. and W.L.; software, S.Z. and X.C.; validation, B.T. and J.W.; formal analysis, J.W. and S.Z.; investigation, X.F. and W.L.; resources, W.X.; data curation, X.C.; writing—original draft preparation, X.F.; writing—review and editing, X.F. and J.P.; visualization, R.Y.; supervision, J.P.; project administration, H.N.; funding acquisition, H.N. All authors have read and agreed to the published version of the manuscript.

Funding: This research was funded by Key-Area Research and Development Program of Guangdong Province (No.2020B010183002), National Natural Science Foundation of China (Grant No.51771074, 62074059 and 22090024), Guangdong Major Project of Basic and Applied Basic Research (No.2019B030302007), Fundamental Research Funds for the Central Universities (No. 2020ZYGXZR060 and 2019MS012), Ji Hua Laboratory scientific research project (X190221TF191), South China University of Technology 100 Step Ladder Climbing Plan Research Project (No.j2tw202004035, j2tw202004034 and j2tw202004095), National College Students' Innovation and Entrepreneurship Training Program (No.202010561001, 202010561004 and 202010561009) and 2020 Guangdong University Student Science and Technology Innovation Special Fund ("Climbing Plan" Special Fund).

Conflicts of Interest: The authors declare no conflict of interest.

References

1. Twyman, N.M.; Tetzner, K.; Anthopoulos, T.D.; Payne, D.J.; Regoutz, A. Rapid photonic curing of solution-processed In_2O_3 layers on flexible substrates. *Appl. Surf. Sci.* **2019**, *479*, 974–979. [CrossRef]
2. Chen, R.; Lan, L. Solution-processed metal-oxide thin-film transistors: A review of recent developments. *Nanotechnology* **2019**, *30*, 312001. [CrossRef]
3. Lee, S.; Shin, J.; Jang, J. Top Interface Engineering of Flexible Oxide Thin-Film Transistors by Splitting Active Layer. *Adv. Funct. Mater.* **2017**, *27*, 1604921. [CrossRef]

4. Xu, W.; Long, M.; Zhang, T.; Liang, L.; Cao, H.; Zhu, D.; Xu, J. Fully solution-processed metal oxide thin-film transistors via a low-temperature aqueous route. *Ceram. Int.* **2017**, *43*, 6130–6137. [CrossRef]

5. Nomura, K.; Ohta, H.; Takagi, A.; Kamiya, T.; Hirano, M.; Hosono, H. Room-temperature fabrication of transparent flexible thin-film transistors using amorphous oxide semiconductors. *Nat. Cell Biol.* **2004**, *432*, 488–492. [CrossRef]

6. Park, J.S.; Maeng, W.J.; Kim, H.S.; Park, J.S. Review of recent developments in amorphous oxide semiconductor thin-film transistor devices. *Thin Solid Films* **2012**, *520*, 1679–1693. [CrossRef]

7. Myny, K. The development of flexible integrated circuits based on thin-film transistors. *Nat. Electron.* **2018**, *1*, 30–39. [CrossRef]

8. Nomura, K.; Ohta, H.; Ueda, K.; Kamiya, T.; Hirano, M.; Hosono, H. Thin-Film Transistor Fabricated in Single-Crystalline Transparent Oxide Semiconductor. *Science* **2003**, *300*, 1269–1272. [CrossRef] [PubMed]

9. Karnaushenko, D.; Münzenrieder, N.; Karnaushenko, D.D.; Koch, B.; Meyer, A.K.; Baunack, S.; Petti, L.; Tröster, G.; Makarov, D.; Schmidt, O.G. Biomimetic Microelectronics for Regenerative Neuronal Cuff Implants. *Adv. Mater.* **2015**, *27*, 6797–6805. [CrossRef] [PubMed]

10. Huang, G.; Duan, L.; Dong, G.; Zhang, D.; Qiu, Y. High-Mobility Solution-Processed Tin Oxide Thin-Film Transistors with High-κ Alumina Dielectric Working in Enhancement Mode. *ACS Appl. Mater. Interfaces* **2014**, *6*, 20786–20794. [CrossRef] [PubMed]

11. Bin Esro, M.; Vourlias, G.; Somerton, C.; Milne, W.I.; Adamopoulos, G. High-Mobility ZnO Thin Film Transistors Based on Solution-processed Hafnium Oxide Gate Dielectrics. *Adv. Funct. Mater.* **2015**, *25*, 134–141. [CrossRef]

12. Cai, W.; Wilson, J.; Zhang, J.; Park, S.; Majewski, L.; Song, A. Low-Voltage, Flexible InGaZnO Thin-Film Transistors Gated with Solution-Processed, Ultra-Thin Al_xO_y. *IEEE Electron Device Lett.* **2018**, *40*, 36–39. [CrossRef]

13. Chen, C.-K.; Hsieh, H.-H.; Shyue, J.-J.; Wu, C.-C. The Influence of Channel Compositions on the Electrical Properties of Solution-Processed Indium-Zinc Oxide Thin-Film Transistors. *J. Disp. Technol.* **2009**, *5*, 509–514. [CrossRef]

14. Jang, J.; Kitsomboonloha, R.; Swisher, S.L.; Park, E.S.; Kang, H.; Subramanian, V. Transparent High-Performance Thin Film Transistors from Solution-Processed SnO_2/ZrO_2 Gel-like Precursors. *Adv. Mater.* **2012**, *25*, 1042–1047. [CrossRef]

15. Lee, C.-G.; Dodabalapur, A. Solution-processed zinc–tin oxide thin-film transistors with low interfacial trap density and improved performance. *Appl. Phys. Lett.* **2010**, *96*, 243501. [CrossRef]

16. Kim, S.J.; Yoon, S.; Kim, H.J. Review of solution-processed oxide thin-film transistors. *Jpn. J. Appl. Phys.* **2014**, *53*, 02BA02. [CrossRef]

17. Zhang, J.; Fu, X.; Zhou, S.; Ning, H.; Wang, Y.; Guo, D.; Cai, W.; Li, J.-A.; Yao, R.; Peng, J. The Effect of Zirconium Doping on Solution-Processed Indium Oxide Thin Films Measured by a Novel Nondestructive Testing Method (Microwave Photoconductivity Decay). *Coatings* **2019**, *9*, 426. [CrossRef]

18. Park, S.; Kim, C.H.; Lee, W.J.; Sung, S.; Yoon, M.H. Sol-gel metal oxide dielectrics for all-solution-processed electronics. *Mater. Eng. R Rep.* **2017**, *114*, 1–22. [CrossRef]

19. Chang, S.B.; Chae, H.U.; Kim, H.S. Structural, optical, electrical and morphological properties of different concentration sol-gel ZnO seeds and consanguineous ZnO nanostructured growth dependence on seeds. *J. Alloys Compd.* **2017**, *729*, 571–582.

20. Yang, J.; Li, C.; Zhang, X.; Quan, Z.; Zhang, C.; Li, H.; Lin, J. Self-Assembled 3D Architectures of $LuBO_3:Eu^{3+}$: Phase-Selective Synthesis, Growth Mechanism, and Tunable Luminescent Properties. *Chemistry* **2008**, *14*, 4336–4345. [CrossRef]

21. Rana, A.U.H.S.; Kang, M.; Kim, H.-S. Microwave-assisted Facile and Ultrafast Growth of ZnO Nanostructures and Proposition of Alternative Microwave-assisted Methods to Address Growth Stoppage. *Sci. Rep.* **2016**, *6*, 24870. [CrossRef] [PubMed]

22. Hafeezullah; Yamani, Z.H.; Iqbal, J.; Qurashi, A.; Hakeem, A.S. Rapid sonochemical synthesis of In_2O_3 nanoparticles their doping optical, electrical and hydrogen gas sensing properties. *J. Alloys Compd.* **2014**, *616*, 76–80. [CrossRef]

23. Rana, A.U.H.S.; Kim, H.S. NH4OH Treatment for an Optimum Morphological Trade-off to Hydrothermal Ga-Doped n-ZnO/p-Si Heterostructure Characteristics. *Materials* **2018**, *11*, 37. [CrossRef] [PubMed]

24. Zhang, G.; Lu, K.; Zhang, X.; Yuan, W.; Shi, M.; Ning, H.; Tao, R.; Liu, X.; Yao, R.; Peng, J. Effects of Annealing Temperature on Optical Band Gap of Sol–gel Tungsten Trioxide Films. *Micromachines* **2018**, *9*, 377. [CrossRef]

25. Zheng, M.J.; Zhang, L.D.; Li, G.H.; Zhang, X.Y.; Wang, X.F. Ordered indium-oxide nanowire arrays and their photoluminescence properties. *Appl. Phys. Lett.* **2001**, *79*, 839–841. [CrossRef]

26. Seo, S.-J.; Choi, C.G.; Hwang, Y.-H.; Bae, B.-S. High performance solution-processed amorphous zinc tin oxide thin film transistor. *J. Phys. D Appl. Phys.* **2008**, *42*, 035106. [CrossRef]

27. Chiang, H.Q.; Wager, J.; Hoffman, R.L.; Jeong, J.; Keszler, D.A. High mobility transparent thin-film transistors with amorphous zinc tin oxide channel layer. *Appl. Phys. Lett.* **2005**, *86*, 013503. [CrossRef]

28. Nguyen, M.-C.; Jang, M.; Lee, D.-H.; Bang, H.-J.; Lee, M.; Jeong, J.K.; Yang, H.; Choi, R. Li-Assisted Low-Temperature Phase Transitions in Solution-Processed Indium Oxide Films for High-Performance Thin Film Transistor. *Sci. Rep.* **2016**, *6*, 25079. [CrossRef]

29. Choi, H.S.; Cho, W.J. Controlling In-Ga-Zn-O Thin-Film Resistance by Vacuum Rapid Thermal Annealing and Application to Transparent Electrode. *Phys. Status Solidi* **2019**, *216*, 1800653.1–1800653.6. [CrossRef]

30. Nguyen, M.-C.; Nguyen, A.H.T.; Ji, H.; Cheon, J.; Kim, J.-H.; Yu, K.-M.; Cho, S.-Y.; Kim, S.-W.; Choi, R. Application of Single-Pulse Charge Pumping Method on Evaluation of Indium Gallium Zinc Oxide Thin-Film Transistors. *IEEE Trans. Electron. Dev.* **2018**, *65*, 3786–3790. [CrossRef]

31. Kim, H.S.; Byrne, P.D.; Facchetti, A.; Marks, T.J. High Performance Solution-Processed Indium Oxide Thin-Film Transistors. *J. Am. Chem. Soc.* **2008**, *130*, 12580–12581. [CrossRef] [PubMed]

32. Troughton, J.; Atkinson, D. Amorphous InGaZnO and metal oxide semiconductor devices: An overview and current status. *J. Mater. Chem. C* **2019**, *7*, 12388–12414. [CrossRef]
33. Jeon, H.; Jin, S.Y.; Park, W.H.; Lee, H.; Kim, H.T.; Ryou, M.H.; Lee, Y.M. Plasma-assisted water-based Al_2O_3 ceramic coating for polyethylene-based microporous separators for lithium metal secondary batteries. *Electrochim. Acta* **2016**, *212*, 649–656. [CrossRef]
34. Huang, K.-M.; Ho, C.-L.; Chang, H.-J.; Wu, M.-C. Fabrication of inverted zinc oxide photonic crystal using sol–gel solution by spin coating method. *Nanoscale Res. Lett.* **2013**, *8*, 306. [CrossRef] [PubMed]
35. Zhuang, L.; Jiang, K.; Zhang, G.; Tang, J.; Lee, S.W.R. O-2 Plasma Treatment in Polymer Insulation Process for Through Silicon Vias. In Proceedings of the 2014 15th International Conference on Electronic Packaging Technology (ICEPT), Chengdu, China, 12–15 August 2014.
36. Tho, S.Y.; Ibrahim, K.; Kamarulazizi, B.I. Plasma Pre-Treatment of Polyethylene Terephthalate Substrate Influence on the Properties of ZnO Thin Film. *Adv. Mater. Res.* **2014**, *895*, 41–44. [CrossRef]
37. Park, J.H.; Yoo, Y.B.; Lee, K.H.; Jang, W.S.; Oh, J.Y.; Chae, S.S.; Lee, H.W.; Han, S.W.; Baik, H.K. Boron-Doped Peroxo-Zirconium Oxide Dielectric for High-Performance, Low-Temperature, Solution-Processed Indium Oxide Thin-Film Transistor. *ACS Appl. Mater. Interfaces* **2013**, *5*, 8067–8075. [CrossRef]
38. Park, J.H.; Kim, K.; Yoo, Y.B.; Park, S.Y.; Lim, K.-H.; Lee, K.H.; Baik, H.K.; Kim, Y.S. Water adsorption effects of nitrate ion coordinated Al_2O_3 dielectric for high performance metal-oxide thin-film transistor. *J. Mater. Chem. C* **2013**, *1*, 7166–7174. [CrossRef]
39. Jiang, S.; Zhang, Q.; Li, Y.; Zhang, Y.; Sun, X.; Jiang, B. Structural characteristics of $SrTiO_3$ thin films processed by rapid thermal annealing. *J. Cryst. Growth* **2005**, *274*, 500–505. [CrossRef]
40. Choi, C.-H.; Han, S.-Y.; Su, Y.-W.; Fang, Z.; Lin, L.-Y.; Cheng, C.-C.; Chang, C.-H. Fabrication of high-performance, low-temperature solution processed amorphous indium oxide thin-film transistors using a volatile nitrate precursor. *J. Mater. Chem. C* **2014**, *3*, 854–860. [CrossRef]
41. Kang, D.; Lim, H.; Kim, C.; Song, I.; Park, J.; Park, Y.; Chung, J. Amorphous gallium indium zinc oxide thin film transistors: Sensitive to oxygen molecules. *Appl. Phys. Lett.* **2007**, *90*, 192101. [CrossRef]

 *micromachines*

Article

Effect of Gas Annealing on the Electrical Properties of Ni/AlN/SiC

Dong-Hyeon Kim, Michael A. Schweitz and Sang-Mo Koo *

Department of Electronic Materials Engineering, Kwangwoon University, 20 Kwangwoon-ro, Nowon-gu, Seoul 01897, Korea; gogomatt@kw.ac.kr (D.-H.K.); michael.schweitz@schweitzlee.com (M.A.S.)
* Correspondence: smkoo@kw.ac.kr; Tel.: +82-2-940-5763

Abstract: It is shown in this work that annealing of Schottky barrier diodes (SBDs) in the form of Ni/AlN/SiC heterojunction devices in an atmosphere of nitrogen and oxygen leads to a significant improvement in the electrical properties of the structures. Compared to the non-annealed device, the on/off ratio of the annealed SBD devices increased by approximately 100 times. The ideality factor, derived from the current-voltage (IV) characterization, decreased by a factor of ~5.1 after annealing, whereas the barrier height increased from ~0.52 to 0.71 eV. The bonding structure of the AlN layer was characterized by X-ray photoelectron spectroscopy. Examination of the N 1 s and O 1 s peaks provided direct indication of the most prevalent chemical bonding states of the elements.

Keywords: aluminum nitride; silicon carbide; Schottky barrier diodes; radio frequency sputtering; X-ray diffraction; X-ray photoelectron spectroscopy

Citation: Kim, D.-H.; Schweitz, M.A.; Koo, S.-M. Effect of Gas Annealing on the Electrical Properties of Ni/AlN/SiC. *Micromachines* **2021**, *12*, 283. https://doi.org/10.3390/mi12030283

Academic Editor: Giovanni Verzellesi

Received: 17 February 2021
Accepted: 5 March 2021
Published: 8 March 2021

Publisher's Note: MDPI stays neutral with regard to jurisdictional claims in published maps and institutional affiliations.

Copyright: © 2021 by the authors. Licensee MDPI, Basel, Switzerland. This article is an open access article distributed under the terms and conditions of the Creative Commons Attribution (CC BY) license (https://creativecommons.org/licenses/by/4.0/).

1. Introduction

Wide bandgap semiconductor materials have superior thermal and electrical properties compared to those of conventional semiconductors, and thus have potential for use in high-power, high-temperature, microwave and optoelectronic applications [1–3]. Next-generation high-temperature electronics will require increasing use of semiconductor materials such as SiC, GaN, AlN, and AlGaN due to their superior electrical properties resulting from their wide band–gap and high thermal conductivity. AlN has the largest bandgap of 6.2 eV and critical electric field of 11.7 MV/cm, as well as the highest thermal conductivity of 320 W/mK among group III materials [4,5]. These material properties make AlN highly versatile for use in high-power and high-temperature applications, including motor drives, energy conversion systems, high-temperature sensors, and space exploration. AlN films can be manufactured by several different processes, including chemical vapor deposition, molecular beam epitaxy, and radio frequency (RF)-sputtering. Compared to high-temperature thin film deposition techniques, RF-sputtering is cheaper and simpler to implement. Importantly, RF-sputtering offers the possibility to manufacture high-quality, large-scale films with desirable material properties, including on temperature-sensitive substrates if required [6–8]. However, RF-sputtering results in a low polar field which reduces the performance of high electron mobility transistors [9]. Furthermore, AlN thin films grown by RF-sputtering contain defects related to oxygen impurities, resulting in impaired electrical and optical properties. The theoretical modeling of oxygen in semiconductor materials remains computationally challenging, largely because traditional empirical or semi-empirical methods fail to satisfactorily explain the large electronegativity of oxygen, along with its chemical binding properties.

Annealing is a potentially important process for manufacturing high-quality compound semiconductors with thin films of materials such as GaAs, SiC, and AlN because the process appears to have the capacity to reduce unintentional defects in these films by orders of magnitude [10–12]. However, previous experimental studies have focused mainly on the sputtering parameters, such as pressure, power, sputtering ambient, and target distance [13,14]. It is known that annealing improves the crystallization of films similar to those mentioned [15]; however, very few researchers have studied the influence of post-annealing treatment on the surface morphology and crystallization orientation of AlN films [16].

While the properties of AlN films are determined by the deposition method, they are also affected by post-deposition treatment parameters, such as annealing temperature and duration [17]. The manufactured SBDs are expected to be of use in high temperature applications. Their behavior and use as temperature sensors for temperatures up to 475 K is explored in [18]. Further investigations of the manufactured device properties are ongoing. Indicative values of, for instance, breakdown voltage of related AlN thin film structures can be found in [19]. In this study, we investigate the influence of post-deposition annealing in nitrogen (N_2) and oxygen (O_2) atmospheres on the electrical properties of AlN thin films. We demonstrate that gas annealing of AlN thin films in either a nitrogen or oxygen atmosphere results in films with lower leakage currents at 300 K than for non-annealed AlN thin films. This effect was demonstrated by characterizing the electrical properties of manufactured heterojunction (Ni/AlN/SiC) Schottky barrier diodes (SBDs).

2. Materials and Methods

A schematic of the manufactured vertical AlN Schottky barrier diode (SBD) is shown in Figure 1. N-type 4H-SiC wafers (base substrate: $N_D = 1 \times 10^{19}$ cm^{-3}; n-type epitaxial layer: $N_D = 5 \times 10^{16}$ cm^{-3}) acted as starting substrates. The SiC substrate was cleaned in a sulfuric peroxide mixture (sulfuric acid (H_2SO_4) to hydrogen peroxide (H_2O_2) ratio of 4:1), after which the native SiO_2 layer was stripped using a buffered oxide etch (BOE) solution. A 150 nm thick Ni-film was then deposited by e-beam evaporation to create a diode cathode on the reverse side of the substrate. The samples were subjected to rapid thermal annealing (RTA) at 1323 K in N_2 for 90 s to form nickel silicide (Ni_2Si) ohmic contacts. AlN films were subsequently deposited by RF sputtering of a 99.9% pure AlN target onto the substrate at 300 K under injection of high purity argon gas (99.999%) with a flow rate of 5.5 sccm, using a mass flow controller. The sputtering power was 150 W, and the target diameter was 5.08 cm. Deposition chamber working pressure was maintained at 10 mTorr during the 120 min deposition step, resulting in a film thickness of approximately 200 nm. An Alpha-Step stylus profilometer and atomic force microscopy (AFM) were used to measure the thickness of the AlN films. AFM thickness measurements are in good agreement with measurements performed by ellipsometry [20]. AlN thin film samples either remained as deposited, i.e., not annealed, or were annealed at 773 K for 30 min, in either an ambient nitrogen or oxygen atmosphere. Top electrode contacts were created by depositing a 150 nm thick nickel layer on the AlN thin films, thus completing the sample SBDs. The mobility of the AlN films were measured using a Ecopia HMS-5000 Hall Effect Measurement System. The four Hall measurement probes were each connected through ohmic contacts located at each upper corner of the AlN film samples.

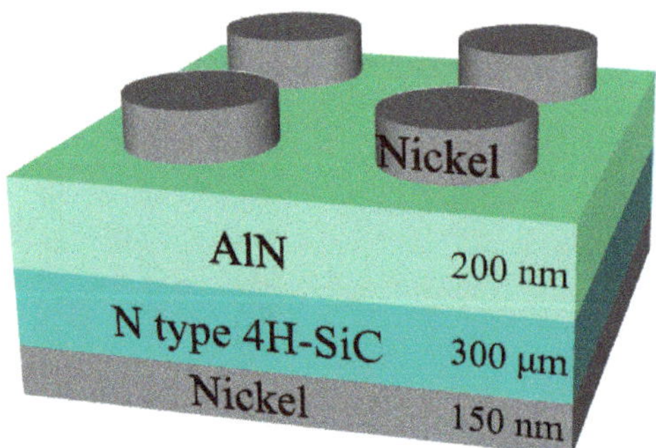

Figure 1. AlN/4H-SiC heterojunction diode structure.

Before deposition of the top electrodes, the samples were subjected to X-ray diffraction (XRD) measurement, X-ray photoelectron spectroscopy (XPS), and Fourier transform infrared spectroscopy (FTIR). XPS was performed to analyze the orientation of the wurtzite crystal structure of the AlN thin films [21]. Annealing atmosphere influence on the AlN film binding energies was measured by X-ray photoelectron spectroscopy (XPS). Al, N, and O contents of the films were estimated from Al 2p, N s, and O s XPS core-level spectra. The AlN film samples were analyzed by Fourier transform infrared spectroscopy (FTIR), and, finally, the finished SBDs were characterized by current voltage (I-V) measurements at 300 K using a semiconductor analyzer (Keithley 4200-SCS, Tektronix, Beaverton, OR 97077, USA).

3. Result and Discussion

Figure 2 shows the XRD patterns of the untreated ("as-grown", i.e., before annealing) and gas (nitrogen or oxygen) annealed AlN films, indicating the main crystal plane orientations of the films. The three graphs in Figure 2 each display two intensity peaks at $2\theta = 38.2°$ and $44.4°$, corresponding to the $(2\bar{1}\bar{1}1)$ and $(2\bar{1}\bar{1}0)$ AlN crystal diffraction planes, respectively. The amplitudes of these peaks decrease after annealing. The different XRD patterns indicate that the samples annealed in either N_2 or O_2 atmospheres were influenced at the crystal domain level to cause the observed decrease in intensity and full width at half maximum (FWHM) [22].

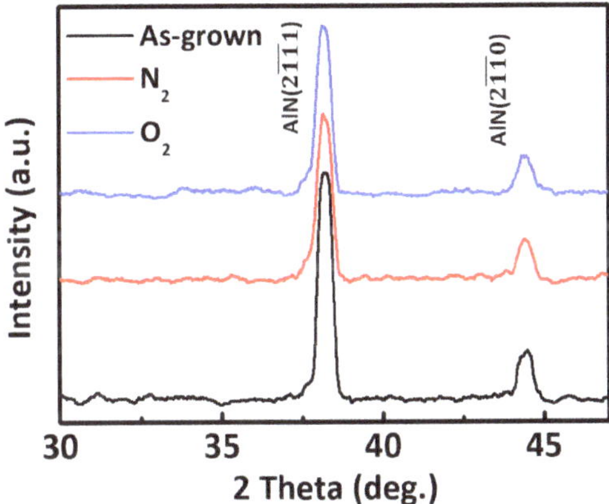

Figure 2. XRD patterns of the as-grown, N_2, and O_2 annealed AlN films.

Figure 3 shows the FWHM of the Al $(2\bar{1}\bar{1}0)$ peak and the corresponding average AlN crystal domain size. The average domain size of the as-grown, N_2 annealed, and O_2 annealed samples was calculated according to the Scherrer equation given by

$$D = \frac{K\lambda}{\beta \sin(\theta)} \tag{1}$$

where K is the shape factor (0.9), λ is the X-ray wavelength (1.5406 Å), β is the FWHM of the peaks in radians, and θ is the Bragg diffraction angle. The results indicate that the FWHM was reduced as a result of the annealing process. The average grain sizes of the as-grown, N_2, and O_2 annealed samples estimated by the Scherrer equation and the XRD data are 87.07 nm, 168.17 nm, and 162.64 nm, respectively. This indicates increases in both uniformity and the level of crystallinity [16]. This result suggests that annealing can to some degree remove or "soften" the grain boundaries in the film. Consequently, the annealed samples contain fewer or less distinct grain boundaries, leading to a reduced number of charge carrier traps or scattering obstacles along grain boundaries [23].

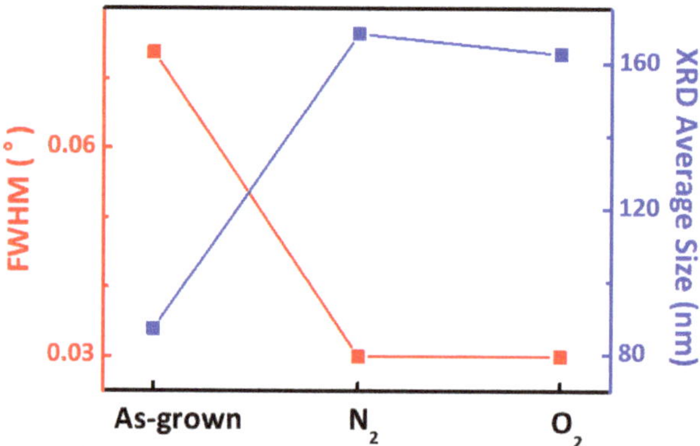

Figure 3. Full width at half maximum (FWHM) of the XRD Al($2\bar{1}\bar{1}0$) peak and the corresponding average crystal size for the as-grown, N_2, and O_2 annealed AlN films.

Figure 4 shows the FTIR spectra of AlN films before and after N_2 and O_2 annealing. Spectra were obtained in the range from 600 to 800 cm^{-1}. The absorption peak at 668 cm^{-1} corresponds to the characteristic value of aluminum nitride. The magnitude of this peak increased with gas annealing, as seen in Figure 4 [24]. Furthermore, the FTIR spectra in Figure 4 show increased overall transmittance of the N_2 and O_2 annealed AlN films, indicating improved film crystallinity [25].

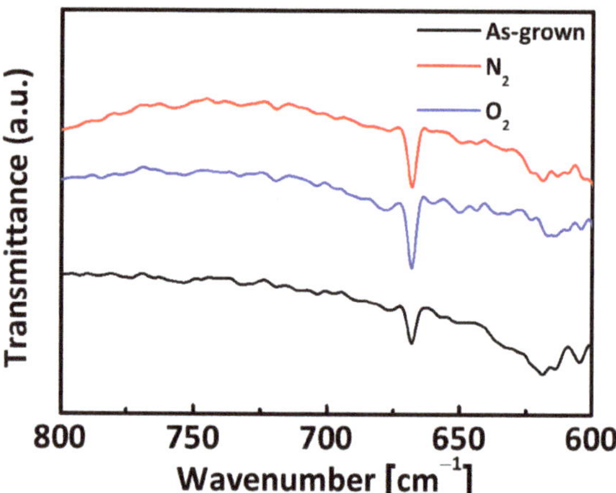

Figure 4. FTIR transmittance spectra of the as-grown, N_2, and O_2 annealed AlN films.

In order to investigate the effect of the annealing process on the chemical bonding state of nitrogen atoms in the AlN layer, we performed an XPS core level measurement. Figure 5a,b show the N 1 s and O 1 s core level XPS spectra for the AlN films. Figure 5a shows these spectra were deconvoluted into three components with peaks at 403.2 ± 0.1, 397.4 ± 0.2, and 396.7 eV, which correspond to the chemical bonding states of Al-$(NO_x)_y$, AlN-O, and Al-N, respectively, as previously reported in the literature [26]. The XPS response after annealing displays a new peak at 403.2 ± 0.1 eV, distinct from the expected one at 397.4 ± 0.2 eV [27]. This indicates increased levels of surface oxygen, which is understood to result from the creation of an oxide film.

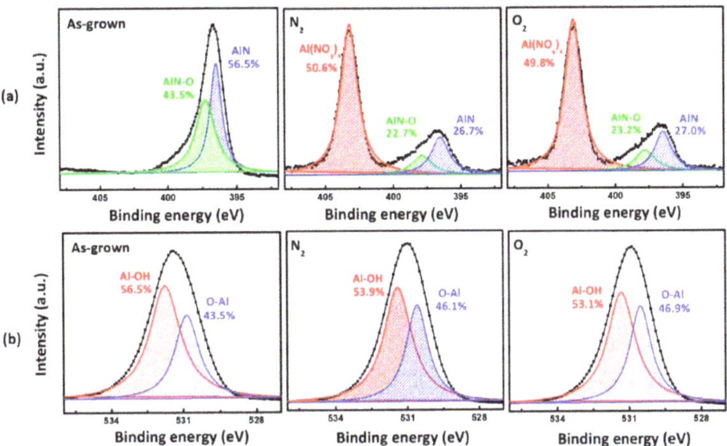

Figure 5. XPS spectra (**a**) (N 1 s) and (**b**) (O 1 s) of the as-grown, N_2, and O_2 annealed AlN films.

When it reducing the relative nitrogen content, AlN chemical bonds break up. This is likely due to the result of the oxidation of the AlN surface barrier that is presumed to occur [28]; also, annealing has related to the creation of nitrogen defects in the form of interstitial N_2 trapped between the surface oxide film and AlN film interlayer. Figure 5b shows the measured XPS core-level O 1s spectra. These can be deconvoluted into two peaks at 531.78 and 530.88 eV of the as-grown sample, corresponding to the Al-OH and O-Al chemical bond states, respectively [29]. After gas annealing, in the N_2 annealed sample, Al–OH and O–Al binding energies shifted to 531.38 eV and 530.58 eV, respectively. Said energies were 0.4 eV and 0.3 eV lower than those of the as-grown sample. In the O_2 annealed sample, the Al–OH and O–Al binding energies shifted to 531.32 eV and 530.50 eV, which in turn were lower than the respective binding energies of the N_2 annealed sample. Additionally, the relative area of the Al–OH peak decreased. Annealing effectively decomposes the OH- species in the AlN film. Simultaneously, the residual OH- bonds release Al atoms to form more Al-O bonds. Then, the OH- groups are removed and the oxygen vacancies are filled [30]. The relationship between the electrical properties of the AlN thin film samples and the film quality was investigated by Hall measurements. Figure 6 shows charge carrier mobility and concentration relative to the different gas annealing conditions. The maximum achieved carrier mobility (528 cm^2/Vs) was observed in the O_2 annealed sample. A reduced number of crystal defects result in higher mobility and decreased charge carrier concentration [31]. From the literature, we infer that also in AlN thin films, grain boundaries are the main source of defects that limit charge carrier mobility and give rise to charge carriers [32,33]. Film crystallinity is proportional to the average crystal grain size, which in turn is inversely proportional to the density of grain boundaries. With the O_2 annealed device having a larger average grain size, it is reasonable to expect that the density of the grain boundaries may have decreased accordingly. This, in turn, would explain the observed reduction in carrier concentration and the increase in mobility.

Figure 7 shows the typical I-V characteristics of the fabricated AlN/4H-SiC SBDs measured on a logarithmic scale. The diode currents were measured for terminal potentials ranging from -5 V to $+5$ V. From 0 V to 2 V, the forward current of the as-grown sample was higher than that of the annealed samples (N_2 annealed, O_2 annealed), while for voltages higher than 2 V, the forward current of the O_2 annealed sample was the highest. For the case of reverse bias, the annealed AlN SBDs exhibited lower leakage currents ($\sim 1.3 \times 10^{-6}$ A) than the as-grown AlN SBD (9.5×10^{-5} A).

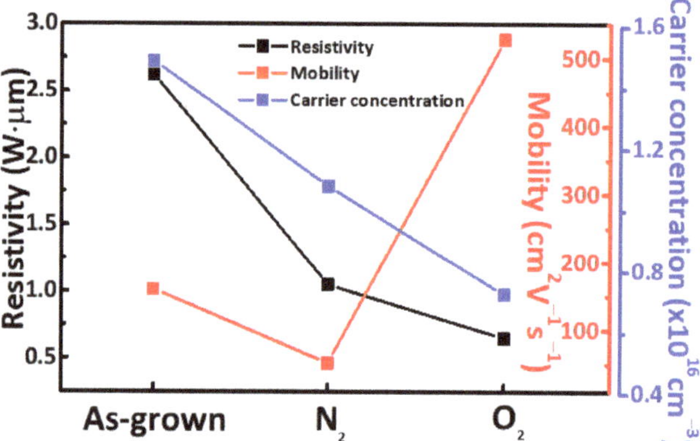

Figure 6. Mobility and carrier concentration of the as-grown, N_2, and O_2 annealed AlN films.

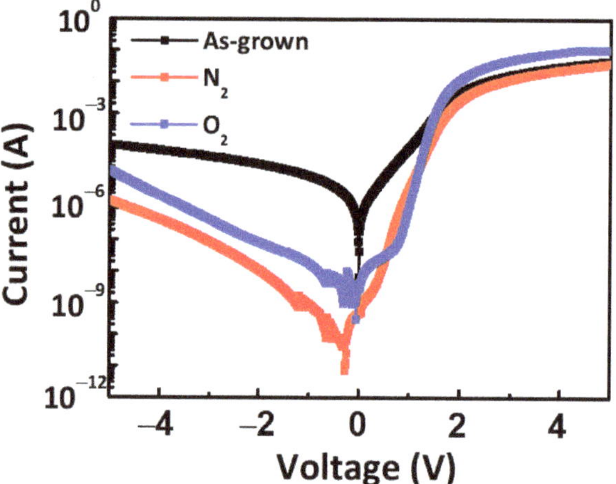

Figure 7. I–V characteristics for the as-grown, N_2, and O_2 annealed AlN/4H-SiC SBDs.

The Schottky barrier height (ϕ_B) of the manufactured diodes was calculated according to Equation (2) and shown in Figure 8:

$$I_S = AA^*T^2 \left[\exp\left(\frac{-q\phi_B}{kT} \right) \right] \tag{2}$$

where ϕ_B is the barrier height, A is the effective area of the diode for current transport, I_S is the saturation current, T is the measurement temperature, and A^* is the Richardson constant (theoretically ~57.6 A cm^{-1} K^{-2} for AlN) [34,35]. The Figure 8 shows the N_2 annealed device exhibited the highest Schottky barrier height ϕ_B = 0.71 eV at reverse bias. From 0 V to 2 V forward bias, the O_2 annealed device had the highest barrier height ϕ_B = 0.59 eV, which can be explained by the filled oxygen vacancies identified from the Al-OH/Al-O peak ratio of the XPS O 1s data. After 2 V forward bias, the O_2 annealed device also had the lowest barrier height ϕ_B = 0.29 eV, which can be explained by the Hall carrier mobility.

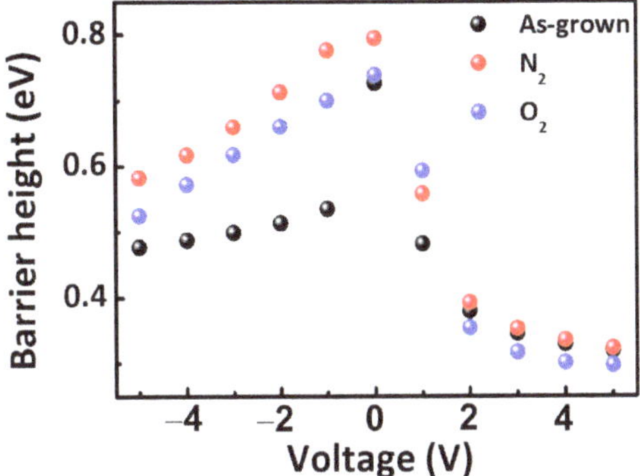

Figure 8. Barrier height of AlN on 4H-SiC SBDs with different gas annealing.

The I_{on}/I_{off} ratio and ideality factor values of the fabricated devices are shown in Figure 9. According to thermionic emission (TE) theory, the SBD I-V curves in forward bias can be expressed in the form of Equations (3)–(5) [36,37].

$$\varnothing_B = \frac{kT}{-q} ln\left(\frac{I_{ST}}{AA^*T^2}\right) \tag{3}$$

$$I = I_s\left[\exp\left(\frac{q(V - IR_s)}{\eta kT}\right) - 1\right] \tag{4}$$

$$\eta = \frac{q}{KT}\left[\frac{dV}{d(lnI)}\right] \tag{5}$$

where η is the ideality factor, q is the elementary electric charge, k is the Boltzmann constant, T is the absolute temperature, and I_s is the saturation current. The ideality factor value is explained assuming a Gaussian distribution of the barrier height around the AlN/4H-SiC interface. As Figure 9 shows, the lower the ideality factor, the greater the barrier height. The on/off ratios at room temperature of the as grown, N_2 annealed, and O_2 annealed samples were calculated to be ~4.5 × 10^2, ~2.2 × 10^4, and ~6.7 × 10^3, respectively. The corresponding ideality factors for the as grown, N_2 annealed, and O_2 annealed samples were 8.5, 4.1, and 5.29, respectively. The on/off ratio of the annealed device was two orders of magnitude (~100 times) higher than the on/off ratio of the as-grown device, while the ideality factor was the lowest (4.1) for the N_2 annealed sample.

In summary, after the annealing process, the electrical conduction properties of the AlN SBDs improved.

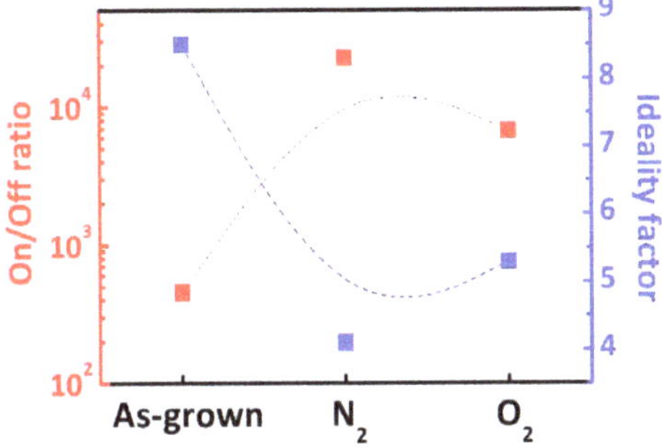

Figure 9. On/off ratio and ideality factor of AlN on 4H-SiC SBDs with different gas annealing.

4. Conclusions

The influence of different annealing atmospheres on AlN thin films was investigated. We measured and analyzed the electrical characteristics of an AlN SBD before and after the gas of nitrogen and oxygen annealing. For the N_2 and O_2 annealed sample, we observed that the AlN thin films were formed with relatively large grain sizes 168.17 nm and 162.62 nm, respectively, and higher magnitude of Al-N bond peaks. These less distinct grain boundaries related to a reduced number of charge carrier traps or scattering obstacles along grain boundaries. Charge carrier mobility was the highest (528 cm^2/Vs) for the O_2 annealed sample. From this sample, we also observed a relatively a reduced barrier height, and increased forward current. The high-temperature annealing process caused the decomposition of the Al-OH bonds; as a result, the relative area of the Al-O peak increased, while the number of oxygen vacancies decreased in O_2 annealed sample. Consequently, the current decreased with the increasing electric resistance until 2 V. One main result of this study is that the characteristic XPS data of the N 1 s region show unusual feature at 404 eV. It is related that the reduced reverse leakage current is a result of the trapped nitrogen defect. The barrier height decreased with improved conductivity, which in turn resulted in an improved on/off ratio in the N_2 annealed devices. In conclusion, the N_2 annealed sample had the lowest reverse leakage current and the O_2 annealed sample displayed the highest forward current level. Gas annealing thus constitutes a method for controlling the electrical properties of manufactured AlN thin films.

Author Contributions: Conceptualization, D.-H.K. and S.-M.K.; writing original draft, D.-H.K.; writing review and editing, D.-H.K., M.A.S. and S.-M.K.; methodology, D.-H.K.; project administration, S.-M.K.; validation, M.A.S. and S.-M.K.; supervision, S.-M.K. All authors have read and agreed to the published version of the manuscript.

Funding: This work was supported by the Technology Innovation Program (20003540), GRDC Program through the National Research Foundation (NRF) funded by the MSIT of Korea (NRF-2018K1A4A3079552), and a Research Grant form Kwangwoon University in 2021.

Institutional Review Board Statement: Not applicable.

Informed Consent Statement: Not applicable.

Data Availability Statement: Data is contained within the article.

Conflicts of Interest: The authors declare no conflict of interest.

References

1. Irokawa, Y.; Víllora, E.A.G.; Shimamura, K. Shottky barrier diodes on AlN free-standing substrates. *Jpn. J. Appl. Phys.* **2012**, *51*, 040206. [CrossRef]
2. Kinoshita, T.; Nagashima, T.; Obata, T.; Takashima, S.; Yamamoto, R.; Togashi, R.; Sitar, Z. Fabrication of vertical Schottky barrier diodes on n-type freestanding AlN substrates grown by hydride vapor phase epitaxy. *Appl. Phys. Express* **2015**, *8*, 061003. [CrossRef]
3. Fu, H.; Baranowski, I.; Huang, X.; Chen, H.; Lu, Z.; Montes, J.; Zhao, Y. Demonstration of AlN Schottky barrier diodes with blocking voltage over 1 kV. *IEEE Electron Device Lett.* **2017**, *38*, 1286–1289. [CrossRef]
4. Roccaforte, F.; Fiorenza, P.; Greco, G.; Nigro, R.L.; Giannazzo, F.; Iucolano, F.; Saggio, M. Emerging trends in wide band gap semiconductors (SiC and GaN) technology for power devices. *Microelectron. Eng.* **2018**, *187*, 66–77. [CrossRef]
5. O'leary, S.K.; Foutz, B.E.; Shur, M.S.; Eastman, L.F. Steady-state and transient electron transport within the III–V nitride semiconductors, GaN, AlN, and InN: A review. *J. Mater. Sci. Mater. Electron.* **2006**, *17*, 87–126. [CrossRef]
6. Prabaswara, A.; Birch, J.; Junaid, M.; Serban, E.A.; Hultman, L.; Hsiao, C.L. Review of GaN Thin Film and Nanorod Growth Using Magnetron Sputter Epitaxy. *Appl. Sci.* **2020**, *10*, 3050. [CrossRef]
7. Cheng, H.; Sun, Y.; Zhang, J.X.; Zhang, Y.B.; Yuan, S.; Hing, P. AlN films deposited under various nitrogen concentrations by RF reactive sputtering. *J. Cryst. Growth* **2003**, *254*, 46–54. [CrossRef]
8. Núñez-Cascajero, A.; Valdueza-Felip, S.; Monteagudo-Lerma, L.; Monroy, E.; Taylor-Shaw, E.; Martin, R.W.; Naranjo, F.B. In-rich $Al_xIn_{1-x}N$ grown by RF-sputtering on sapphire: From closely-packed columnar to high-surface quality compact layers. *J. Phys. D Appl. Phys.* **2017**, *50*, 065101. [CrossRef]
9. Jiang, F.; Zheng, C.; Changda, W.; Fang, L.; Wenqing, P.; Yong, D.; Dai, J. The growth and properties of ZnO film on Si (111) substrate with an AlN buffer by AP-MOCVD. *J. Lumin.* **2007**, *122*, 905–907. [CrossRef]

10. Xu, X.H.; Wu, H.S.; Zhang, C.J.; Jin, Z.H. Morphological properties of AlN piezoelectric thin films deposited by DC reactive magnetron sputtering. *Thin Solid Films* **2001**, *388*, 62–67. [CrossRef]
11. Mattila, T.; Nieminen, R.M. Ab initio study of oxygen point defects in GaAs, GaN, and AlN. *Phys. Rev. B* **1996**, *54*, 16676. [CrossRef] [PubMed]
12. Singh, R.; Christiansen, S.H.; Moutanabbir, O.; Gösele, U. The phenomenology of ion implantation-induced blistering and thin-layer splitting in compound semiconductors. *J. Electron. Mater.* **2010**, *39*, 2177–2189. [CrossRef]
13. Liu, H.Y.; Tang, G.S.; Zeng, F.; Pan, F. Influence of sputtering parameters on structures and residual stress of AlN films deposited by DC reactive magnetron sputtering at room temperature. *J. Cryst. Growth* **2013**, *363*, 80–85. [CrossRef]
14. Ekpe, S.D.; Jimenez, F.J.; Dew, S.K. Effect of process conditions on the microstructural formation of dc reactively sputter deposited AlN. *J. Vac. Sci. Technol. A Vac. Surf. Films* **2010**, *28*, 1210–1214. [CrossRef]
15. Kumar, V.; Kumar, V.; Som, S.; Yousif, A.; Singh, N.; Ntwaeaborwa, O.M.; Swart, H.C. Effect of annealing on the structural, morphological and photoluminescence properties of ZnO thin films prepared by spin coating. *J. Colloid Interface Sci.* **2014**, *428*, 8–15. [CrossRef]
16. Fu, B.; Wang, F.; Cao, R.; Han, Y.; Miao, Y.; Feng, Y.; Zhang, K. Optimization of the annealing process and nanoscale piezoelectric properties of (002) AlN thin films. *J. Mater. Sci. Mater. Electron.* **2017**, *28*, 9295–9300. [CrossRef]
17. Kar, J.P.; Bose, G.; Tuli, S. Effect of annealing on DC sputtered aluminum nitride films. *Surf. Coat. Technol.* **2005**, *198*, 64–67. [CrossRef]
18. Jung, S.W.; Shin, M.C.; Schweitz, M.A.; Oh, J.M.; Koo, S.M. Influence of Gas Annealing on Sensitivity of AlN/4H-Sic-Based Temperature Sensors. *Materials* **2021**, *14*, 683. [CrossRef]
19. Fu, H.; Huang, X.; Chen, H.; Lu, Z.; Zhao, Y. Fabrication and characterization of ultra-wide bandgap AlN-based Schottky diodes on sapphire by MOCVD. *IEEE J. Electron Devices Soc.* **2017**, *5*, 518–524. [CrossRef]
20. Jung, J.C.; Koo, S.M. The Effect of Catalytic Metal Work Functions and Interface States on the High Temperature SiC-based Gas Sensors. *J. Korean Inst. Electr. Electron. Mater. Eng.* **2011**, *24*, 280–284. [CrossRef]
21. Clement, M.; Vergara, L.; Sangrador, J.; Iborra, E.; Sanz-Hervás, A. SAW characteristics of AlN films sputtered on silicon substrates. *Ultrasonics* **2004**, *42*, 403–407. [CrossRef] [PubMed]
22. Wang, M.X.; Xu, F.J.; Xie, N.; Sun, Y.H.; Liu, B.Y.; Qin, Z.X.; Shen, B. Crystal quality evolution of AlN films via high-temperature annealing under ambient N_2 conditions. *CrystEngComm* **2018**, *20*, 6613–6617. [CrossRef]
23. Chen, F.H.; Hung, M.N.; Yang, J.F.; Kuo, S.Y.; Her, J.L.; Matsuda, Y.H.; Pan, T.M. Effect of surface roughness on electrical characteristics in amorphous InGaZnO thin-film transistors with high-κ Sm_2O_3 dielectrics. *J. Phys. Chem. Solids* **2013**, *74*, 570–574. [CrossRef]
24. Hajakbari, F.; Hojabri, A.; Mojtahedzadeh Larijani, M. Effect of Thickness on Structural and Morphological Properties of AlN Films Prepared Using Single Ion Beam Sputtering. *J. Appl. Chem. Res.* **2014**, *8*, 45–52.
25. Khan, S.; Shahid, M.; Mahmood, A.; Shah, A.; Ahmed, I.; Mehmood, M.; Alam, M. Texture of the nano-crystalline AlN thin films and the growth conditions in DC magnetron sputtering. *Prog. Nat. Sci. Mater. Int.* **2015**, *25*, 282–290. [CrossRef]
26. Baltrusaitis, J.; Jayaweera, P.M.; Grassian, V.H. XPS study of nitrogen dioxide adsorption on metal oxide particle surfaces under different environmental conditions. *Phys. Chem. Chem. Phys.* **2009**, *11*, 8295–8305. [CrossRef]
27. Rosenberger, L.; Baird, R.; McCullen, E.; Auner, G.; Shreve, G. XPS analysis of aluminum nitride films deposited by plasma source molecular beam epitaxy. *Surf. Interface Anal. Int. J. Devoted Dev. Appl. Tech. Anal. Surf. Interfaces Thin Films* **2008**, *40*, 1254–1261. [CrossRef]
28. Bourlier, Y.; Bouttemy, M.; Patard, O.; Gamarra, P.; Piotrowicz, S.; Vigneron, J.; Etcheberry, A. Investigation of InAlN Layers Surface Reactivity after Thermal Annealings: A Complete XPS Study for HEMT. *ECS J. Solid State Sci. Technol.* **2018**, *7*, P329. [CrossRef]
29. Cao, D.; Cheng, X.; Xie, Y.H.; Zheng, L.; Wang, Z.; Yu, X.; Yu, Y. Effects of rapid thermal annealing on the properties of AlN films deposited by PEALD on AlGaN/GaN heterostructures. *RSC Adv.* **2015**, *5*, 37881–37886. [CrossRef]
30. Raja, J.; Nguyen, C.P.T.; Lee, C.; Balaji, N.; Chatterjee, S.; Jang, K.; Yi, J. Improved data retention of InSnZnO nonvolatile memory by H_2O_2 treated Al_2O_3 tunneling layer: A cost-effective method. *IEEE Electron Device Lett.* **2016**, *37*, 1272–1275. [CrossRef]
31. Taniyasu, Y.; Kasu, M.; Makimoto, T. Increased electron mobility in n-type Si-doped AlN by reducing dislocation density. *Appl. Phys. Lett.* **2006**, *89*, 182112. [CrossRef]
32. Vladimirov, I.; Kühn, M.; Geßner, T.; May, F.; Weitz, R.T. Energy barriers at grain boundaries dominate charge carrier transport in an electron-conductive organic semiconductor. *Sci. Rep.* **2018**, *8*, 1–10. [CrossRef] [PubMed]
33. Horowitz, G.; Hajlaoui, M.E. Grain size dependent mobility in polycrystalline organic field-effect transistors. *Synth. Met.* **2001**, *122*, 185–189. [CrossRef]
34. Bozack, M.J. Surface studies on SiC as related to contacts. *Phys. Status Solidi* **1997**, *202*, 549–580. [CrossRef]
35. Han, S.M.; Takasawa, T.; Yasutomi, K.; Aoyama, S.; Kagawa, K.; Kawahito, S. A time-of-flight range image sensor with background canceling lock-in pixels based on lateral electric field charge modulation. *IEEE J. Electron Devices Soc.* **2014**, *3*, 267–275. [CrossRef]
36. Cetin, H.; Ayyildiz, E.N.İ.S.E. On barrier height inhomogeneities of Au and Cu/n-InP Schottky contacts. *Phys. B Condens. Matter* **2010**, *405*, 559–563. [CrossRef]
37. Bouzid, F.; Pezzimenti, F.; Dehimi, L.; Megherbi, M.L.; Della Corte, F.G. Numerical simulations of the electrical transport characteristics of a Pt/n-GaN Schottky diode. *Jpn. J. Appl. Phys.* **2017**, *56*, 094301. [CrossRef]

 *micromachines*

Article

First-Principles Studies for Electronic Structure and Optical Properties of Strontium Doped β-Ga$_2$O$_3$

Loh Kean Ping [1], Mohd Ambri Mohamed [1,*], Abhay Kumar Mondal [1], Mohamad Fariz Mohamad Taib [2], Mohd Hazrie Samat [2,3], Dilla Duryha Berhanuddin [1], P. Susthitha Menon [1] and Raihana Bahru [1]

[1] Institute of Microengineering and Nanoelectronics (IMEN), Universiti Kebangsaan Malaysia (UKM), Bangi 43600, Selangor, Malaysia; p103776@siswa.ukm.edu.my (L.K.P.); abhay.nano17@gmail.com (A.K.M.); dduryha@ukm.edu.my (D.D.B.); susi@ukm.edu.my (P.S.M.); raihanabahru@ukm.edu.my (R.B.)

[2] Faculty of Applied Sciences, Universiti Teknologi MARA (UiTM), Shah Alam 40450, Selangor, Malaysia; mfariz@uitm.edu.my (M.F.M.T.); mohdhazrie@uitm.edu.my (M.H.S.)

[3] Ionic Materials & Devices (iMADE) Research Laboratory, Institute of Science, Universiti Teknologi MARA (UiTM), Shah Alam 40450, Selangor, Malaysia

* Correspondence: ambri@ukm.edu.my; Tel.: +60-3-8911-8558

Citation: Kean Ping, L.; Mohamed, M.A.; Kumar Mondal, A.; Mohamad Taib, M.F.; Samat, M.H.; Berhanuddin, D.D.; Menon, P.S.; Bahru, R. First-Principles Studies for Electronic Structure and Optical Properties of Strontium Doped β-Ga$_2$O$_3$. *Micromachines* **2021**, *12*, 348. https://doi.org/10.3390/mi12040348

Academic Editor: Giovanni Verzellesi

Received: 3 February 2021
Accepted: 22 March 2021
Published: 24 March 2021

Publisher's Note: MDPI stays neutral with regard to jurisdictional claims in published maps and institutional affiliations.

Copyright: © 2021 by the authors. Licensee MDPI, Basel, Switzerland. This article is an open access article distributed under the terms and conditions of the Creative Commons Attribution (CC BY) license (https://creativecommons.org/licenses/by/4.0/).

Abstract: The crystal structure, electron charge density, band structure, density of states, and optical properties of pure and strontium (Sr)-doped β-Ga$_2$O$_3$ were studied using the first-principles calculation based on the density functional theory (DFT) within the generalized-gradient approximation (GGA) with the Perdew–Burke–Ernzerhof (PBE). The reason for choosing strontium as a dopant is due to its p-type doping behavior, which is expected to boost the material's electrical and optical properties and maximize the devices' efficiency. The structural parameter for pure β-Ga$_2$O$_3$ crystal structure is in the monoclinic space group (C2/m), which shows good agreement with the previous studies from experimental work. Bandgap energy from both pure and Sr-doped β-Ga$_2$O$_3$ is lower than the experimental bandgap value due to the limitation of DFT, which will ignore the calculation of exchange-correlation potential. To counterbalance the current incompatibilities, the better way to complete the theoretical calculations is to refine the theoretical predictions using the scissor operator's working principle, according to literature published in the past and present. Therefore, the scissor operator was used to overcome the limitation of DFT. The density of states (DOS) shows the hybridization state of Ga 3d, O 2p, and Sr 5s orbital. The bonding population analysis exhibits the bonding characteristics for both pure and Sr-doped β-Ga$_2$O$_3$. The calculated optical properties for the absorption coefficient in Sr doping causes red-shift of the absorption spectrum, thus, strengthening visible light absorption. The reflectivity, refractive index, dielectric function, and loss function were obtained to understand further this novel work on Sr-doped β-Ga$_2$O$_3$ from the first-principles calculation.

Keywords: first-principles; density functional theory; pure β-Ga$_2$O$_3$; Sr-doped β-Ga$_2$O$_3$; p-type doping; band structure; density of states; optical absorption

1. Introduction

Gallium oxide (Ga$_2$O$_3$) embraces five different kinds of polymorphism, such as α, β, γ, δ, and ε phase [1,2]. Other examples of metal oxide structures are lead oxide (Pb$_2$O$_3$) [3], molybdenum dioxide (MoO$_2$) [4], aluminium oxide (Al$_2$O$_3$) [5], and zirconium oxide (ZrO$_2$) [6], which have a variety of polymorph phase similar to Ga$_2$O$_3$. Among all of these polymorphs of gallium oxide, β-Ga$_2$O$_3$ plays an essential role in ultrawide bandgap (UWBG) applications, with a bandgap energy of 4.8 eV between the valence band and conduction band [7,8]. This UWBG material has excellent heat thermal stability utilized in power electronics applications [9,10]. Perhaps, history of the monoclinic β-Ga$_2$O$_3$ can be traced back in the recent past few years due to its stable properties that eventually draw scientists' profound interest [11–13].

One of the significant applications of β-Ga$_2$O$_3$ which is still being used now is its metastable state crystal structure compared to other Ga$_2$O$_3$ polymorphs [14]. This is because β-Ga$_2$O$_3$ is a wide bandgap semiconductor with a bandgap (E_g) of about 4.8 eV [15]. The wide-bandgap material can sustain high incoming voltage resulting from large electrical breakdown strength when it comes to short circuit situation. Recently, β-Ga$_2$O$_3$ has attracted much attention for its potential use in the next-generation optoelectronic devices in the near UV wavelength region. It shows the shortest cut-off wavelength at around 270 nm, whereas conventional transparent conductive oxides (TCOs) belongs to the visible wavelength region. The applications for a monoclinic β-Ga$_2$O$_3$ structure are such as in solar energy devices [16], passivation coating [17], optoelectronic devices [18], gas sensor [19], and deep ultraviolet radiation devices [20].

The strontium element has the same chemical properties as alkaline earth metals such as beryllium, magnesium, calcium, and barium from group II elements. It is generally used for fireworks and flares purposes [21]. It is also used to produce ferrite magnets and is responsible for zinc's refining process [22]. One famous application of the strontium-90 radio-isotope is in the military field, especially nuclear weapons [23] and nuclear reactors [24]. The radio-isotope of the strontium-90 is the by-product of a nuclear explosion that initially comes from the uranium element's nuclear fission [25]. It has a half-life of 29 years [26], which was considered a long duration to degenerate its radioactivity. As for the application of semiconductors, strontium ions can be reused for the color television cathode ray tube (CRTs) to avoid X-ray emission [26,27].

There are no reports on the strontium (Sr^{2+}) doping, to date with the pure β-Ga$_2$O$_3$ based on density functional theory (DFT) in the present first-principles study. The importance of this paper is to determine the effect of Sr doping in Ga$_2$O$_3$ as p-type doping based on its material properties in the simulation structure. Therefore, this paper will focus on the theoretical investigation of the electronic band structure, total and partial density of states, and optical properties of pure and Sr-doped β-Ga$_2$O$_3$. Besides, this simulation process allow researchers to discover more about the material's theoretical part.

2. Materials and Methods

The calculations were carried out using Cambridge Serial Total Energy Package (CASTEP) code. This simulation can be traced back to the application of density functional theory (DFT), which utilizes the total-energy plane-wave pseudopotential method [28,29]. The exchange-correlation potential effects were handled by the generalized gradient approximation (GGA) with the Perdew–Burke–Ernzerhof (PBE) functional [30–32]. Theoretically, the DFT is based on the ground state, which causes the exchange-correlation potential between the excited electrons to be underestimated [33]. Thus, the calculations from GGA-PBE results in lower energy levels above the valence band than that of experimental results. The utilization of the GGA-PBE in this simulation work was used to compare the parameters, cell volume, and cell angle. β-Ga$_2$O$_3$ crystal structure belongs to the C2/m group, which is monoclinic. Figure 1 shows the crystal structure of the unit cell ($1 \times 1 \times 1$) of pure β-Ga$_2$O$_3$ and Sr-doped β-Ga$_2$O$_3$, while Figure 2 show the crystal structure of supercell ($1 \times 2 \times 2$) of pure β-Ga$_2$O$_3$ and Sr-doped β-Ga$_2$O$_3$. The unit cell crystal structure consists of 20 atoms (8 Ga atoms and 12 O atoms), and the supercell crystal structure consists of 80 atoms (32 Ga atoms and 48 O atoms). Ga atoms occupy 4i Wyckoff position at Ga1(0.09050, 0, 0.79460) and Ga2(0.15866, 0.5, 0.31402), whereas the O atoms occupy 4i Wyckoff positions defined by O1(0.1645, 0, 0.1098), O2(0.1733, 0, 0.5632), and O3($-$0.0041, 0.5, 0.2566) [34]. For obtaining exact band gaps and optical properties, the scissors operator has been carried out.

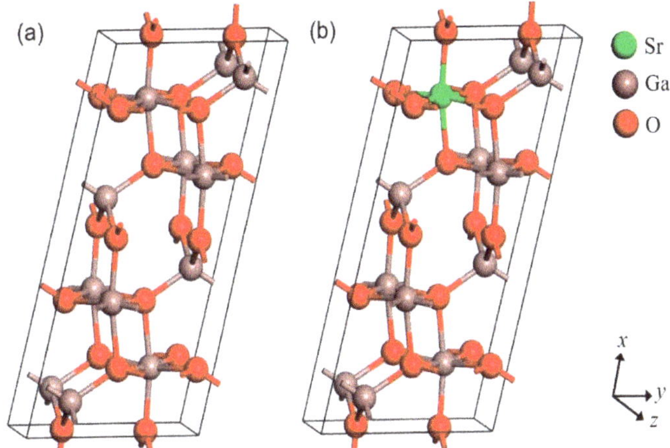

Figure 1. Crystal structure of $1 \times 1 \times 1$ unit cell of (**a**) pure β-Ga$_2$O$_3$ and (**b**) Sr-doped β-Ga$_2$O$_3$.

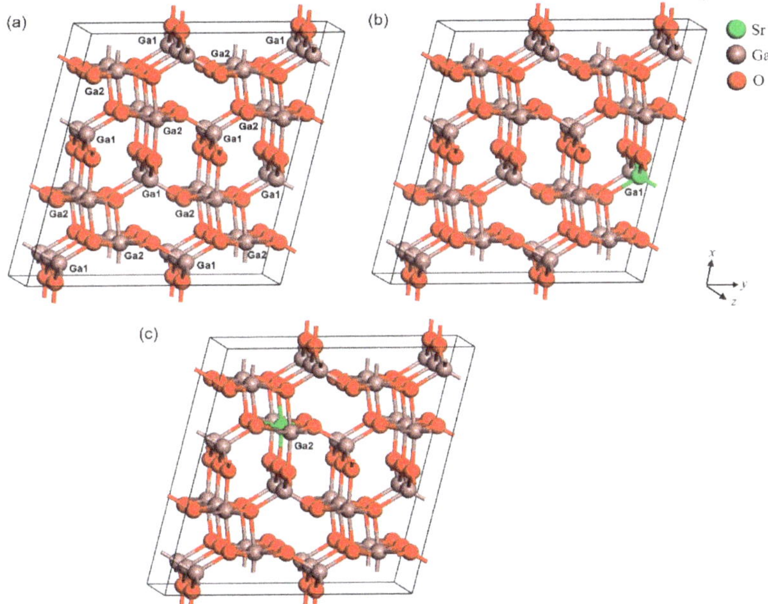

Figure 2. Crystal structure of $1 \times 2 \times 2$ supercell of (**a**) pure β-Ga$_2$O$_3$, (**b**) Sr-doped β-Ga$_2$O$_3$ at Ga1, and (**c**) Sr-doped Ga$_2$O$_3$ at Ga2.

The electronic interaction between the valence electrons and conduction holes was modeled using ultra-fine quality pseudopotentials with an energy cut-off of 380 eV. This cut-off energy was able to bring out the optimized results of the band structure, density of states, electron density, and optical properties calculations. As for the Monhorst Pack scheme k-point grid sampling of the reduced Brillouin zone, $1 \times 1 \times 1$ and $1 \times 2 \times 2$ k-points were set for pure β-Ga$_2$O$_3$ and Sr-doped β-Ga$_2$O$_3$, respectively. The valence electronic configurations for Ga, O, and Sr are 3d^{10}4s^{2}4p^1, 2s^{2}2p^4, and 3d^{10}4p^{6}5s^2, respectively. During the geometry optimization, the cut-off energy for both the structure model was

380 eV, the energy convergence for this structure is 1.209×10^{-6} eV/atom, the maximum displacement is 5×10^{-4} Å, maximum stress is 0.02 GPa, and maximum force is 0.01 eV/Å.

3. Results and Discussion

3.1. Structural Properties

The structure of β-Ga_2O_3 and Sr-doped β-Ga_2O_3 were investigated, and the geometry optimized crystal structure is shown in Figures 1 and 2. Table 1 shows the list of the optimized lattice parameters, cell volume, and angle from GGA-PBE functional. The theoretical value for β-Ga_2O_3 from GGA-PBE calculation has a small difference in lattice parameter, cell volume, and angle value compared to the experimental value. The lattice parameter difference is less than 2.29% while for the cell volume, it gives a 6.14% error. As for the Sr-doped β-Ga_2O_3 part, the error percentage indicates the differences in theoretical results between β-Ga_2O_3 and Sr-doped β-Ga_2O_3. The lattice parameter and volume of Ga_2O_3 increase after Sr doping. This is similar to the other reports using Mg-doped Ga_2O_3. The ionic radius of Mg^{2+} of 0.72 Å is larger than that of Ga^{3+} of 0.62 Å. Thus, it is reasonable that the lattice parameters of Ga_2O_3 increases after Mg doping [35]. Furthermore, Zn-doped Ga_2O_3 also shows the increase of structural parameter after Zn doping because of the ionic radius of Zn^{2+} of 0.74 Å is larger than that of Ga^{3+} of 0.62 Å, which resulted in the lattice spacing gradually being enlarged [36]. However, to the best of our knowledge, there is no theoretical and experimental data available for Sr-doped β-Ga_2O_3 for comparison with this work.

Table 1. Calculated lattice parameter (a, b and c), cell volume, and cell angle for pure and Sr-doped β-Ga_2O_3 crystal structure in $1 \times 1 \times 1$ unit cell.

Structures	Parameters	GGA-PBE	Experiment [37]
β-Ga_2O_3	a (Å)	12.494 (+2.29%)	12.214
	b (Å)	3.096 (+1.94%)	3.037
	c (Å)	5.898 (+1.72%)	5.798
	Cell Volume (Å^3)	221.647 (+6.14%)	208.835
		$\alpha = 90°$	$\alpha = 90°$
	Cell Angle (°)	$\beta = 103.705°$	$\beta = 103.830°$
		$\gamma = 90°$	$\gamma = 90°$
Sr-doped β-Ga_2O_3	a (Å)	12.506 (+0.10%)	-
	b (Å)	3.158 (+2.00%)	-
	c (Å)	5.794 (−1.76%)	-
	Cell Volume (Å^3)	222.317 (+0.30%)	-
	Cell Angle (°)	$\alpha = 90°$	-
		$\beta = 103.701°$	-
		$\gamma = 90°$	-

As for Table 2, different doping sites were taken at Ga1 and Ga2 for $1 \times 2 \times 2$ supercells. Sr-doped β-Ga_2O_3 has a greater lattice parameter and cell volume compared to pure β-Ga_2O_3. This also means that Sr-dopant enlarged the original size of pure β-Ga_2O_3. The highest error of estimation for the lattice parameter between Sr-doped β-Ga_2O_3 at Ga1 and Ga2 are about 1.11 % indicates that the results are nearly close to each other.

The average bond length for pure and Sr-doped β-Ga_2O_3 is shown in Table 3, while the average bond length for pure and Sr-doped β-Ga_2O_3 at Ga1 and Ga2 in $1 \times 2 \times 2$ supercell were listed in Table 4. It is shown that the atomic radius of the Ga atom (1.36 Å) is smaller than the Sr (2.19 Å), according to the periodic table. Therefore, Ga-O bond length is shorter than Sr-O and O-O bonds in Sr-doped β-Ga_2O_3. This situation is the same as the pure β-Ga_2O_3, where Ga-O bond length is shorter than O-O bonds. After Sr doping, the overall bond length for Sr-doped β-Ga_2O_3 is increased more than pure β-Ga_2O_3 due to the presence of Sr^{2+} ions.

Table 2. Calculated lattice parameter (a, b, and c), cell volume, and cell angle of pure and Sr-doped β-Ga_2O_3 crystal structure in $1 \times 2 \times 2$ supercell at Ga1 and Ga2.

Parameters	Pure β-Ga_2O_3	Sr-Doped β-Ga_2O_3 at Ga1	Sr-Doped β-Ga_2O_3 at Ga2
a (Å)	12.497	12.545 (+0.38%)	12.599 (+0.82%)
b (Å)	6.187	6.218 (+0.50%)	6.233 (+0.74%)
c (Å)	11.806	11.937 (+1.11%)	11.857 (+0.43%)
Cell Volume (Å^3)	886.770	903.865	904.761
Cell Angle (°)	$\alpha = 90°$ $\beta = 103.723°$ $\gamma = 90°$	$\alpha = 90°$ $\beta = 103.903°$ $\gamma = 90°$	$\alpha = 90°$ $\beta = 103.667°$ $\gamma = 90°$

Table 3. Calculated average bond length of pure β-Ga_2O_3 and Sr-doped β-Ga_2O_3 for $1 \times 1 \times 1$ unit cell.

Bond Length	$1 \times 1 \times 1$ Unit Cell	
	Pure β-Ga_2O_3	Sr-Doped β-Ga_2O_3
Ga-O (Å)	1.964	1.965
O-O (Å)	2.861	2.864
Sr-O (Å)	-	2.426

Table 4. Calculated average bond length of pure β-Ga_2O_3 and Sr-doped β-Ga_2O_3 for $1 \times 2 \times 2$ unit cell.

Bond Length	$1 \times 2 \times 2$ Supercell		
	Pure β-Ga_2O_3	Sr-Doped β-Ga_2O_3 at Ga1	Sr-Doped β-Ga_2O_3 at Ga2
Ga-O (Å)	1.971	1.979	1.973
O-O (Å)	2.882	2.873	2.876
Sr-O (Å)	-	2.229	2.366

The spatial electron density maps determine whether the structure model belongs to ionic or covalent bonds. Figure 3 exhibits the distribution of the different structure's electron density in pure and Sr-doped β-Ga_2O_3. It shows that pure β-Ga_2O_3 has strong covalent bonding characteristics before doping. On the other hand, the bond population analysis for Sr-doped β-Ga_2O_3 exhibits a weak ionic bonding effect compared to pure β-Ga_2O_3. This is because, the Sr^{2+} dopant had changed the bonding characteristics of β-Ga_2O_3 after doping. Table 5 shows the bond population analysis indicator for electron density distribution in pure and Sr-doped β-Ga_2O_3. It can be determined from Figure 3a that pure β-Ga_2O_3 shows strong covalent bonding characteristics. This is because the Ga and O atoms were located at the strong covalent region nearby to the red color region. On the other hand, there were weak ionic bonding characteristics in Sr-doped β-Ga_2O_3, where some of the Ga and O atoms are located in the green color indicator region.

Table 5. Indicator of population analysis for electron density.

Bonding Characteristics	*Ionic*	*Weak Ionic*	*Weak Covalent*	*Covalent*
Bond Population	0–0.32	0.32–0.50	0.50–0.75	0.75–1.0

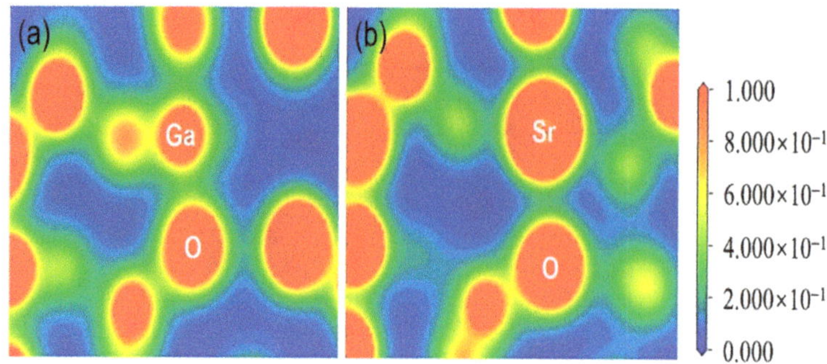

Figure 3. Distribution of electron density of (**a**) pure β-Ga$_2$O$_3$ and (**b**) Sr-doped β-Ga$_2$O$_3$.

3.2. Properties of Electronic Structure

This section will discuss the electronic structure properties used to determine the band structure, total and partial density of states (DOS) of β-Ga$_2$O$_3$. Gallium oxide displays an indirect bandgap of 4.8–4.9 eV, both for the experimental and simulation work [38,39]. Besides, the application of Ga$_2$O$_3$ is applied in power MOSFETs, where high breakdown electric field and large Balinga's figure of merit occur [40]. The (100) plane of Ga$_2$O$_3$ is mostly taken in experimental work because it gives a high resolution of result analysis which takes places at the specific Brillouin zone, Γ-Z and A-M directions [41]. This unique electronic structure properties of Ga$_2$O$_3$ bring out new hope in future applications, especially in electronics, optoelectronics, and sensing systems [7].

The bandgap is usually measured between the conduction band minimum (CBM) and the valence band maximum (VBM), located at the Fermi level. The high-symmetry direction of the Brillouin zone of pure β-Ga$_2$O$_3$ along with the G-F-Q-Z-G path, is illustrated in Figure 4. Figure 5a shows that pure β-Ga$_2$O$_3$ has a bandgap energy of 1.939 eV, lower than the bandgap energy from the experimental work but consistent with other calculated results from DFT [42,43]. The measured bandgap energy for experimental work is 4.8 eV [15]. This phenomenon can be explained by the underestimation of density functional theory (DFT) limitations. Therefore, the calculations of band structures with scissor operator were considered to overcome bandgap underestimation from the DFT method. The scissor operator was introduced to shift all the conduction levels to agree with the band gap's measured value [44]. In our case, the scissor operator's value for unit cell $1 \times 1 \times 1$ was taken to be $4.8 - 1.939 = 2.861$ eV, accounting for the difference between the experimental band gap (4.8 eV) [15] and the calculated GGA bandgap (1.939 eV) for β-Ga$_2$O$_3$. For supercell $1 \times 2 \times 2$, the scissor operator's value is 2.868 eV. The band structure of pure β-Ga$_2$O$_3$ in P1 symmetry along with G-F-Q-Z-Q path is shown in Figure 5 to compare with the band structure of Sr-doped β-Ga$_2$O$_3$ also in P1 symmetry. Figure 5b shows that the bandgap energy decreases to 1.879 eV after Sr doping. This indicates that Sr^{2+} ions possess p-type doping behavior, which creates more holes to accept electrons that allow the semiconductor to perform efficiently during the presence of conducting current. Figure 6 presents the band structures of pure β-Ga$_2$O$_3$ and Sr-doped β-Ga$_2$O$_3$ with a scissor operator. It shows that the bandgap of pure β-Ga$_2$O$_3$ was corrected to match the experimental band gap at 4.8 eV [15] while the bandgap of Sr-doped β-Ga$_2$O$_3$ is 4.740 eV which is lower than pure β-Ga$_2$O$_3$.

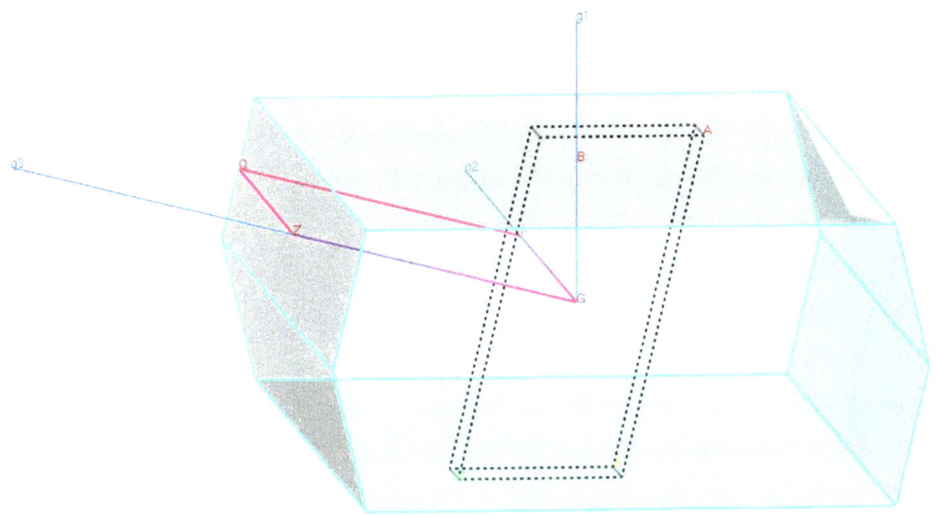

Figure 4. Brillouin zone path at G-F-G-Z-Q direction for β-Ga$_2$O$_3$ in P1 symmetry.

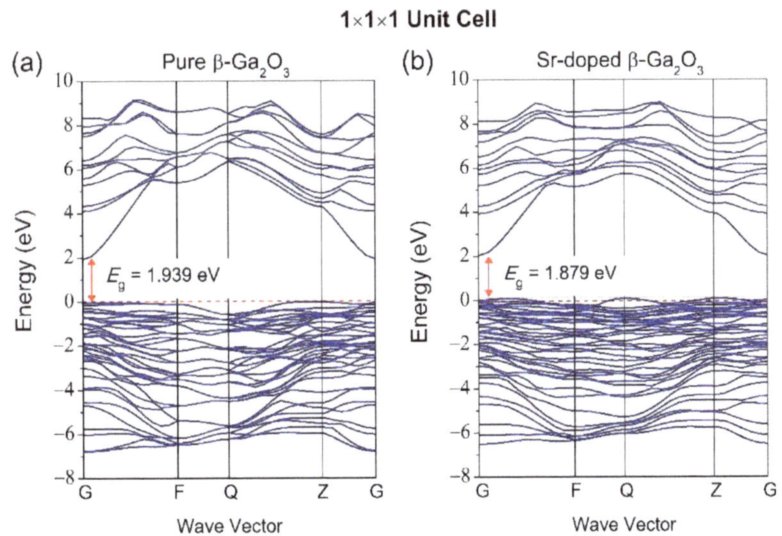

Figure 5. Band structure of (**a**) pure β-Ga$_2$O$_3$ and (**b**) Sr-doped β-Ga$_2$O$_3$ without scissor operator.

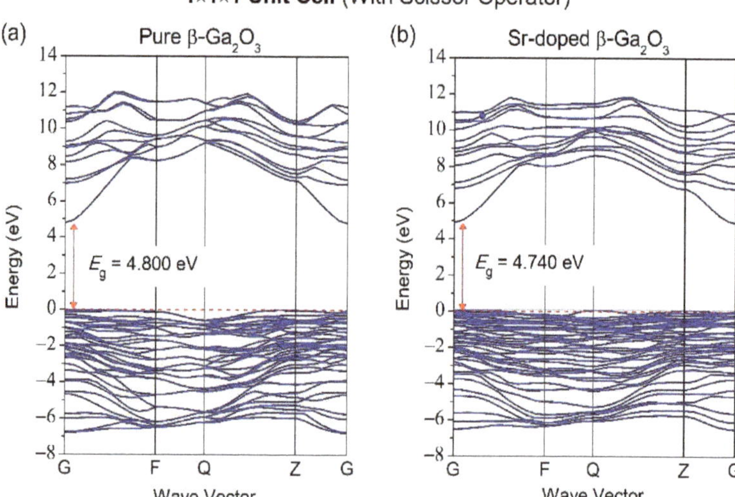

Figure 6. Band structure of (**a**) pure β-Ga$_2$O$_3$ and (**b**) Sr-doped β-Ga$_2$O$_3$ with scissor operator.

As for Figure 7, it is the band structure for $1 \times 2 \times 2$ supercell for pure and Sr-doped β-Ga$_2$O$_3$. Pure β-Ga$_2$O$_3$ has a bandgap of 1.932 eV. On the other hand, the Sr-doped at different doping sites at Ga1 and Ga2 have bandgaps of 1.826 and 1.840 eV, respectively. The band structure for unit cell and supercell show slight differences in bandgap, according to this investigation. For band structures with scissor operator of $1 \times 2 \times 2$ supercell for pure and Sr-doped β-Ga$_2$O$_3$ in Figure 8, the bandgap of pure β-Ga$_2$O$_3$ in $1 \times 2 \times 2$ supercell is 4.8 eV, and the bandgap of Sr-doped β-Ga$_2$O$_3$ at Ga1 and Ga2 is 4.694 and 4.708 eV, respectively.

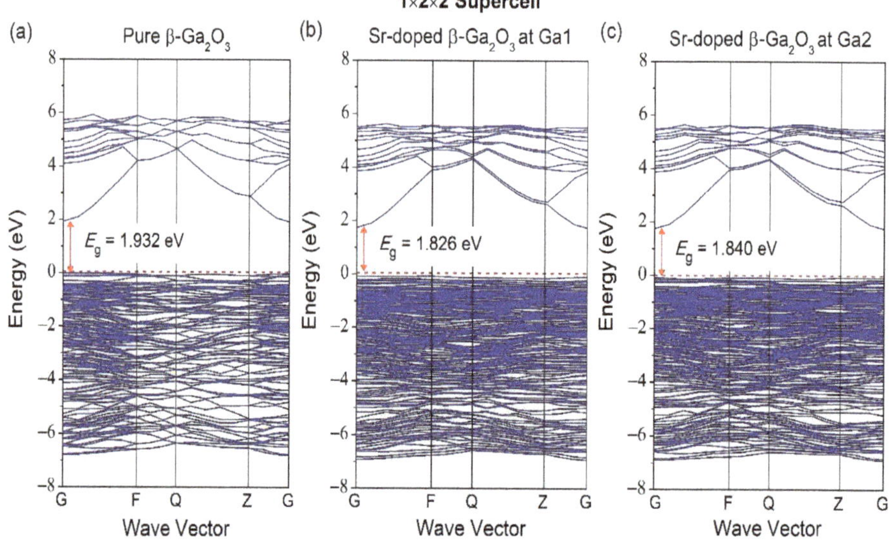

Figure 7. Band structure of (**a**) pure β-Ga$_2$O$_3$, (**b**) Sr-doped β-Ga$_2$O$_3$ at Ga1, and (**c**) Sr-doped β-Ga$_2$O$_3$ at Ga2 without scissor operator.

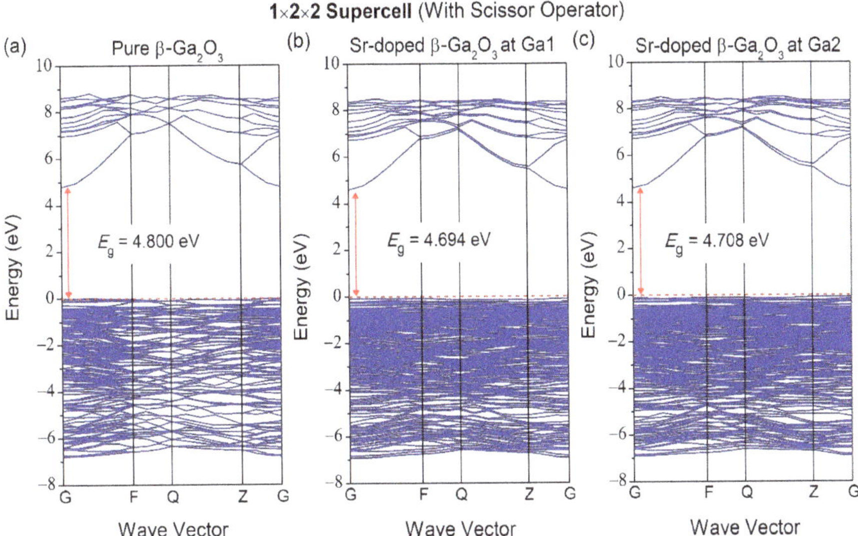

Figure 8. Band structure of (**a**) pure β-Ga$_2$O$_3$, (**b**) Sr-doped β-Ga$_2$O$_3$ at Ga1, and (**c**) Sr-doped β-Ga$_2$O$_3$ at Ga2 with scissor operator.

Figure 9 shows the total and partial DOS of pure and Sr-doped β-Ga$_2$O$_3$. Pure β-Ga$_2$O$_3$ comprises of Ga 4s at the top of the valence band and O 2p located at the bottom of the conduction band. As for the Sr-doped β-Ga$_2$O$_3$, the Sr atom has an atom of s orbital and introduces dopant energy levels in the pure β-Ga$_2$O$_3$. The valence band mainly consist of Sr 5s, Sr 4p, O 2p, and Ga 3d at -32, -12.5, -2, and -13 eV, respectively. The conduction band was dominated by Sr 3d, O 2p, and Ga 4p. Therefore, Sr dopant changes the covalent bonding characteristics of pure β-Ga$_2$O$_3$. This also indicates that weak ionic bonding characteristic appears between Ga, O, and Sr atoms. Figure 10 shows the total and partial DOS for $1 \times 2 \times 2$ supercell of pure and Sr-doped β-Ga$_2$O$_3$ at different doping sites Ga1 and Ga2. The same hybridization states occur in the valence band, such as Sr 5s, Sr 4p, O 2p, and Ga 3d. The conduction band is also dominated by the same hybridization states of Sr 3d, O 2p, and Ga 4p. All these results could be determined when comparing $1 \times 1 \times 1$ unit cell and $1 \times 2 \times 2$ supercell.

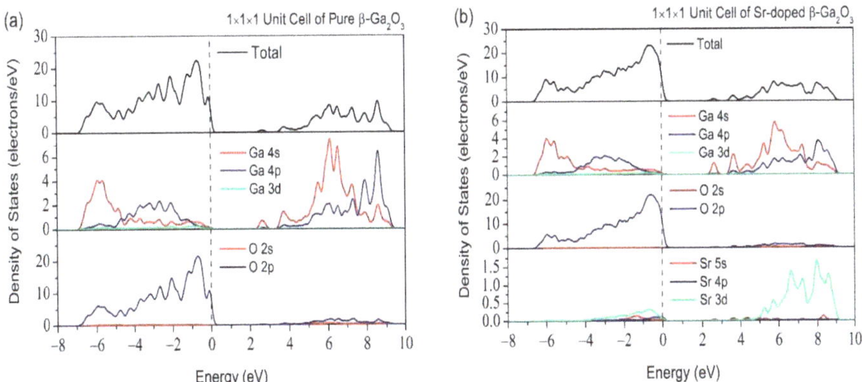

Figure 9. Total and partial density of states of (**a**) pure β-Ga$_2$O$_3$ and (**b**) Sr-doped β-Ga$_2$O$_3$ in $1 \times 1 \times 1$ supercell.

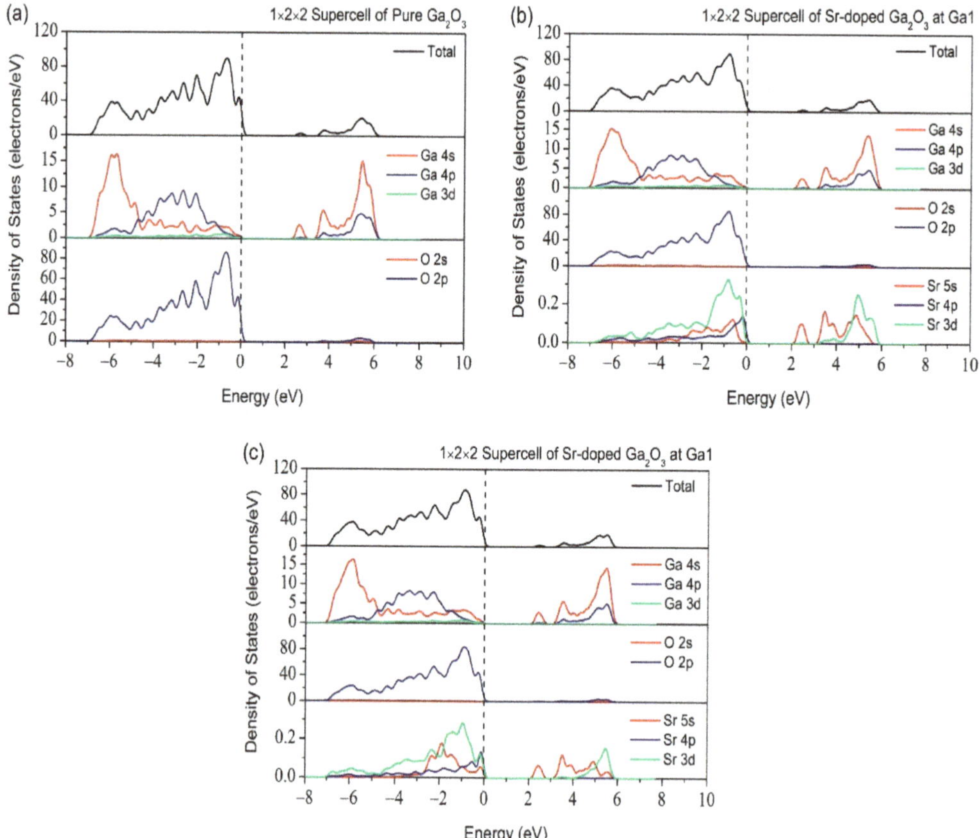

Figure 10. Total and Partial density of states of (**a**) pure β-Ga₂O₃, (**b**) Sr-doped β-Ga₂O₃ at Ga1, and (**c**) Sr-doped β-Ga₂O₃ at Ga2 in 1 × 2 × 2 supercell.

3.3. Optical Properties

Optical properties' importance is usually highlighted in the absorption coefficient, reflectivity, refractive index, dielectric function, and loss function. All these optical properties features are related to the complex dielectric function formula, which is written as in Equation (1):

$$\varepsilon(\omega) = \varepsilon_1(\omega) + i\varepsilon_2(\omega). \tag{1}$$

The $\varepsilon_1(\omega)$ and $\varepsilon_2(\omega)$ are the real and imaginary part, respectively. The real part is correlated to the degree of electronic polarization and calculated from the Kramers–Kronig relation. On the other hand, the imaginary part is associated with the material's dielectric losses. All other optical properties can be derived from $\varepsilon_1(\omega)$ and $\varepsilon_2(\omega)$ by the Kramer–Kronig relation.

The other well-known formula for optical properties such as absorption ($\alpha(\omega)$), reflectivity ($R(\omega)$), refractive index ($n(\omega)$), and loss function ($L(\omega)$) is defined as follows in Equations (2)–(5):

$$\alpha(\omega) = \frac{4\pi k}{\lambda} = \frac{2k\omega}{c} = \sqrt{2}\omega \left[\sqrt{\varepsilon_1^2(\omega) + \varepsilon_2^2(\omega)} - \varepsilon_1(\omega) \right]^{1/2}, \tag{2}$$

$$R(\omega) = \left| \frac{1 - \sqrt{\varepsilon(\omega)}}{1 + \varepsilon(\omega)} \right|^2, \tag{3}$$

$$n(\omega) = \sqrt{\frac{|\varepsilon(\omega)| + \varepsilon_1(\omega)}{2}}, \tag{4}$$

$$L(\omega) = Im\left[\frac{-1}{\varepsilon(\omega)}\right] = \frac{\varepsilon_2(\omega)}{\varepsilon_1^2(\omega) + \varepsilon_2^2(\omega)}. \tag{5}$$

The band structure of pure β-Ga$_2$O$_3$ shows an underestimated bandgap value of 1.939 eV compared to the experimental bandgap. Such underestimation of calculated bandgap values is a common feature of the DFT calculations and can be overcome by applying the so-called scissor operator [45]. Such a correction is significant for calculations of the optical properties. To facilitate comparison with the experimental results, we utilized a scissor operator to match the calculated optical gap determined via experimental techniques. The calculated optical properties of pure Ga$_2$O$_3$ and Sr-doped β-Ga$_2$O$_3$ without and with scissor operator are presented in Tables 6 and 7. For the absorption coefficient, the absorption edge with the scissor operator shifts the light absorption towards the UV light region, corresponding to the bandgap value of β-Ga$_2$O$_3$. This result is consistent with other experimental results for absorption spectra in a deep UV-Vis range of 200 to 300 nm using UV-Vis spectroscopy [46]. The scissor operator's shifts the major peak of reflectivity and loss function towards higher photon energy. For the refractive index of pure Ga$_2$O$_3$ and Sr-doped β-Ga$_2$O$_3$, the scissor operator decreases its value by 11.4% and 31.8%, while the dielectric constant decreases by 21.6% and 52.9% as compared to the calculations without scissor operator. Figures 11–13 present the optical properties using a scissor operator.

Table 6. Calculated optical properties without and with scissor operator (SO) of pure Ga$_2$O$_3$ and Sr-doped β-Ga$_2$O$_3$.

	Pure Ga$_2$O$_3$		Sr-Doped β-Ga$_2$O$_3$	
	without SO	with SO	without SO	with SO
Absorption edge	400–700 nm	100–300 nm	400–700 nm	100–300 nm
Dielectric constant	3.05	2.39	5.37	2.53
Refractive index	1.75	1.55	2.33	1.59
Reflectivity peak	13.3 eV	16.3 eV	11.1 eV	15.5 eV
Loss function peak	14.3 eV	17.4 eV	13.3 eV	16.3 eV

Table 7. Calculated optical properties without and with scissor operator (SO).

	Pure Ga$_2$O$_3$		Sr-Doped β-Ga$_2$O$_3$ at Ga1		Sr-Doped β-Ga$_2$O$_3$ at Ga2	
	without SO	with SO	without SO	with SO	without SO	with SO
Absorption edge	400–700 nm	100–300 nm	400–700 nm	100–300 nm	400–700 nm	100–300 nm
Dielectric constant	2.18	1.74	14.7	3.11	9.63	2.55
Refractive index	1.48	1.32	3.88	1.76	3.13	1.60
Reflectivity peak	9.46 eV	12.3 eV	9.46 eV	3.54 eV	9.33 eV	12.3 eV
Loss function peak	9.69 eV	12.7 eV	9.69 eV	12.8 eV	9.57 eV	12.5 eV

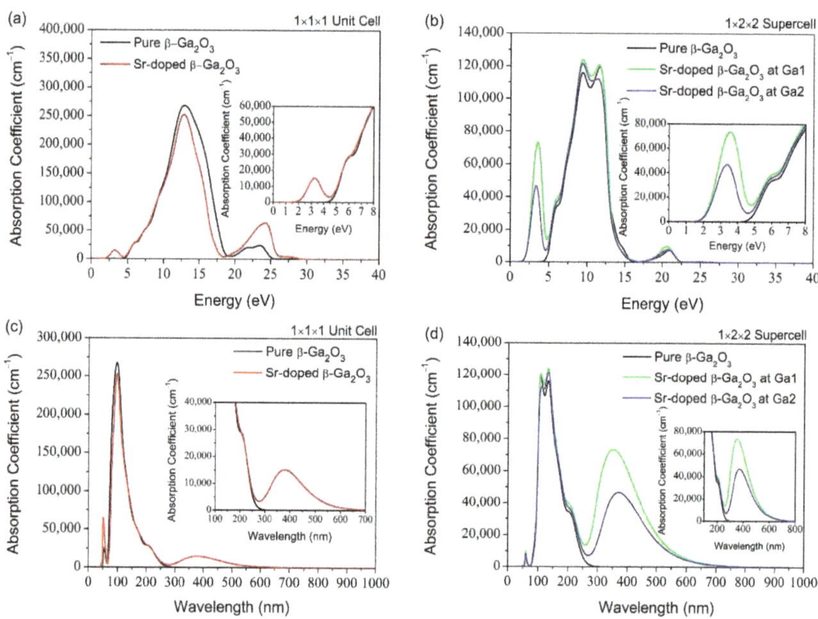

Figure 11. Absorption coefficient against (**a**) Energy (eV) for $1 \times 1 \times 1$ pure and Sr-doped β-Ga$_2$O$_3$, (**b**) Energy (eV) for $1 \times 2 \times 2$ supercell pure and Sr-doped β-Ga$_2$O$_3$, (**c**) Wavelength (nm) for $1 \times 1 \times 1$ unit cell pure and Sr-doped β-Ga$_2$O$_3$, (**d**) Wavelength (nm) for $1 \times 2 \times 2$ supercell for pure and Sr-doped β-Ga$_2$O$_3$.

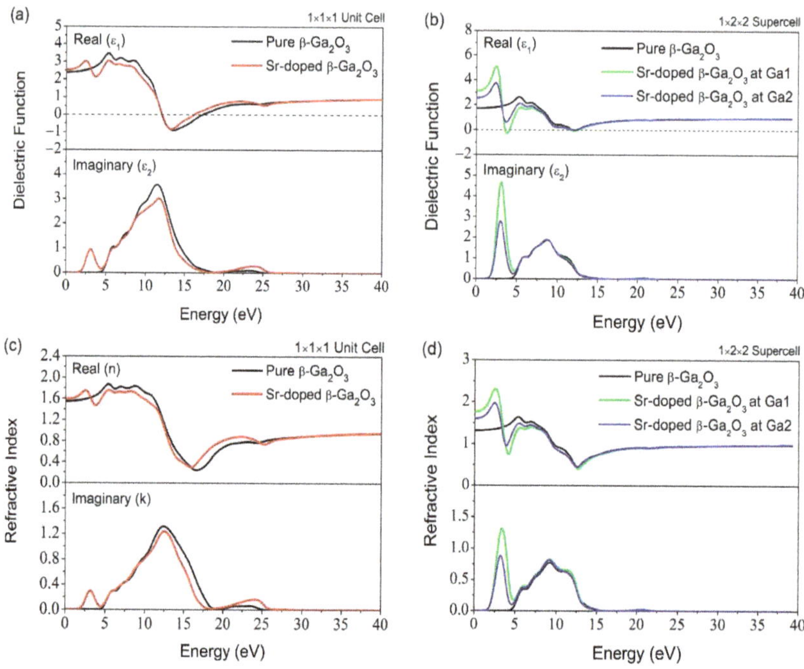

Figure 12. (**a**) Dielectric Function $1 \times 1 \times 1$ unit cell, (**b**) Dielectric Function $1 \times 2 \times 2$ supercell, (**c**) Refractive Index $1 \times 1 \times 1$ unit cell, (**d**) Refractive Index $1 \times 2 \times 2$ supercell.

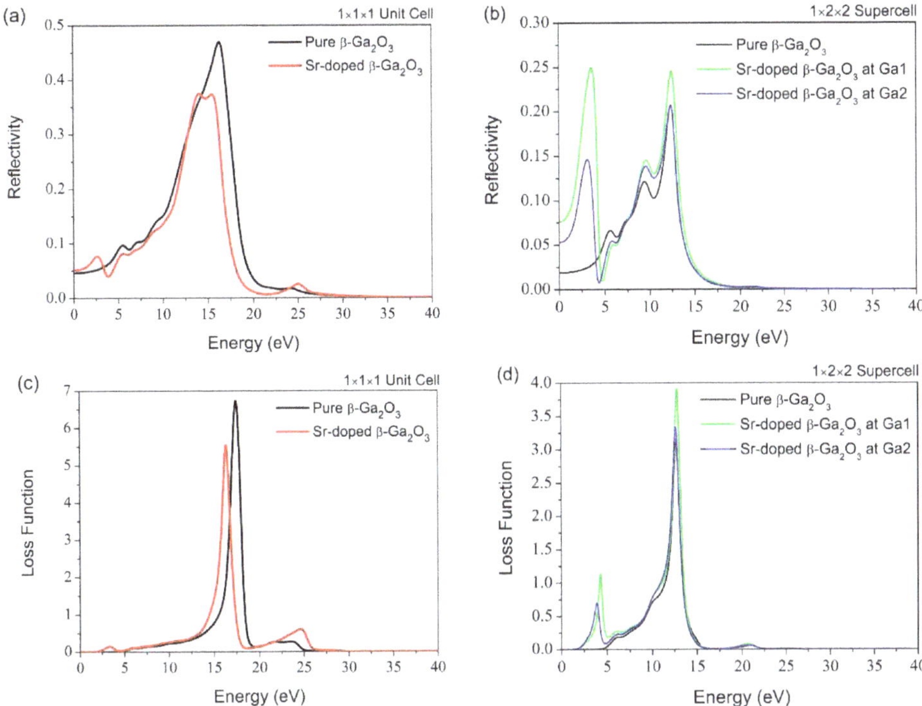

Figure 13. (**a**) Reflectivity of $1 \times 1 \times 1$ unit cell, (**b**) Reflectivity of $1 \times 2 \times 2$ supercell, (**c**) Loss Function of $1 \times 1 \times 1$ unit cell, (**d**) Loss Function of $1 \times 2 \times 2$ supercell.

Figure 11 presents the absorption spectrum of pure and Sr-doped β-Ga$_2$O$_3$ structure in $1 \times 1 \times 1$ unit cell and $1 \times 2 \times 2$ supercell. It can be observed that the absorption region is a broad spectrum. The most important wavelength emission for pure and Sr-doped β-Ga$_2$O$_3$ system mainly resides in the deep UV region, as shown in Figure 11d. After Sr-doping, the wavelength emission increases as the bandgap energy of Sr-doped β-Ga$_2$O$_3$ decreases. This simulation result matches with the principle of the Einstein–Planck relation: $E = hf = hc/\lambda$, where its bandgap energy decreases and increases the wavelength emission spectrum.

Meanwhile, this also indicates that the emission spectrum has been red-shifted after Sr-doping, proving that Sr dopant possesses p-type doping behavior. There is no available absorption spectrum report for Sr-doped β-Ga$_2$O$_3$ for comparison with this theoretical work. However, this finding is similar to other dopants in Ga$_2$O$_3$, such as Mg-doped Ga$_2$O$_3$ [46] and Zn-doped Ga$_2$O$_3$ from experimental work, which decreases the bandgap after doping and also possesses p-type doping behavior. The results for Mg-doped Ga$_2$O$_3$ suggested that it is a promising material candidate for solar-blind photodetector due to its lower dark current, higher sensitivity, and faster decay time which can be attributed to the high insulating and low defect concentration. For Zn-doped Ga$_2$O$_3$, its bandgap is 4.90–4.93 eV for different Zn doping contents which is reduced by 0.20–0.81% compared to pure β-Ga$_2$O$_3$ (4.94 eV). This is agreeable with our theoretical work, which decreases the bandgap after Sr doping by 1.25–5.49%.

Figures 12 and 13 shows the reflectivity, refractive index, dielectric function, and loss function of pure and Sr-doped β-Ga$_2$O$_3$ for both the $1 \times 1 \times 1$ unit cell and $1 \times 2 \times 2$ supercell. The energy range is shown as ~0–25 eV in this simulation work. For the energy spectrum of reflectivity, the energy increases slightly to 16 eV. Simultaneously, this incident causes a decrease of reflectivity after the Sr-doping, which eventually affects surface material effectiveness to reflect the electromagnetic radiation energy. More input electromagnetic radiation energy cannot be reflected completely and finally reside in the Sr-doped β-Ga$_2$O$_3$ structure compared to pure β-Ga$_2$O$_3$.

The refractive index exhibits a different growing emission spectrum trend compared to reflectivity. It has a real (n) and imaginary (k) part for both pure and Sr-doped β-Ga$_2$O$_3$. The emission spectrum releases energy in the range of ~0–25 eV for both systems. However, there are still differences in its refractive index spectrum in both real (n) and the imaginary (k) parts between pure and Sr-doped β-Ga$_2$O$_3$. The refractive index of Sr-doped β-Ga$_2$O$_3$ is 1.59, which is slightly higher than pure β-Ga$_2$O$_3$ (1.55).

As for the dielectric function, this feature of optical properties is usually observed in the spectrum's imaginary part (ε_2). It has dielectric constant, $\varepsilon_0 = 2.39$ for pure Ga$_2$O$_3$ and $\varepsilon_0 = 2.53$ for Sr-doped Ga$_2$O$_3$. The imaginary part's major peaks are located at 11.5 and 11.8 eV for pure and Sr-doped β-Ga$_2$O$_3$. This result indicates that the Sr dopant could affect the optical properties to determine the energy range. After the Sr-doping, the peak energy spectrum increases compared to pure β-Ga$_2$O$_3$. The peak energy shift also means a shift in the localized degree of a free electron and holes between the conduction and valence band in the impurity doping structure.

Lastly, the loss function feature in the optical properties is often used to determine the energy loss of the free electrons crossing along with the material. The major peak energy for the Sr-doped β-Ga$_2$O$_3$ is located at 16.3 eV, and the pure β-Ga$_2$O$_3$ is located at 17.4 eV, which is lower than Sr-doped β-Ga$_2$O$_3$. The Sr doping causes a decrease in the energy loss function of β-Ga$_2$O$_3$. This indicates that the Sr dopant tends to reduce its energy loss and improve its emission of peak energy efficiency for better performance in the material, matching with the characteristic of p-type doping material. It can withstand high voltage and current with minimum energy loss.

4. Conclusions

In conclusion, the first-principles studies provided calculations of the structural, electronic, and optical properties of pure and Sr-doped β-Ga$_2$O$_3$. There are not many differences in the calculated lattice parameter, cell volume, and angle of the pure β-Ga$_2$O$_3$ structure at GGA-PBE with experimental data. GGA-PBE is still preferable for the calculated structural parameter as it has good theoretical results, which almost matches the experimental work. Pure β-Ga$_2$O$_3$ has an indirect bandgap of 1.939 eV while for Sr-doped β-Ga$_2$O$_3$, the bandgap is 1.879 eV for $1 \times 1 \times 1$ unit cell. Meanwhile, the $1 \times 2 \times 2$ supercell for pure β-Ga$_2$O$_3$ has a bandgap of 1.932 eV and Sr-doped β-Ga$_2$O$_3$ at different doping sites Ga1 and Ga2 have bandgaps of 1.826 and 1.840 eV, respectively. This bandgap was underestimated compared to the experimental value, so the scissor operator was used to correct the bandgap. The decrease in bandgap energy was due to the creation of more holes to accept more incoming free electrons, indicating a p-type doping behavior for Sr dopant. The electronic interaction between the valence band and conduction band contains O 2p and Ga 4s orbital for pure β-Ga$_2$O$_3$. On the other hand, Sr-doped β-Ga$_2$O$_3$ consists of Sr 5s, Sr 4p, O 2p, and Ga 3d for the electronic interaction between valence and conduction band. These hybridization states can be observed from the total and partial DOS for both $1 \times 1 \times 1$ unit cell and $1 \times 2 \times 2$ supercell. The population analysis for pure β-Ga$_2$O$_3$ was considered as strong covalent bonding characteristics. As for the Sr-doped β-Ga$_2$O$_3$, it changes its bonding characteristics to weak ionic bonds. The optical absorption for Sr-doped β-Ga$_2$O$_3$ exhibited a red-shifted spectrum compared to pure β-Ga$_2$O$_3$. This matches the optical behavior for p-type doping, where the material emits broader wavelength emission, and a red-shifted spectrum occurred. The optical absorption for both pure and Sr-doped β-Ga$_2$O$_3$ was found in the deep ultraviolet light (DUV) region according to the absorption coefficient.

Author Contributions: Formal analysis, Investigation, Writing—original draft, Writing—review and editing, L.K.P.; Assistance of analysis, give opinions regarding the paper, A.K.M.; Conceptualization, Methodology, Validation, Correction, Supervision, Project administration. M.A.M.; Co-supervision,

Project administration, D.D.B., P.S.M. and R.B.; Provide facilities for conducting simulation analysis work, Validation, Correction, M.F.M.T. and M.H.S. All authors have read and agreed to the published version of the manuscript.

Funding: This work is supported in part by the Universiti Kebangsaan Malaysia grant No. DPK-2020-010.

Data Availability Statement: No new data were created or analyzed in this study. Data sharing does not apply to this article.

Conflicts of Interest: The authors declare no conflict of interest.

References

1. Roy, R.; Hill, V.G.; Osborn, E.F. Polymorphism of Ga_2O_3 and the system Ga_2O_3—H_2O. *J. Am. Chem. Soc.* **1952**, *74*, 719–722. [CrossRef]
2. Stepanov, S.I.; Nikolaev, V.I.; Bougrov, V.E.; Romanov, A.E. Gallium oxide: Properties and applications—A review. *Rev. Adv. Mater. Sci.* **2016**, *44*, 63–86.
3. Bakhtatou, A.; Ersan, F. Effects of the number of layers on the vibrational, electronic and optical properties of alpha lead oxide. *Phys. Chem. Chem. Phys.* **2019**, *21*, 3868–3876. [CrossRef]
4. Ersan, F.; Sarıkurt, S.; Sarikurt, S. Monitoring the electronic, thermal and optical properties of two-dimensional MoO2 under strain via vibrational spectroscopies: A first-principles investigation. *Phys. Chem. Chem. Phys.* **2019**, *21*, 19904–19914. [CrossRef] [PubMed]
5. Xie, T.; Edwards, T.E.J.; Della Ventura, N.M.; Casari, D.; Huszár, E.; Fu, L.; Zhou, L.; Maeder, X.; Schwiedrzik, J.J.; Utke, I.; et al. Synthesis of model Al-Al2O3 multilayer systems with monolayer oxide thickness control by circumventing native oxidation. *Thin Solid Films* **2020**, *711*, 138287. [CrossRef]
6. Zywotko, D.R.; Zandi, O.; Faguet, J.; Abel, P.R.; George, S.M. ZrO2 monolayer as a removable etch stop layer for thermal Al2O3 atomic layer etching using hydrogen fluoride and trimethylaluminum. *Chem. Mater.* **2020**. [CrossRef]
7. Galazka, Z. β-Ga_2O_3 for wide-bandgap electronics and optoelectronics. *Semicond. Sci. Technol.* **2018**, *33*, 113001. [CrossRef]
8. Ghadi, H.; McGlone, J.F.; Feng, Z.; Bhuiyan, A.F.M.A.U.; Zhao, H.; Arehart, A.R.; Ringel, S.A. Influence of growth temperature on defect states throughout the bandgap of MOCVD-grown β-Ga_2O_3. *Appl. Phys. Lett.* **2020**, *117*. [CrossRef]
9. Lee, I.; Kumar, A.; Zeng, K.; Singisetti, U.; Yao, X. Modeling and Power Loss Evaluation of Ultra Wide Band Gap Ga_2O_3 Device for High Power Applications. In Proceedings of the 2017 IEEE Energy Conversion Congress and Exposition (ECCE 2017), Cincinnati, OH, USA, 1–5 October 2017; pp. 4377–4382. [CrossRef]
10. Green, A.J.; Chabak, K.D.; Heller, E.; Mccandless, J.P.; Moser, N.A.; Fitch, R.C.; Walker, D.E.; Tetlak, S.E.; Crespo, A.; Leedy, K.D.; et al. Recent progress of β-Ga_2O_3 MOSFETs for power electronic applications. *Appl. Phys. Lett.* **2016**, *109*, 391–394.
11. Pearton, S.J.; Yang, J.; Iv, P.H.C.; Ren, F.; Kim, J.; Tadjer, M.J.; Mastro, M.A. A review of Ga_2O_3 materials, processing, and devices. *Appl. Phys. Rev.* **2018**, *5*. [CrossRef]
12. Li, P.; Shi, H.; Chen, K.; Guo, D.; Cui, W.; Zhi, Y.; Wang, S.; Wu, Z.; Chen, Z.; Tang, W. Construction of GaN/Ga_2O_3 p–n junction for an extremely high responsivity self-powered UV photodetector. *J. Mater. Chem. C* **2017**, *5*, 10562–10570. [CrossRef]
13. Jinno, R.; Uchida, T.; Kaneko, K.; Fujita, S. Control of crystal structure of Ga2O3 on sapphire substrate by introduction of α-(AlxGa1−x)2O3 buffer layer. *Phys. Status Solidi Basic Res.* **2018**, *255*, 3–7. [CrossRef]
14. Wheeler, V.D.; Nepal, N.; Boris, D.R.; Qadri, S.B.; Nyakiti, L.O.; Lang, A.; Koehler, A.; Foster, G.; Walton, S.G.; Eddy, C.R.; et al. Phase control of crystalline Ga_2O_3 films by plasma-enhanced atomic layer deposition. *Chem. Mater.* **2020**, *32*, 1140–1152. [CrossRef]
15. Lorenz, M.; Woods, J.; Gambino, R. Gambino, Some electrical properties of the semiconductor β-Ga_2O_3. *J. Phys. Chem. Solids* **1967**, *28*, 403–404. [CrossRef]
16. Ma, J.; Zheng, M.; Chen, C.; Zhu, Z.; Zheng, X.; Chen, Z.; Guo, Y.; Liu, C.; Yan, Y.; Fang, G. Efficient and stable nonfullerene-graded heterojunction inverted perovskite solar cells with inorganic Ga_2O_3 tunneling protective nanolayer. *Adv. Funct. Mater.* **2018**, *28*, 1–11. [CrossRef]
17. Wurdack, M.; Yun, T.; Estrecho, E.; Syed, N.; Bhattacharyya, S.; Pieczarka, M.; Zavabeti, A.; Chen, S.; Haas, B.; Müller, J.; et al. Ultrathin Ga_2O_3 glass: A large-scale passivation and protection material for monolayer WS 2. *Adv. Mater.* **2020**, 1–6. [CrossRef]
18. Choi, Y.H.; Baik, K.H.; Kim, S.; Kim, J. Photoelectrochemical etching of ultra-wide bandgap β-Ga_2O_3 semiconductor in phosphoric acid and its optoelectronic device application. *Appl. Surf. Sci.* **2021**, *539*, 148130. [CrossRef]
19. Afzal, A. β-Ga2O3 nanowires and thin films for metal oxide semiconductor gas sensors: Sensing mechanisms and performance enhancement strategies. *J. Mater.* **2019**, *5*, 542–557. [CrossRef]
20. Ayhan, M.E.; Shinde, M.; Todankar, B.; Desai, P.; Ranade, A.K.; Tanemura, M.; Kalita, G. Ultraviolet radiation-induced photovoltaic action in γ-CuI/β-Ga_2O_3 heterojunction. *Mater. Lett.* **2020**, *262*, 127074. [CrossRef]
21. Negi, G.; Goswami-Giri, A.S. Potentiometric graphite coated electrode based on ditertbutyl-cyclohexano18crown6 for detection of strontium (II) ions. *Anal. Chem. Lett.* **2018**, *8*, 25–34. [CrossRef]

22. Boanini, E.; Gazzano, M.; Nervi, C.; Chierotti, M.R.; Rubini, K.; Gobetto, R.; Bigi, A. Strontium and zinc substitution in β-tricalcium phosphate: An X-ray diffraction, solid state NMR and ATR-FTIR study. *J. Funct. Biomater.* **2019**, *10*, 20. [CrossRef]

23. Sasa, K.; Honda, M.; Hosoya, S.; Takahashi, T.; Takano, K.; Ochiai, Y.; Sakaguchi, A.; Kurita, S.; Satou, Y.; Sueki, K. A sensitive method for Sr-90 analysis by accelerator mass spectrometry. *J. Nucl. Sci. Technol.* **2021**, *58*, 72–79. [CrossRef]

24. Bikit, K.; Knezevic, J.; Mrdja, D.; Todorovic, N.; Kuzmanovic, P.; Forkapic, S.; Nikolov, J.; Bikit, I. Application of 90Sr for industrial purposes and dose assessment. *Radiat. Phys. Chem.* **2021**, *179*, 109260. [CrossRef]

25. Turkington, G.; Gamage, K.A.A.; Graham, J. Detection of Strontium-90, a Review and the Potential for Direct In Situ Detection. In Proceedings of the 2018 IEEE Nuclear Science Symposium and Medical Imaging Conference (NSS/MIC), Sydney, Australia, 10–17 November 2018. [CrossRef]

26. Semenishchev, V.S.; Voronina, A.V. Isotopes of strontium: Properties and applications. *Handb. Environ. Chem.* **2020**, *88*, 25–42. [CrossRef]

27. Jin, Y.; Lei, Z.; Tong, X.; Zhilin, Z.; Shuo, L.; Wenjie, T. Research Progress on Recycling Technology of Lead-containing Glass in Waste Cathode Ray Tubes. In Proceedings of the IOP Conference Series: Earth Environmental Science, Bucharest, Romania, 16–18 October 2020; Volume 446, p. 5. [CrossRef]

28. Cabrera, M.; Pérez, P.; Rosales, J.; Agrela, F. Feasible use of cathode ray tube glass (CRT) and recycled aggregates as unbound and cement-treated granular materials for road sub-bases. *Materials* **2020**, *13*, 748. [CrossRef]

29. Iasir, A.R.M.; Hammond, K.D. Pseudopotential for plane-wave density functional theory studies of metallic uranium. *Comput. Mater. Sci.* **2020**, *171*, 109221. [CrossRef]

30. Torres, E.; Kaloni, T. Projector augmented-wave pseudopotentials for uranium-based compounds. *Comput. Mater. Sci.* **2020**, *171*, 109237. [CrossRef]

31. Mondal, A.; Mohamed, M.; Ping, L.; Taib, M.M.; Samat, M.; Haniff, M.M.; Bahru, R. First-principles studies for electronic structure and optical properties of p-type calcium doped α-Ga$_2$O$_3$. *Materials* **2021**, *14*, 604. [CrossRef]

32. Guo, R.; Su, J.; Yuan, H.; Zhang, P.; Lin, Z.; Zhang, J.; Chang, J.; Hao, Y. Surface functionalization modulates the structural and optoelectronic properties of two-dimensional Ga$_2$O$_3$. *Mater. Today Phys.* **2020**, *12*. [CrossRef]

33. Cohen, A.J.; Mori-Sánchez, P.; Yang, W. Challenges for density functional theory. *Chem. Rev.* **2012**, *112*, 289–320. [CrossRef]

34. Roehrens, D.; Brendt, J.; Samuelis, D.; Martin, M. On the ammonolysis of Ga$_2$O$_3$: An XRD, neutron diffraction and XAS investigation of the oxygen-rich part of the system Ga$_2$O$_3$–GaN. *J. Solid State Chem.* **2010**, *183*, 532–541. [CrossRef]

35. Qian, Y.; Guo, D.; Chu, X.; Shi, H.; Zhu, W.; Wang, K.; Huang, X.; Wang, H.; Wang, S.; Li, P.; et al. Mg-doped p-type β-Ga$_2$O$_3$ thin film for solar-blind ultraviolet photodetector. *Mater. Lett.* **2017**, *209*, 558–561. [CrossRef]

36. Feng, Q.; Liu, J.; Yang, Y.; Pan, D.; Xing, Y.; Shi, X.; Xia, X.; Liang, H. Catalytic growth and characterization of single crystalline Zn doped p-type β-Ga$_2$O$_3$ nanowires. *J. Alloy. Compd.* **2016**, *687*, 964–968. [CrossRef]

37. Mykhaylyk, V.B.; Kraus, H.; Kapustianyk, V.; Rudko, M. Low temperature scintillation properties of Ga$_2$O$_3$. *Appl. Phys. Lett.* **2019**, *115*, 081103. [CrossRef]

38. Liao, C.-H.; Li, K.-H.; Torres-Castanedo, C.G.; Zhang, G.; Li, X. Wide range tunable bandgap and composition β-phase (AlGa)$_2$O$_3$ thin film by thermal annealing. *Appl. Phys. Lett.* **2021**, *118*, 032103. [CrossRef]

39. Zhang, C.; Liao, F.; Liang, X.; Gong, H.; Liu, Q.; Li, L.; Qin, X.; Huang, X.; Huang, C. Electronic transport properties in metal doped beta-Ga$_2$O$_3$: A first principles study. *Phys. B Condens. Matter* **2019**, *562*, 124–130. [CrossRef]

40. Zhang, H.; Yuan, L.; Tang, X.; Hu, J.; Sun, J.; Zhang, Y.; Zhang, Y.; Jia, R. Progress of ultra-wide bandgap Ga$_2$O$_3$ semiconductor materials in power MOSFETs. *IEEE Trans. Power Electron.* **2020**, *35*, 5157–5179. [CrossRef]

41. Mohamed, M.; Unger, I.; Janowitz, C.; Manzke, R.; Galazka, Z.; Uecker, R.; Fornari, R. The surface band structure of β-Ga$_2$O$_3$. *J. Phys. Conf. Ser.* **2011**, *286*. [CrossRef]

42. Zhang, Y.; Yan, J.; Zhao, G.; Xie, W. First-principles study on electronic structure and optical properties of Sn-doped β-Ga$_2$O$_3$. *Phys. B Condens. Matter* **2010**, *405*, 3899–3903. [CrossRef]

43. Yan, J.; Qu, C.; Jinliang, Y.; Chong, Q. Electronic structure and optical properties of F-doped β-Ga$_2$O$_3$ from first principles cal culations. *J. Semicond.* **2016**, *37*. [CrossRef]

44. Vincenzo, F. Dielectric scaling of the self-energy scissor operator in semiconductors and insulators. *Phys. Rev. B* **1995**, *51*, 196–198.

45. Godby, R.W.; Schlüter, M.; Sham, L.J. Self-energy operators and exchange-correlation potentials in semiconductors. *Phys. Rev. B* **1988**, *37*, 10159–10175. [CrossRef]

46. Zhang, J.; Li, B.; Xia, C.; Pei, G.; Deng, Q.; Yang, Z.; Xu, W.; Shi, H.; Wu, F.; Wu, Y.; et al. Growth and spectral characterization of β-Ga$_2$O$_3$ single crystals. *J. Phys. Chem. Solids* **2006**, *67*, 2448–2451. [CrossRef]

micromachines

Article

Effect of High-Temperature Nitridation and Buffer Layer on Semi-Polar (10–13) AlN Grown on Sapphire by HVPE

Qian Zhang [1,†], Xu Li [1,†], Jianyun Zhao [1], Zhifei Sun [2], Yong Lu [1], Ting Liu [1,*] and Jicai Zhang [1,3,*]

[1] College of Mathematics and Physics, Beijing University of Chemical Technology, Beijing 100029, China; zhang2020@buct.edu.cn (Q.Z.); xuli@mail.buct.edu.cn (X.L.); jyzhao@mail.buct.edu.cn (J.Z.); luy@mail.buct.edu.cn (Y.L.)

[2] School of Physical Education and Health Management, Guangxi Normal University, Guilin 541001, China; szfszfffbb@163.com

[3] State Key Laboratory of Chemical Resource Engineering, Beijing University of Chemical Technology, Beijing 100029, China

* Correspondence: liuting2021@buct.edu.cn (T.L.); jczhang@mail.buct.edu.cn (J.Z.)

† These authors contributed equally to this work.

Abstract: We have investigated the effect of high-temperature nitridation and buffer layer on the semi-polar aluminum nitride (AlN) films grown on sapphire by hydride vapor phase epitaxy (HVPE). It is found the high-temperature nitridation and buffer layer at 1300 °C are favorable for the formation of single (10–13) AlN film. Furthermore, the compressive stress of the (10–13) single-oriented AlN film is smaller than polycrystalline samples which have the low-temperature nitridation layer and buffer layer. On the one hand, the improvement of (10–13) AlN crystalline quality is possibly due to the high-temperature nitridation that promotes the coalescence of crystal grains. On the other hand, as the temperature of nitridation and buffer layer increases, the contents of N-Al-O and Al-O bonds in the AlN film are significantly reduced, resulting in an increase in the proportion of Al-N bonds.

Keywords: HVPE; AlN; high-temperature; buffer layer; nitridation

Citation: Zhang, Q.; Li, X.; Zhao, J.; Sun, Z.; Lu, Y.; Liu, T.; Zhang, J. Effect of High-Temperature Nitridation and Buffer Layer on Semi-Polar (10–13) AlN Grown on Sapphire by HVPE. *Micromachines* **2021**, *12*, 1153. https://doi.org/10.3390/mi12101153

Academic Editor: Giovanni Verzellesi

Received: 21 August 2021
Accepted: 23 September 2021
Published: 25 September 2021

Publisher's Note: MDPI stays neutral with regard to jurisdictional claims in published maps and institutional affiliations.

Copyright: © 2021 by the authors. Licensee MDPI, Basel, Switzerland. This article is an open access article distributed under the terms and conditions of the Creative Commons Attribution (CC BY) license (https://creativecommons.org/licenses/by/4.0/).

1. Introduction

AlN is a potential material for the deep ultraviolet (DUV) optical devices, such as light-emitting diodes (LEDs) and laser diodes (LDs) [1], since it has wide direct band gap of 6.2 eV and excellent physical and chemical stability. These optoelectronic devices are generally grown along AlN c-axis [2,3]. However, the strong spontaneous and piezoelectric polarization along c-axis will weaken the recombination of carriers in the quantum wells [4], which in turn decrease the luminescence efficiency of devices. The use of semi-polar substrates and epitaxial layers [5–9], such as (10–11), (10–12) and (10–13) AlN, can effectively solve this problem since the polarization field is greatly weakened [10,11]. According to the energy band structure, other semi-polar plane, such as (11–22), can produce a negative polarization field across multiple quantum wells used in optoelectronic devices, which can reduce the confinement of hole states and increase the carrier loss. In contrast, the (10–13) plane has a positive polarization field, which is favorable for the devices [12].

Semi-polar nitride films were generally grown on M-plane sapphire [5,6], silicon [7,8] and ZnO [13,14]. It is also reported that native semi-polar AlN substrate were used for homoepitaxial semi-polar AlN [15]. Shen et al. obtained high quality (10–13) AlN by ammonia-free metalorganic vapor phase at high temperature of 1650 °C [5]. Kukushkin er al. investigated the semi-polar AlN without cracks on Si (001) and hybrid SiC/Si (001) substrates [7]. Bessolov et al. prepared the hexagonal AlN layer on Si with V-groove nanostructured surface by HVPE at 1080 °C [8]. Ueno et al. obtained high quality semi-polar AlN and AlGaN on ZnO substrates with annealing in the air by growing a room temperature epitaxial AlN buffer layer [13,14].

Due to the transparency and low cost, sapphire substrate is normally used for preparation of AlN substrates and AlN-based deep ultraviolet optoelectronic devices. For (10–13) AlN on m-plane sapphire, the in-plane epitaxial relationship between (10–13) AlN and m-plane sapphire is $[30\text{-}3\text{-}2]_{AlN}//[1\text{-}210]_{sapphire}$ and $[1\text{-}210]AlN//[0001]sa_{pphire}$ [5]. According to the epitaxial relationship, it can be determined that the lattice mismatch between m-plane sapphire and (10–13) AlN is the smallest [12]. When growing AlN on sapphire [16–19], nitridation is a common method to improve crystal quality [20,21]. Moreover, nitridation has a large impact on the crystal orientation of semi-polar AlN [22,23]. Furthermore, it is well known that buffer layer is an efficient approach to reduce the lattice mismatch between III-V nitrides and foreign substrates [24]. At present, the effect of the buffer layer on the quality of polar AlN crystals has been widely confirmed [24–27]. In addition, compared with metalorganic chemical vapor deposition (MOCVD), hydride vapor phase epitaxy (HVPE) is more suitable for AlN substrate due to the high growth rate and low cost [28]. However, the comprehensive influence of high-temperature nitridation and high-temperature buffer layer on the semi-polar AlN film grown by HVPE has not yet been clarified.

In this work, we use HVPE to grow semi-polar AlN on m-plane sapphire substrate. The influence of high-temperature nitridation and buffer layer on the semi-polar AlN films has been carefully studied.

2. Experiment

The growth of semi-polar (10–13) AlN sample was performed in a home-made horizontal HVPE system. 2-inch m-plane sapphire with miscut-angle of $\pm0.1°$ was used as substrate, and HCl and NH_3 were used as input active gases. The mixture of H_2 and N_2 were utilized as carrier gas under 40 Torr. At first, the m-plane sapphire substrate was heated to the nitridation temperature in the carrier gas and kept for 10 min in H_2 ambient to remove the surface pollution and achieve thermal stability. The sapphire substrate was then nitrided in NH_3 ambient about 10 min. Next, the buffer layer was grown for 1 min. Finally, the temperature was raised to 1500 °C to grow AlN film for 30 min. Table 1 shows the growth conditions of four samples with different nitridation temperature and buffer layer temperature. X-ray diffractometer (XRD, Philips, X'pert MRD PIXcel, Amsterdam, The Netherlands) was used to characterize the orientation and quality of AlN film. Scanning electron microscope (SEM, HITACHI, SU8020, Tokyo, Japan) was used to study the cross-section morphology. Raman spectrometer (HORIBA, LabRam HR Evolution, Kyoto, Japan) was used to study the stress distribution of AlN film. X-ray photoelectron spectroscopy (XPS, Thermo Scientific, Escalab 250Xi, Waltham, MA, USA) was used to analyze the chemical composition of the AlN film.

Table 1. Growth conditions of sample A, B, C and D.

Sample	Nitridation Temperature/°C	Temperature of Buffer Layer/°C	Growth Temperature/°C
A	1050	800	1500
B	1050	1300	1500
C	1300	800	1500
D	1300	1300	1500

3. Results and Discussion

Nitridation and buffer layer are introduced to grow semi-polar (10–13) AlN by HVPE. As shown in Figure 1, the growth orientation of the semi-polar AlN films is characterized by XRD ω-2θ scan. Low-temperature (1050 °C) nitridation and low-temperature (800 °C) buffer layer are first tried to grow semi-polar (10–13) AlN. In Figure 1a, (10–11) and (20–22) diffraction peaks are also observed except the strong (10–13) diffraction peak, indicating that sample A is not the single crystal. In sample B, the buffer layer temperature is kept at 800 °C and the nitridation temperature is increased to 1300 °C. Although the additional diffraction peaks still exist, the (10–13) diffraction peak is narrowed as shown in Figure 1b. In contrast, for sample C, the nitridation temperature is 1050 °C, while the buffer layer temperature rises to 1300 °C. From Figure 1c, it is clearly to see that the intensity of impurity peaks of (10–11) and (20–22) is significantly reduced. Comparing with sample A, B and C, we can conclude that high-temperature nitridation and high-temperature buffer layer are promising to improve the crystal quality of the semi-polar (10–13) AlN. Therefore, both high-temperature (1300 °C) nitridation and high-temperature buffer (1300 °C) layer are used in sample D. As expected, in Figure 1d, there are no other impurity peaks but only sharp (10–13) AlN diffraction peak, implying that sample D is (10–13)-oriented single crystal. The weak peak between (10–13) AlN and (30–30) sapphire is most probably a shoulder peak due to the twin structures, which are common small-angle grains in (10–13) AlN [29]. Compared with sample D, an additional broad peak around 44.4° exits in samples A-C. Unfortunately, the origin of the peak is unknown currently. But from the characterizations of this peak, we believe it should come from AlN.

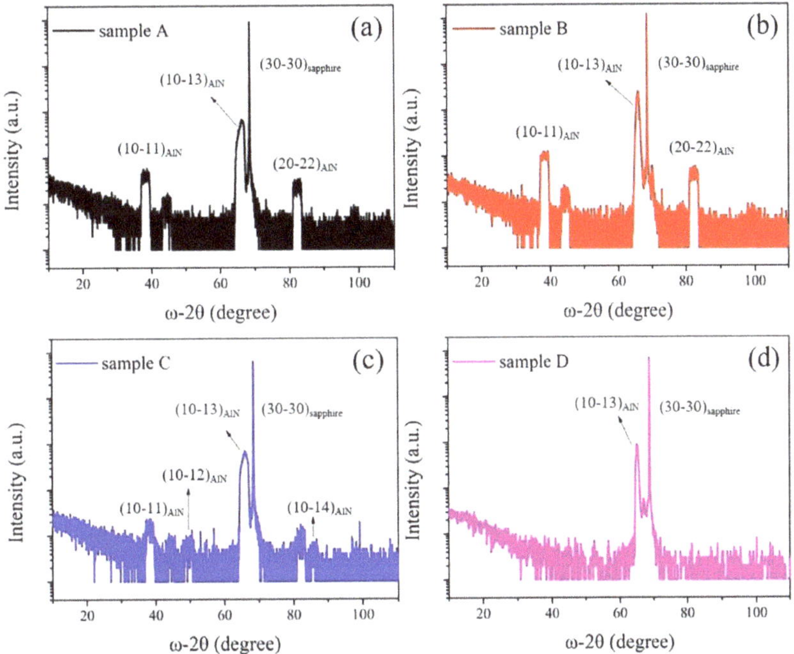

Figure 1. (a–d) corresponding to XRD $\omega-2\theta$ scan of sample A, sample B, sample C and sample D, respectively.

Figure 2 shows the variations of full width at half maximum (FWHM) of (10–13) ω scans at different angle ϕ in the range of 0° and 360° for sample A, B, C and D, respectively. There are two FWHM values at the same φ angle due to the effect of stress. It can be clearly observed that the change of FWHM has an M-shaped curve, indicating that it has obvious anisotropy characteristics. The FWHM value of sample B is significantly lower than that of sample A, suggesting that the crystal quality has been significantly improved after increasing the temperature of the buffer layer. All the FWHM values for sample D are smaller than those of other samples, demonstrating that high-temperature nitridation and high-temperature buffer layer can improve the crystal quality of (10–13) AlN films significantly. The insert of Figure 2d shows the rocking curves of (0002) and (10–13) diffraction. The FWHM values are 0.359° and 0.356° for (0002) and (10–13) diffraction, respectively. It is very close to the value of literature [6].

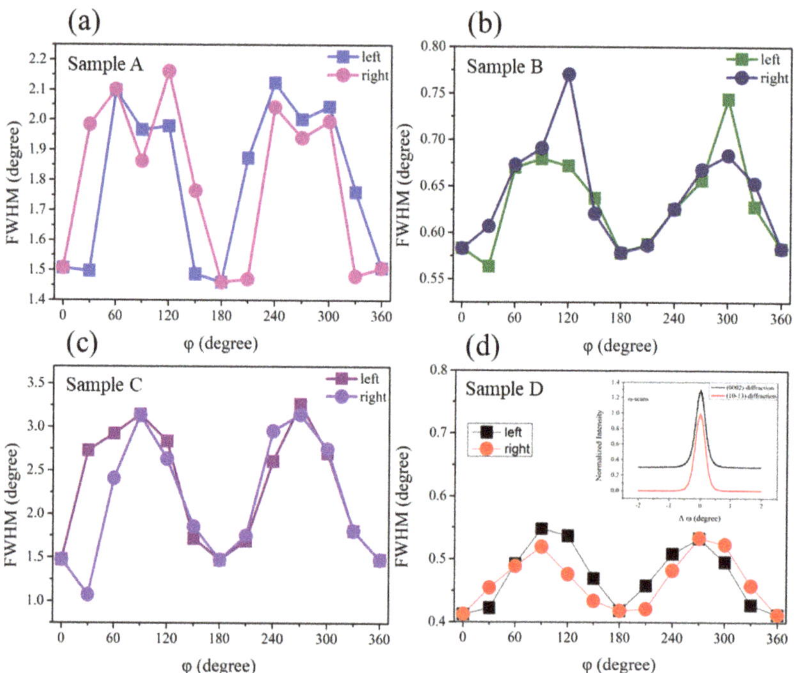

Figure 2. (**a–d**) is the variations of FWHM values of XRC at different φ angle for (10–13) AlN films of sample A, sample B, sample C and sample D, respectively. The left and right corresponding to different FWHM values of XRC at the same φ angle. The insert of figure (**d**) shows the FWHM of (0002) and (10–13) diffraction.

Figure 3 shows the cross-sectional SEM images of the four samples. In sample D (Figure 3d), the crystal column extends from the growth interface to the surface, and the orientation of the crystal column is almost the same, which is consistent with the result of XRD. However, the crystal columns are still not coalescent, as shown in the insert of Figure 3d. While in the other three samples, the crystal columns become smaller and their orientation looks disordered. Moreover, in sample A, there are many holes as large as 200 nm at the growth interface. In the case of low-temperature nitridation and buffer layer, high density of AlN defects appear and AlN grains are not easy to merge at the initial growth stage [6], resulting in that sapphire can't be completely covered by AlN film. During the high-temperature (1500 °C) growth, due to lack of AlN protective layer, the exposed sapphire surface is decomposed, leading to large holes in sample A as shown in Figure 1a.

Raman spectrum is then introduced to evaluate the residual stress in the semi-polar AlN films. In the x(yy)z scattering configuration [30], E_2(low), A_1(TO), E_2(high), E_1(TO) and E_1(LO) Raman peaks of AlN can be observed in Figure 4a. Among these peaks, E_2(high) peak could reflect the residual stress of AlN film and the value of E_2(high) peak under stress free is 657.4 cm^{-1} [7]. The enlarged image (Figure 4b) of E_2(high) peak demonstrates that the peak values of the four samples are larger than 657.4 cm^{-1}, which means they are all in the compressive state. Furthermore, the E_2(high) peak value of sample B and D is smaller than A and C, indicating that the buffer layer at high temperature is beneficial to reduce the compressive stress of (10–13) AlN film.

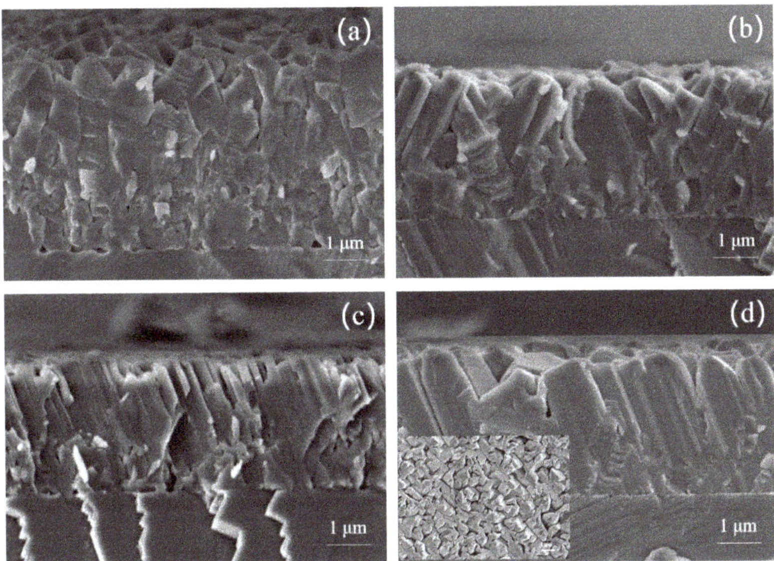

Figure 3. (**a–d**) corresponding to the cross-sectional SEM image of sample A, sample B, sample C and sample D, respectively. The insert of figure (**d**) shows the surface SEM image of sample D, in which the scale bar is 1 um.

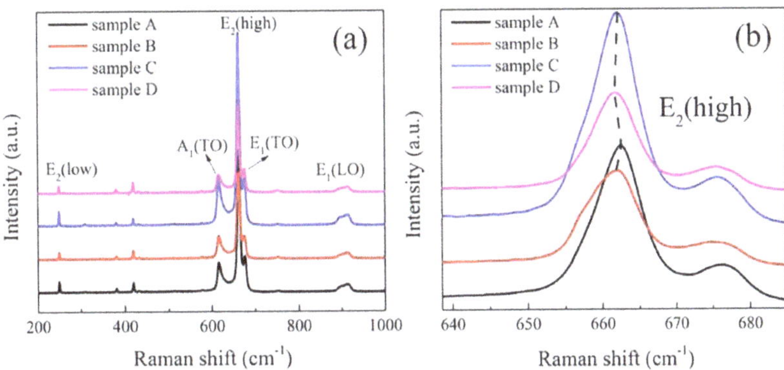

Figure 4. (**a**) Raman spectra of AlN samples, including E_2(low), A_1(TO), E_2(high), E_1(TO) and E_1(LO) peaks of AlN. (**b**) The enlarged view of the E_2(high) peaks in (**a**). The black, red, blue and pink curves corresponding to sample A, B, C and D, respectively. The black dash line is guide for eyes.

In order to investigate the effect of nitridation and buffer layer on the chemical states of AlN films, we perform an XPS core level measurement. Figure 5 shows the XPS core level spectra of N 1s and Al 2p. In Figure 5a, the N 1s spectrum in all samples can be deconvoluted into two peaks at 396.8 ± 0.2 eV and 397.7 ± 0.3 eV, corresponding to the N−Al bond and N−Al−O bond, respectively [31]. In sample A, the relative area of the N−Al peak is only 25.51% while the relative area of the N−Al−O peak is as large as 74.49%. With the increasing temperature of nitridation and buffer layer, the relative area of the N−Al peak is enhanced obviously, increasing to 77.06% in sample D. As a result, comparing with sample A, the center of the N 1s spectra shifts from 398.1 eV to 396.7 eV in sample D due to the expanded content of N−Al peak [31]. Furthermore, in Figure 5b, the Al−N and Al−O peaks in the Al 2p spectrum are centered at 73.5 ± 0.2 eV and 74.3 ± 0.1 eV, respectively [21]. It is obvious that sample D has the larger Al−N content and the smaller

Al−O content than other samples. Therefore, both N 1s and Al 2p spectrum show that the high-temperature nitridation and buffer layer are beneficial to reduce the combination of oxygen impurity with Al atom or N atom, resulting in a better AlN film.

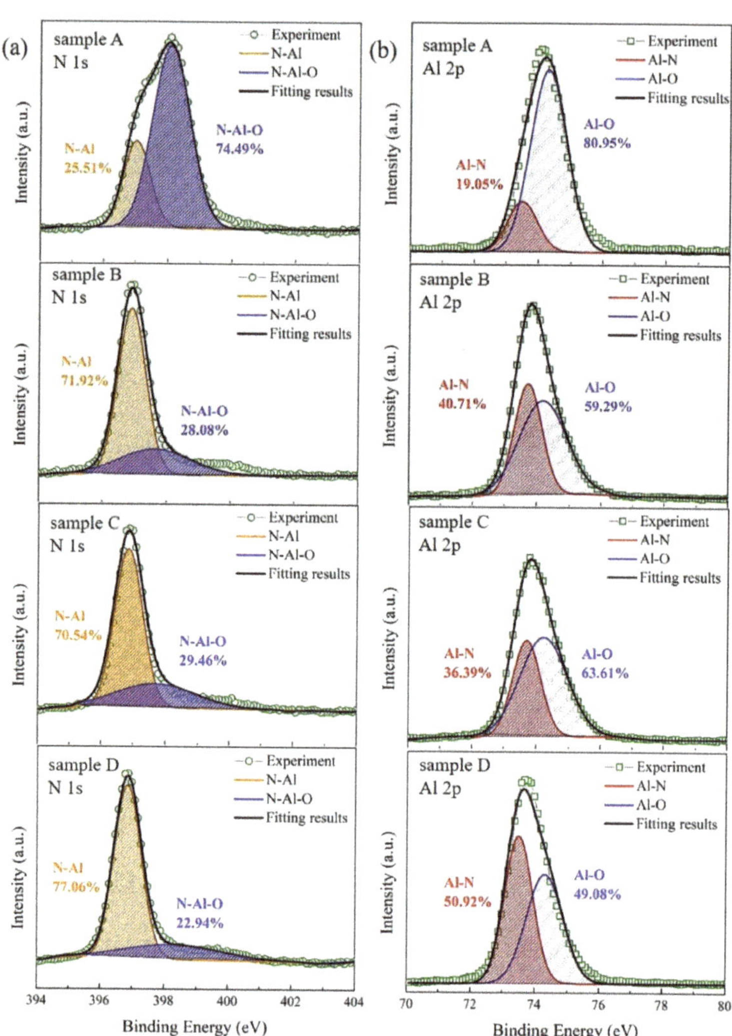

Figure 5. (a) N 1s high-resolution XPS scans of sample A, B, C and D. The green circle, orange, blue and black curves corresponding to the experiment result, N−Al bond, N−Al−O bond and fitting result, respectively. (b) Al 2p high-resolution XPS scans of sample A, B, C and D. The green square, red, blue and black curves corresponding to the experiment result, Al−N bond, Al−O bond and fitting result, respectively.

4. Conclusions

In conclusion, with the high-temperature nitridation and buffer layer, the single (10–13) AlN film is successfully obtained on m-plane sapphire by HVPE. Comparing with the polycrystalline samples which have the low-temperature nitridation layer and buffer layer, there is the smallest compressive stress in the (10–13) single-oriented AlN film. The simultaneous introduction of high-temperature nitridation and buffer layer is beneficial to promote the coalescence of crystal grains and reduce the content of impurity components, like $N-Al-O$ and $Al-O$, thus improving the crystal quality of semi-polar (10–13) AlN.

Author Contributions: Conceptualization, Q.Z., X.L., T.L. and J.Z. (Jicai Zhang); data collection, Q.Z., X.L., J.Z. (Jianyun Zhao) and Z.S.; supervision, T.L. and J.Z. (Jicai Zhang); data analysis, Q.Z., X.L., J.Z. (Jianyun Zhao), Y.L., T.L. and J.Z. (Jicai Zhang); writing-review and editing, Q.Z., X.L., T.L. and J.Z. (Jicai Zhang). All authors have read and agreed to the published version of the manuscript.

Funding: This work was partly supported by the Beijing Municipal Natural Science Foundation (4182046), the National Natural Science Foundation of China (61874007, 12074028, 52102152), Shandong Provincial Major Scientific and Technological Innovation Project (2019JZZY010209), Key-area research and development program of Guangdong Province (2020B010172001), and the Fundamental Research Funds for the Central Universities (buctrc201802, buctrc201830, buctrc202127).

Institutional Review Board Statement: Not applicable.

Informed Consent Statement: Not applicable.

Data Availability Statement: Data sharing is not applicable to this article.

Conflicts of Interest: The authors declare no conflict of interest.

References

1. Li, D.; Jiang, K.; Sun, X.; Guo, C. AlGaN photonics: Recent advances in materials and ultraviolet devices. *Adv. Opt. Photonics* **2018**, *10*, 43–110. [CrossRef]
2. Jo, M.; Hirayama, H. Growth of non-polar a-plane AlN on r-plane sapphire. *Jpn. J. Appl. Phys.* **2016**, *55*. [CrossRef]
3. Le, D.D.; Kim, D.Y.; Hong, S.-K. Crystal orientation variation of nonpolar AlN films with III/V ratio on r-plane sapphire substrates by plasma-assisted molecular beam epitaxy. *Electron. Mater. Lett.* **2014**, *10*, 1109–1114. [CrossRef]
4. Leathersich, J.; Suvarna, P.; Tungare, M.; Shahedipour-Sandvik, F. Homoepitaxial growth of non-polar AlN crystals using molecular dynamics simulations. *Surf. Sci.* **2013**, *617*, 36–41. [CrossRef]
5. Shen, X.-Q.; Kojima, K.; Okumura, H. Single-phase high-quality semipolar (10–13) AlN epilayers on m-plane (10–10) sapphire substrates. *Appl. Phys. Express* **2020**, *13*, 035502. [CrossRef]
6. Li, X.; Zhao, J.; Liu, T.; Lu, Y.; Zhang, J. Growth of semi-polar (10–13) AlN film on m-plane sapphire with high-temperature nitridation by HVPE. *Materials* **2021**, *14*, 1722. [CrossRef] [PubMed]
7. Kukushkin, S.A.; Osipov, A.V.; Redkov, A.V.; Sharofidinov, S.S. Epitaxial growth of bulk semipolar AlN films on Si (001) and Hybrid SiC/Si (001) Substrates. *Tech. Phys. Lett.* **2020**, *46*, 539–542. [CrossRef]
8. Bessolov, V.N.; Kompan, M.E.; Konenkova, E.V.; Panteleev, V.N. Hydride vapor-phase epitaxy of a semipolar AlN(10–12) layer on a nanostructured Si(100) substrate. *Tech. Phys. Lett.* **2020**, *46*, 59–61. [CrossRef]
9. Jo, M.; Itokazu, Y.; Kuwaba, S.; Hirayama, H. Improved crystal quality of semipolar AlN by employing a thermal annealing technique with MOVPE. *J. Cryst. Growth* **2019**, *507*, 307–309. [CrossRef]
10. Tajima, J.; Murakami, H.; Kumagai, Y.; Takada, K.; Koukitu, A. Preparation of a crack-free AlN template layer on sapphire substrate by hydride vapor-phase epitaxy at 1450 °C. *J. Cryst. Growth* **2009**, *311*, 2837–2839. [CrossRef]
11. Ranalli, F.; Parbrook, P.J.; Bai, J.; Lee, K.B.; Wang, T.; Cullis, A.G. Non-polar AlN and GaN/AlN on r-plane sapphire. *Phys. Status Solidi C* **2009**, *6*, 780–783. [CrossRef]
12. Hu, N.; Dinh, D.V.; Pristovsek, M.; Honda, Y.; Amano, H. How to obtain metal-polar untwinned high-quality (10–13) GaN on m-plane sapphire. *J. Cryst. Growth* **2019**, *507*, 205–208. [CrossRef]
13. Ueno, K.; Kobayashi, A.; Ohta, J.; Fujioka, H. Improvement in the Crystalline Quality of Semipolar AlN(1–102) Films by Using ZnO Substrates with Self-Organized Nanostripes. *Appl. Phys. Express* **2010**, *3*, 041002. [CrossRef]
14. Ueno, K.; Kobayashi, A.; Ohta, J.; Fujioka, H. Structural properties of semipolar $Al_xGa_{1-x}N$ (1–103) films grown on ZnO substrates using room temperature epitaxial buffer layers. *Phys. Status Solidi A* **2010**, *207*, 2149–2152. [CrossRef]
15. Ichikawa, S.; Funato, M.; Kawakami, Y. Metalorganic vapor phase epitaxy of pit-free AlN homoepitaxial films on various semipolar substrates. *J. Cryst. Growth* **2019**, *522*, 68–77. [CrossRef]
16. Wu, Z.; Yan, J.; Guo, Y.; Zhang, L.; Lu, Y.; Wei, X.; Wang, J.; Li, J. Study of the morphology evolution of AlN grown on nano-patterned sapphire substrate. *J. Semicond.* **2019**, *40*, 122803. [CrossRef]
17. Sun, X.; Li, D.; Chen, Y.; Song, H.; Jiang, H.; Li, Z.; Miao, G.; Zhang, Z. In situ observation of two-step growth of AlN on sapphire using high-temperature metal–organic chemical vapour deposition. *CrystEngComm* **2013**, *15*, 6066–6073. [CrossRef]
18. Lahourcade, L.; Bellet-Amalric, E.; Monroy, E.; Chauvat, M.P.; Ruterana, P. Molecular beam epitaxy of semipolar AlN(11–22) and GaN(11–22) on m-sapphire. *J. Mater. Sci.: Mater. Electron.* **2008**, *19*, 805–809. [CrossRef]
19. Nagashima, T.; Harada, M.; Yanagi, H.; Fukuyama, H.; Kumagai, Y.; Koukitu, A.; Takada, K. Improvement of AlN crystalline quality with high epitaxial growth rates by hydride vapor phase epitaxy. *J. Cryst. Growth* **2007**, *305*, 355–359. [CrossRef]

20. Ma, Z.-C.; Chiu, K.-A.; Wei, L.-L.; Chang, L. Formation of m-plane AlN on plasma-nitrided m-plane sapphire. *Jpn. J. Appl. Phys.* **2019**, *58*, Sc1033. [CrossRef]
21. Won, Y.; So, B.; Woo, S.; Lee, D.; Kim, M.; Nam, K.; Im, S.; Shim, K.B.; Nam, O. Effect of nitridation on the orientation of GaN layer grown on m-sapphire substrates using hydride vapor phase epitaxy. *J. Ceram. Process. Res.* **2014**, *15*, 61–65.
22. Stellmach, J.; Frentrup, M.; Mehnke, F.; Pristovsek, M.; Wernicke, T.; Kneissl, M. MOVPE growth of semipolar (11–22) AlN on m-plane (10–10) sapphire. *J. Cryst. Growth* **2012**, *355*, 59–62. [CrossRef]
23. Jo, M.; Itokazu, Y.; Kuwaba, S.; Hirayama, H. Controlled crystal orientations of semipolar AlN grown on an m-plane sapphire by MOCVD. *Jpn. J. Appl. Phys.* **2019**, *58*, SC1031. [CrossRef]
24. Zhao, D.G.; Jiang, D.S.; Wu, L.L.; Le, L.C.; Li, L.; Chen, P.; Liu, Z.S.; Zhu, J.J.; Wang, H.; Zhang, S.M.; et al. Effect of dual buffer layer structure on the epitaxial growth of AlN on sapphire. *J. Alloys Compd.* **2012**, *544*, 94–98. [CrossRef]
25. Lin, C.H.; Yamashita, Y.; Miyake, H.; Hiramatsu, K. Fabrication of high-crystallinity a-plane AlN films grown on r-plane sapphire substrates by modulating buffer-layer growth temperature and thermal annealing conditions. *J. Cryst. Growth* **2017**, *468*, 845–850. [CrossRef]
26. Miyake, H.; Nishio, G.; Suzuki, S.; Hiramatsu, K.; Fukuyama, H.; Kaur, J.; Kuwano, N. Annealing of an AlN buffer layer in N_2-CO for growth of a high-quality AlN film on sapphire. *Appl. Phys. Express* **2016**, *9*, 025501. [CrossRef]
27. Jiang, K.; Sun, X.; Ben, J.; Jia, Y.; Liu, H.; Wang, Y.; Wu, Y.; Kai, C.; Li, D. The defect evolution in homoepitaxial AlN layers grown by high-temperature metal-organic chemical vapor deposition. *CrystEngComm* **2018**, *20*, 2720–2728. [CrossRef]
28. Sun, M.; Li, J.; Zhang, J.; Sun, W. The fabrication of AlN by hydride vapor phase epitaxy. *J. Semicond.* **2019**, *40*, 121803. [CrossRef]
29. Okojie, R.S.; Holzheu, T.; Huang, X.; Dudley, M. X-ray diffraction measurement of doping induced lattice mismatch in n-type 4H-SiC epilayers grown on p-type substrates. *Appl. Phys. Lett.* **2003**, *83*, 1971–1973. [CrossRef]
30. Kallel, T.; Dammak, M.; Wang, J.; Jadwisienczak, W.M. Raman characterization and stress analysis of AlN: Er^{3+} epilayers grown on sapphire and silicon substrates. *Mater. Sci. Eng. B* **2014**, *187*, 46–52. [CrossRef]
31. Kim, D.H.; Schweitz, M.A.; Koo, S.M. Effect of gas annealing on the electrical properties of Ni/AlN/SiC. *Micromachines* **2021**, *12*, 283. [CrossRef]

 *micromachines*

Article

Crystal Growth of Cubic and Hexagonal GaN Bulk Alloys and Their Thermal-Vacuum-Evaporated Nano-Thin Films

Marwa Fathy [1,*], Sara Gad [1,*], Badawi Anis [2,3] and Abd El-Hady B. Kashyout [1]

1   Electronic Materials Department, Advanced Technology and New Materials Research Institute, City of Scientific Research and Technological Applications (SRTA-City), Alexandria 21934, Egypt; akashyout@srtacity.sci.eg

2   Spectroscopy Department, Physics Division, National Research Centre, 33 El Bohouth St., Dokki, Giza 12622, Egypt; badawi.ali@gmail.com

3   Molecular and Fluorescence Lab., Central Laboratories Network, National Research Centre, 33 El Bohouth Str., Dokki, Giza 12622, Egypt

*   Correspondence: mbahnase@srtacity.sci.eg (M.F.); sgad@srtacity.sci.eg (S.G.)

Abstract: In this study, we investigate a novel simple methodology to synthesize gallium nitride nanoparticles (GaN) that could be used as an active layer in light-emitting diode (LED) devices by combining the crystal growth technique with thermal vacuum evaporation. The characterizations of structural and optical properties are carried out with different techniques to investigate the main featured properties of GaN bulk alloys and their thin films. Field emission scanning electron microscopy (FESEM) delivered images in bulk structures that show micro rods with an average diameter of 0.98 μm, while their thin films show regular microspheres with diameter ranging from 0.13 μm to 0.22 μm. X-ray diffraction (XRD) of the bulk crystals reveals a combination of 20% hexagonal and 80% cubic structure, and in thin films, it shows the orientation of the hexagonal phase. For HRTEM, these microspheres are composed of nanoparticles of GaN with diameter of 8–10 nm. For the optical behavior, a band gap of about from 2.33 to 3.1 eV is observed in both cases as alloy and thin film, respectively. This article highlights the fabrication of the major cubic structure of GaN bulk alloy with its thin films of high electron lifetime.

Keywords: GaN; crystal growth; cubic and hexagonal structure; blue and yellow luminescence; electron lifetime

Citation: Fathy, M.; Gad, S.; Anis, B.; Kashyout, A.E.-H.B. Crystal Growth of Cubic and Hexagonal GaN Bulk Alloys and Their Thermal-Vacuum-Evaporated Nano-Thin Films. *Micromachines* **2021**, *12*, 1240. https://doi.org/10.3390/mi12101240

Academic Editors: Giovanni Verzellesi and Aiqun Liu

Received: 29 August 2021
Accepted: 8 October 2021
Published: 13 October 2021

Publisher's Note: MDPI stays neutral with regard to jurisdictional claims in published maps and institutional affiliations.

Copyright: © 2021 by the authors. Licensee MDPI, Basel, Switzerland. This article is an open access article distributed under the terms and conditions of the Creative Commons Attribution (CC BY) license (https://creativecommons.org/licenses/by/4.0/).

1. Introduction

Generally, gallium nitride (GaN) is one of the group III-V nitrides, with a band gap ranging between 3.27 eV and 3.47 eV [1]. At room temperature, it has two different phases of structural properties and dislocations: wurtzite (2H) and zinc-blende (3C). The cubic phase (c-GaN) is metastable, making its growth a difficult challenge. It is *p*-type doping and has higher mobility than hexagonal structure. Cubic or zinc-blende structure is easily cleavable, which, combined with the evidence for optical gain, offers attractive possibilities such as blue light emission and promises to achieve improved efficiencies for green-wavelength LEDs [2]. Hexagonal phase (h-GaN) is a promising candidate for creating high power devices due to its large band gap (3.47 eV) with high saturation velocity. Furthermore, spontaneous piezoelectric polarization occurs in h-GaN-induced internal electric fields, which causes energy band tilting [3] and influence optoelectronic devices' performance [4].

Many studies fabricated GaN using different physical growth techniques such as metal organic chemical vapor deposition (MOCVD) [5], ion-beam-assisted molecular beam epitaxy [6], reactive molecular beam epitaxy [7], thermal ammonization [8], physical vacuum vapor deposition [9], chemical vapor deposition (CVD) [10,11], thermal vapor deposition [12], and combustion method [13–15].

MOCVD is the most useful technique in the preparation of high-quality h-GaN wurtzite phase, although it is not suitable for c-GaN because the process is done at high temperature. By using lower-temperature techniques such as molecular beam epitaxy, the percentage of h-GaN in the thermodynamically stable phase could be dominant. Furthermore, the structural and optical properties of the deposited GaN depend on the type of substrate. [1]

This work involves the synthesis of high-quality GaN bulk alloy as a mixture of cubic and hexagonal microstructure using a simple technique, "crystal growth", under a temperature of 850 °C from Ga metal with pure ammonia gas, and demonstrates the structure transformation process due to the effect of the deposition process using high thermal vacuum evaporation on a glass substrate.

2. Materials and Methods

The following chart explains the experimental procedure carried out through this work (Figure 1).

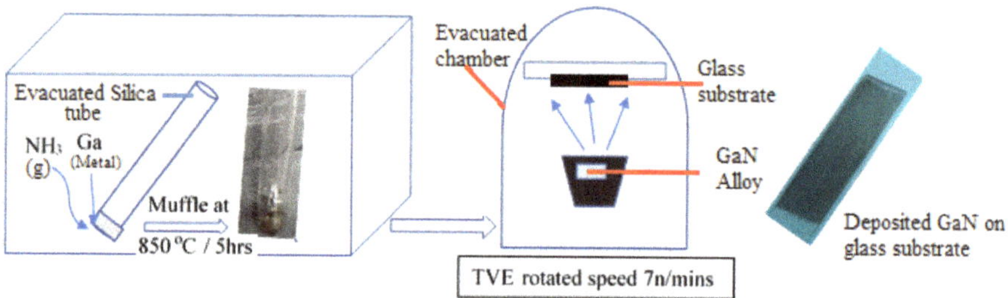

Figure 1. Scheme diagram of the experimental work.

For GaN bulk alloy preparation, gallium metal of 99.999% purity (Aldrich-Sigma, Saint Louis, MO, USA) is introduced in a silica tube, which is first vacuumed at 10^{-4} Torr and then filled with pure NH_3 gas (99.9%). The tube is finally sealed by welding a narrow nick, which is made to insert the NH_3 gas. The ammonia flow rate is kept at 0.1 sccm. The tube temperature reaches 850 °C for 2 h in a muffle furnace (Carbolite AAF117) and then is reduced inside the furnace in order to prevent the segregations previously detected in InGaN alloys [16,17].

For GaN thin film using thermal vacuum evaporator, the vapor faces the glass substrates and condenses as a thin film. The glass substrate is fixed at the substrate holder, which is rotated at a fixed speed during the evaporation process. Finally, the GaN is deposited on the glass substrate with a thickness of 119 nm (measured from SEM cross section). The structural and morphological properties of the GaN thin film are systematically analyzed by X-ray diffraction (XRD-Shimadzu XRD 7000 maxima powder diffractometer, Kyoto, Japan), field emission scanning electron microscopy (FESEM-Quanta 250, USGS, Laurel, MD, USA), photoluminescence (PL-Perkin Elmer Luminescence Spectrometer Model LSS, Shared Instrumentations, Richmond, CA, USA), transmission electron microscopy (JEM-2100, JEOL, Japan) running at 200 KV. Also, we report triggered single-photon emission from gallium nitride with nanostructure by using fluorescence lifetime imaging microscopy (FLIM system Alba with v5 from ISS). A laser diode of 640 nm was used for the excitation, coupled with a scanning module of (ISS) through multi-band dichroic filter to epifluorescence microscope (Model IX73, Olympus, Tokyo, Japan) with UPLSA 60X objective 1.2 NA and 0.28 mm width. Emission is observed and detected by cooled low noise (below 100 counts/s) with a detector of GaAs fast PMTs for time-correlated single-photon counting (TCSPC). FLIM data are acquired using ISS A330 Fast FLIM module with n harmonics of 20 MHz laser repetition frequency. FLIM data are analyzed with Vista Vision Suite software (Vista v.204 from ISS). The FLIM analysis is carried out using the fitting indicating the formation of cubic structure. The XRD peaks at 60.8° (301), 57.33° (110), 37.9° (101) and 33.39° (100) are for JCPDS Card No. [01-079-2499], which presents the hexagonal structure. In general, the common structure algorithm and the phasor analysis are used with the hexagonal phase.

3. Results and Discussion

3.1. XRD

Figure 2 shows the XRD spectra of the bulk alloy and thin film of GaN. The direct reaction of Ga-melt with NH_3-gas at a temperature of 850 °C gives rise to the formation of micro-crystalline GaN. As shown in Figure 2a, the XRD peaks at 37.9° (101) and 33.39° (100) are for JCPDS Card No. [01-076-0703] [1], and the cubic phase is prepared with special conditions.

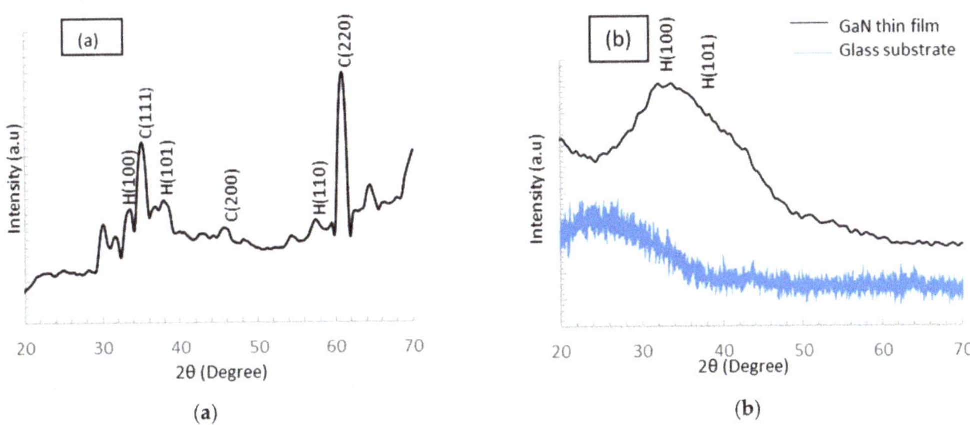

(a) **(b)**

Figure 2. The XRD spectra of GaN (**a**) bulk alloy and (**b**) thin film on glass substrate.

For bulk alloy, the structural quality of cubic GaN with low hexagonal phase was determined, and the percentage of cubic to hexagonal structure is 80% according to Equation (1) below [18]:

$$X_C = 1 - [1/(1+1.26(I_c/I_h))]^{-1} \tag{1}$$

where Xc is the weight fraction of cubic structure in the mixture, and I_c (220) and I_h (110) are the diffraction peak intensities of the cubic and hexagonal structure, respectively.

There is a shift in XRD theoretical value peaks from the experimental peaks due to the strain from scattered intensity distribution [2]. According to Reference [2], the lattice parameters of GaN are; a = 0.3232 nm and c = 0.5269 nm, so the unit cell c/a = 1.630, but in this work, c/a = 1.3; lattice constants a and c can be obtained from the Bragg diffraction formula (Equations (2)–(4)) [2]:

$$2d \sin \theta = n\lambda \tag{2}$$

$$a = d(hkl)[((3/4)(h2 + hk + k2) + l2 \ (a/c)^2]0.5 \tag{3}$$

$$c = d(hkl)[((3/4)(h2 + hk + k2) + l2 \ (c/a)^2]0.5 \tag{4}$$

where n as an integer is the "order" of the reflection, λ is the incident X-rays of the wavelength in nm, d is the inter-planar spacing of the crystal in Å, θ is the incidence angle in radians, $d(hkl)$ is the inter-planar distance, and h, k, l are the values of Miller Indices.

Figure 2b shows the diffraction pattern of thin film that is deposited on glass substrate (XRD of glass substrate shown in Figure 2b as blue line) using thermal vacuum evaporation. A broad diffraction peak with semi-crystalline structure appears at 2θ of 31° to 33.88° (100) and another peak at 38.40° (101); JCPDS Card No. [01-079-2499] refers to GaN with hexagonal structure with no appearance of cubic structure. This result is dependent on the vapor–solid (VS) mechanism in the vacuumed chamber of the thermal vacuum evaporator [19]. The high temperature helps in the transformation of cubic structure into h-GaN layers [19].

3.2. EDX Analysis

Table 1 shows the EDX analysis for the produced GaN bulk alloy prepared by crystal growth technique at 850 °C and GaN thin film. For the GaN bulk alloy, EDX data confirm the existence of Ga and N elements, indicating the formation of GaN structure (the reaction mechanism is shown in Equations (5) and (6)) and form an alloy with Ga:N composition of 1:1.

$$2NH_3(g) \rightarrow N_2(g) + 3H_2(g) \tag{5}$$

$$2Ga(s) + N_2(g) \rightarrow 2GaN(s) \tag{6}$$

Table 1. The EDX analysis of GaN bulk alloy and thin film.

Sample	Ga%	N%
Bulk alloy	50.8	49.2
Thin film	44.13	55.87

After the deposition of GaN alloy on glass substrate using thermal evaporation technique, it can be seen that the atomic percentage of nitrogen increases due to thermal decomposition of GaN at high deposition temperature [20].

3.3. FESEM of GaN

GaN Bulk Alloy

The morphology of the GaN bulk alloy and its thermal-vacuum-evaporated thin film is investigated using field emission scanning electron microscopy (as shown in Figure 3). For GaN bulk alloy (Figure 3a), uniform micro-rods with an average diame-ter of 0.98 μm are observed. In some places, these microrods clearly coalesce to form few micron-size GaN bundles with lengths up to 30 μm. It shows that the growth of GaN alloy on silica tube yields rough and regular surface morphologies.

Figure 3. FESEM of GaN; (**a**,**a′**) bulk alloy, (**b**) thin film, and (**c**) cross-section area of the thin film.

After the thermal evaporation deposition process, microrods convert to uniform microsphere particles with diameter ranging from 0.13 μm to 0.22 μm (as shown in Figure 3b). The mechanism of the thin film deposition may occur in the following stag-es: nucleation stage, fast attachment to form micro-rods network, growth of mi-cro-wires accompanied by fragmentation of network, and cleaving of spherical-like particles [21]. The cross-sectional view shown in Figure 3c of the thin film shows two parallel surface structures. Furthermore, there are no peels and cracks on the surface. The average thickness of the thin film as shown in Figure 3c is about 100 nm. A well-adhered and high film coverage on the glass substrate is clearly evident, as shown in Figure 1 for the GaN thin film.

3.4. Transmission Electron Microscopy

The microstructural properties of GaN thin film deposited on glass substrate are further investigated by TEM (as shown in Figure 4). We detached the GaN film to prepare a TEM specimen by scratching a thin layer and dispersing it in ethanol. The surface morphology of GaN thin film shows nanoparticles with spherical nature with size around 8-10 nm. The nano-crystallization process may be explained by the "propagation" of the SFs network. It probably occurred because of the highly strained area indicated in the XRD data of the thin film. This strained area is below the SFs that are sensitive to the lattice disorder, as they appeared in HRTEM. This growth of GaN nanoparticles resembles overgrowth by MOCVD and PLD [22–24]. The lattice spacing is 0.223, which is a reflection of the (100) plane of h-GaN, as shown in Figure 4b. Ga and N elements are distributed regularly across the particles, as shown in Figure 4d,d′,d″. Selected-area electron diffraction (SAED) shows (Figure 4e) semi-crystalline behavior, as indicated by the regular rings.

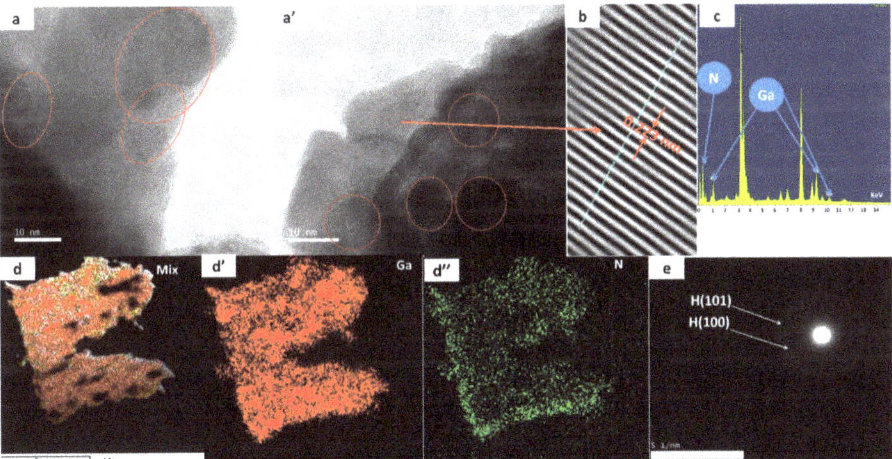

Figure 4. TEM and HRTEM of GaN; (**a,a′**) thin film, (**b,b′**) d-spacing, (**c**) EDAX spectrum, (**d,d′,d″**) Ga and N mapping, and (**e**) SAED patterns.

3.5. Optical Properties of GaN as a Bulk and as Thin Film

Photoluminescence (PL) measurements are commonly applied techniques for the qual-itative investigation of both GaN bulk alloys and thin films to detect material defects [25]. These native defects, which are present in semiconductor materials, arise from either non-stoichiometric crystal growth or an annealing process and consequently affect the electrical and optical properties of these materials [22]. The dissociation of nitrogen element during the reaction procedures as shown in Equations (5) and (6) may lead to the generation of the crystallographic defects and results in either nitrogen vacancy VN or gallium vacancy VGa, rather than a change in impurities [26]. GaN has two peaks of luminescence emissions; the yellow luminescence (YL) band centered at 2.1 eV (590 nm)–2.3 eV (539.13 nm) and the blue luminescence (BL) band centered at about 3.1 eV (401.3 nm) are generally considered related to defects in GaN [25].

The rate of dissociation is low enough to be compensated for by the NH_3 ambient gas, yielding a small amount of native defects in the GaN microcrystals, which is responsible for the difference in emission peaks. PL measurements for the GaN bulk alloy and thin film deposited on glass substrate are presented in Figure 5. As shown in this figure, the PL graph of the GaN bulk alloy has near-wavelength emission (NBE); blue (BL) and yellow emission (YL) at wavelengths of 404.4 nm (3.1 eV) to 533.5 nm (2.32 eV), respectively. The luminescence of (BL) and (YL) may be due to the presence of Ga vacancies, dislocations [17], and amorphous phases [25]. The intensity of the 3.1 eV band is related to a significant presence of stacking faults or dislocations and point defects in the samples [27].

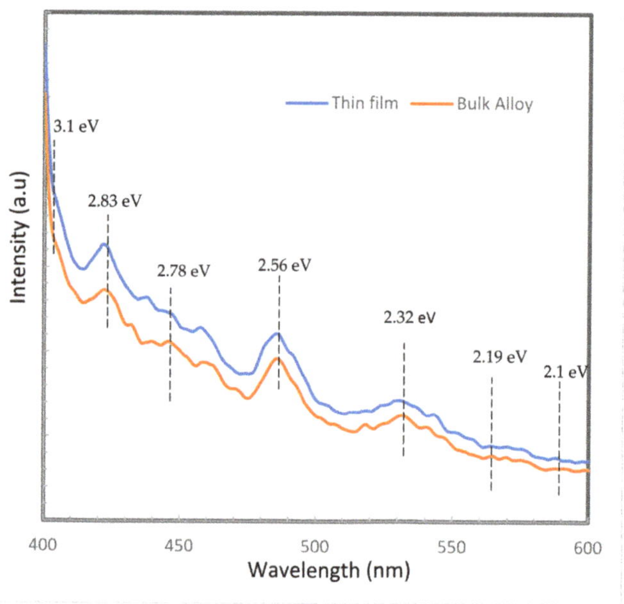

Figure 5. Bandgap of GaN bulk alloy and thin film.

The peak at 438.5 nm (2.83 eV) may originate from the luminescence coming from the dissociation of excitons bound to neutral donors, according to Reshchikov and Morkoc [22]. The PL spectrum of the bulk alloy also reveals the presence of a BL at both 457.48 nm (2.71 eV) and 438.5 nm, which are be attributed to gallium and nitrogen vacancies and deep-level impurities [25].

For GaN thin film, the sample shows many emission peaks ranging from 533.5 nm to 404.4 nm with a slightly blue shift. The reason for the blue shift is the ionization of the bound excitons as the deposition temperature increases, resulting in recombination of free excitons, causing the resultant blue shift [28]. A luminescence centered at around 446.48 nm (2.78 eV) is associated with a broad shoulder at 457.48 nm (2.71 eV). The peak at 485.07 nm (2.56 eV) is a broad and intense green band that is associated with the luminescent center produced at the dislocation edges originating from both Ga and N vacancies. It also could be explained by the gallium vacancy, as seen in the bulk that appears in hexagonal and cubic phases.

In general, the intensity of BL decreases quickly, while the intensity of YL shows a slight decrease, and its peak energy declines greatly. BL intensity slightly decreases, and its energy peak falls slightly with temperature increase through the deposition process. However, YL intensity is increased when the peak energy is reduced (as shown in Figure 5).

The optical transient behaviors of BL or YL in GaN have been reported in the literature before [17,29]. There may be two main explanations for their optical transient phenomena. First, the Coulomb fields, which are caused by the charge-trapping centers, may block the diffusion of carries to BL-related (or YL-related) defects [30,31]. This enhancement in the shield effect will cause decreases in BL and YL as a result of the evaporated high temperature. A second explanation may arise from the transformation of the meta-stable point defects through the recombination process, enhancing the defect reaction mechanism, which appeared according to the cubic phase's meta-stable property, as shown in the XRD data [25].

3.6. Time-Resolved Analysis

Figure 6 shows the raw FLIM data for GaN, with photoluminescence intensity decay curve of the nano-crystals. The decay curve is fitted by the sum of two exponential functions, with more than 97% of counts having 1.12 ns lifetime and the rest having 4.2 ns. The right panel of Figure 6 shows the phasor plot for the GaN. The intensity of the phasor plot decays for the corresponding FLIM image, as shown in the left panel of Figure 6. In this plot, every pixel in the image is normally represented in a 2D diagram with two coordinates; namely, S and G. These two coordinates are based on the phase shift (φ) between the transmitted wave and the resulting PL wave and demodulation factor (m) in the lase source. The S and G components are given by the following Equations (7) and (8) [32]:

$$S = m\sin(\varphi) = \frac{\omega\tau}{1 + \omega^2\,\tau^2} \tag{7}$$

$$G = m\cos(\varphi) = \frac{1}{1 + \omega^2\,\tau^2} \tag{8}$$

where τ is the electron lifetime and $\omega = 2\pi f$ is the laser modulation angular frequency (20 MHz). From Equations (7) and (8), the lifetime $\tau = \left(\frac{1}{\omega}\right)\frac{S}{G}$. The phasor plot for the GaN shows a cluster of points located at the edge of the phasor plot semicircle, indicating that the GaN decay is single-exponential decay with an average life-time of 1.12 ns as calculated from Equations (7) and (8).

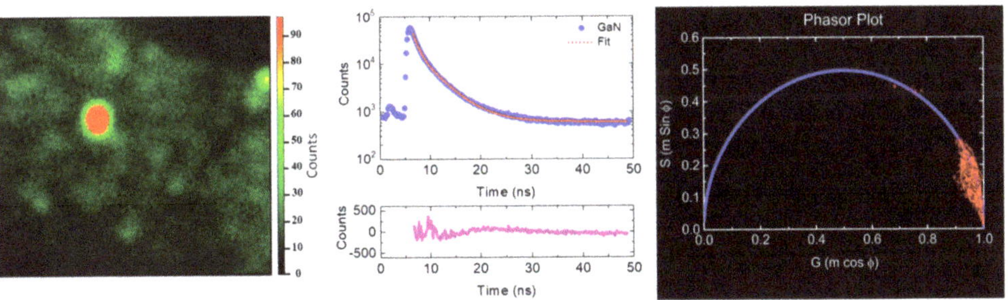

Figure 6. The raw FLIM data for GaN. The middle panel shows the photoluminescence intensity decay with the fitting curve. The curve at the bottom panels is the fitting residual. The left panel in is the phasor plot representations from fluorescence FLIM data.

4. Conclusions

In this study, a new line in GaN emission shifted to red light. The preparation of the bulk and thin film is very simple and economic. The bulk alloy preparation is a new technique for the generation of a III-nitride group depending on the crystal growth temperature. Furthermore, it was simple to use thermal vacuum evaporation to achieve this the target, and it is also dependent on the passing current and the deposition time. Furthermore, it plays an important role of eliminating the oxidation to obtain a pure GaN thin film. The crystal growth occurs in two phases apparent in XRD diffraction peaks and from PL. This thin film could be applied as a window material or emitter in some solar devices. From XRD and TEM investigation, we detected the formation of a strained region beyond the SFs network, which may be due to the point defect clustering and a band of planar defects that appeared in the surface. Time-resolved analysis reveals that PL intensity decays mainly in 1.12 ns, while the rest of the counts decay in 4.2 ns. Finally, these results support the use of a thermal evaporator for the growth of nano GaN particles used in short light-emitting diodes.

Author Contributions: Conceptualization, A.E.-H.B.K., M.F. and S.G.; methodology, A.E.-H.B.K., M.F. and S.G.; Validation, A.E.-H.B.K., M.F. and S.G.; formal analysis, A.E.-H.B.K., M.F., B.A. and S.G.; investigation, A.E.-H.B.K., M.F. and S.G.; resources, A.E.-H.B.K., M.F., B.A. and S.G.; data curation, A.E.-H.B.K., M.F., B.A. and S.G.; writing—original draft preparation, A.E.-H.B.K., M.F., B.A. and S.G.; writing—review and editing, A.E.-H.B.K., M.F., B.A. and S.G.; supervision, A.E.-H.B.K. and M.F. All authors have read and agreed to the published version of the manuscript.

Funding: This research received no external funding.

Conflicts of Interest: The authors declare no conflict of interest.

References

1. Daudin, B.; Feuillet, G.; Hübner, J.; Samson, Y.; Widmann, F.; Philippe, A.; Bru-Chevallier, C.; Guillot, G.; Bustarret, E.; Bentoumi, G.; et al. How to grow cubic GaN with low hexagonal phase content on (001) SiC by molecular beam epitaxy. *J. Appl. Phys.* **1998**, *84*, 2295–2300. [CrossRef]
2. Frentrup, M.; Lee, L.Y.; Sahonta, S.-L.; Kappers, M.J.; Massabuau, F.; Gupta, P.; Oliver, R.A.; Humphreys, C.J.; Wallis, D.J. X-ray diffraction analysis of cubic zincblende III-nitrides. *J. Phys. D Appl. Phys.* **2017**, *50*, 433002. [CrossRef]
3. Fatahilah, M.F.; Strempel, K.; Yu, F.; Vodapally, S.; Waag, A.; Wasisto, H.S. 3D GaN nanoarchitecture for field-effect transistors. *Micro Nano Eng.* **2019**, *3*, 59–81. [CrossRef]
4. Hernández-Gutiérrez, C.A.; Casallas-Moreno, Y.L.; Rangel-Kuoppa, V.-T.; Cardona, D.; Hu, Y.; Kudriatsev, Y.; Zambrano-Serrano, M.A.; Gallardo-Hernandez, S.; Lopez-Lopez, M. Study of the heavily p-type doping of cubic GaN with Mg. *Sci. Rep.* **2020**, *10*, 1–7. [CrossRef]
5. Liu, Z.; Lin, C.-H.; Hyun, B.-R.; Sher, C.-W.; Lv, Z.; Luo, B.; Jiang, F.; Wu, T.; Ho, C.-H.; Kuo, H.-C.; et al. Micro-light-emitting diodes with quantum dots in display technology. *Light. Sci. Appl.* **2020**, *9*, 1–23. [CrossRef]
6. Poppitz, D.; Lotnyk, A.; Gerlach, J.W.; Rauschenbach, B. Microstructure of porous gallium nitride nanowall networks. *Acta Mater.* **2014**, *65*, 98–105. [CrossRef]
7. Poppitz, D.; Lotnyk, A.; Gerlach, J.W.; Lenzner, J.; Grundmann, M.; Rauschenbach, B. An aberration-corrected STEM study of structural defects in epitaxial GaN thin films grown by ion beam assisted MBE. *Micron* **2015**, *73*, 1–8. [CrossRef]
8. Xue, S.; Zhang, X.; Huang, R.; Tian, D.; Zhuang, H.; Xue, C. A simple method for the growth of high-quality GaN nanobelts. *Mater. Lett.* **2008**, *62*, 2743–2745. [CrossRef]
9. Saron, K.; Hashim, M. Broad visible emission from GaN nanowires grown on n-Si (111) substrate by PVD for solar cell application. *Superlattices Microstruct.* **2013**, *56*, 55–63. [CrossRef]
10. Tang, Y.B.; Chen, Z.; Song, H.S.; Lee, C.-S.; Cong, H.T.; Cheng, H.-M.; Zhang, W.; Bello, I.; Lee, S.T. Vertically Aligned p-Type Single-Crystalline GaN Nanorod Arrays on n-Type Si for Heterojunction Photovoltaic Cells. *Nano Lett.* **2008**, *8*, 4191–4195. [CrossRef]
11. Abdullah, Q.; Ahmed, A.; Ali, A.; Yam, F.; Hassan, Z.; Bououdina, M.; Almessiere, M. Growth and characterization of GaN nanostructures under various ammoniating time with fabricated Schottky gas sensor based on Si substrate. *Superlattices Microstruct.* **2018**, *117*, 92–104. [CrossRef]
12. Saron, K.M.A.; Hashim, M.R.; Farrukh, M.A. Growth of GaN films on silicon (1 1 1) by thermal vapor deposition method: Optical functions and MSM UV photodetector applications. *Superlattices Microstruct.* **2013**, *64*, 88–97. [CrossRef]
13. Micic, O.I.; Ahrenkiel, S.P.; Bertram, D.; Nozik, A.J. Synthesis, structure, and optical properties of colloidal GaN quantum dots. *Appl. Phys. Lett.* **1999**, *75*, 478–480. [CrossRef]
14. Kuo, T.-J.; Kuo, C.-L.; Kuo, C.-H.; Huang, M.H. Growth of Core–Shell Ga–GaN Nanostructures via a Conventional Reflux Method and the Formation of Hollow GaN Spheres. *J. Phys. Chem. C* **2009**, *113*, 3625–3630. [CrossRef]
15. Saron, K.; Hashim, M. Study of using aqueous NH3 to synthesize GaN nanowires on Si(111) by thermal chemical vapor deposition. *Mater. Sci. Eng. B* **2013**, *178*, 330–335. [CrossRef]
16. Kashyout, A.E.-H.B.; Fathy, M.; Gad, S.; Badr, Y.; Bishara, A.A. Synthesis of Nanostructure InxGa1-xN Bulk Alloys and Thin Films for LED Devices. *Photonics* **2019**, *6*, 44. [CrossRef]

17. Gad, S.; Fathy, M.; Badr, Y.; Kashyout, A.E.-H.B. Pulsed Laser Deposition of $In_{0.1}Ga_{0.9}N$ Nanoshapes by Nd:YAG Technique. *Coatings* **2020**, *10*, 465. [CrossRef]
18. Kashyout, A.; Soliman, M.; Fathy, M. Effect of preparation parameters on the properties of TiO_2 nanoparticles for dye sensitized solar cells. *Renew. Energy* **2010**, *35*, 2914–2920. [CrossRef]
19. Tsai, C.-Y.; Su, Y.-Z.; Yu, I.-S. Effects of temperature and nitradition on phase transformation of GaN quantum dots grown by droplet epitaxy. *Surf. Coatings Technol.* **2019**, *358*, 182–189. [CrossRef]
20. Musiał, M.; Gosk, J.; Twardowski, A.; Janik, J.F.; Drygaś, M. Nanopowders of gallium nitride GaN surface functionalized with manganese. *J. Mater. Sci.* **2017**, *52*, 145–161. [CrossRef]
21. Pong, B.-K.; Elim, H.I.; Chong, J.-X.; Ji, W.; Trout, B.L.; Lee, J.-Y. New Insights on the Nanoparticle Growth Mechanism in the Citrate Reduction of Gold(III) Salt: Formation of the Au Nanowire Intermediate and Its Nonlinear Optical Properties. *J. Phys. Chem. C* **2007**, *111*, 6281–6287. [CrossRef]
22. Reshchikov, M.A.; Morkoç, H. Luminescence properties of defects in GaN. *J. Appl. Phys.* **2005**, *97*, 061301. [CrossRef]
23. Elsner, J.; Jones, R.; Heggie, M.I.; Sitch, P.K.; Haugk, M.; Frauenheim, T.; Öberg, S.; Briddon, P.R. Deep acceptors trapped at threading-edge dislocations in GaN. *Phys. Rev. B* **1998**, *58*, 12571–12574. [CrossRef]
24. Ravash, R.; Bläsing, J.; Hempel, T.; Noltemeyer, M.; Dadgar, A.; Christen, J.; Krost, A. Metal organic vapor phase epitaxy growth of single crystalline GaN on planar Si(211) substrates. *Appl. Phys. Lett.* **2009**, *95*, 242101. [CrossRef]
25. Wang, B.; Liang, F.; Zhao, D.; Ben, Y.; Yang, J.; Chen, P.; Liu, Z. Transient behaviours of yellow and blue luminescence bands in unintentionally doped GaN. *Opt. Express* **2021**, *29*, 3685–3693. [CrossRef] [PubMed]
26. Ganchenkova, M.G.; Nieminen, R.M. Nitrogen Vacancies as Major Point Defects in Gallium Nitride. *Phys. Rev. Lett.* **2006**, *96*, 196402. [CrossRef] [PubMed]
27. Santana, G.; De Melo, O.; Aguilar-Hernández, J.; Mendoza-Pérez, R.; Monroy, B.M.; Escamilla-Esquivel, A.; López-López, M.; De Moure, F.; Hernández, L.A.; Contreras-Puente, G. Photoluminescence Study of Gallium Nitride Thin Films Obtained by Infrared Close Space Vapor Transport. *Materials* **2013**, *6*, 1050–1060. [CrossRef]
28. Church, S.A.; Hammersley, S.; Mitchell, P.W.; Kappers, M.J.; Sahonta, S.L.; Frentrup, M.; Nilsson, D.; Ward, P.J.; Shaw, L.J.; Wallis, D.J.; et al. Photoluminescence studies of cubic GaN epilayers. *Phys. Status Solidi A* **2017**, *254*, 1600733. [CrossRef]
29. Al-Heuseen, K.; Hashim, M.; Ali, N. Synthesis of hexagonal and cubic GaN thin film on Si (111) using a low-cost electrochemical deposition technique. *Mater. Lett.* **2010**, *64*, 1604–1606. [CrossRef]
30. Tokuda, Y.; Matsuoka, Y.; Ueda, H.; Ishiguro, O.; Soejima, N.; Kachi, T. DLTS study of n-type GaN grown by MOCVD on GaN substrates. *Superlattices Microstruct.* **2006**, *40*, 268–273. [CrossRef]
31. Nayak, S.K.; Shamoon, D.; Ghatak, J.; Shivaprasad, S.M. Nanostructuring GaN thin film for enhanced light emission and extraction. *Phys. Status Solidi A* **2016**, *214*. [CrossRef]
32. Szmacinski, H.; Toshchakov, V.; Lakowicz, J.R. Application of phasor plot and autofluorescence correction for study of heterogeneous cell population. *J. Biomed. Opt.* **2014**, *19*, 46017. [CrossRef] [PubMed]

 *micromachines*

Article

Improvement of Crystal Quality of AlN Films with Different Polarities by Annealing at High Temperature

Yang Yue [1], Maosong Sun [1], Jie Chen [1], Xuejun Yan [1], Zhuokun He [1], Jicai Zhang [1,2,3,*] and Wenhong Sun [1,4,5,*]

[1] Research Center for Optoelectronic Materials and Devices, School of Physical Science and Technology, Guangxi University, Nanning 530004, China; 1907301107@st.gxu.edu.cn (Y.Y.); 1907401011@st.gxu.edu.cn (M.S.); a709161389@163.com (J.C.); xjuny100@163.com (X.Y.); hzkgmy@hotmail.com (Z.H.)
[2] College of Mathematics and Physics, Beijing University of Chemical Technology, Beijing 100029, China
[3] State Key Laboratory of Chemical Resource Engineering, Beijing University of Chemical Technology, Beijing 100029, China
[4] Guangxi Key Laboratory of Processing for Non-Ferrous Metal and Featured Materials, Guangxi University, Nanning 530004, China
[5] Guangxi Key Laboratory for Relativistic Astrophysics, School of Physical Science & Technology, Guangxi University, Nanning 530004, China
* Correspondence: jczhang@mail.buct.edu.cn (J.Z.); 20180001@gxu.edu.cn (W.S.)

Abstract: High-quality AlN film is a key factor affecting the performance of deep-ultraviolet optoelectronic devices. In this work, high-temperature annealing technology in a nitrogen atmosphere was used to improve the quality of AlN films with different polarities grown by magnetron sputtering. After annealing at 1400–1650 °C, the crystal quality of the AlN films was improved. However, there was a gap between the quality of non-polar and polar films. In addition, compared with the semi-polar film, the quality of the non-polar film was more easily improved by annealing. The anisotropy of both the semi-polar and non-polar films decreased with increasing annealing temperature. The results of Raman spectroscopy, scanning electron microscopy and X-ray photoelectron spectroscopy revealed that the annihilation of impurities and grain boundaries during the annealing process were responsible for the improvement of crystal quality and the differences between the films with different polarities.

Keywords: AlN; polar; semi-polar; non-polar; magnetron sputtering; HTA

Citation: Yue, Y.; Sun, M.; Chen, J.; Yan, X.; He, Z.; Zhang, J.; Sun, W. Improvement of Crystal Quality of AlN Films with Different Polarities by Annealing at High Temperature. *Micromachines* **2022**, *13*, 129. https://doi.org/10.3390/mi13010129

Academic Editor: Giovanni Verzellesi

Received: 14 December 2021
Accepted: 10 January 2022
Published: 14 January 2022

Publisher's Note: MDPI stays neutral with regard to jurisdictional claims in published maps and institutional affiliations.

Copyright: © 2022 by the authors. Licensee MDPI, Basel, Switzerland. This article is an open access article distributed under the terms and conditions of the Creative Commons Attribution (CC BY) license (https://creativecommons.org/licenses/by/4.0/).

1. Introduction

The III-nitride semiconductor materials are a popular research topic because of their wide applications in visible and ultraviolet light-emitting diodes, laser diodes and high-power electronic devices [1–4]. Among them is AlN, a promising material with many advantages such as high thermal conductivity, high temperature resistance and stable chemical properties [5–8]. On account of the limitations of wafer size and the production cost, AlN materials are currently mostly epitaxial films grown on sapphire substrates [9–11]. However, the lattice and thermal mismatch between the AlN film and the substrate will result in a high dislocation density in the epilayer [12]. High-quality AlN materials are extremely important for improving the related device performance. Additionally, when the material is grown along a [0001] direction, the large polarization field influences its properties significantly [13–15]. Semi-polar and non-polar materials are usually used to avoid the influence of the built-in electric field. Yusuke Yoshizumi et al. fabricated high-performance green laser diodes by using a semi-polar thin film [16]. Hwang, S.M. et al. [17] prepared non-polar a-plane light-emitting diodes (LED) on an r-plane sapphire substrate and proved that the crystallization quality of non-polar materials strongly affects the device performance. The preparation of high-quality semi-polar and non-polar materials is extremely important. However, due to the anisotropy of the atomic structure on the surface of semi-polar and non-polar thin films, there are a lot of dislocations and stacking faults formed during growth [18,19]. These defects make the quality of semi-polar and non-polar AlN thin films much lower than the quality of the polar AlN films and limit the practical application of semi-polar and non-polar films. In recent years, some studies have pointed out that the quality of polar thin films can be significantly improved by high-temperature annealing (HTA) technology [20]. Unfortunately, there are still few studies on the annealing of semi-polar and non-polar thin films with great application potential. In this work, thin AlN films were deposited on a-, m- and r-sapphire substrates and HTA was used to improve the crystal quality of the films. The influence of HTA on AlN films with different polarities was studied and the different influences were explained by the annihilation of impurities and grain boundaries during the annealing process.

2. Materials and Methods

Polar (0001), semi-polar (11−22) and non-polar (11−20) AlN films were deposited on a-, m- and r-sapphire substrates, respectively, by magnetron sputtering at 650 °C and 700 W. Oxygen, nitrogen and argon were employed for the plasma that reacts to aluminum target. In the initial stage, a small flux of O_2 was introduced to form a very thin AlON layer (about several nanometers) on the substrate to prepare high-quality AlN films. The thickness of all sputtered AlN films was 300 nm. The samples were treated with HTA under 150 Torr in a nitrogen atmosphere at 1400, 1500, 1600 and 1650 °C for 30 min. To inhibit the thermal decomposition of the AlN film during thermal annealing, the surface of the AlN film was covered with another sample or sapphire in a face-to-face setup, as shown in Figure 1.

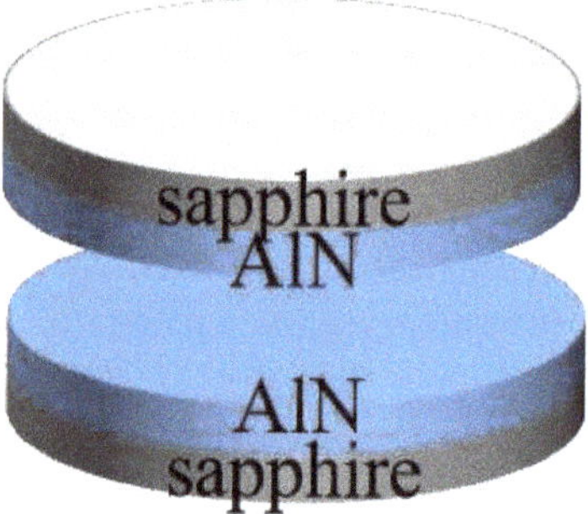

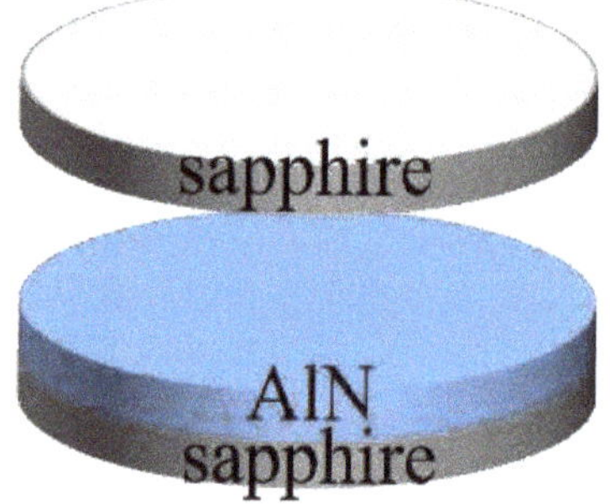

Figure 1. Schematic diagram of face-to-face setup.

The crystallinity of the AlN films was measured by high-resolution X-ray diffraction (XRD). The XRD results were obtained by using a PANalytical instrument with a Cu $K_{\alpha 1}$ = 1.5406 Å radiation. Raman spectroscopy, scanning electron microscopy (SEM) and X-ray photoelectron spectroscopy (XPS) were used to further characterize the quality, stress and surface morphology of the films. Raman spectra were obtained using a Labram HR Evolution instrument with a laser of wavelength 532 nm. SEM images were obtained using a VEECO Dimension 3100 instrument. XPS was performed using an ESCALAB 250XI+ instrument with a monochromatic Al target X-ray source.

3. Results and Discussion

Firstly, we investigated the effect of annealing temperature on film quality with XRD. Figure 2a shows the relationship between the full-width at half-maximum (FWHM) of the XRD rocking curves (XRC) of the polar, semi-polar and non-polar films, and the annealing temperature. With the increase in annealing temperature, the FWHM of the films decreased and the film quality was improved. However, there was still a certain gap between the quality of the semi-polar and non-polar films and that of the polar film after HTA. The polar films exhibited the highest quality after annealing. Additionally, the quality of the non-polar film was better than that of the semi-polar film after HTA, which may be related to the content of impurities in the films before and after annealing.

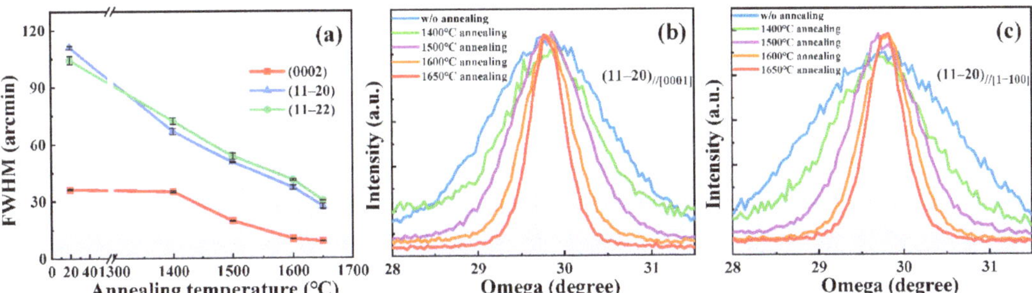

Figure 2. (**a**) The FWHM of rocking curves of polar (0002), semi-polar (11–22) and non-polar (11–20) films after annealing at 1400–1650 °C as a function of annealing temperature. The XRC of the non-polar (11–20) film along (**b**) [0001] direction and (**c**) [1−100] direction without and with annealing.

The non-polar thin film had structural anisotropy due to the asymmetry of the crystals in the plane. The anisotropy of the thin-film structure was caused by the anisotropy of atomic diffusion length or growth rate on the growth surface [21]. Figure 2b,c show XRC from two different incident beam directions, the c-axis [0001] and the m-axis [1−100] of AlN. It can be seen that the rocking curve of the non-polar film was wide before annealing, which indicated that the sputtered film had a high density of dislocation. As the annealing temperature was increased, the rocking curves of the non-polar films became smooth and sharp, and the FWHM decreased, which indicated that HTA significantly reduced the dislocations in the films. Though the crystal quality along the c-axis and the m-axis was improved, the FWHM values of the XRCs in the two directions were different. The FWHM along the c-axis [0001] direction was smaller than that in the [1−100] direction, which was mainly attributed to the smaller stacking fault density and the faster growth rate along the c-axis [0001]. Due to the difference in the in-plane growth rates, the crystal mosaic degree also exhibited anisotropy. With the increase in annealing temperature, the difference in FWHM between the two perpendicular directions decreased, and the anisotropy of the films decreased gradually, as shown in Table 1. The anisotropy was related to the stacking fault density in the film; HTA resulted in the decrease in the growth rate difference between the two directions, which reduced the stacking fault density and improved the film quality.

Table 1. The FWHM (arcmin) of non-polar film along two perpendicular directions without and with annealing.

Direction	W/O	1400 °C HTA	1500 °C HTA	1600 °C HTA	1650 °C HTA
[0001]	111	73	57	37	26
[1−100]	159	87	60	38	27

The formation of dislocations was often related to impurities. We carried out high-resolution XPS to analyze the changes in the impurity concentration in the films. The Al 2p, N 1s, O 1s and C 1s in the films were scanned with high resolution to analyze atomic concentrations (at. %). The surface element concentrations were calculated according to the element characteristic peak area and the corresponding element sensitivity factor [22]. The element concentrations of the unannealed samples are shown in Table 2. On account of AlN exhibiting a high affinity for oxygen, oxygen atoms easily entered the lattice to form defects [23]. When oxygen atoms entered the AlN lattice, oxygen replaced nitrogen atoms in the lattice to form aluminum vacancies [24]. As the oxygen content increased, the defects agglomerated and evolved into extended defects [24,25]. The impurity model was used to explain the difference in the oxygen impurity levels and the quality of the films with different polarities (shown in Figure 3). It can be seen that when an oxygen atom tried to enter the lattice of the AlN film, the aluminum atom of the polar structure film prevented the oxygen impurity from entering the lattice to replace the nitrogen atom. The structure of the semi-polar film was inclined at a certain angle compared with the polar film, which weakened the blocking effect of the aluminum atoms on oxygen contamination. In contrast, the non-polar film exhibited no blocking effect of aluminum atoms on impurities; therefore, non-polar films may contain more oxygen impurities and defects. Moreover, this was consistent with the XPS test results (as shown in Table 2), which showed that the non-polar film was more vulnerable to impurities than the polar and semi-polar films. This may be the reason why the quality of the non-polar film was lower than that of the polar and semi-polar films before annealing.

Table 2. The elemental concentrations of the unannealed polar, semi-polar and non-polar films.

Sample	Al (at. %)	N (at. %)	C (at. %)	O (at. %)
Polar	15.68	8.41	49.88	26.03
Semi-polar	15.62	5.49	49.93	28.96
Non-polar	15.43	4.93	50.03	29.61

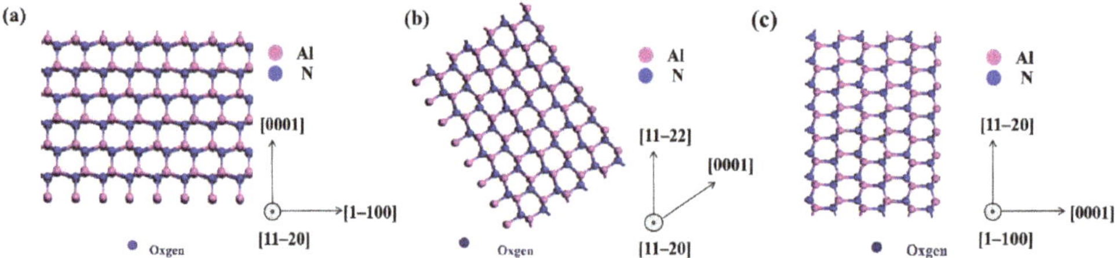

Figure 3. Impurity-binding model for (**a**) polar (0001), (**b**) semi-polar (11–22) and (**c**) non-polar (11–20) films.

In order to further investigate the micromechanism of the quality improvement of the non-polar film after annealing, high-resolution XPS tests were carried out. Figure 4 shows the XPS results of non-polar films with and without annealing. The concentrations of O and C impurities in the films after thermal annealing changed to 17.64% and 40.85%, respectively, and those of Al and N increased to 24.29% and 17.22%, respectively (Figure 4a,b). This means that HTA reduced the concentration of impurities in the non-polar film. A linear combination of Gaussian and Lorentz functions was used to fit the subpeaks of Al 2p and N 1s. The subpeaks within each element peak represented different chemical states. The phase information of non-polar films can be obtained by analyzing the area ratio of subpeaks during annealing. The Al 2p peak contained two subpeaks, A 1 and A 2, which appeared at 73.5 eV and 74.7 eV, respectively (shown in Figure 4c) [26]. The A1 subpeak originated from the Al-N bond in wurtzite and existed in the form of aluminum nitride phase [22]. The A 2 subpeak was derived from the Al-O bond and could be attributed to the alumina domains and the oxidation of AlN grain boundaries [22]. The proportion of Al-N and Al-O bonds changed from 46.7% and 46.1% for the as-grown sample to 66.4% and 27.3% after annealing at 1650 °C, respectively. These results suggested that HTA enhanced the Al-N interaction in the non-polar film and strengthened the AlN phase. The N 1s peak of the non-polar film before and after annealing contained two subpeaks, N 1 and N 2, which were related to the interaction between the N and Al atoms (shown in Figure 4e). In the as-grown non-polar film, in addition to N 1 and N 2 subpeaks, the N 1s peak also had a subpeak at higher binding energies (marked as N 3). This subpeak was the result of the interaction between nitrogen and oxygen atoms on the film surface [22]. The subpeak N 3 disappeared and the areas of the N 1 and N 2 subpeaks increased after HTA (as shown in Figure 4e,f), which meant the N-O bond broke and the interaction between N and Al increased. Therefore, the crystallinity of the film improved.

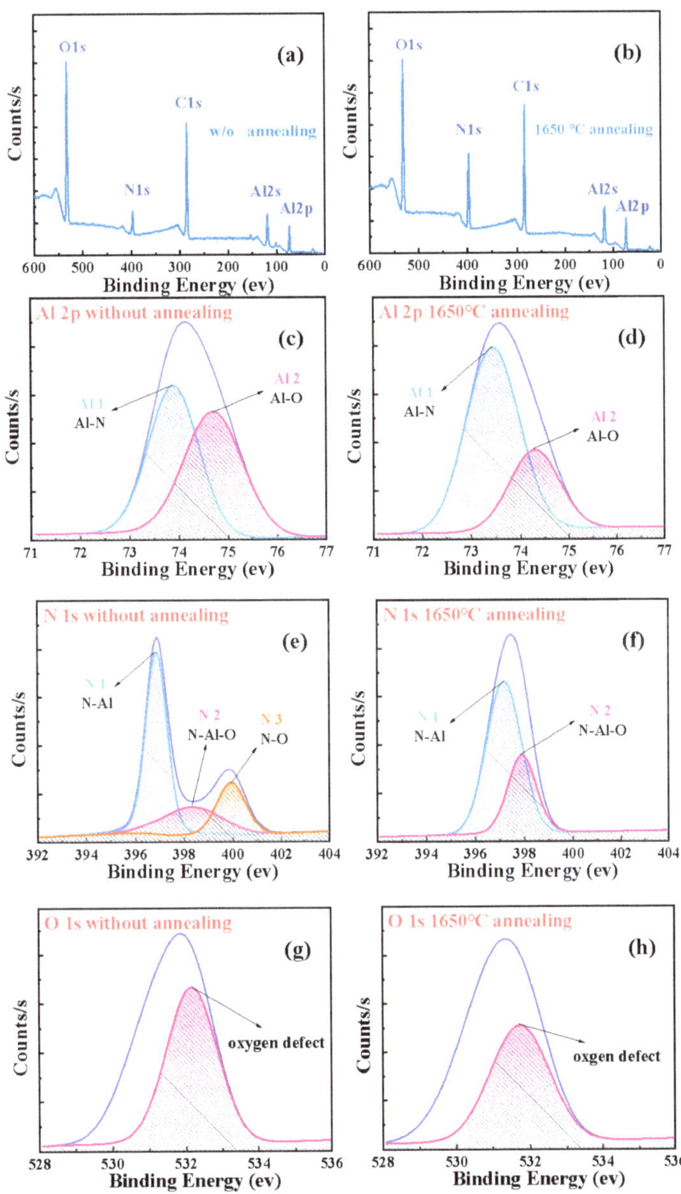

Figure 4. The XPS high-resolution scans for the non-polar film (**a**) without annealing and (**b**) with annealing at 1650 °C. The subpeak spectra of (**c**) Al 2p, (**e**) N 1s and (**g**) O 1s peaks on the surface of the non-polar film without annealing, and (**d**) Al 2p, (**f**) N 1s and (**h**) O 1s peaks with annealing at 1650 °C.

Table 3 shows the concentrations of different atoms in all the samples after annealing at 1650 °C. After annealing at high temperature, the concentration of oxygen impurities in the non-polar film decreased from 29.61% to 17.64%, while that in polar film decreased from 26.03% to 14.49% and in the semi-polar film decreased from 28.96% to 23.81%. This indicated that the non-polar film was strongly affected by HTA and the impurity concentration was greatly reduced. The aggregation of screw dislocations was conducive to the adsorption of oxygen impurities [27]. Consequently, the reduction in the oxygen impurity concentration proved that HTA effectively reduced the dislocations in the films, such that the quality of the non-polar film after annealing was higher than that of the semi-polar film, which was consistent with the XRC tests. Through the fitting of the O 1s spectra, a subpeak was obtained near 531.8 eV, which originated from the Al-O octahedral complex in the grain and was related to oxygen defects [24,26]. After HTA, the proportion of this subpeak decreased from 53.4% to 42.6%. This should be assigned to HTA, which reduced the concentration of oxygen impurities, inhibited the formation of defects, repaired the lattice damage of the film and improved the quality of the non-polar film.

Table 3. The elemental concentrations of the polar, semi-polar and non-polar films with annealing at 1650 °C.

Sample	Al (at. %)	N (at. %)	C (at. %)	O (at. %)
Polar	25.29	20.03	40.19	14.49
Semi-polar	16.73	17.19	42.37	23.81
Non-polar	24.29	17.22	40.85	17.64

The microstructure of the AlN films was characterized by SEM to clarify the mechanism of HTA of the films, as shown in Figure 5. Before annealing, the surfaces of the films were covered with a large number of small particles, which indicated that the films deposited by sputtering were uniform. In addition, the surfaces of the films showed discrete grains and a large number of grain boundaries, showing that the film had a high density of dislocation. The surface grains of the semi-polar and non-polar films were tilted long strips. This was due to anisotropic growth resulting in thin-film stripe characteristics. After annealing at high temperature, the grains of the polar, semi-polar and non-polar films increased obviously and the grain boundaries decreased. The grain boundary contained dislocations, impurity atoms and other defects. This indicated that the recrystallization process reduced the grain boundaries of the films, thus reducing the defects and improving the crystal quality of the films. The surfaces of the semi-polar and non-polar films after annealing no longer exhibited obvious grain characteristics of long stripes. The crystal quality of the non-polar film after annealing was still a degree lower than that of the polar film. However, compared with the semi-polar film, the non-polar film had a higher crystal quality with a larger grain size and fewer holes on the surface. This confirmed that the anisotropy properties of the semi-polar and non-polar films were reduced after HTA, and that the quality of the non-polar film could be improved more easily through thermal annealing than that of the semi-polar film.

Raman spectra were used to analyze the effect of HTA on the stress state and quality of the films. For an AlN crystal, the frequency shift of the E_2 (high) phonon peak position is generally used to reflect the stress of the AlN film. The E_2 (high) peak in stress-free bulk AlN was 657.4 cm^{-1} [28,29]. The E_2 (high) peak of the sputtered AlN films was located at 656.0 cm^{-1}, which indicated that the stress state of the sputtered film was tensile stress. After annealing at high temperature, the E_2 (high) Raman peak of the non-polar film was blue shifted, as shown in Figure 6a. The blue shift increased with the increase in annealing temperature and the shifted peak was greater than 657.4 cm^{-1}. This phenomenon showed that after annealing, the stress of the film was released, and the film changed from tensile stress to compressive stress. The change of stress state may be attributed to the recrystallization process in the film during annealing. In the recrystallization process, there was lattice mismatch and thermal-expansion mismatch between the epitaxial AlN film and the sapphire substrate, which led to compressive strain [30]. The dislocation contributed to the relaxation of the stress in the film [31]. The compressive stress inside the AlN films gradually increased with the enhanced annealing temperature, proving that annealing reduced the dislocation in the film and weakened the relaxation effect of dislocation. Raman peak linewidth was used as a measure of phonon lifetime [32]. When the film contained a large number of grain boundaries and other defects, phonons will decay at grain boundaries or defects during the Raman scattering process, resulting in a shortened lifetime and a widening of Raman linewidth. Thus, the film quality could be obtained from the variation of the linewidth of the Raman phonon peak. It can be seen from the Raman spectra that the linewidth of the E_2 (high) peak decreased gradually with the increase in annealing temperature. In addition to the E_2 (high) phonon modes, A_1 (TO) and E_1 (TO) modes were observed in our Raman spectra. Simultaneously with the increase in annealing temperature, the FWHM of the A_1 (TO) and E_1 (TO) modes also decreased. The atomic vibrations of the A_1 (TO) and E_1 (TO) modes were along the c-axis and m-axis, respectively [33]. This proved that the crystal quality of the non-polar film along the c-axis and the m-axis was improved after HTA. In addition, Figure 6b illustrates that at the same annealing temperature, the E_2 (high) peak offset of films with different polarities was different and the stress of the films was different. This may be due to different degrees of dislocation reduction in the recrystallization process of the films with different polarities, resulting in the different stress and crystallization qualities of the films.

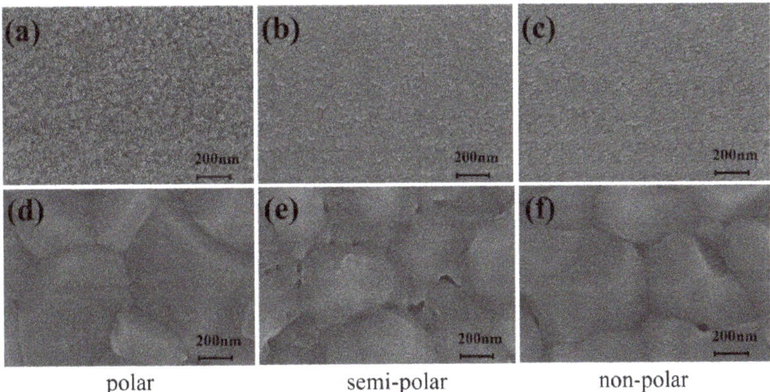

Figure 5. (**a–c**) The surface morphology images of the films without annealing; (**d–f**) the surface morphology images of the films with annealing at 1650 °C.

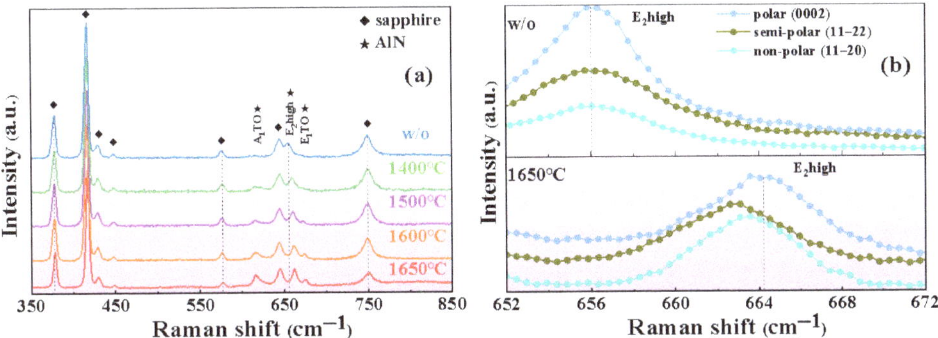

Figure 6. (**a**) Raman spectra of the non-polar film without and with annealing. (**b**) Raman spectra of polar, semi-polar and non-polar films without and with annealing at 1650 °C.

4. Conclusions

In summary, the thermal annealing of AlN films with different polarities was studied in this paper. The quality of polar, semi-polar and non-polar films was improved by thermal annealing. The films with different polarities had different degrees of quality improvement. Although there was a gap between the quality of the non-polar film and that of the polar film after annealing, compared with the semi-polar film, the non-polar film was easier to treat to achieve quality improvement. XPS tests showed that the quality difference between the films was related to the concentration of impurities. Compared with the semi-polar film, the concentration of impurities in the non-polar film was greatly reduced. The thermal annealing process could effectively reduce impurities and inhibit the formation of defects in the film, and the non-polar film was greatly affected by the annealing temperature. Combined with the surface morphology images, we can draw the conclusion that under the treatment conditions used in this study, the essence of crystal quality improvement lies in the annihilation of grain boundaries during the annealing process. In addition, HTA reduced the anisotropy of the non-polar and semi-polar films, improving the quality of these films. The blue shift of the Raman phonon peak and the reduction in the linewidth further revealed that the stress state and quality of the films changed during annealing. The films with different polarities exhibited different stresses after HTA. During the annealing process, the stress state of the films changed from tensile stress to compressive stress, the dislocation and defects of the films were reduced and the quality of the films improved.

Author Contributions: Conceptualization, Y.Y., M.S., J.Z. and W.S.; data collection, Y.Y., J.C., X.Y. and Z.H.; supervision, J.Z. and W.S.; data analysis, Y.Y., J.Z. and W.S.; writing—review and editing, Y.Y., J.Z. and W.S. All authors have read and agreed to the published version of the manuscript.

Funding: This work was supported in part by the Beijing Municipal Natural Science Foundation (4182046), the National Natural Science Foundation of China (61874007, 12074028), the Shandong Provincial Major Scientific and Technological Innovation Project (2019JZZY010209), the Key-Area Research and Development Program of Guangdong Province (2020B010172001) and the Fundamental Research Funds for the Central Universities (buctrc201802, buctrc201830), the funding for the Bagui Talent of Guangxi province (No. T3120099202 and T3120097921) and Talent Model Base (AD19110157), Disinfection Robot Based on High Power AlGaN-based UVLEDs (No. BB31200014).

Institutional Review Board Statement: Not applicable.

Informed Consent Statement: Not applicable.

Data Availability Statement: Data sharing is not applicable to this article.

Conflicts of Interest: The authors declare no conflict of interest.

References

1. Mukai, T. Recent progress in group-III nitride light-emitting diodes. *IEEE J. Sel. Top. Quantum Electron.* **2002**, *8*, 264–270. [CrossRef]
2. Kneissl, M.; Rass, J. *III-Nitride Ultraviolet Emitters: Technology and Applications*; Springer: Berlin/Heidelberg, Germany, 2016; Volume 227.
3. Strite, S.; Morkoç, H. GaN, AlN, and InN: A review. *J. Vac. Sci. Technol. B Microelectron. Nanometer Struct.* **1992**, *10*, 1237–1266. [CrossRef]
4. Mohammad, S.N.; Morkoç, H. Progress and prospects of group-III nitride semiconductors. *Prog. Quantum Electron.* **1996**, *20*, 361–525. [CrossRef]
5. Yim, W.; Stofko, E.; Zanzucchi, P.; Pankove, J.; Ettenberg, M.; Gilbert, S. Epitaxially Grown AlN and its Optical Band Gap. *J. Appl. Phys.* **1973**, *44*, 292–296. [CrossRef]
6. Li, J.; Nam, K.; Nakarmi, M.; Lin, J.-Y.; Jiang, H.; Carrier, P.; Wei, S.-H. Band structure and fundamental optical transitions in wurtzite AlN. *Appl. Phys. Lett.* **2004**, *83*, 5163–5165. [CrossRef]
7. Goldberg, Y.; Levinshtein, M.; Rumyantsev, S. (Eds.) *Properties of Advanced Semiconductor Materials: GaN, AIN, InN, BN, SiC, SiGe*; SciTech Book News; Wiley-Interscience: Hoboken, NJ, USA, 2001; Volume 25, pp. 93–146.
8. Sun, M.; Li, J.; Zhang, J.; Sun, W. The fabrication of AlN by hydride vapor phase epitaxy. *J. Semicond.* **2019**, *40*, 121803. [CrossRef]

9. Jeschke, J.; Martens, M.; Knauer, A.; Kueller, V.; Zeimer, U.; Netzel, C.; Kuhn, C.; Krueger, F.; Reich, C.; Wernicke, T.; et al. UV-C Lasing from AlGaN Multiple Quantum Wells on Different Types of AlN/Sapphire Templates. *IEEE Photonics Technol. Lett.* **2015**, *27*, 1969–1972. [CrossRef]

10. Wu, Z.; Yan, J.; Guo, Y.; Zhang, L.; Lu, Y.; Wei, X.; Wang, J.; Li, J. Study of the morphology evolution of AlN grown on nano-patterned sapphire substrate. *J. Semicond.* **2019**, *40*, 122803. [CrossRef]

11. Zhang, B.; Egawa, T.; Ishikawa, H.; Liu, Y.; Jimbo, T. High-Bright InGaN Multiple-Quantum-Well Blue Light-Emitting Diodes on Si (111) Using AlN/GaN Multilayers with a Thin AlN/AlGaN Buffer Layer. *Jpn. J. Appl. Phys.* **2003**, *42*, L226–L228. [CrossRef]

12. Kneissl, M.; Kolbe, T.; Chua, C.; Kueller, V.; Lobo, N.; Stellmach, J.; Knauer, A.; Rodriguez, H.; Einfeldt, S.; Yang, Z.; et al. Advances in group III-nitride-based deep UV light-emitting diode technology. *Semicond. Sci. Technol.* **2010**, *26*, 014036. [CrossRef]

13. Deguchi, T.; Sekiguchi, K.; Nakamura, A.; Sota, T.; Matsuo, R.; Chichibu, S.; Nakamura, S. Quantum-Confined Stark Effect in an AlGaN/GaN/AlGaN Single Quantum Well Structure. *Jpn. J. Appl. Phys.* **1999**, *38*, L914–L916. [CrossRef]

14. Craven, M.; Lim, S.; Wu, F.; Speck, J.; Denbaars, S. Structural characterization of nonpolar (110) a-plane GaN thin films grown on (102) r-plane sapphire. *Appl. Phys. Lett.* **2002**, *81*, 469–471. [CrossRef]

15. Smorchkova, I.; Elsass, C.; Ibbetson, J.; Vetury, R.; Heying, B.; Fini, P.; Haus, E.; Denbaars, S.; Speck, J.; Mishra, U.K. Polarization-induced charge and electron mobility in AlGaN/GaN heterostructures grown by plasma-assisted molecular-beam epitaxy. *J. Appl. Phys.* **1999**, *86*, 4520–4526. [CrossRef]

16. Yoshizumi, Y.; Adachi, M.; Enya, Y.; Kyono, T.; Tokuyama, S.; Sumitomo, T.; Akita, K.; Ikegami, T.; Ueno, M.; Katayama, K.; et al. Continuous-Wave Operation of 520 nm Green InGaN-Based Laser Diodes on Semi-Polar {2021} GaN Substrates. *Appl. Phys. Express* **2009**, *2*, 092101. [CrossRef]

17. Hwang, S.M.; Seo, Y.G.; Baik, K.; Cho, I.-S.; Baek, J.-H.; Jung, S.; Kim, T.; Cho, M. Demonstration of nonpolar a-plane InGaN/GaN light emitting diode on r-plane sapphire substrate. *Appl. Phys. Lett.* **2009**, *95*, 071101. [CrossRef]

18. Pristovsek, M.; Frentrup, M.; Han, Y.; Humphreys, C.J. Optimizing GaN (11–22) hetero-epitaxial templates grown on (10–10) sapphire. *Phys. Status Solidi* **2016**, *253*, 61–66. [CrossRef]

19. Rajpalke, M.K.; Roul, B.; Kumar, M.; Bhat, T.N.; Sinha, N.; Krupanidhi, S.B. Structural and optical properties of nonpolar (11−20) a-plane GaN grown on (1−102) r-plane sapphire substrate by plasma-assisted molecular beam epitaxy. *Scr. Mater.* **2011**, *65*, 33–36. [CrossRef]

20. Miyake, H.; Lin, C.H.; Tokoro, K.; Hiramatsu, K. Preparation of high-quality AlN on sapphire by high-temperature face-to-face annealing. *J. Cryst. Growth* **2016**, *456*, 155–159. [CrossRef]

21. Wang, H.; Chen, C.; Gong, Z.; Zhang, J.; Gaevski, M.; Su, M.; Yang, J.; Khan, M.A. Anisotropic structural characteristics of (11$\bar{2}$0) GaN templates and coalesced epitaxial lateral overgrown films deposited on (10$\bar{1}$2) sapphire. *Appl. Phys. Lett.* **2004**, *84*, 499–501. [CrossRef]

22. Sharma, N.; Ilango, S.; Dash, S.; Tyagi, A.K. X-ray photoelectron spectroscopy studies on AlN thin films grown by ion beam sputtering in reactive assistance of N^+/N_2^+ ions: Substrate temperature induced compositional variations. *Thin Solid Film.* **2017**, *636*, 626–633. [CrossRef]

23. Youngman, R.A.; Harris, J.H. Luminescence Studies of Oxygen-Related Defects in Aluminum Nitride. *J. Am. Ceram. Soc.* **1990**, *73*, 3238–3246. [CrossRef]

24. Harris, J.; Youngman, R.; Teller, R. On the Nature of the Oxygen Related Defect in Aluminum Nitride. *J. Mater. Res.* **1990**, *5*, 1763–1773. [CrossRef]

25. Westwood, A.D.; Youngman, R.A.; McCartney, M.R.; Cormack, A.N.; Notis, M.R. Oxygen incorporation in aluminum nitride via extended defects: Part II. Structure of curved inversion domain boundaries and defect formation. *J. Mater. Res.* **1995**, *10*, 1287–1300. [CrossRef]

26. Rosenberger, L.; Baird, R.; McCullen, E.; Auner, G.; Shreve, G. XPS Analysis of Aluminum Nitride Films Deposited by Plasma Source Molecular Beam Epitaxy. *Surf. Interface Anal.* **2008**, *40*, 1254–1261. [CrossRef]

27. Heinke, H.; Kirchner, V.; Einfeldt, S.; Hommel, D. X-ray diffraction analysis of the defect structure in epitaxial GaN. *Appl. Phys. Lett.* **2000**, *77*, 2145–2147. [CrossRef]

28. Holtz, M.; Nikishin, S.; Faleev, N.; Temkin, H.; Zollner, S. Vibrational Properties of AlN Grown on (111)-Oriented Silicon. *Phys. Rev. B* **2001**, *63*, 125313.

29. Yang, S.; Miyagawa, R.; Miyake, H.; Hiramatsu, K.; Harima, H. Raman Scattering Spectroscopy of Residual Stresses in Epitaxial AlN Films. *Appl. Phys. Express* **2011**, *4*, 031001. [CrossRef]

30. Claudel, A.; Fellmann, V.; Gélard, I.; Coudurier, N.; Sauvage, D.; Balaji, M.; Blanquet, E.; Boichot, R.; Beutier, G.; Coindeau, S.; et al. Influence of the V/III ratio in the gas phase on thin epitaxial AlN layers grown on (0001) sapphire by high temperature hydride vapor phase epitaxy. *Thin Solid Films* **2014**, *573*, 140–147. [CrossRef]

31. Jain, S.C.; Harker, A.H.; Cowley, R.A. Misfit strain and misfit dislocations in lattice mismatched epitaxial layers and other systems. *Philos. Mag. A* **1997**, *75*, 1461–1515. [CrossRef]

32. Liu, M.S.; Nugent, K.W.; Prawer, S.; Bursill, L.A.; Peng, J.L.; Tong, Y.Z.; Jewsbury, P. Micro-Raman Scattering Properties of Highly Oriented AlN Films. *Int. J. Mod. Phys. B* **1998**, *12*, 1963–1974. [CrossRef]

33. Wagner, J.-M.; Bechstedt, F. Properties of strained wurtzite GaN and AlN: Ab initio studies. *Phys. Rev. B* **2002**, *66*, 115202. [CrossRef]

MDPI

St. Alban-Anlage 66

4052 Basel

Switzerland

Tel. +41 61 683 77 34

Fax +41 61 302 89 18

www.mdpi.com

Micromachines Editorial Office

E-mail: micromachines@mdpi.com

www.mdpi.com/journal/micromachines

www.ingramcontent.com/pod-product-compliance
Lightning Source LLC
LaVergne TN
LVHW071534170726
843515LV00008B/2140